Cracking the AP® PHYSICS B Exam

2013 Edition

Cracking the AP® PHYSICS B Exam

2013 Edition

Steven A. Leduc and John Miller

PrincetonReview.com

Random House, Inc. New York

The Princeton Review, Inc.
111 Speen Street, Suite 550
Framingham, MA 01701
E-mail: editorialsupport@review.com

ISBN: 978-0-307-94515-0
ISSN: 1937-6383

Editor: Calvin S. Cato
Production Editor: Michelle Krapf
Production Coordinator: Deborah A. Silvestrini

Printed in the United States of America on partially recycled paper.

10 9 8 7 6 5 4 3 2 1

2013 Edition

ACKNOWLEDGMENTS

My thanks and appreciation to John Katzman, Steve Quattrociocchi, Paul Maniscalco, Kris Gamache, Tricia McCloskey, Andy Lutz, and Suellen Glasser for making me feel at home. Thanks also to Art Brown for his thoughtful and valuable input.

Special thanks to Paul Kanarek for his friendship, counsel, and encouragement.

—S.L.

Thanks to Heather Brady, Meave Shelton, Kim Howie, and Al Mercado for making my words and diagrams make more sense. Thanks to Pete Insley, Scott Welty, and Jeff Rylander—the guys who taught me how to teach physics. Thanks to my wife and three children who continue to teach me about living. The Princeton Review would also like to thank Douglas Laurence for his hard work in reviewing this title for the latest edition.

—J.M.

DEDICATION

This work is dedicated to the memory of my great aunt, Norma Perron Lamb Piette.

CONTENTS

INTRODUCTION

WHAT IS THE PRINCETON REVIEW?

The Princeton Review is an international test-preparation company with branches in all major U.S. cities and several abroad. In 1981, John Katzman started teaching an SAT prep course in his parents' living room. Within five years, The Princeton Review had become the largest SAT prep program in the country.

Our phenomenal success in improving students' scores on standardized tests is due to a simple, innovative, and radically effective philosophy: Study the test, not just what the test claims to test. This approach has led to the development of techniques for taking standardized tests based on the principles the test writers themselves use to write the tests.

The Princeton Review has found that its methods work not just for cracking the SAT, but for any standardized test. We've already successfully applied our system to the GMAT, LSAT, MCAT, and GRE, to name just a few. Obviously, you need to be well versed in physics to do well on the AP Physics B Exam, but you should remember that any standardized test is partly a measure of your ability to think like the people who write standardized tests. This book will help you brush up on your AP Physics and prepare for the exam using our time-tested principle: Crack the system based on how the test is created.

We also offer books and online services that cover an enormous variety of education and career-related topics. If you're interested, check out our website at **PrincetonReview.com**.

The AP (Advanced Placement) Examinations are given in May of each year and offer students the opportunity to earn advanced placement or college credit for work they've done in high school or independently.

There are two versions of the AP Physics Exam; they're called Physics B and Physics C. The Physics B Exam covers a wide range of topics, but the questions are not too mathematically sophisticated. The Physics C Exam covers a smaller range of topics, but the questions are more challenging and calculus is used widely. In case you're curious, the Physics C Exam is actually composed of two separate exams: one in Mechanics and one in Electricity and Magnetism (E & M). You can take just the Mechanics, just the E & M, or both. Separate scores are reported for the Mechanics and E & M sections.

The Physics B exam consists of two sections: a multiple-choice section and a free-response section. Questions in the multiple-choice section are each followed by five possible responses (only one of which is correct), and your job, of course, is to choose the right answer. Each correct answer is worth one point, and there is no penalty for an incorrect answer. There are 70 multiple-choice questions and the time limit is 90 minutes. You may *not* use a calculator on the multiple-choice section.

The free-response section consists of six multi-part questions, which require you to actually write out your solutions, showing your work. The total amount of time for this section is 90 minutes, so you have an average of 15 minutes per question. Unlike the multiple choice section, which is scored by computer, the free-response section is graded by high school and college teachers. They have guidelines for awarding partial credit, so you don't need to correctly answer every part to get points. You are allowed to use a calculator (programmable or graphing calculators are okay, but ones with a typewriter-style keyboard are not) on the free-response section, and a table of equations is provided for your use on this section. The two sections—multiple choice and free response—are weighed equally, so each is worth 50 percent of your grade.

Grades on the AP Physics Exam are reported as a number: either 1, 2, 3, 4, or 5. The descriptions for each of these five numerical scores are as follows:

AP Exam Grade	Description
5	Extremely well qualified
4	Well qualified
3	Qualified
2	Possibly qualified
1	No recommendation

Colleges are generally looking for a 4 or 5, but some may grant credit for a 3. How well do you have to do to earn such a grade? Each test is curved, and specific cut-offs for each grade vary a little from year to year, but here's a rough idea of how many points you must earn—as a percentage of the maximum possible raw score—to achieve each of the grades 2 through 5:

AP Exam Grade	Percentage Needed
5	$\geq$ 75%
4	$\geq$ 60%
3	$\geq$ 45%
2	$\geq$ 35%

So, what's on the exams and how do you prepare for them? Here's a list of the major topics covered on the AP Physics B Exam, along with an approximate percentage of the questions in each topic (what the College Board calls the *percentage goal*):

Topic	Physics B
Newtonian Mechanics	**35%**
Kinematics	7%
Newton's Laws	9%
Work, Energy, and Power	5%
Linear Momentum	4%
Circular Motion and Rotation	4%
Oscillations and Gravitation	6%
Fluid Mechanics and Thermal Physics	**15%**
Fluid Mechanics	6%
Temperature and Heat	2%
Kinetic Theory and Thermodynamics	7%
Electricity and Magnetism	**25%**
Electrostatics	5%
Conductors and Capacitors	4%
Electric Circuits	7%
Magnetic Fields	4%
Electromagnetism	5%
Waves	**5%**
Optics	**10%**
Atomic and Nuclear Physics	**10%**

Naturally, it's important to be familiar with the topics—to understand the basics of the theory, to know the definitions of the fundamental quantities, and to recognize and be able to use the equations. Then, you must acquire practice at applying what you've learned to answering questions like you'll see on the exam. This book is designed to review all of the content areas covered on the exam, illustrated by hundreds of examples. Also, each chapter (except the first) is followed by practice multiple-choice and free-response questions, and perhaps even more important, *answers and explanations are provided for every example and question in this book*. You'll learn as much—if not more—from actively reading the solutions as you will from reading the text and examples. Also, two full-length practice tests (with solutions) are provided at the back of the book. The difficulty level of the examples and questions in this book is at or slightly above AP level, so if you have the time and motivation to attack these questions and learn from the solutions, you should feel confident that you can do your very best on the real thing.

SOME FRIENDLY ADVICE

Here are a few simple strategies: On the multiple-choice section, do not linger over any one question. Go through the exam and answer the questions on the topics you know well, leaving the tough ones for later when you make another pass through the section. All the questions are worth the same amount, so you don't want to run out of time and not get to questions you could have answered because you spent too much time agonizing over a few complex questions. No one is expected to answer all of them, so maximize the number you get right.

Make a copy of the table on page 437 and use it when you work on the problems in this book. A similar table will be provided when you take the exam, so you do not need to memorize everything on the page. However, it will help to be familiar with the table when you take the exam. Also, you should work the multiple-choice questions at the end of the chapters and the practice tests without a calculator. Students are not permitted calculators for the multiple-choice portion of the AP Physics test, so it is important that you practice mental arithmetic.

On the free-response section, be sure to show the graders what you're thinking. Write clearly—that is *very* important—and show your steps. If you make a mistake in one part and carry an incorrect result to a later part of the question, you can still earn valuable points if your method is correct. But the graders cannot give you credit for work they can't follow or can't read. And, where appropriate, be sure to include units on your final answers.

The most important advice we can give you for the free-response section of the AP Physics Exam is to read the questions carefully and answer according to exactly what the questions are asking you to do. Credit for the answers depends not only on the quality of the solutions but also on how they are explained. On the AP Physics Exam, the words "justify," "explain," "calculate," "what is," "determine," and "derive" have specific meanings, and the graders are looking for very precise approaches in your explanations in order to get maximum credit.

Questions that ask you to "justify" are looking for you both to show an understanding in words of the principles underlying physical phenomena and to perform the mathematical operations needed to arrive at the correct answer. The word "justify" as well as the word "explain" requires that you support your answers with text, equations, calculations, diagrams, or graphs. In some cases, the text or equations must elucidate physics fundamentals or laws, while in other cases they will serve to analyze the behavior of different values or different types of variables in the equation.

The word "calculate" requires you to show numerical or algebraic work to arrive at the final answer. In contrast, "what is" and "determine" questions signify that full credit may be given without showing mathematical work. Just remember, showing work that leads to the correct answer is always a good idea when possible, especially since showing work may still earn you partial credit even if the answer is not correct.

"Derive" questions are looking for a more specific approach, which entails beginning the solution with one or more fundamental equations and then arriving at the final answer through the proper use of mathematics, usually involving some algebra.

Your answers should be concise and focused and should not contain irrelevant or off-the-point information. If you make a mistake, you may either cross it out or erase it. The graders will not score crossed-out work. And, as we mentioned before, partial solutions may receive partial credit, so you should definitely show all your work, especially since correct answers without supporting work may lose credit. This is particularly true when you are asked to "justify" your answer, as graders are looking for some evidence of how you arrived at your solution. Finally, make sure that all of your numerical answers are in the appropriate units.

Sample test questions are also available directly from the College Board. You can request information about ordering such materials by contacting them by mail, phone, or on the web:

Advanced Placement Program
PO Box 6671
Princeton, NJ 08541-6671
(609) 771-7300/(888) 225-5427
www.collegeboard.com

We wish you all the best as you study for the AP Physics B Exam. Good luck!

Vectors

INTRODUCTION

Vectors will show up all over the place in our study of physics. Some physical quantities that are represented as vectors are displacement, velocity, acceleration, force, momentum, and electric and magnetic fields. Since vectors play such a recurring role, it's important to become comfortable working with them; the purpose of this chapter is to provide you with a mastery of the fundamental vector algebra we'll use in subsequent chapters. For now, we'll restrict our study to two-dimensional vectors (that is, ones that lie flat in a plane).

DEFINITION

A **vector** is a quantity that involves both magnitude and direction and obeys the **commutative law for addition,** which we'll explain in a moment. A quantity that does not involve direction is a **scalar**. For example, the quantity *55 miles per hour* is a scalar, while the quantity *55 miles per hour to the north* is a vector. Other examples of scalars include: mass, work, energy, power, temperature, and electric charge.

Vectors can be denoted in several ways, including:

$$\mathbf{A},\ A,\ \vec{\mathrm{A}},\ \bar{\mathrm{A}}$$

In textbooks, you'll usually see one of the first two, but when it's handwritten, you'll see one of the last two.

Displacement (which is net distance traveled plus direction) is the prototypical example of a vector:

$$\underbrace{\mathbf{A}}_{\text{displacement}} = \underbrace{4\text{ miles}}_{\text{magnitude}}\ \underbrace{\text{to the north}}_{\text{direction}}$$

When we say that vectors obey the commutative law for addition, we mean that if we have two vectors of the same type, for example, another displacement,

$$\mathbf{B} = \underbrace{3\text{ miles}}_{\text{magnitude}}\ \underbrace{\text{to the east}}_{\text{direction}}$$

then **A** + **B** must equal **B** + **A**. The vector sum **A** + **B** means *the vector A followed by* **B**, while the vector sum **B** + **A** means *the vector* **B** *followed by A*. That these two sums are indeed identical is shown in the following figure:

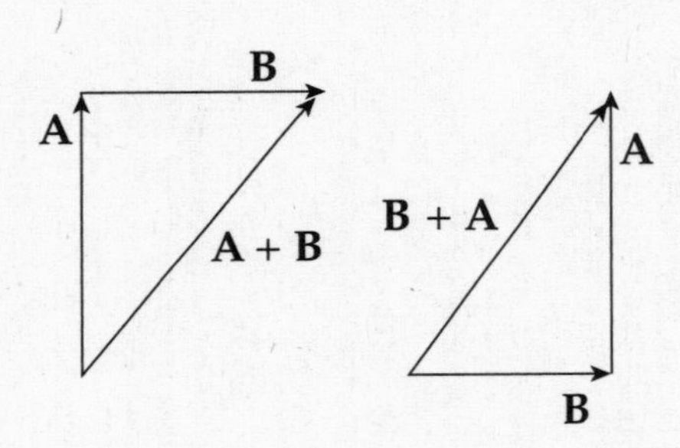

Two vectors are equal if they have the same magnitude and the same direction.

VECTOR ADDITION (GEOMETRIC)

The figure above illustrates how vectors are added to each other geometrically. Place the tail (the initial point) of one vector at the tip of the other vector, then connect the exposed tail to the exposed tip. It is essential that the original magnitude and direction of each vector be preserved. The vector formed is the sum of the first two. This is called the "tip-to-tail" method of vector addition.

Example 1.1 Add the following two vectors:

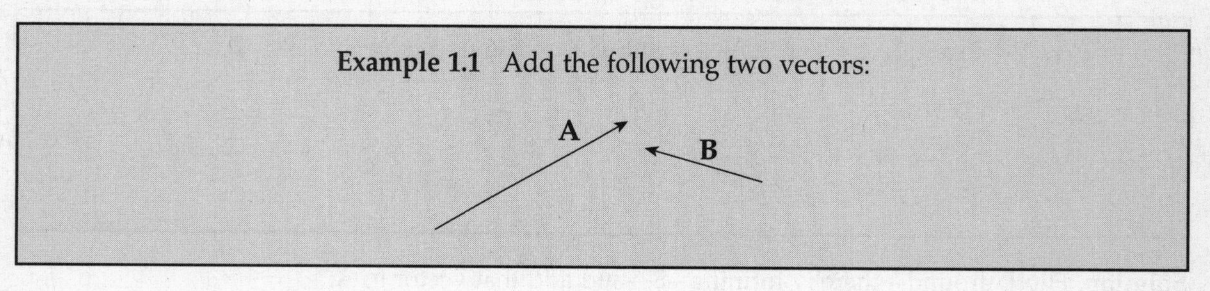

Solution. Place the tail of **B** at the tip of **A** and connect them:

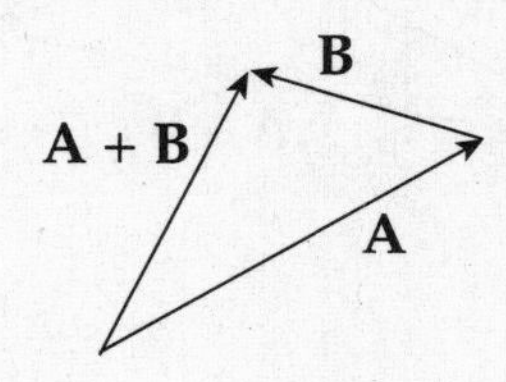

SCALAR MULTIPLICATION

A vector can be multiplied by a scalar (that is, by a number), and the result is a vector. If the original vector is **A** and the scalar is *k*, then the scalar multiple *k***A** is as follows:

$$\text{magnitude of } k\mathbf{A} = |k| \times (\text{magnitude of } \mathbf{A})$$

$$\text{direction of } k\mathbf{A} = \begin{cases} \text{the same as } \mathbf{A} \text{ if } k \text{ is positive} \\ \text{the opposite of } \mathbf{A} \text{ if } k \text{ is negative} \end{cases}$$

Example 1.2 Sketch the scalar multiples 2**A**, $\frac{1}{2}$**A**, –**A**, and –3**A** of the vector **A**:

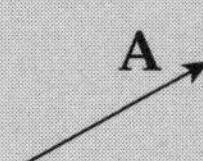

Solution.

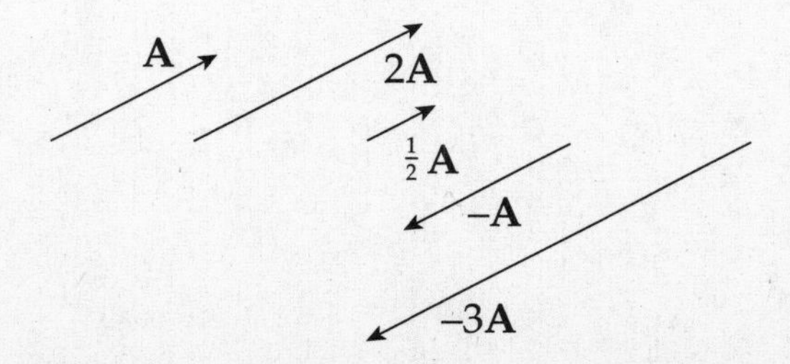

VECTOR SUBTRACTION (GEOMETRIC)

To subtract one vector from another, for example, to get **A** – **B**, simply form the vector –**B**, which is the scalar multiple (–1)**B**, and add it to **A**:

$$\mathbf{A} - \mathbf{B} = \mathbf{A} + (-\mathbf{B})$$

Example 1.3 For the two vectors **A** and **B**, find the vector **A** – **B**.

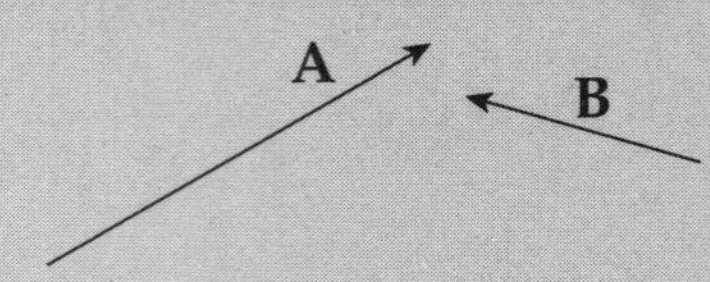

Solution. Flip **B** around—thereby forming –**B**—and add that vector to **A**:

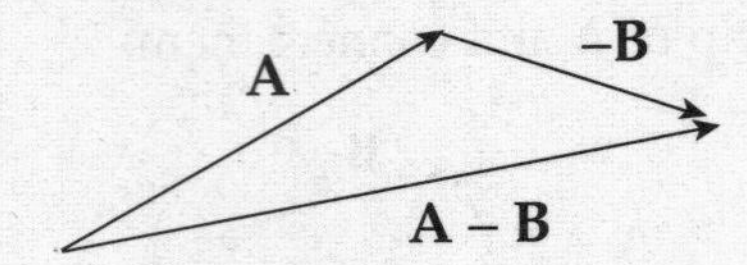

STANDARD BASIS VECTORS

Two-dimensional vectors, that is, vectors that lie flat in a plane, can be written as the sum of a horizontal vector and a vertical vector. For example, in the following diagram, the vector **A** is equal to the horizontal vector **B** plus the vertical vector **C**:

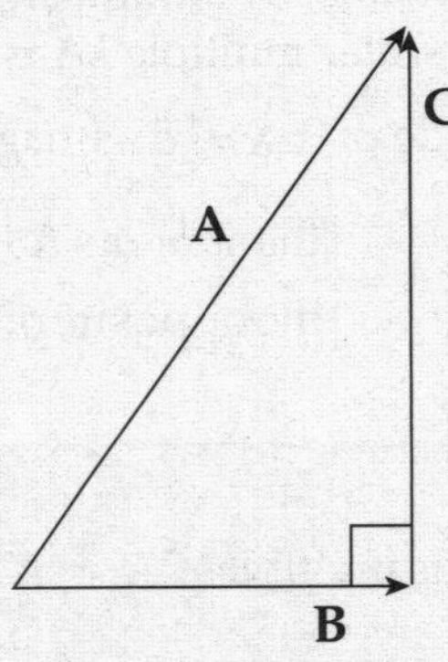

The horizontal vector is always considered a scalar multiple of what's called the **horizontal basis vector**, **i**, and the vertical vector is a scalar multiple of the **vertical basis vector**, **j**. Both of these special vectors have a magnitude of 1, and for this reason, they're called **unit vectors.** Unit vectors are often represented by placing a hat (caret) over the vector; for example, the unit **vectors i** and **j** are sometimes denoted $\hat{\mathbf{i}}$ and $\hat{\mathbf{j}}$.

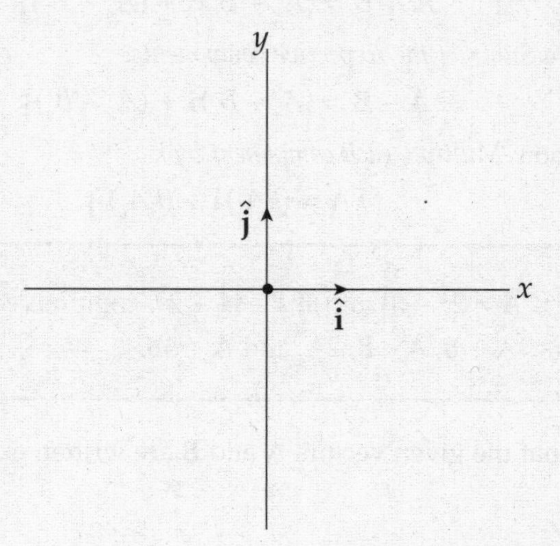

For instance, the vector **A** in the figure below is the sum of the horizontal vector $\mathbf{B} = 3\hat{\mathbf{i}}$ and the vertical vector $\mathbf{C} = 4\hat{\mathbf{j}}$.

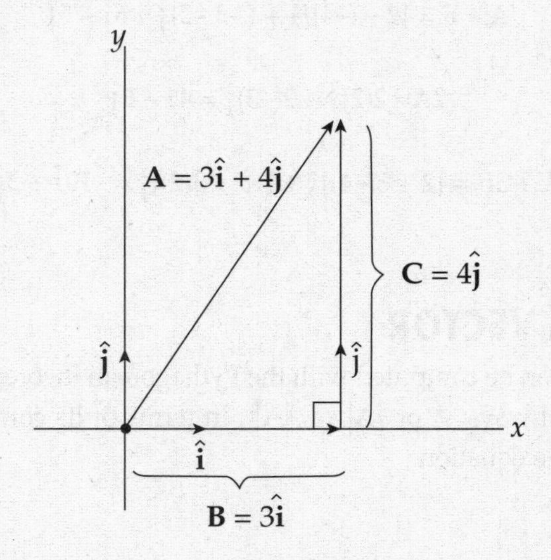

The vectors **B** and **C** are called the **vector components** of **A**, and the scalar multiples of $\hat{\mathbf{i}}$ and $\hat{\mathbf{j}}$ which give **A**—in this case, 3 and 4—are called the **scalar components** of **A**. So vector **A** can be written as the sum $A_x\hat{\mathbf{i}} + A_y\hat{\mathbf{j}}$, where A_x and A_y are the scalar components of **A**. The component A_x is called the **horizontal** scalar component of **A**, and A_y is called the **vertical** scalar component of **A**. In general, any vector in a plane can be described in this manner.

VECTOR OPERATIONS USING COMPONENTS

The use of components makes the vector operations of addition, subtraction, and scalar multiplication pretty straightforward:

Vector addition: *Add the respective components.*

$$\mathbf{A} + \mathbf{B} = (A_x + B_x)\hat{\mathbf{i}} + (A_y + B_y)\hat{\mathbf{j}}$$

Vector subtraction: *Subtract the respective components.*

$$\mathbf{A} - \mathbf{B} = (A_x - B_x)\hat{\mathbf{i}} + (A_y - B_y)\hat{\mathbf{j}}$$

Scalar multiplication: *Multiply each component by k.*

$$k\mathbf{A} = (kA_x)\hat{\mathbf{i}} + (kA_y)\hat{\mathbf{j}}$$

Example 1.4 If $\mathbf{A} = 2\hat{\mathbf{i}} - 3\hat{\mathbf{j}}$ and $\mathbf{B} = -4\hat{\mathbf{i}} + 2\hat{\mathbf{j}}$, compute each of the following vectors: **A + B, A – B, 2A,** and **A + 3B.**

Solution. It's very helpful that the given vectors **A** and **B** are written explicitly in terms of the standard basis vectors $\hat{\mathbf{i}}$ and $\hat{\mathbf{j}}$:

$$\mathbf{A} + \mathbf{B} = (2 - 4)\hat{\mathbf{i}} + (-3 + 2)\hat{\mathbf{j}} = -2\hat{\mathbf{i}} - \hat{\mathbf{j}}$$

$$\mathbf{A} - \mathbf{B} = [2 - (-4)]\hat{\mathbf{i}} + (-3 - 2)\hat{\mathbf{j}} = 6\hat{\mathbf{i}} - 5\hat{\mathbf{j}}$$

$$2\mathbf{A} = 2(2)\hat{\mathbf{i}} + 2(-3)\hat{\mathbf{j}} = 4\hat{\mathbf{i}} - 6\hat{\mathbf{j}}$$

$$\mathbf{A} + 3\mathbf{B} = [2 + 3(-4)]\hat{\mathbf{i}} + [-3 + 3(2)]\hat{\mathbf{j}} = -10\hat{\mathbf{i}} + 3\hat{\mathbf{j}}$$

MAGNITUDE OF A VECTOR

The magnitude of a vector can be computed with the Pythagorean theorem. The magnitude of vector **A** can be denoted in several ways: A or $|\mathbf{A}|$ or $\|\mathbf{A}\|$. In terms of its components, the magnitude of $\mathbf{A} = A_x\hat{\mathbf{i}} + A_y\hat{\mathbf{j}}$ is given by the equation

$$A = \sqrt{(A_x)^2 + (A_y)^2}$$

which is the formula for the length of the hypotenuse of a right triangle with sides of lengths A_x and A_y.

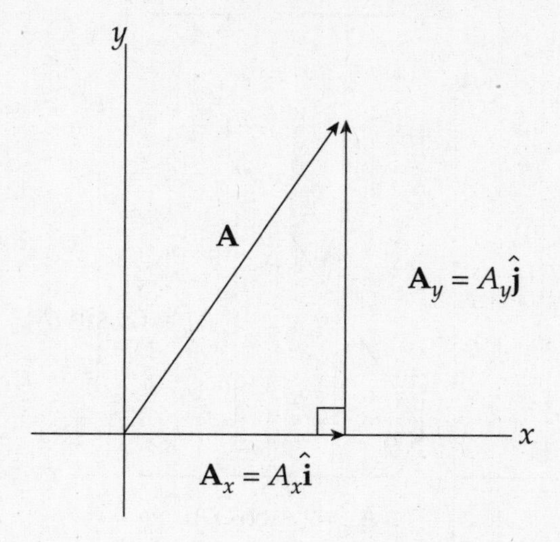

DIRECTION OF A VECTOR

The direction of a vector can be specified by the angle it makes with the positive x axis. You can sketch the vector and use its components (and an inverse trig function) to determine the angle. For example, if θ denotes the angle that the vector $\mathbf{A} = 3\hat{\mathbf{i}} + 4\hat{\mathbf{j}}$ makes with the $+x$ axis, then $\tan\theta = 4/3$, so $\theta = \tan^{-1}(4/3) = 53.1°$.

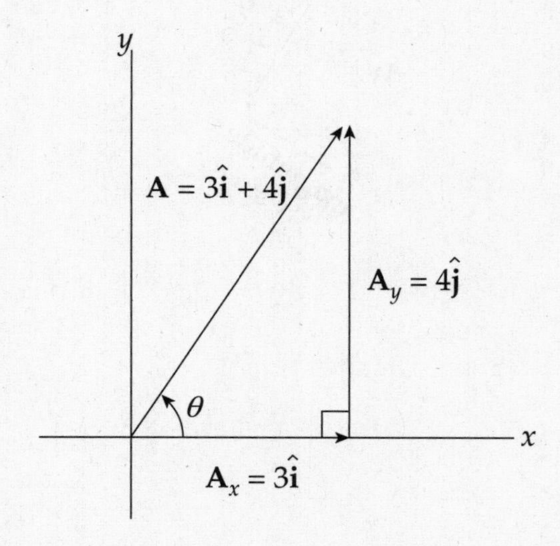

In general, the axis that θ is made to is known as the adjacent axis. The adjacent component is always going to get the cos θ. For example, if **A** makes the angle θ with the $+x$ axis, then its x- and y-components are $A \cos \theta$ and $A \sin \theta$, respectively (where A is the magnitude of **A**).

$$\mathbf{A} = \underbrace{(A\cos\theta)}_{A_x}\hat{\mathbf{i}} + \underbrace{(A\sin\theta)}_{A_y}\hat{\mathbf{j}}$$

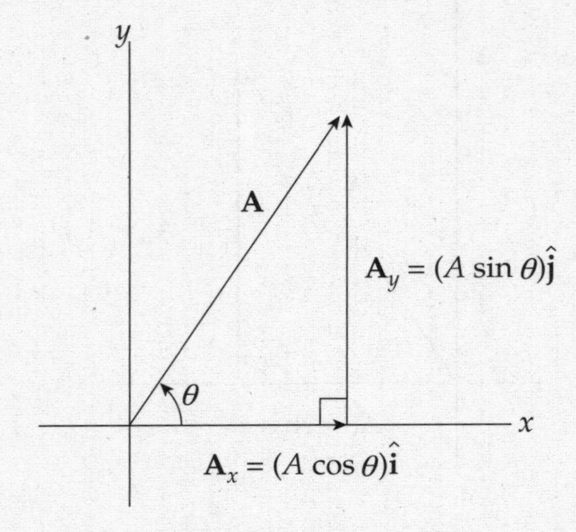

In general, any vector in the plane can be written in terms of two perpendicular component vectors. For example, vector **W** (shown below) is the sum of two component vectors whose magnitudes are $W \cos \theta$ and $W \sin \theta$:

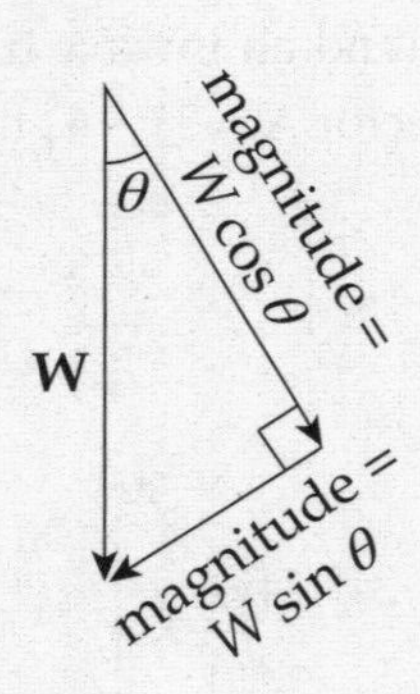

Kinematics

INTRODUCTION

Kinematics is the study of an object's motion in terms of its displacement, velocity, and acceleration. Questions such as, *How far does this object travel?* Or, *How fast and in what direction does it move?* Or, *At what rate does its speed change?* All properly belong to kinematics. In the next chapter, we will study **dynamics**, which delves more deeply into *why* objects move the way they do.

POSITION

Any object exists in some location somewhere in the universe. We call that location the object's position, but the position is meaningless without some reference point. We can arbitrarily choose one location and call it the origin. Many times it makes sense to let the object in question start at the origin, but this is not required and the laws of physics work no matter where you place the origin or whether you choose to make the up-direction positive and down-direction negative or vice versa. As we go more in depth in the problem-solving sections, you will notice that the mathematics sometimes simplifies greatly with a cleverly chosen coordinate system and origin.

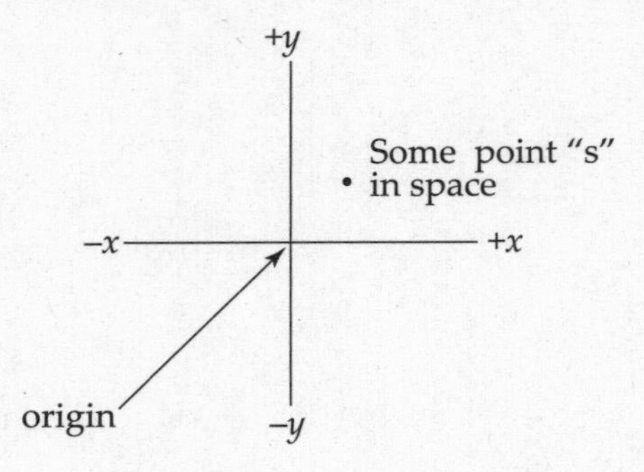

DISPLACEMENT

Displacement is an object's change in position. It's the vector that points from the object's initial position to its final position, regardless of the path actually taken. Since displacement means *change in position*, it is generically denoted $\Delta\mathbf{s}$, where Δ denotes *change in* and **s** means spatial location. (The letter **p** is not used for position because it's reserved for another quantity called **momentum**. We'll discuss momentum in Chapter 5.) If it's known that the displacement is horizontal, then it can be called $\Delta\mathbf{x}$; if the displacement is vertical, then it's $\Delta\mathbf{y}$. The magnitude of this vector is the *net* distance traveled and, sometimes, the word *displacement* refers just to this scalar quantity. Since a distance is being measured, the SI unit for displacement is the meter $[\Delta s] = \text{m}$.

> **Example 2.1** A rock is thrown straight upward from the edge of a 30 m cliff, rising 10 m then falling all the way down to the base of the cliff. Find the rock's displacement.

Solution. Displacement only refers to the object's initial position and final position, not the details of its journey. Since the rock started on the edge of the cliff and ended up on the ground 30 m below, its displacement is 30 m, downward.

Example 2.2 An infant crawls 5 m east, then 3 m north, then 1 m east. Find the magnitude of the infant's displacement.

Solution. Although the infant crawled a *total* distance of 5 + 3 + 1 = 9 m, this is not the displacement, which is merely the *net* distance traveled.

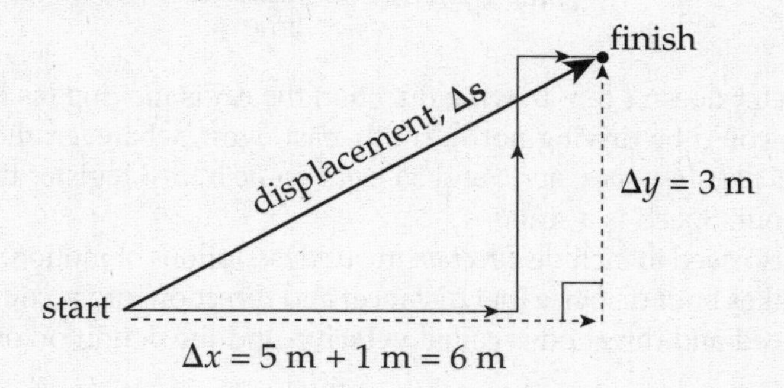

Using the Pythagorean theorem, we can calculate that the magnitude of the displacement is

$$\Delta s=\sqrt{(\Delta x)^2+(\Delta(y))^2}=\sqrt{(6\text{ m})^2+(3\text{ m})^2}=\sqrt{45\text{ m}^2}=6.7\text{ m}$$

Example 2.3 In a track-and-field event, an athlete runs exactly once around an oval track, a total distance of 500 m. Find the runner's displacement for the race.

Solution. If the runner returns to the same position from which she left, then her displacement is zero.

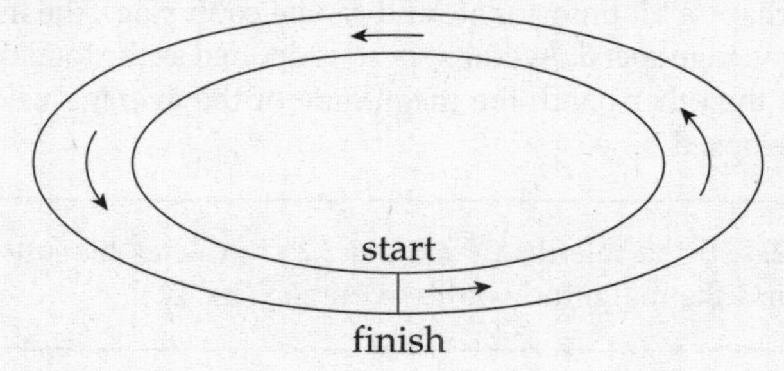

The *total* distance covered is 500 m, but the net distance—the displacement—is 0.

A Note about Notation

Δs is a more general term that works in space. The term x or "$\Delta x = x_f - x_i$" has a specific meaning that is defined in the x direction. However to be consistent with AP notation, and to avoid confusion between spacial location and speed, from this point on we will use x in our development of the concepts of speed, velocity, and acceleration. It should be noted that the concepts developed work equally well in the y direction.

SPEED AND VELOCITY

When we're in a moving car, the speedometer tells us how fast we're going; it gives us our speed. But what does it mean to have a speed of say, 10 m/s? It means that we're covering a distance of 10 meters every second. By definition, **average speed** is the ratio of the total distance traveled to the time required to cover that distance:

$$\text{average speed} = \frac{\text{total distance}}{\text{time}}$$

The car's speedometer doesn't care in what direction the car is moving (as long as the wheels are moving forward). You could be driving north, south, east, west, whatever; the speedometer would make no distinction. *55 miles per hour, north* and *55 miles per hour, east* register the same on the speedometer: 55 miles per hour. Speed is a scalar.

However, we will also need to include *direction* in our descriptions of motion. We just learned about displacement, which takes both distance (net distance) and direction into account. The single concept that embodies both speed and direction is called **velocity**, and the definition of average velocity is:

$$\text{average velocity} = \frac{\text{displacement}}{\text{time}}$$

$$\bar{\mathbf{v}} = \frac{\Delta \mathbf{x}}{\Delta t}$$

(The bar over the **v** means *average*.) Because Δx is a vector, $\bar{\mathbf{v}}$ is also a vector, and because Δt is a *positive* scalar, the direction of $\bar{\mathbf{v}}$ is the same as the direction of Δx. The magnitude of the velocity vector is called the object's **speed**, and is expressed in units of meters per second (m/s).

Note the distinction between speed and velocity. In everyday language, they're often used interchangeably. However, in physics, *speed* and *velocity* are technical terms whose definitions are not the same. *Velocity is speed plus direction.* An important note: The magnitude of the velocity is the speed. However (and this is perhaps a bit unfortunate and can be confusing), the magnitude of the average velocity is not called the average speed. Average speed is defined as the total distance traveled divided by the elapsed time. On the other hand, the magnitude of the average velocity is the net distance traveled divided by the elapsed time.

Example 2.4 If the infant in Example 2.2 completes his journey in 20 seconds, find the magnitude of his average velocity.

Solution. Since the displacement is 6.7 m, the magnitude of his average velocity is

$$\bar{v} = \Delta x/\Delta t = (6.7\text{ m})/(20\text{ s}) = 0.34\text{ m/s}$$

Example 2.5 Assume that the runner in Example 2.3 completes the race in 1 minute and 18 seconds. Find her average speed and the magnitude of her average velocity.

Solution. *Average speed is total distance divided by elapsed time.* Since the length of the track is 500 m, the runner's average speed was (500 m) / (78 s) = 6.4 m/s. However, since her displacement was zero, her average velocity was zero also: $\bar{v} = \Delta x/\Delta t = (0\text{ m}) / (78\text{ s}) = 0\text{ m/s}$.

Example 2.6 Is it possible to move with constant speed but not constant velocity? Is it possible to move with constant velocity but not constant speed?

Solution. The answer to the first question is *yes*. For example, if you set your car's cruise control at 55 miles per hour but turn the steering wheel to follow a curved section of road, then the direction of your velocity changes (which means your velocity is not constant), even though your speed doesn't change.

The answer to the second question is *no*. Velocity means speed and direction; if the velocity is constant, then that means both speed and direction are constant. If speed were to change, then the velocity vector's magnitude would change (by definition), which immediately implies that the vector changes.

ACCELERATION

When you step on the gas pedal in your car, the car's speed increases; step on the brake and the car's speed decreases. Turn the wheel, and the car's direction of motion changes. In all of these cases, the velocity changes. To describe this change in velocity, we need a new term: **acceleration**. In the same way that velocity measures the rate-of-change of an object's position, acceleration measures the rate-of-change of an object's velocity. An object's average acceleration is defined as follows:

$$\text{average acceleration} = \frac{\text{change in velocity}}{\text{time}}$$

$$\bar{\mathbf{a}} = \frac{\Delta \mathbf{v}}{\Delta t}$$

The units of acceleration are meters per second, per second: $[a] = \text{m/s}^2$. Because $\Delta\mathbf{v}$ is a vector, $\bar{\mathbf{a}}$ is also a vector; and because Δt is a *positive* scalar, the direction of $\bar{\mathbf{a}}$ is the same as the direction of $\Delta\mathbf{v}$.

Furthermore, if we take an object's original direction of motion to be positive, then an increase in speed corresponds to a positive acceleration, while a decrease in speed corresponds to a negative acceleration (deceleration). However, if an object's original direction is negative, then an increase in speed corresponds to a negative acceleration, while a decrease in speed corresponds to a positive acceleration. So, when the velocity and acceleration are in the same direction (they have the same sign), the object speeds up. When the velocity and acceleration are in the opposite direction (one is negative and one is positive, no matter which is which) the object slows down.

Note that an object can accelerate even if its speed doesn't change. (Again, it's a matter of not allowing the everyday usage of the word *accelerate* to interfere with its technical, physics usage.) This is because acceleration depends on $\Delta\mathbf{v}$, and the velocity vector $\mathbf{v}$ changes if (1) speed changes, or (2) direction changes, or (3) both speed and direction change. For instance, a car traveling around a circular racetrack is constantly accelerating even if the car's *speed* is constant, because the direction of the car's velocity vector is constantly changing.

> **Example 2.7** A car is traveling in a straight line along a highway at a constant speed of 80 miles per hour for 10 seconds. Find its acceleration.

Solution. Since the car is traveling at a constant velocity, its acceleration is zero. If there's no change in velocity, then there's no acceleration.

> **Example 2.8** A car is traveling along a straight highway at a speed of 20 m/s. The driver steps on the gas pedal and, 3 seconds later, the car's speed is 32 m/s. Find its average acceleration.

Solution. Assuming that the direction of the velocity doesn't change, it's simply a matter of dividing the change in velocity, 32 m/s – 20 m/s = 12 m/s, by the time interval during which the change occurred: $\bar{a} = \Delta v/\Delta t = (12 \text{ m/s}) / (3 \text{ s}) = 4 \text{ m/s}^2$.

> **Example 2.9** Spotting a police car ahead, the driver of the car in the previous example slows from 32 m/s to 20 m/s in 2 seconds. Find the car's average acceleration.

Solution. Dividing the change in velocity, 20 m/s – 32 m/s = –12 m/s, by the time interval during which the change occurred, 2 s, give us $\bar{\mathbf{a}} = \Delta \text{v}/\Delta t = (-12 \text{ m/s}) / (2 \text{ s}) = -6 \text{ m/s}^2$. The negative sign here means that the direction of the acceleration is opposite the direction of the velocity, which describes slowing down.

UNIFORMLY ACCELERATED MOTION AND THE BIG FIVE

The simplest type of motion to analyze is motion in which the acceleration is *constant* (possibly equal to zero). Although true uniform acceleration is rarely achieved in the real world, many common motions are governed by approximately constant acceleration and, in these cases, the kinematics of uniformly accelerated motion provide a pretty good description of what's happening. Notice that if the acceleration is constant, then taking an average yields nothing new, so $\bar{a} = a$.

Another restriction that will make our analysis easier is to consider only motion that takes place along a straight line. In these cases, there are only two possible directions of motion. One is positive, and the opposite direction is negative. Most of the quantities we've been dealing with—displacement, velocity, and acceleration—are vectors, which means that they include both a magnitude and a direction. With straight-line motion, direction can be specified simply by attaching a + or – sign to the magnitude of the quantity. Therefore, although we will often abandon the use of bold letters to denote the vector quantities of displacement, velocity, and acceleration, the fact that these quantities include direction will still be indicated by a positive or negative sign.

Let's review the quantities we've seen so far. The fundamental quantities are position (x), velocity (v), and acceleration (a). Acceleration is a change in velocity, from an initial velocity (v_i or v_0) to a final velocity (v_f or simply v—with no subscript). And, finally, the motion takes place during some elapsed time interval, Δt. Also, if we agree to start our clocks at time $t_i = 0$, then $\Delta t = t_f - t_i = t - 0 = t$, so we can just write t instead of Δt in the first four equations. This simplification in notation makes these equations a little easier to write down. Therefore, we have five kinematics quantities: Δx, v_0, v, a, and Δt.

These five quantities are related by a group of five equations that we call the *Big Five*. They work in cases where acceleration is uniform, which are the cases we're considering.

		Variable that's missing
Big Five #1*:	$\Delta x = \bar{v}t$	a
Big Five #2**:	$v = v_o + at$	x
Big Five #3**:	$x = x_o + v_o t + \frac{1}{2}at^2$	v
Big Five #4:	$x = x_o + vt - \frac{1}{2}at^2$	v_o
Big Five #5**:	$v^2 = v_o{}^2 + 2a(x - x_o)$	t

* Because the acceleration is uniform, this may also be written $\Delta x = \frac{1}{2}(v_o + v)t$

** Historically, these three equations have been given on the AP B equation sheet for the free-response section. Again, no equations are provided for the multiple-choice section.

Each of the Big Five equations is missing one of the five kinematic quantities. The way you decide which equation to use when solving a problem is to determine which of the kinematic quantities is missing from the problem—that is, which quantity is neither given nor asked for—and then use the equation that doesn't contain that variable. For example, if the problem never mentions the final velocity—v is neither given nor asked for—the equation that will work is the one that's missing v. That's Big Five #3.

Big Five #1 and #2 are simply the definitions of $\bar{v}$ and $\bar{a}$ written in forms that don't involve fractions. The other Big Five equations can be derived from these two definitions and the equation $\bar{v} = \frac{1}{2}(v_0 + v)$, using a bit of algebra.

Example 2.10 An object with an initial velocity of 4 m/s moves along a straight axis under constant acceleration. Three seconds later, its velocity is 14 m/s. How far did it travel during this time?

Solution. We're given v_0, Δt, and v, and we're asked for x. So a is missing; it isn't given and it isn't asked for, and we use Big Five #1:

$$x = \bar{v}t = \frac{1}{2}(v_0 + v)t = \frac{1}{2}(4\text{ m/s} + 14\text{ m/s})(3\text{ s}) = 27\text{ m}$$

Note: This last equation could also have been written as

$$x = \bar{v}t = \frac{1}{2}(v_0 + v)t = \frac{1}{2}(4 + 14)(3) = 27\text{ m}$$

That is, it's okay to leave off the units in the middle of the calculation *as long as you remember to include them in your final answer*. Leaving units off of your final answer will cost you points on the AP exam.

Example 2.11 A car that's initially traveling at 10 m/s accelerates uniformly for 4 seconds at a rate of 2 m/s², in a straight line. How far does the car travel during this time?

Solution. We're given v_0, t, and a, and we're asked for x. So, v is missing; it isn't given and it isn't asked for, and we use Big Five #3:

$$x = x_0 + v_0\Delta t + \frac{1}{2}a(\Delta t)^2 = (10\text{ m/s})(4\text{ s}) + \frac{1}{2}(2\text{ m/s}^2)(4\text{ s})^2 = 56\text{ m}$$

Example 2.12 A rock is dropped off a cliff that's 80 m high. If it strikes the ground with an impact velocity of 40 m/s, what acceleration did it experience during its descent?

Solution. If something is dropped, then that means it has no initial velocity: $v_0 = 0$. So, we're given v_0, Δx, and v, and we're asked for a. Since t is missing, we use Big Five #5:

$$v^2 = v_0^2 + 2a(x - x_0) \Rightarrow v^2 = 2a(x - x_0) \quad (\text{since } v_0 = 0)$$

$$a = \frac{v^2}{2(x - x_0)} = \frac{(40\text{ m/s})^2}{2(80\text{ m})} = 10\text{ m/s}^2$$

Note that since a has the same sign as $(x - x_0)$, the acceleration vector points in the same direction as the displacement vector. This makes sense here, since the object moves downward and the acceleration it experiences is due to gravity, which also points downward.

KINEMATICS WITH GRAPHS

So far, we have dealt with kinematics problems algebraically, but you should also be able to handle kinematics questions in which information is given graphically. The two most popular graphs in kinematics are position-vs.-time graphs and velocity-vs.-time graphs. For example, consider an object that's moving along an axis in such a way that its position x as a function of time t is given by the following position-vs.-time graph:

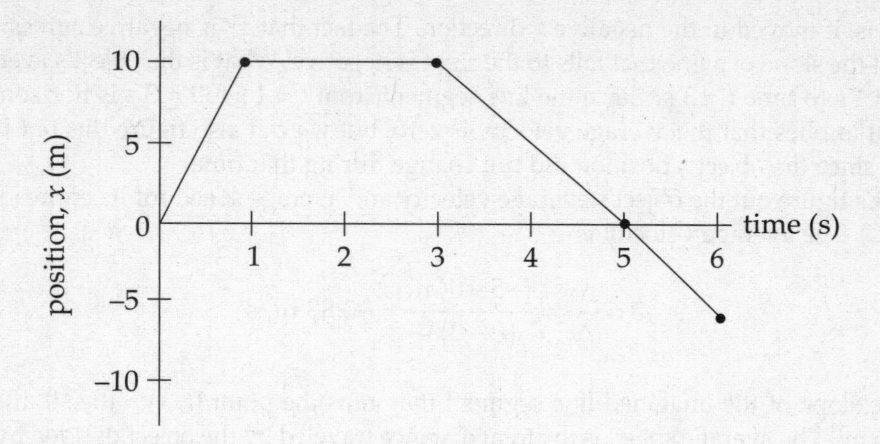

What does this graph tell us? It says that at time $t = 0$, the object was at position $x = 0$. Then, in the first second, its position changed from $x = 0$ to $x = 10$ m. Then, at time $t = 1$ s to 3 s it stopped. From $t = 3$ s to t = 6 s, it reversed direction, reaching $x = 0$ at time $t = 5$ s, and continued, reaching position $x = -5$ m at time $t = 6$ s. Notice how economically the graph embodies all this information!

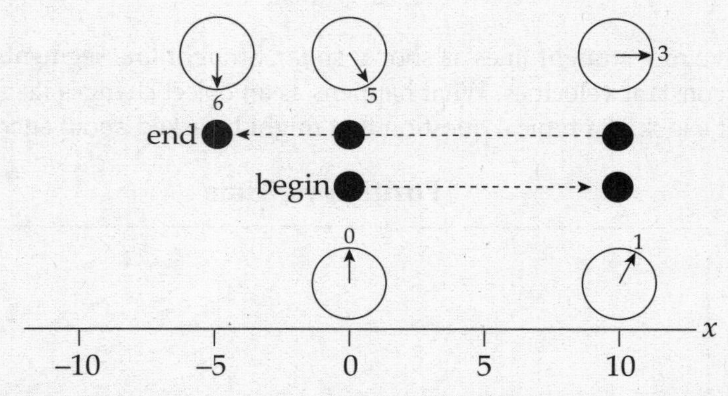

We can also determine the object's average velocity (and average speed) during particular time intervals. For example, its average velocity from time $t = 0$ to time $t = 1$ s is equal to the object's displacement, $10 - 0 = 10$ m, divided by the elapsed time, 2 s:

$$\bar{v} = \frac{\Delta x}{\Delta t} = \frac{(10-0)\text{ m}}{(1-0)\text{ s}} = 10 \text{ m/s}$$

Note, however, that the ratio that defines the average velocity, $\Delta x/\Delta t$, also defines the slope of the x vs. t graph. Therefore, we know the following important fact:

The slope of a position-vs.-time graph gives the velocity.

What was the average velocity from time $t = 3$ s to time $t = 6$ s? The slope of the line segment joining the point $(t, x) = (3\text{ s}, 10\text{ m})$ to the point $(t, x) = (6\text{ s}, -5\text{m})$ is

$$\bar{v} = \frac{\Delta x}{\Delta t} = \frac{(-5-10)\text{ m}}{(6-3)\text{ s}} = -5 \text{ m/s}$$

The fact that $\bar{v}$ is negative tells us that the object's displacement was negative during this time interval; that is, it moved in the negative x direction. The fact that $\overline{v}$ is negative agrees with the observation that the slope of a line that falls to the right is negative. What is the object's average velocity from time $t = 1$ s to time $t = 3$ s? Since the line segment from $t = 1$ s to $t = 3$ s is horizontal, its slope is zero, which implies that the average velocity is zero, but we can also figure this out from looking at the graph, since the object's position did not change during that time.

Finally, let's figure out the object's average velocity and average speed for its entire journey (from $t = 0$ to $t = 6$ s). The average velocity is

$$\overline{v} = \frac{\Delta x}{\Delta t} = \frac{(-5-0)\text{ m}}{(6-0)\text{ s}} = -0.83\text{ m/s}$$

This is the slope of the imagined line segment that joins the point $(t, x) = (0\text{ s}, 0\text{ m})$ to the point $(t, x) = (6\text{ s}, -5\text{ m})$. The average speed is the total distance traveled by the object divided by the elapsed time. In this case, notice that the object traveled 10 m in the first 1 s, then 15 m (albeit backwards) in the next 3 s. It covered no additional distance from $t = 1$ s to $t = 3$ s. Therefore, the total distance traveled by the object is $d = 10 + 15 = 25$ m, which took 6 s, so

$$\text{average speed} = \frac{d}{\Delta t} = \frac{25\text{ m}}{6\text{ s}} = 4.2\text{ m/s}$$

Not all graphs have nice straight lines as shown so far. Straight line segments represent constant slopes and therefore constant velocities. What happens as an object changes its velocity? The "lines" become "curves." Let's look at a typical question that might be asked about such a curve.

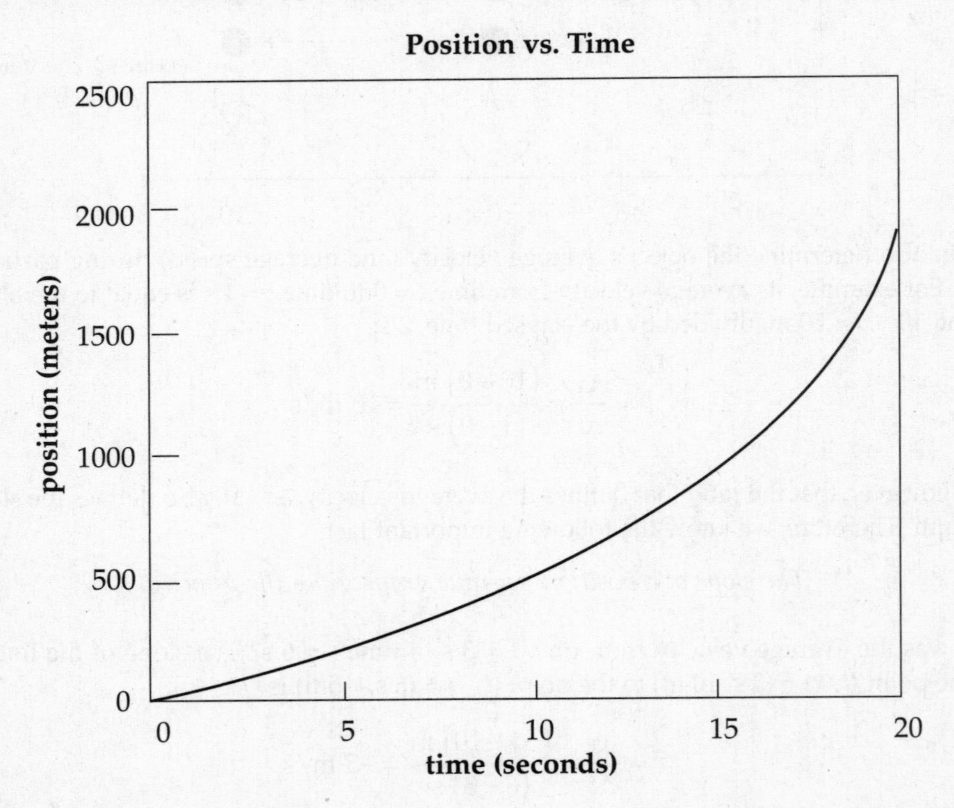

Example 2.13

a) What is the average velocity from 0 to 10 seconds?

b) What is the average velocity from 10 to 20 seconds?

c) What is the average velocity from 0 to 20 seconds?

Solution. This is familiar territory. To find the average velocity use

a) $v_{ave} = \dfrac{\Delta x}{\Delta t} \Rightarrow \dfrac{(500-0)}{(10-0)} \Rightarrow 50 \text{ m/s}$

b) $v_{ave} = \dfrac{\Delta x}{\Delta t} \Rightarrow \dfrac{(2000-500)}{(20-10)} \Rightarrow 150 \text{ m/s}$

c) $v_{ave} = \dfrac{\Delta x}{\Delta t} \Rightarrow \dfrac{(2000-0)}{(20-0)} \Rightarrow 100 \text{ m/s}$

The instantaneous velocity is the velocity at a given moment in time. When you drive in a car and look down at the speedometer, you see your instantaneous velocity at that time. To find the true instantaneous velocity from a position-vs.-time graph you need some calculus. However, we can get a very good approximation by finding the position a very small amount of time before the time in question and the position a very small time after the time in question.

For example, what if you wish to find the instantaneous velocity at 10 seconds? The velocity from 9–10 seconds is close to the velocity at 10 seconds, but it is still a bit too slow. The velocity from 10–11 seconds is also close to the velocity at 10 seconds, but it is still a bit too fast. You can find the middle ground between these two ideas, or the slope of the line that connects the point before and the point after 10 seconds. This is very close to the instantaneous velocity. A true tangent line touches the curve at only one point, but this line is close enough for our purposes.

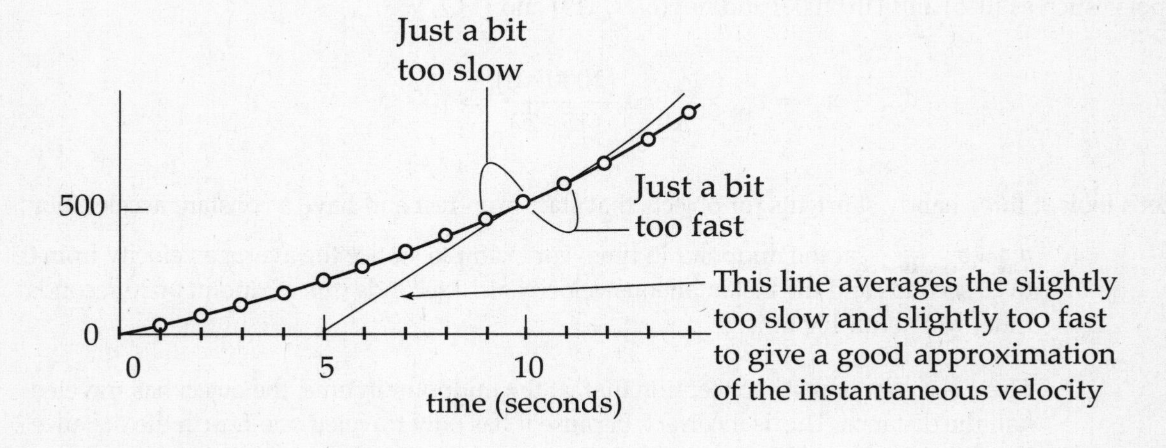

Example 2.14 Using the graph below, what is the instantaneous velocity at 10 seconds?

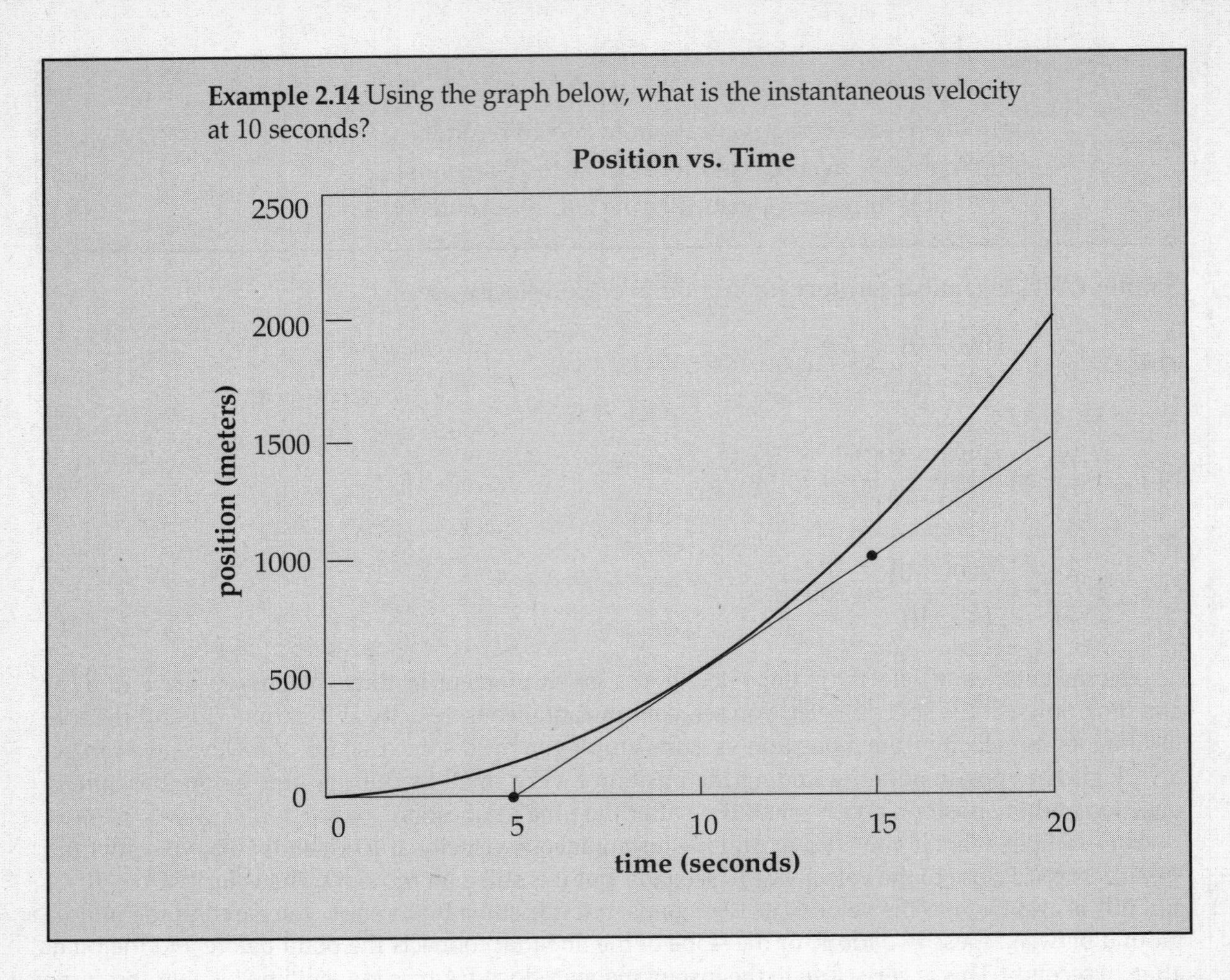

Solution. Draw a tangent line. Find the slope of the tangent by picking any two points on the tangent line. It usually helps if the points are kind of far apart and it also helps if points can be found at "easy spots" such as (0, 5) and (15, 1000) and not (6.27, 113) and (14.7, 983)

$$v_{ins} = v_{\tan} = \frac{\Delta x}{\Delta t} \Rightarrow \frac{(1000-0)}{(15-5)} \Rightarrow 100 \text{ m/s}$$

Let's look at three handy shortcuts for objects that start from rest and have a constant acceleration:

1. $v_{\text{ave}} = v_{\text{instantaneous}}$ at the midpoint in time. For example, notice the average velocity from 0 to 20 seconds and the instantaneous velocity at 10 seconds (the midpoint of 20 seconds time) were both 100 m/s.

2. There is a common misconception that, at the midpoint in time, the object has traveled half the distance. This is incorrect, because it has only traveled one-fourth the distance. For this example, at 10 seconds (the midpoint of 20 seconds time) the object has only traveled 500 m out of the 2000 m total or one-fourth the distance.

3. The instantaneous velocity at any point is twice the average velocity from $t = 0$ to that point. For example, the average velocity from 0 to 10 seconds is 50 m/s. The instantaneous velocity at 10 seconds was 100 m/s, or twice the average velocity of 50 m/s.

Please note: Shortcuts 2 and 3 are true only for objects that start from rest and have a constant acceleration.

Let's next consider an object moving along a straight axis in such a way that its velocity, v, as a function of time, t, is given by the following velocity-vs.-time graph:

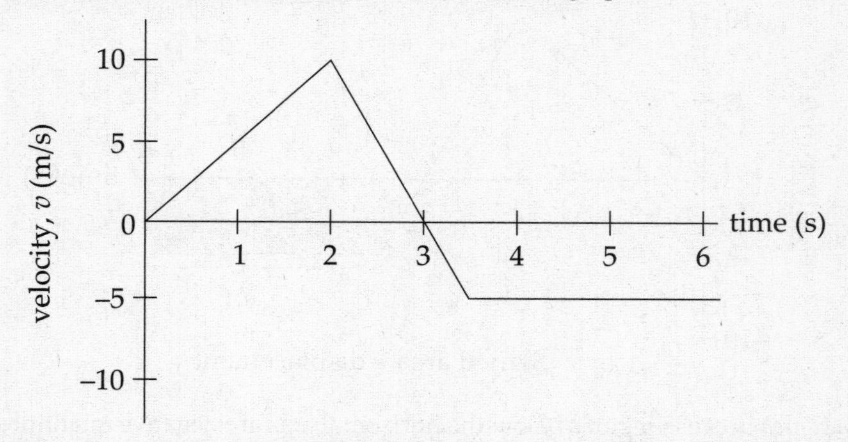

What does this graph tell us? It says that, at time $t = 0$, the object's velocity was $v = 0$. Over the first two seconds, its velocity increased steadily to 10 m/s. At time $t = 2$ s, the velocity then began to decrease (eventually becoming $v = 0$, at time $t = 3$ s). The velocity then became negative after $t =$ 3 s, reaching $v = -5$ m/s at time $t = 3.5$ s. From $t = 3.5$ s on, the velocity remained a steady –5 m/s.

What can we ask about this motion? First, the fact that the velocity changed from $t = 0$ to $t = 2$ s tells us that the object accelerated. The acceleration during this tme was

$$a = \frac{\Delta v}{\Delta t} = \frac{(10-0)\text{ m/s}}{(2-0)\text{ s}} = 5\text{ m/s}^2$$

Note, however, that the ratio that defines the acceleration, $\Delta v/\Delta t$, also defines the slope of the v vs. t graph. Therefore,

The slope of a velocity-vs.-time graph gives the acceleration.

What was the acceleration from time $t = 2$ s to time $t = 3.5$ s? The slope of the line segment joining the point (t, v) = (2 s, 10 m/s) to the point (t, v) = (3.5 s, –5 m/s) is

$$a = \frac{\Delta v}{\Delta t} = \frac{(-5-10)\text{ m/s}}{(3.5-2)\text{ s}} = -10\text{ m/s}^2$$

The fact that a is negative tells us that the object's velocity change was negative during this time interval; that is, the object accelerated in the negative direction. In fact, after time $t = 3$ s, the velocity became more negative, indicating that the direction of motion was negative at increasing speed. What is the object's acceleration from time $t = 3.5$ s to time $t = 6$ s? Since the line segment from $t = 3.5$ s to $t = 6$ s is horizontal, its slope is zero, which implies that the acceleration is zero, but you can also see this from looking at the graph; the object's velocity did not change during this time interval.

Another question can be asked when a velocity-vs.-time graph is given: How far did the object travel during a particular time interval? For example, let's figure out the displacement of the object from time $t = 4$ s to time $t = 6$ s. During this time interval, the velocity was a constant -5 m/s, so the displacement was $\Delta x = v\Delta t = (-5 \text{ m/s})(2 \text{ s}) = -10$ m.

Geometrically, we've determined the area between the graph and the horizontal axis. After all, the area of a rectangle is *base* × *height* and, for the shaded rectangle shown below, the *base* is Δt, and the *height* is v. So, *base* × *height* equals $\Delta t \times v$, which is displacement.

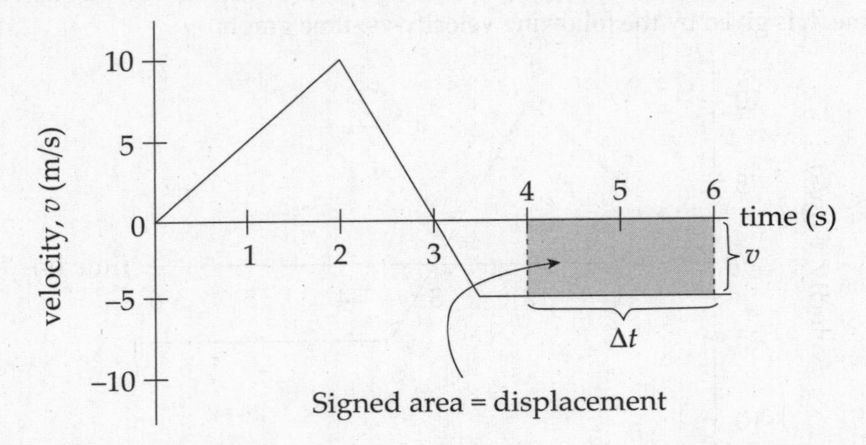

We say *signed area* because regions below the horizontal axis are negative quantities (since the object's velocity is negative, its displacement is negative). Therefore, counting areas above the horizontal axis as positive and areas below the horizontal axis as negative, we can make the following claim:

Given a velocity-vs.-time graph, the area between the graph and the t axis equals the object's displacement.

What is the object's displacement from time $t = 0$ to $t = 3$ s? Using the fact that displacement is the area bounded by the velocity graph, we figure out the area of the triangle shown below:

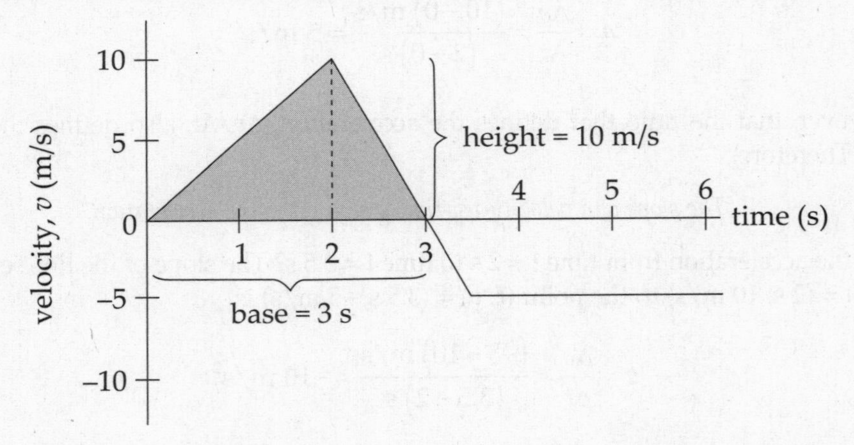

Since the area of a triangle is $\left(\frac{1}{2}\right)$ × base × height, we find that $\Delta x = \frac{1}{2}(3 \text{ s})(10 \text{ m/s}) = 15 \text{ m}$.

QUALITATIVE GRAPHING

Beyond all the math, you are at a clear advantage when you start to recognize that position-vs.-time and velocity-vs.-time graphs have a few basic shapes, and that all the graphs you will see will be some form of these basic shapes. Having a feel for these building blocks will go a long way toward understanding kinematics graphs in physics.

Either of the following two graphs represents something that is not moving.

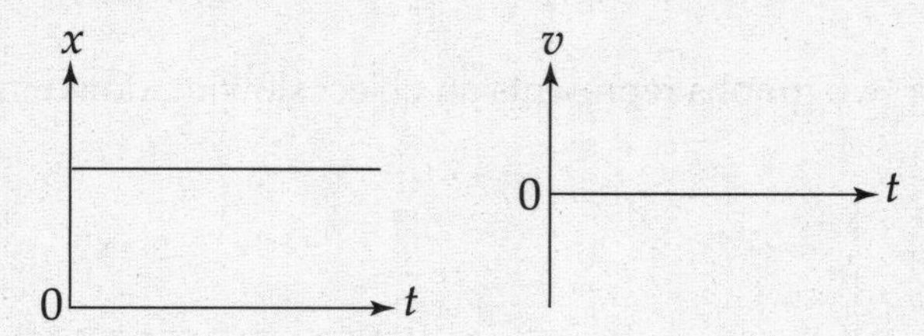

Either of the following two graphs represents an object moving at a constant velocity in the positive direction.

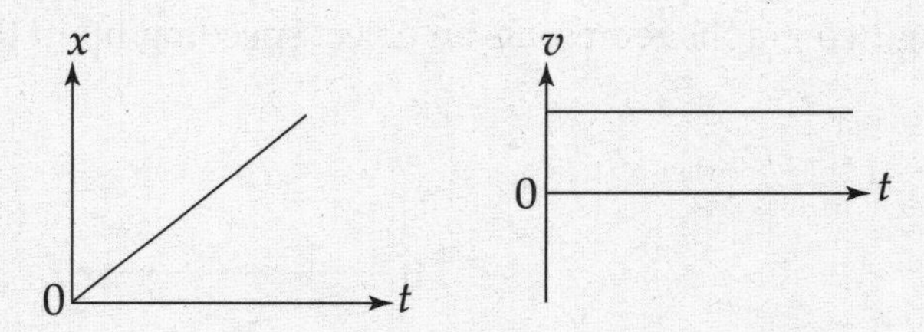

Either of the following two graphs represents an object moving at a constant velocity in the negative direction.

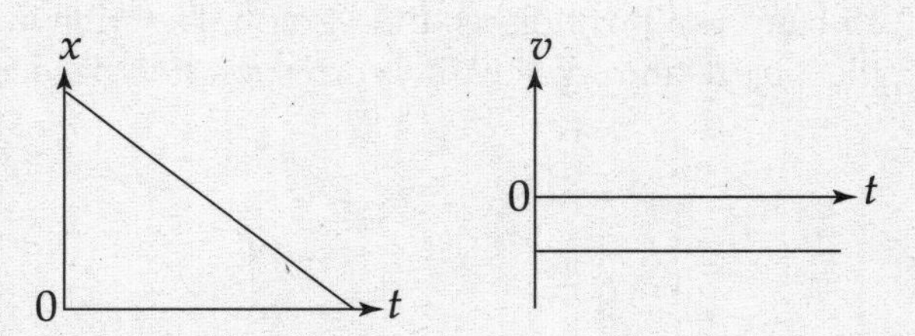

Either of the following two graphs represents an object speeding up in the positive direction.

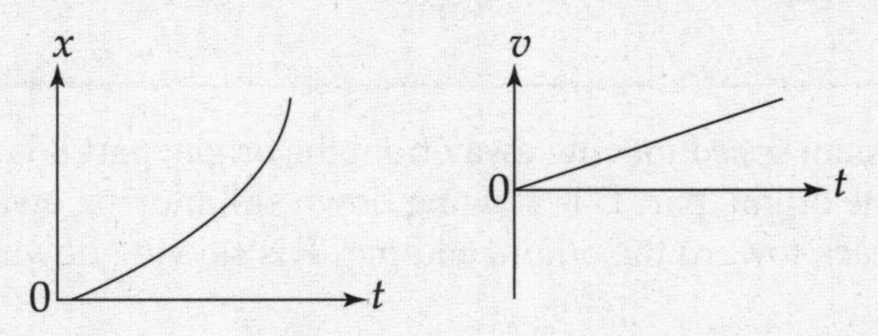

Either of the following two graphs represents an object slowing down in the positive direction.

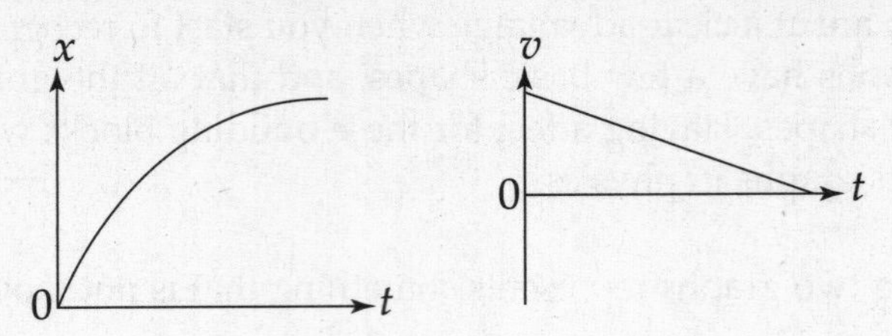

Either of the following two graphs represents an object slowing down in the negative direction.

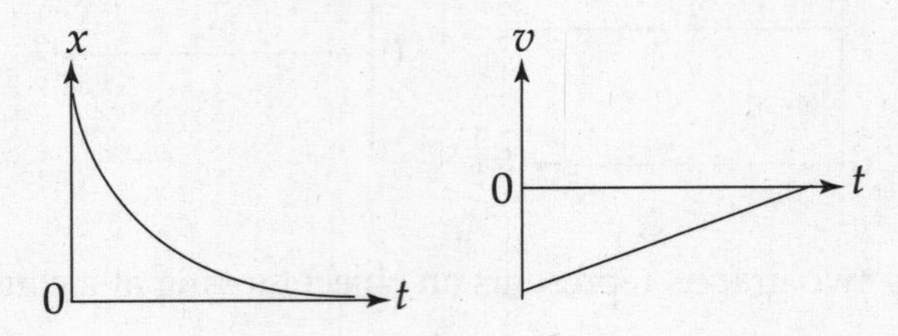

Either of the following two graphs represents an object speeding up in the negative direction.

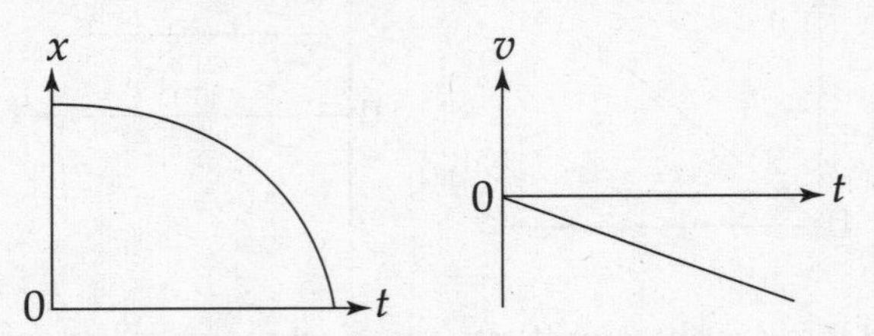

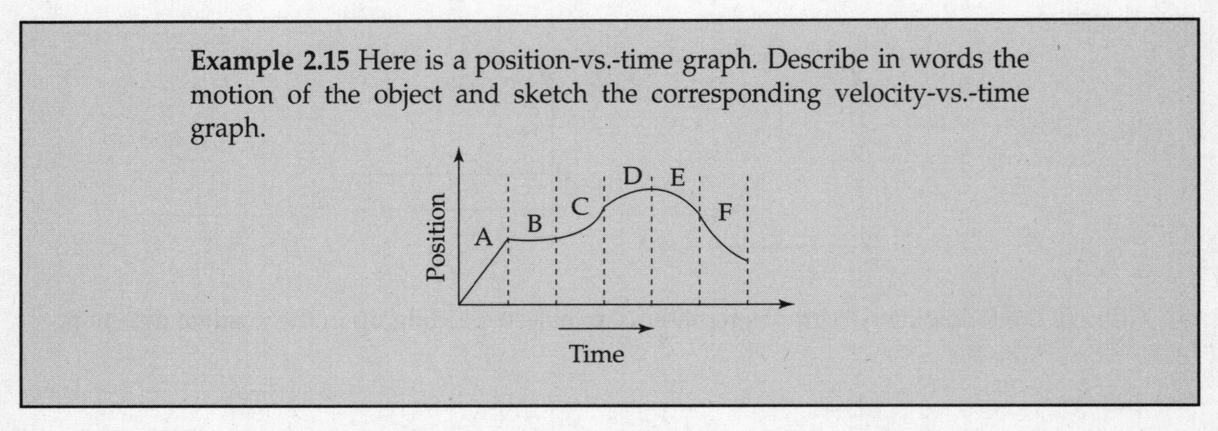

Example 2.15 Here is a position-vs.-time graph. Describe in words the motion of the object and sketch the corresponding velocity-vs.-time graph.

Solution. Part A is a constant speed moving away from the origin, part B is at rest, part C is speeding up moving away from the origin, part D is slowing down still moving away from the origin, part E is speeding up moving back toward the origin, and part F is slowing down moving back toward the origin

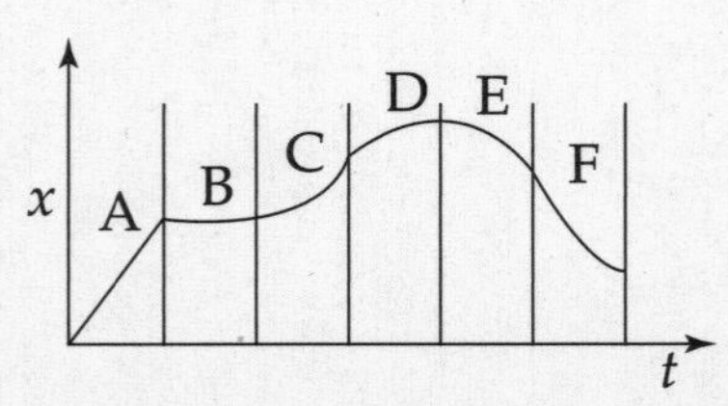

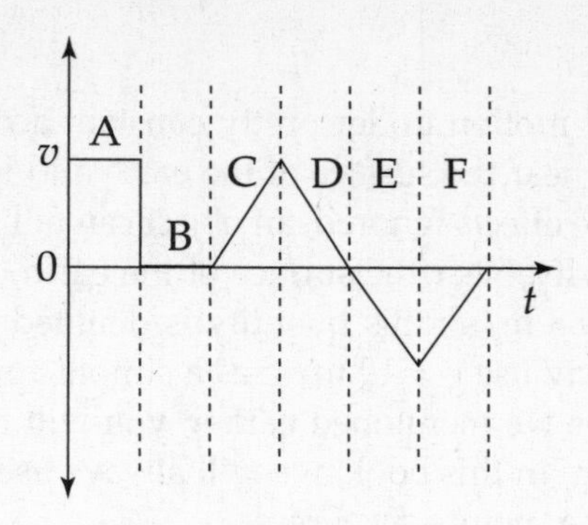

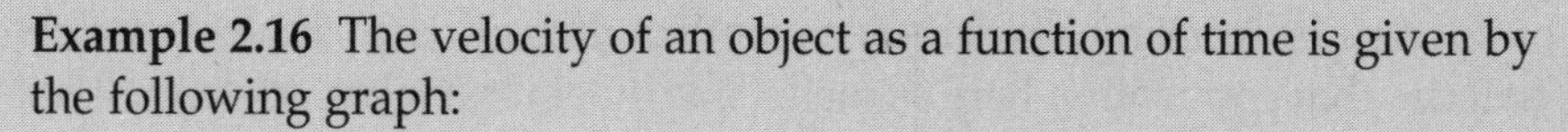

Example 2.16 The velocity of an object as a function of time is given by the following graph:

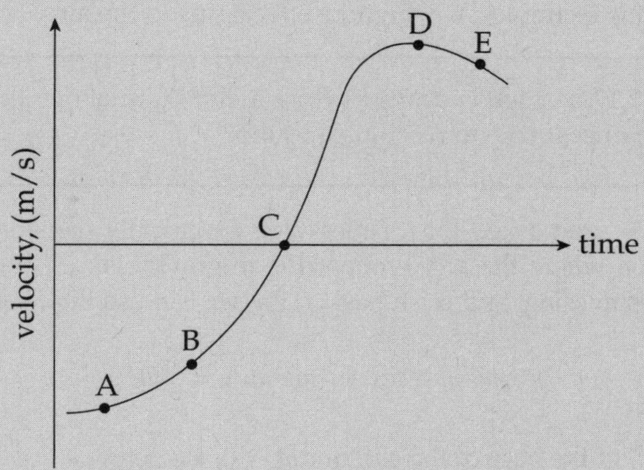

At which point (A, B, C, D, or E) is the magnitude of the acceleration the greatest? How would you answer this same question if the graph shown were a position-vs.-time graph?

Solution. The acceleration is the slope of the velocity-vs.-time graph. Although this graph is not composed of straight lines, the concept of slope still applies; at each point, the slope of the curve is the slope of the tangent line to the curve. The slope is essentially zero at Points A and D (where the curve is flat), small and positive at B, and small and negative at E. The slope at Point C is large and positive, so this is where the object's acceleration is the greatest.

If the graph shown were a position-vs.-time graph, then the slope would be the velocity. The slope of the given graph starts at zero (around Point A), slowly increases to a small positive value at B, continues to slowly increase to a large positive value at C, then, at around Point D, this large positive slope decreases quickly to zero. Of the points designated on the graph, Point D is the location of the greatest slope change, which means that this is the point of the greatest velocity change. Therefore, this is the point at which the magnitude of the acceleration is greatest.

FREE FALL

The simplest real-life example of motion under pretty constant acceleration is the motion of objects in the earth's gravitational field, near the surface of the earth and ignoring any effects due to the air (mainly air resistance). With these effects ignored, an object can fall *freely*, that is, it can fall experiencing only acceleration due to gravity. Near the surface of the earth, the gravitational acceleration has a constant magnitude of about 9.8 m/s^2; this quantity is denoted g (for gravitational acceleration). On the AP Physics Exam, you may use g = 10 m/s^2 as a simple approximation to g = 9.8 m/s^2. This is particularly helpful because, as we mentioned earlier, you will not be permitted to use a calculator on the multiple-choice section. In this book, we will always use g = 10 m/s^2. And, of course, the gravitational acceleration vector, **g**, points *downward*.

Although you are allowed to choose any coordinate system you wish, most students will increase their degree of success by sticking with the convention that up is positive and down is negative. Typically, we will use that convention here as well. Gravity points down, so we will use **g** = –10 m/s^2.

In each of the following examples, we'll ignore effects due to the air.

Example 2.17 A rock is dropped from a cliff 80 m above the ground. How long does it take to reach the ground?

Solution. We are given v_0, and asked for t. Unless you specifically see words to the contrary (such as, "you are on the moon where the acceleration due to gravity is....") assume you are also given a = –10m/s^2. Because v is missing, and it isn't asked for, we can use Big Five equation #3.

$$y = y_0 + v_0 t + \frac{1}{2}at^2 \Rightarrow y = \frac{1}{2}at^2$$

If we set the origin y_0 = 0 at the base of the cliff and v_0 = 0, we get:

$$t = \sqrt{\frac{2y}{a}}$$

$$t = \sqrt{\frac{2(-80\text{ m})}{(-10\text{ m/s}^2)}} = 4\text{ s}$$

Note: The negative in front of the 80 is inserted because the rock fell in the down direction.

Example 2.18 A baseball is thrown straight upward with an initial speed of 20 m/s. How high will it go?

Solution. We are given v_0, a = –10 m/s^2 is implied, and we are asked for y. Now, neither t nor v is expressly given; however, we know the vertical velocity at the top is 0 (otherwise the baseball would still rise). Consequently, we use Big Five equation #5.

$$v^2 = v_0^{\,2} + 2a(y - y_0) \Rightarrow -2ay = v_0^{\,2}$$

We set $y_0 = 0$ and we know that $v = 0$, so that leaves us with:

$$y = -\frac{v_0^{\ 2}}{2a}$$

$$y = -\frac{(20 \text{ m/s})^2}{2(-10 \text{ m/s}^2)} = 20 \text{ m}$$

Example 2.19 One second after being thrown straight down, an object is falling with a speed of 20 m/s. How fast will it be falling 2 seconds later?

Solution. We're given v_0, a, and t, and asked for v. Since x is missing, we use Big Five #2:

$$v = v_0 + at = (-20 \text{ m/s}) + (-10 \text{ m/s}^2)(2 \text{ s}) = -40 \text{ m/s}$$

The negative sign in front of the 40 simply indicates that the object is traveling in the down direction.

Example 2.20 If an object is thrown straight upward with an initial speed of 8 m/s and takes 3 seconds to strike the ground, from what height was the object thrown?

Solution. We're given a, v_0, and t, and we need to find y_0. Because v is missing, we use Big Five #3:

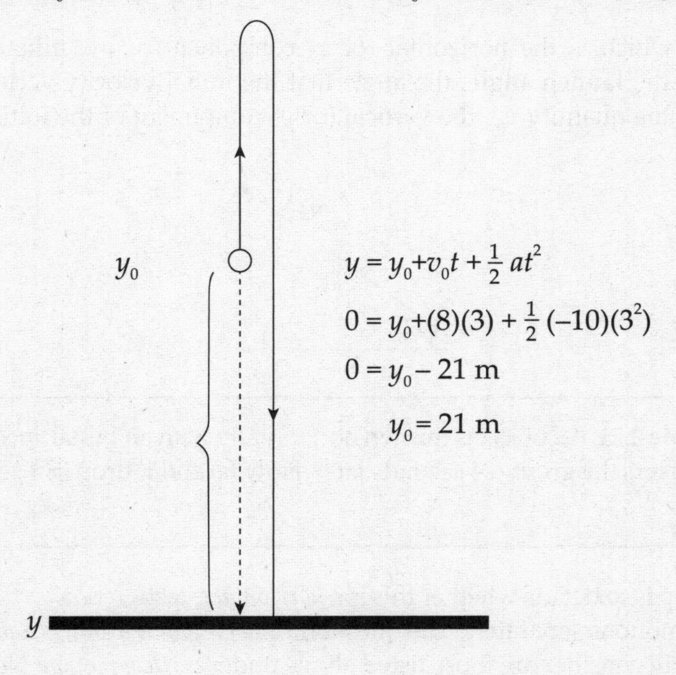

PROJECTILE MOTION

In general, an object that moves near the surface of the earth will not follow a straight-line path (for example, a baseball hit by a bat, a golf ball struck by a club, or a tennis ball hit from the baseline). If we launch an object at an angle other than straight upward and consider only the effect of acceleration due to gravity, then the object will travel along a parabolic trajectory.

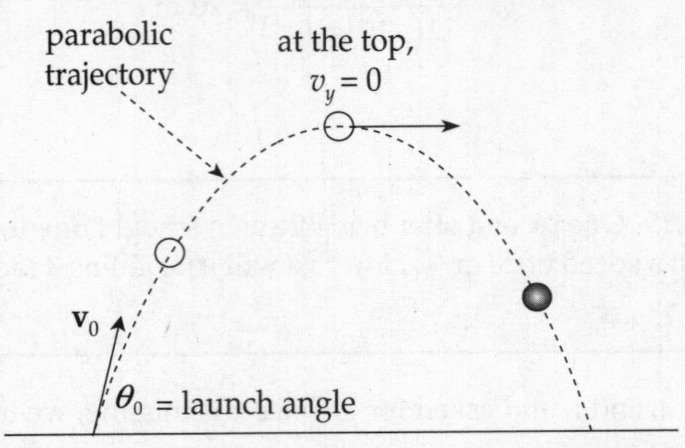

To simplify the analysis of parabolic motion, *we analyze the horizontal and vertical motions* ***separately***, using the Big Five. This is the key to doing projectile motion problems. Calling *down* the negative direction, we have

Horizontal motion:	**Vertical motion:**
$\Delta x = v_{0x}t$	$y = y_0 + v_{0y}t - \frac{1}{2}gt^2$
$v_x = v_{0x}$ (constant!)	$v_y = v_{0y} - gt$
$a_x = 0$	$a_y = -g = -10 \text{ m/s}^2$

The quantity v_{0x}, which is the horizontal (or x) component of the initial velocity, is equal to $v_0 \cos \theta_0$, where θ_0 is the **launch angle**, the angle that the initial velocity vector, $\mathbf{v}_0$, makes with the horizontal. Similarly, the quantity v_{0y}, the vertical (or y) component of the initial velocity, is equal to $v_0 \sin \theta_0$.

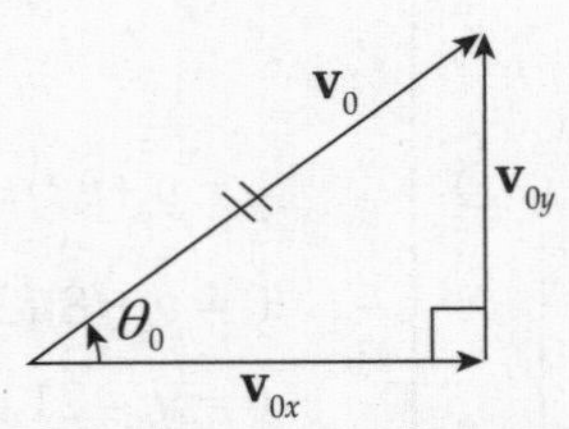

> **Example 2.21** An object is thrown horizontally with an initial speed of 10 m/s. It hits the ground 4 seconds later. How far did it drop in 4 seconds?

Solution. The first step is to decide whether this is a *horizontal* question or a *vertical* question, since you must consider these motions separately. The question *How far will it drop?* is a *vertical* question, so the set of equations we will consider are those listed above under *vertical motion*. Next, *How far...?* implies that we will use the first of the vertical-motion equations, the one that gives vertical displacement, Δy.

Now, since the object is thrown horizontally, there is no vertical component to its initial velocity vector $\mathbf{v}_0$; that is, $v_{0y} = 0$. Therefore,

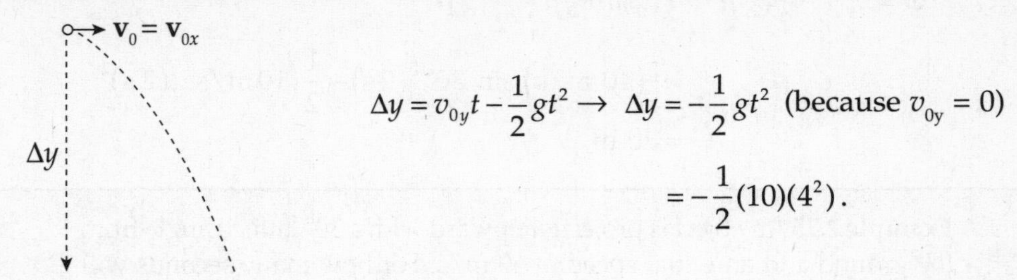

$$\Delta y = v_{0y}t - \frac{1}{2}gt^2 \rightarrow \Delta y = -\frac{1}{2}gt^2 \text{ (because } v_{0y} = 0\text{)}$$

$$= -\frac{1}{2}(10)(4^2).$$

The fact that Δy is negative means that the displacement is *down*. Also, notice that the information given about v_{0x} is irrelevant to the question.

> **Example 2.22** From a height of 100 m, a ball is thrown horizontally with an initial speed of 15 m/s. How far does it travel horizontally in the first 2 seconds?

Solution. The question, *How far does it travel horizontally...?*, immediately tells us that we should use the first of the horizontal-motion equations listed above:

$$\Delta x = v_{0x}t = (15 \text{ m/s})(2 \text{ s}) = 30 \text{ m}$$

The information that the initial vertical position is 100 m above the ground is irrelevant (except for the fact that it's high enough that the ball doesn't strike the ground before the two seconds have elapsed).

> **Example 2.23** A projectile is traveling in a parabolic path for a total of 6 seconds. How does its horizontal velocity 1 s after launch compare to its horizontal velocity 4 s after launch?

Solution. The only acceleration experienced by the projectile is due to gravity, which is purely vertical, so that there is no horizontal acceleration. If there's no horizontal acceleration, then the horizontal velocity cannot change during flight, and the projectile's horizontal velocity 1 s after it's launched is the same as its horizontal velocity 3 s later.

> **Example 2.24** An object is projected upward with a 30° launch angle and an initial speed of 40 m/s. How long will it take for the object to reach the top of its trajectory? How high is this?

Solution. When the projectile reaches the top of its trajectory, its velocity vector is momentarily horizontal; that is, $v_y = 0$. Using the vertical-motion equation for v_y, we can set it equal to 0 and solve for t:

$$v_y \overset{\text{set}}{=} 0 \Rightarrow v_{0y} - gt = 0$$

$$t = \frac{v_{0y}}{g} = \frac{v_0 \sin\theta_0}{g} = \frac{(40 \text{ m/s})\sin 30^\circ}{10 \text{ m/s}^2} = 2 \text{ s}$$

At this time, the projectile's vertical displacement is

$$\Delta y = v_{0y}t - \frac{1}{2}(g)t^2 = (v_0 \sin\theta_0)t - \frac{1}{2}(g)t^2$$

$$= [(40 \text{ m/s}) \sin 30^\circ](2 \text{ s}) - \frac{1}{2}(10 \text{ m/s}^2)(2 \text{ s})^2$$

$$= 20 \text{ m}$$

Example 2.25 An object is projected upward with a 30° launch angle from the ground and an initial speed of 60 m/s. For how many seconds will it be in the air? How far will it travel horizontally? Assume it returns to its original height.

Solution. The total time the object spends in the air is equal to twice the time required to reach the top of the trajectory (because the parabola is symmetrical). So, as we did in the previous example, we find the time required to reach the top by setting v_y equal to 0, and now double that amount of time:

$$v_y \overset{\text{set}}{=} 0 \Rightarrow v_{0y} - gt = 0$$

$$t = \frac{v_{0y}}{g} = \frac{v_0 \sin\theta_0}{g} = \frac{(60 \text{ m/s})\sin 30^\circ}{10 \text{ m/s}^2} = 3 \text{ s}$$

Therefore, the *total* flight time (that is, up and down) is $t_t = 2t = 2 \times (3 \text{ s}) = 6$ s.

Now, using the first horizontal-motion equation, we can calculate the horizontal displacement after 6 seconds:

$$\Delta x = v_{0x}t_t = (v_0 \cos\theta_0)t_t = [(60 \text{ m/s})\cos 30^\circ](6 \text{ s}) = 312 \text{ m}$$

By the way, assuming it lands back at its original height, the full horizontal displacement of a projectile is called the projectile's **range**.

CHAPTER 2 REVIEW QUESTIONS

Solutions can be found in Chapter 18.

SECTION I: MULTIPLE CHOICE

1. An object that's moving with constant speed travels once around a circular path. Which of the following is/are true concerning this motion?

   I. The displacement is zero.
   II. The average speed is zero.
   III. The acceleration is zero.

   (A) I only
   (B) I and II only
   (C) I and III only
   (D) III only
   (E) II and III only

2. 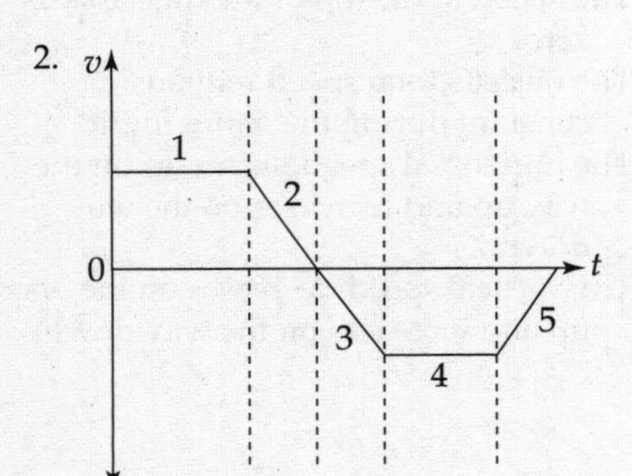

   In section 5 of the velocity-time graph the object is

   (A) speeding up moving in the positive direction
   (B) slowing down moving in the positive direction
   (C) speeding up moving in the negative direction
   (D) slowing down moving in the negative direction
   (E) this can not be determined without the position-time graph

3. Which of the following is/are true?

   I. If an object's acceleration is constant, then it must move in a straight line.
   II. If an object's acceleration is zero, then its speed must remain constant.
   III. If an object's speed remains constant, then its acceleration must be zero.

   (A) I and II only
   (B) I and III only
   (C) II only
   (D) III only
   (E) II and III only

4. A baseball is thrown straight upward. What is the ball's acceleration at its highest point?

   (A) 0
   (B) $\frac{1}{2}g$, downward
   (C) g, downward
   (D) $\frac{1}{2}g$, upward
   (E) g, upward

5. How long would it take a car, starting from rest and accelerating uniformly in a straight line at 5 m/s^2, to cover a distance of 200 m?

   (A) 9.0 s
   (B) 10.5 s
   (C) 12.0 s
   (D) 15.5 s
   (E) 20.0 s

6. A rock is dropped off a cliff and strikes the ground with an impact velocity of 30 m/s. How high was the cliff?

(A) 15 m
(B) 20 m
(C) 30 m
(D) 45 m
(E) 60 m

7. A stone is thrown horizontally with an initial speed of 10 m/s from a bridge. If air resistance could be ignored, how long would it take the stone to strike the water 80 m below the bridge?

(A) 1 s
(B) 2 s
(C) 4 s
(D) 6 s
(E) 8 s

8. A soccer ball, at rest on the ground, is kicked with an initial velocity of 10 m/s at a launch angle of 30°. Calculate its total flight time, assuming that air resistance is negligible.

(A) 0.5 s
(B) 1 s
(C) 1.7 s
(D) 2 s
(E) 4 s

9. A stone is thrown horizontally with an initial speed of 30 m/s from a bridge. Find the stone's total speed when it enters the water 4 seconds later. (Ignore air resistance.)

(A) 30 m/s
(B) 40 m/s
(C) 50 m/s
(D) 60 m/s
(E) 70 m/s

10. Which one of the following statements is true concerning the motion of an ideal projectile launched at an angle of 45° to the horizontal?

(A) The acceleration vector points opposite to the velocity vector on the way up and in the same direction as the velocity vector on the way down.
(B) The speed at the top of the trajectory is zero.
(C) The object's total speed remains constant during the entire flight.
(D) The horizontal speed decreases on the way up and increases on the way down.
(E) The vertical speed decreases on the way up and increases on the way down.

SECTION II: FREE RESPONSE

1. This question concerns the motion of a car on a straight track; the car's velocity as a function of time is plotted below.

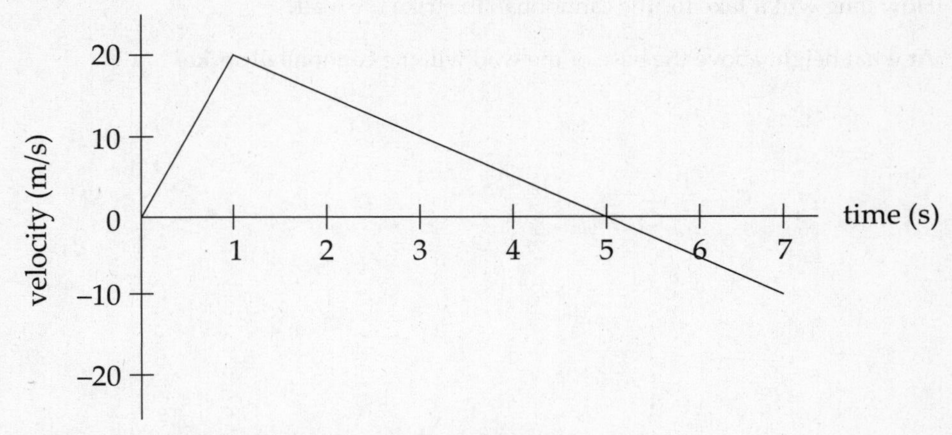

   (a) Describe what happened to the car at time $t = 1$ s.

   (b) How does the car's average velocity between time $t = 0$ and $t = 1$ s compare to its average velocity between times $t = 1$ s and $t = 5$ s?

   (c) What is the displacement of the car from time $t = 0$ to time $t = 7$ s?

   (d) Plot the car's acceleration during this interval as a function of time.

   (e) Make a sketch of the object's position during this interval as a function of time. Assume that the car begins at $x = 0$.

2. Consider a projectile moving in a parabolic trajectory under constant gravitational acceleration. Its initial velocity has magnitude v_0, and its launch angle (with the horizontal) is θ_0.

   (a) Calculate the maximum height, H, of the projectile.

   (b) Calculate the (horizontal) range, R, of the projectile.

   (c) For what value of θ_0 will the range be maximized?

   (d) If $0 < h < H$, compute the time that elapses between passing through the horizontal line of height h in both directions (ascending and descending); that is, compute the time required for the projectile to pass through the two points shown in this figure:

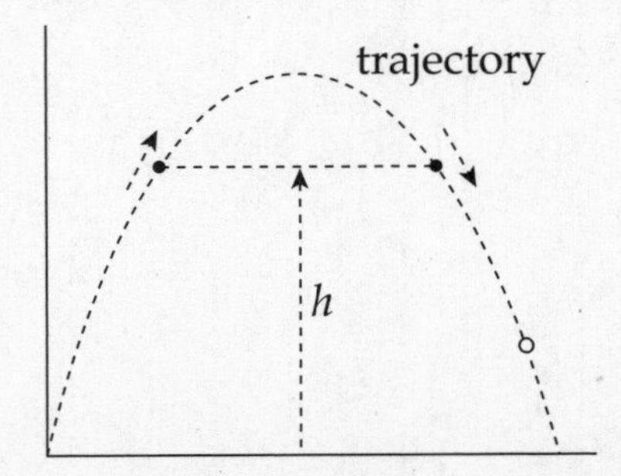

3. A cannonball is shot with an initial speed of 50 m/s at a launch angle of 40° toward a castle wall 220 m away. The height of the wall is 30 m. Assume that effects due to the air are negligible. (For this problem, use $g = 9.8 \text{ m/s}^2$.)

   (a) Show that the cannonball will strike the castle wall.

   (b) How long will it take for the cannonball to strike the wall?

   (c) At what height above the base of the wall will the cannonball strike?

SUMMARY

Graphs are very useful tools to help visualize the motion of an object. They can also help you solve problems once you learn how to translate from one graph to another. As you work with graphs, keep the following things in mind:

- Always look at a graph's axes first. This sounds obvious, but one of the most common mistakes students make is looking at a velocity-vs.-time graph, thinking about it as if it were a position-vs.-time graph.

- Don't ever assume one box is one unit. Look at the numbers on the axis.

- Lining up position-vs.-time graphs *directly above* velocity-vs.-time graphs and *directly above* acceleration-vs.-time graphs is a must. This way you can match up key points from one graph to the next.

- The slope of an x vs. t graph gives velocity. The slope of a v vs. t graph gives acceleration.

- The area under an a vs. t graph gives the change in velocity. The area under a v vs. t graph gives the displacement.

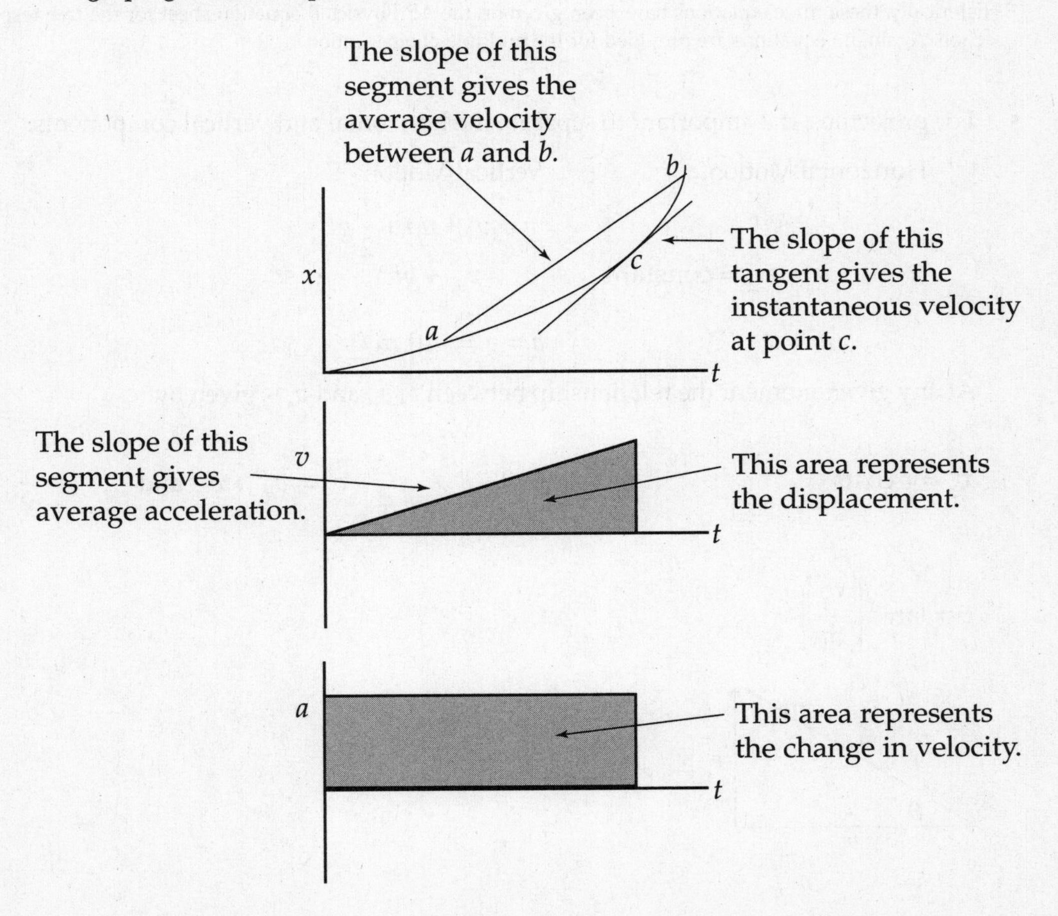

- The motion of an object in one dimension can be described using the Big Five equations. Look for what is given, determine what you're looking for, and use the equation that has those variables in them. Remember that sometimes there is hidden (assumed) information in the problem such as a = –10m/s^2.

		Variable that's missing
Big Five #1*:	$\Delta x = \bar{v}t$	a
Big Five #2**:	$v = v_o + at$	x
Big Five #3**:	$x = x_o + v_o t + \frac{1}{2}at^2$	v
Big Five #4:	$x = x_o + vt - \frac{1}{2}at^2$	v_o
Big Five #5**:	$v^2 = v_o{}^2 + 2a(x - x_o)$	t

* Because the acceleration is uniform, this may also be written as $\Delta x = \frac{1}{2}(v_o + v)t$.

** Historically, these three equations have been given on the AP Physics B equation sheet for the free-response section. Again, no equations are provided for the multiple-choice section.

- For projectiles, it is important to separate the horizontal and vertical components.

Horizontal Motion:	Vertical Motion:
$x = v_x t$	$y = y_o + v_o t + \frac{1}{2}gt^2$
$v_x = v_{ox}$ = constant	$v_y = v_{oy} + gt$
$a_x = 0$	$a = g = -10$ m/s^2

At any given moment the relationship between v, v_x, and v_y is given by

$v_x = v\cos\theta$ $\qquad$ $v_y = v\sin\theta$ $\qquad$ $v^2 = v_x{}^2 + v_y{}^2$ and

$$\theta = \tan^{-1}\left(\frac{v_y}{v_x}\right)$$

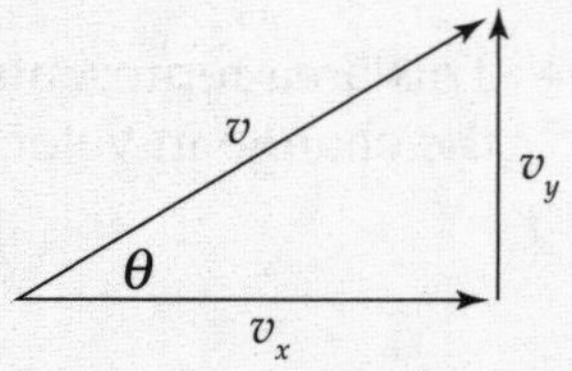

3

Newton's Laws

INTRODUCTION

In the previous chapter we studied the vocabulary and equations that describe motion. Now we will learn why things move the way they do; this is the subject of **dynamics**.

An interaction between two bodies—a push or a pull—is called a **force**. If you lift a book, you exert an upward force (created by your muscles) on it. If you pull on a rope that's attached to a crate, you create a *tension* in the rope that pulls the crate. When a skydiver is falling through the air, the Earth is exerting a downward pull called *gravitational force*, and the air exerts an upward force called *air resistance*. When you stand on the floor, the floor provides an upward, supporting force called the *normal force*. If you slide a book across a table, the table exerts a *frictional force* against the book, so the book slows down and then stops. Static cling provides a directly observable example of the *electrostatic force*. Protons and neutrons are held together in the nuclei of atoms by the *strong nuclear force* and radioactive nuclei decay through the action of the *weak nuclear force*.

The Englishman Sir Isaac Newton published a book in 1687 called *Philosophiae Naturalis Principa Mathematica* (*The Mathematical Principles of Natural Philosophy*)—referred to nowadays as simply *The Principia*—which began the modern study of physics as a scientific discipline. Three of the laws that Newton stated in *The Principia* form the basis for dynamics and are known simply as *Newton's Laws of Motion*.

THE FIRST LAW

Newton's First Law says that *an object will continue in its state of motion unless compelled to change by a force impressed upon it*. That is, unless an unbalanced force acts on an object, the object's velocity will not change: If the object is at rest, then it will stay at rest; and if it is moving, then it will continue to move at a constant speed in a straight line.

This property of objects, their natural resistance to changes in their state of motion, is called **inertia**. In fact, the First Law is often referred to as the **Law of Inertia**.

THE SECOND LAW

Newton's Second Law predicts what will happen when an unbalanced force *does* act on an object: The object's velocity will change; the object will accelerate. More precisely, it says that its acceleration, $\mathbf{a}$, will be directly proportional to the strength of the total—or *net*—force ($\mathbf{F}_{net}$) and inversely proportional to the object's mass, m:

$$\mathbf{F}_{net} = m\mathbf{a} \text{ or } \Sigma F = ma$$

This is the most important equation in mechanics!

The **mass** of an object is the quantitative measure of its inertia; intuitively, it measures how much matter is contained in an object. Two identical boxes, one empty and one full, have different masses. The box that's full has the greater mass, because it contains more stuff; more stuff, more mass. Mass is measured in *kilograms*, abbreviated as kg. (Note: An object whose mass is 1 kg weighs about 2.2 pounds.) It takes twice as much force to produce the same change in velocity of a 2 kg object than of a 1 kg object. Mass is a measure of an object's inertia, its resistance to acceleration.

Forces are represented by vectors; they have magnitude and direction. If several different forces act on an object simultaneously, then the net force, $\mathbf{F}_{net}$, is the vector sum of all these forces. (The phrase *resultant force* is also used to mean *net force*.)

Since $\mathbf{F}_{net} = m\mathbf{a}$, and m is a *positive* scalar, the direction of $\mathbf{a}$ always matches the direction of $\mathbf{F}_{net}$. Finally, since $F = ma$, the units for F equal the units of m times the units of a:

$$\begin{aligned}[F] &= [m][a] \\ &= \text{kg}\cdot\text{m/s}^2\end{aligned}$$

A force of 1 kg·m/s² is renamed 1 **newton** (abbreviated as N). A medium-size apple weighs about 1 N.

THE THIRD LAW

This is the law that's commonly remembered as, *to every action, there is an equal, but opposite, reaction.* More precisely, if Object 1 exerts a force on Object 2, then Object 2 exerts a force back on Object 1, equal in strength but opposite in direction. These two forces, $\mathbf{F}_{1\text{-on-}2}$ and $\mathbf{F}_{2\text{-on-}1}$, are called an **action/reaction pair**.

> **Example 3.1** What net force is required to maintain a 5000 kg object moving at a constant velocity of magnitude 7500 m/s?

Solution. The First Law says that any object will continue in its state of motion unless an unbalanced force acts on it. Therefore, no net force is required to maintain a 5000 kg object moving at a constant velocity of magnitude 7500 m/s.

You might be asking, "If no net force is needed to keep a car moving at a constant speed, why does the driver need to press down on the gas pedal in order to maintain a constant speed?" There is a big difference between *force* and *net force*. As the car moves forward, there is a frictional force (more on that shortly) opposite the direction of motion that would be slowing the car down. The gas supplies energy to the engine to spin the tires so they exert a forward force on the car to counteract friction and make the net force zero, which maintains a constant speed.

Here's another way to look at it: Constant velocity means $\mathbf{a} = 0$, so the equation $F_{net} = ma$ immediately gives $F_{net} = 0$.

> **Example 3.2** How much force is required to cause an object of mass 2 kg to have an acceleration of 4 m/s^2?

Solution. According to the Second Law, $F_{net} = m\mathbf{a} = (2 \text{ kg})(4 \text{ m/s}^2) = 8 \text{ N}$.

> **Example 3.3** An object feels two forces; one of strength 8 N pulling to the left and one of strength 20 N pulling to the right. If the object's mass is 4 kg, what is its acceleration?

Solution. Forces are represented by vectors and can be added and subtracted. Therefore, an 8 N force to the left added to a 20 N force to the right yields a net force of $20 - 8 = 12$ N to the right. Then Newton's Second Law gives $a = F_{net}/m = (12 \text{ N to the right})/(4 \text{ kg}) = 3 \text{ m/s}^2$ to the right.

WEIGHT

Mass and weight are not the same thing—there is a clear distinction between them in physics—but they are often used interchangeably in everyday life. The **weight** of an object is the gravitational force exerted on it by the Earth (or by whatever planet it happens to be on). Mass, by contrast, is an intrinsic property of an object that measures its inertia. An object's mass does not change with location. Put a baseball in a rocket and send it to the Moon. The baseball's weight on the Moon is less than its weight here on Earth (because the Moon's gravitational pull is weaker than the Earth's due to its much smaller mass), but the baseball's mass would be the same.

Since weight is a force, we can use $\mathbf{F} = m\mathbf{a}$ to compute it. What acceleration would the gravitational force impose on an object? The gravitational acceleration, of course! Therefore, setting $\mathbf{a} = \mathbf{g}$, the equation $\mathbf{F} = m\mathbf{a}$ becomes

$$\mathbf{F}_{\mathrm{w}} = m\mathbf{g} \text{ or } F_g = mg$$

This is the equation for the weight of an object of mass m. (Weight is often symbolized as F_g, rather than $\mathbf{F}_{\mathrm{w}}$.) Notice that mass and weight are proportional but not identical. Furthermore, mass is measured in kilograms, while weight is measured in newtons.

Example 3.4 What is the mass of an object that weighs 500 N?

Solution. Since weight is m multiplied by g, mass is F_w (weight) divided by g. Therefore,

$$m = F_w/g = (500 \text{ N})/(10 \text{ m/s}^2) = 50 \text{ kg}$$

Example 3.5 A person weighs 150 pounds. Given that a pound is a unit of weight equal to 4.45 N, what is this person's mass?

Solution. This person's weight in newtons is (150 lb)(4.45 N/lb) = 667.5 N, so his mass is

$$m = F_w/g = (667.5 \text{ N})/(10 \text{ m/s}^2) = 66.75 \text{ kg}$$

Example 3.6 A book whose mass is 2 kg rests on a table. Find the magnitude of the force exerted by the table on the book.

Solution. The book experiences two forces: The downward pull of the Earth's gravity and the upward, supporting force exerted by the table. Since the book is at rest on the table, its acceleration is zero, so the net force on the book must be zero. Therefore the magnitude of the support force must equal the magnitude of the book's weight, which is $F_w = mg = (2 \text{ kg})(10 \text{ m/s}^2) = 20$ N.

AN OVERALL STRATEGY

The above examples are the lowest level of understanding Newton's laws. They are pretty straightforward thinking sometimes referred to as "plug and chug." Most of physics is not that simple. Frequently there is more than one force acting on an object and many times angles are involved. Following the below strategy can greatly increase your chance of success for all but the most trivial of Newton's Second Law problems.

I. You must be able to visualize what's going on. Make a sketch if it helps, but definitely make a free-body diagram by doing the following:

   A. Draw a dot to represent the object. Draw arrows going away from the dot to represent any (all) forces acting on the object.

      i. Anything touching the object exerts a force.

         a. If the thing touching the object is a rope, ropes can only pull. Draw in the force accordingly.

         b. If the thing touching the object is a table, ramp, floor, or some other flat surface, a surface can exert two forces.

            1. The surface exerts a force perpendicular to itself toward the object. This force is always present if two things are in contact and is called the normal force.

            2. If there is kinetic friction present, then the surface exerts a force on the object that is parallel to the surface and opposite to the direction of motion.

      ii. Some things can exert a force without touching an object. For example, the Earth pulls down on everything via the mystery of gravity. Electricity and magnetism also exert their influences without actually touching. Unless you hear otherwise, gravity points down!

      iii. If you know one force is bigger than another, you should draw that arrow longer than the smaller force.

      iv. Don't draw a velocity and mistake it for a force. No self-respecting velocity vector hangs out in a free-body diagram! Oh, and there is no such thing as the force of inertia.

II. Clearly define an appropriate coordinate system. Be sure to break up each force that does not lie on an axis into its x and y components.

III. Write out Newton's Second Law in the form of $\sum F_x = ma_x$ and/or $\sum F_y = ma_y$ using the forces identified in the free-body diagram to fill in the appropriate forces.

IV. Do the math.

As you go though the following examples, notice how this strategy is used.

Example 3.7 Draw a free-body diagram for each of the following situations:

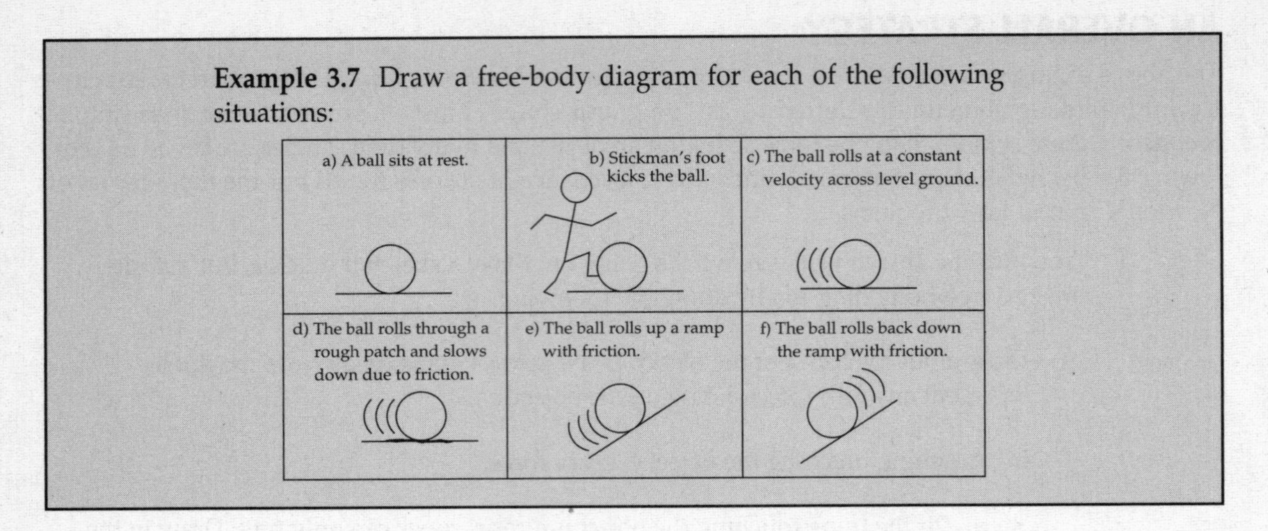

Solution. Notice that gravity always points down (even on ramps). Also the normal force is perpendicular to the surface (even on ramps). Do not always put the normal opposite the direction of gravity, because the normal is relative to the surface, which may be tilted. Finally, friction is always parallel to the surface (or perpendicular to the normal) and tends to point in the opposite direction from motion.

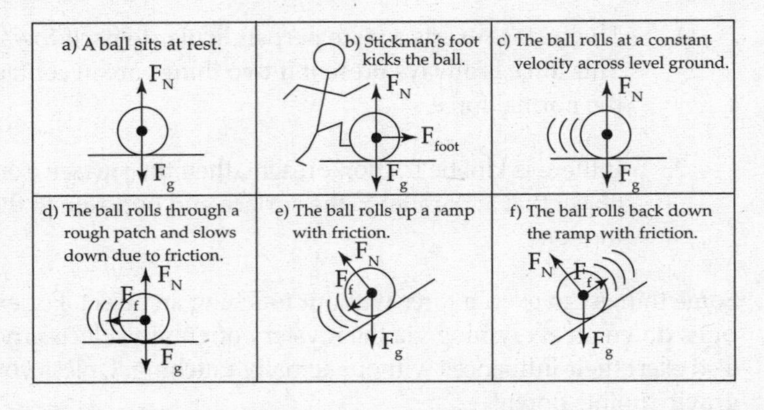

Example 3.8 A can of paint with a mass of 6 kg hangs from a rope. If the can is to be pulled up to a rooftop with an acceleration of 1 m/s^2, what must the tension in the rope be?

Solution. First draw a picture. Represent the object of interest (the can of paint) as a heavy dot, and draw the forces that act on the object as arrows connected to the dot. This is called a **free-body** (or force) **diagram**.

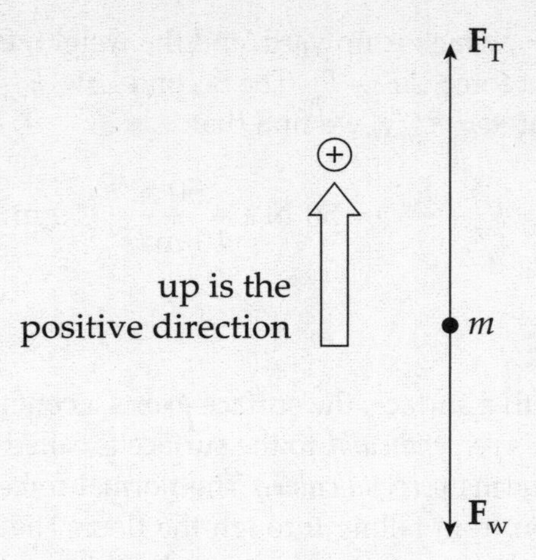

We have the tension force in the rope, $\mathbf{F}_T$ (also symbolized merely by **T**), which is upward, and the weight, $\mathbf{F}_w$, which is downward. Calling *up* the positive direction, the net force is $F_T - F_w$. The Second Law, $F_{net} = ma$, becomes $F_T - F_w = ma$, so

$$F_T = F_w + ma = mg + ma = m(g + a) = 6(10 + 1) = 66 \text{ N}$$

> **Example 3.9** A can of paint with a mass of 6 kg hangs from a rope. If the can is to be pulled up to a rooftop with a constant velocity of 1 m/s, what must the tension in the rope be?

Solution. The phrase "constant velocity" automatically means $a = 0$ and, therefore, $F_{net} = 0$. In the diagram above, $\mathbf{F}_T$ would need to have the same magnitude as $\mathbf{F}_w$ in order to keep the can moving at a constant velocity. Thus, in this case, $F_T = F_w = mg = (6)(10) = 60$ N.

> **Example 3.10** How much tension must a rope have to lift a 50 N object with an acceleration of 10 m/s^2?

Solution. First draw a free-body diagram:

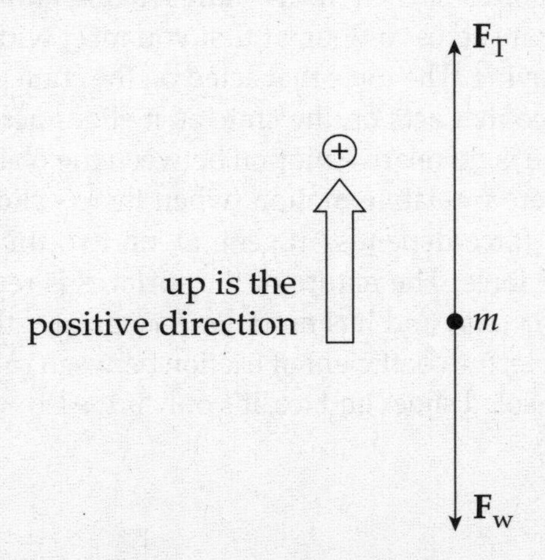

We have the tension force, $\mathbf{F}_T$, which is upward, and the weight, $\mathbf{F}_w$, which is downward. Calling *up* the positive direction, the net force is $F_T - F_w$. The Second Law, $F_{net} = ma$, becomes $F_T - F_w = ma$, so $F_T = F_w + ma$. Remembering that $m = F_w/g$, we find that

$$F_T = F_w + ma = F_w + \frac{F_w}{g}a = 50\text{ N} + \frac{50\text{ N}}{10\text{ m/s}^2}(10\text{ m/s}^2) = 100\text{ N}$$

THE NORMAL FORCE

When an object is in contact with a surface, the surface exerts a contact force on the object. The component of the contact force that's *perpendicular* to the surface is called the **normal force** on the object. (In physics, the word *normal* means *perpendicular*.) The normal force is what prevents objects from falling through tabletops or you from falling through the floor. The normal force is denoted by $\mathbf{F}_N$, or simply by **N**. (If you use the latter notation, be careful not to confuse it with N, the abbreviation for the newton.)

Example 3.11 A book whose mass is 2 kg rests on a table. Find the magnitude of the normal force exerted by the table on the book.

Solution. The book experiences two forces: The downward pull of Earth's gravity and the upward, supporting force exerted by the table. Since the book is at rest on the table, its acceleration is zero, so the net force on the book must be zero. Therefore, the magnitude of the support force must equal the magnitude of the book's weight, which is $F_w = mg = (2)(10) = 20$ N. This means the normal force must be 20 N as well: $F_N = 20$ N. (Note that this is a repeat of Example 3.6, except now we have a name for the "upward, supporting force exerted by the table"; it's called the normal force.)

FRICTION

When an object is in contact with a surface, the surface exerts a contact force on the object. The component of the contact force that's *parallel* to the surface is called the **friction force** on the object. Friction, like the normal force, arises from electrical interactions between atoms of which the object is composed and those of which the surface is composed.

We'll look at two main categories of friction: (1) **static friction** and (2) **kinetic (sliding) friction**. If you attempt to push a heavy crate across a floor, at first you meet with resistance, but then you push hard enough to get the crate moving. The force that acted on the crate to cancel out your initial pushes was static friction, and the force that acts on the crate as it slides across the floor is kinetic friction. Static friction occurs when there is no relative motion between the object and the surface (no sliding); kinetic friction occurs when there *is* relative motion (when there's sliding).

The strength of the friction force depends, in general, on two things: The nature of the surfaces and the strength of the normal force. The nature of the surfaces is represented by the **coefficient of friction,** which is denoted by μ (*mu*) and has no units. The greater this number is, the stronger the friction force will be. For example, the coefficient of friction between rubber-soled shoes and a wooden floor is 0.7, but between rubber-soled shoes and ice, it's only 0.1. Also, since kinetic friction is generally weaker than static friction (it's easier to keep an object sliding once it's sliding than it is to start the object sliding in the first place), there are two coefficients of friction; one for static friction (μ_s) and one for kinetic friction (μ_k). For a given pair of surfaces, it's virtually always true that $\mu_k < \mu_s$. The strengths of these two types of friction forces are given by the following equations:

$$F_{\text{static friction, max}} = \mu_s F_N$$

$$F_{\text{kinetic friction}} = \mu_k F_N$$

Note that the equation for the strength of the static friction force is for the *maximum* value only. This is because static friction can vary, precisely counteracting weaker forces that attempt to move an object. For example, suppose an object experiences a normal force of F_N = 100 N and the coefficient of static friction between it and the surface it's on is 0.5. Then, the *maximum* force that static friction can exert is (0.5)(100 N) = 50 N. However, if you push on the object with a force of, say, 20 N, then the static friction force will be 20 N (in the opposite direction), *not* 50 N; the object won't move. The net force on a stationary object must be zero. Static friction can take on all values, up to a certain maximum, and you must overcome the maximum static friction force to get the object to slide. The direction of $\mathbf{F}_{\text{kinetic friction}} = \mathbf{F}_{\text{f (kinetic)}}$ is opposite to that of motion (sliding), and the direction of $\mathbf{F}_{\text{static friction}} = \mathbf{F}_{\text{f (static)}}$ is usually, but not always, opposite to that of the intended motion.

Example 3.12 A crate of mass 20 kg is sliding across a wooden floor. The coefficient of kinetic friction between the crate and the floor is 0.3.

(a) Determine the strength of the friction force acting on the crate.

(b) If the crate is being pulled by a force of 90 N (parallel to the floor), find the acceleration of the crate.

Solution. First draw a free-body diagram:

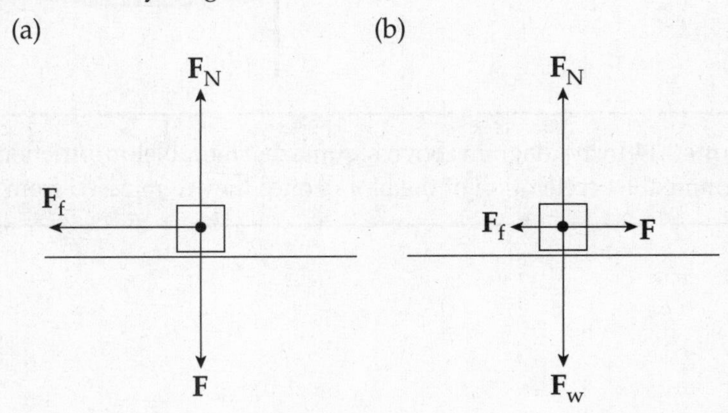

(a) The normal force on the object balances the object's weight, so $F_N = mg$ = (20 kg)(10 m/s²) = 200 N. Therefore, $F_{(\text{kinetic})} = \mu_k F_N$ = (0.3)(200 N) = 60 N.

(b) The net horizontal force that acts on the crate is $F - F_f$ = 90 N – 60 N = 30 N, so the acceleration of the crate is $a = F_{\text{net}}/m$ = (30 N)/(20 kg)= 1.5 m/s².

Example 3.13 A crate of mass 100 kg rests on the floor. The coefficient of static friction is 0.4. If a force of 250 N (parallel to the floor) is applied to the crate, what's the magnitude of the force of static friction on the crate?

Solution. The normal force on the object balances its weight, so $F_N = mg = (100\text{ kg})(10\text{ m/s}^2) = 1{,}000\text{ N}$. Therefore, $F_{\text{static friction, max}} = F_{\text{f (static), max}} = \mu_s F_N = (0.4)(1{,}000\text{ N}) = 400\text{ N}$. This is the *maximum* force that static friction can exert, but in this case it's not the actual value of the static friction force. Since the applied force on the crate is only 250 N, which is less than the $F_{\text{f (static), max}}$, the force of static friction will be less also: $F_{\text{f (static)}} = 250\text{ N}$, and the crate will not slide.

PULLEYS

Pulleys are devices that change the direction of the tension force in the cords that slide over them. Here we'll consider each pulley to be frictionless and massless, which means that their masses are so much smaller than the objects of interest in the problem that they can be ignored.

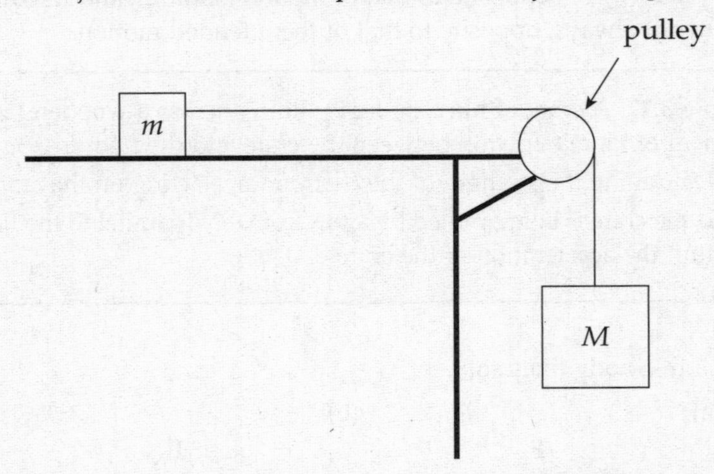

Example 3.14 In the diagram above, assume that the tabletop is frictionless. Determine the acceleration of the blocks once they're released from rest.

Solution. There are two blocks, so we draw two free-body diagrams:

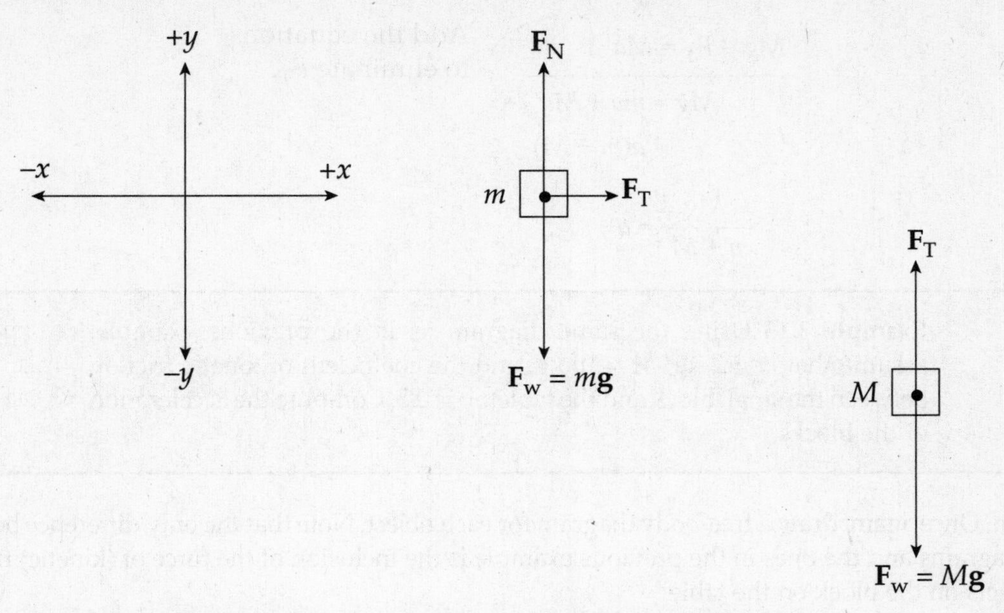

To get the acceleration of each one, we use Newton's Second Law, $\mathbf{F}_{net} = m\mathbf{a}$.

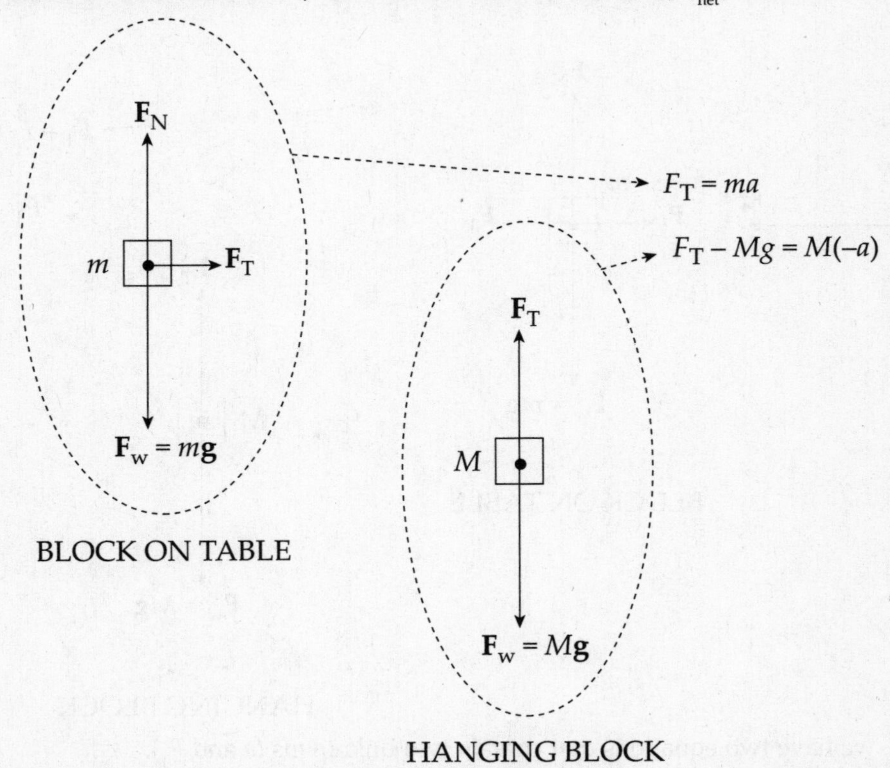

Note that there are two unknowns, F_T and a, but we can eliminate F_T by adding the two equations, and then we can solve for a.

$$F_T = ma$$

$$Mg - F_T = Ma$$

Add the equations to eliminate F_T.

$$Mg = ma + Ma$$

$$= a(m + M)$$

$$\frac{Mg}{m + M} = a$$

Example 3.15 Using the same diagram as in the previous example, assume that m = 2 kg, M = 10 kg, and the coefficient of kinetic friction between the small block and the tabletop is 0.5. Compute the acceleration of the blocks.

Solution. Once again, draw a free-body diagram for each object. Note that the only difference between these diagrams and the ones in the previous example is the inclusion of the force of (kinetic) friction, F_f, that acts on the block on the table.

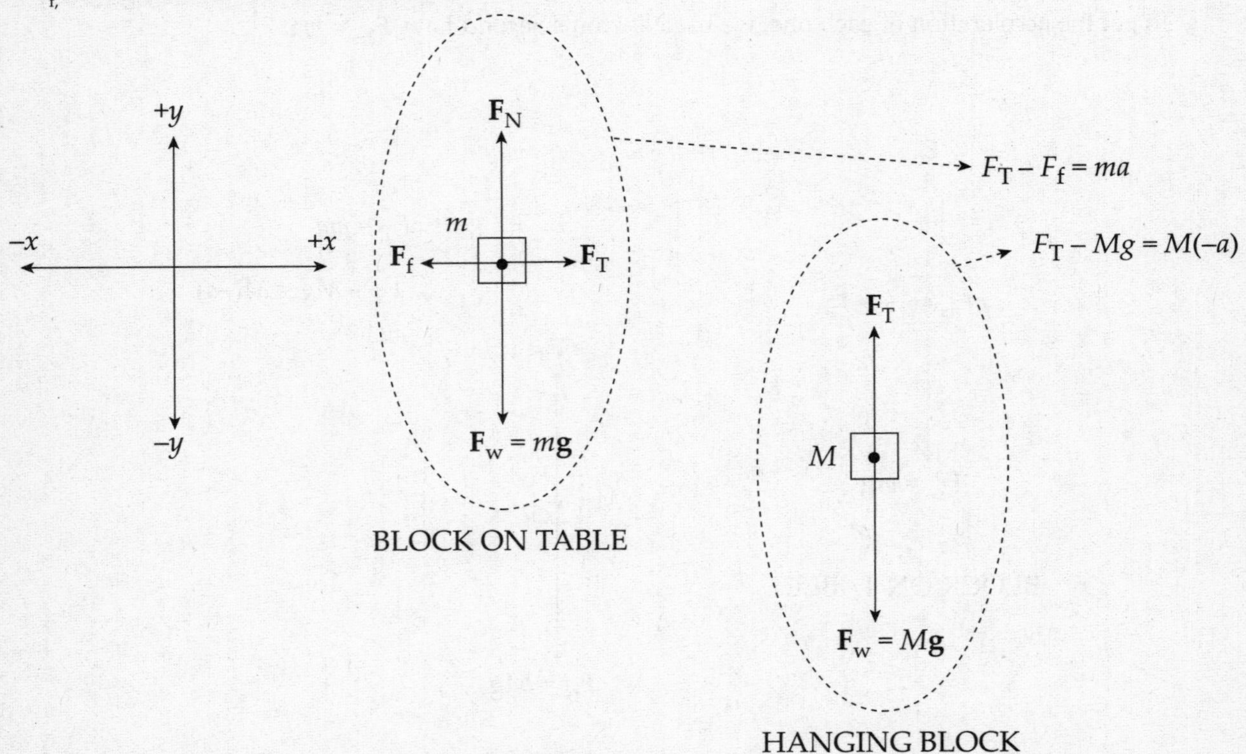

As before, we have two equations that contain two unknowns (a and F_T):

$$F_T - F_f = ma \quad (1)$$

$$F_T - Mg = M(-a) \quad (2)$$

Subtract the equations (thereby eliminating F_T) and solve for a. Note that, by definition, $F_f = \mu F_N$, and from the free-body diagram for m, we see that $F_N = mg$, so $F_f = \mu mg$:

$$Mg - F_f = ma + Ma$$

$$Mg - \mu mg = a(m + M)$$

$$\frac{M - \mu m}{m + M} g = a$$

Substituting in the numerical values given for m, M, and μ, we find that $a = \frac{3}{4}g$ (or 7.5 m/s²).

Example 3.16 In the previous example, calculate the tension in the cord.

Solution. Since the value of a has been determined, we can use either of the two original equations to calculate F_T. Using Equation (2), $F_T - Mg = M(-a)$ (because it's simpler), we find

$$F_T = Mg - Ma = Mg - M \cdot \frac{3}{4}g = \frac{1}{4}Mg = \frac{1}{4}(10)(10) = 25 \text{ N}$$

As you can see, we would have found the same answer if Equation (1) had been used:

$$F_T - F_f = ma \Rightarrow F_T = F_f + ma = \mu mg + ma = \mu mg + m \cdot \frac{3}{4}g = mg\left(\mu + \frac{3}{4}\right)$$
$$= (2)(10)(0.5 + 0.75)$$
$$= 25 \text{ N}$$

INCLINED PLANES

An inclined plane is basically a ramp. If you look at the forces acting on a block that sits on a ramp using a standard coordinate system it initially looks straight forward. However, part of the normal force acts in the x direction, part acts in the y direction, and the block has acceleration in both the x and y direction. If friction is present it also has components in both the x and y direction. The math has the potential to be quite cumbersome.

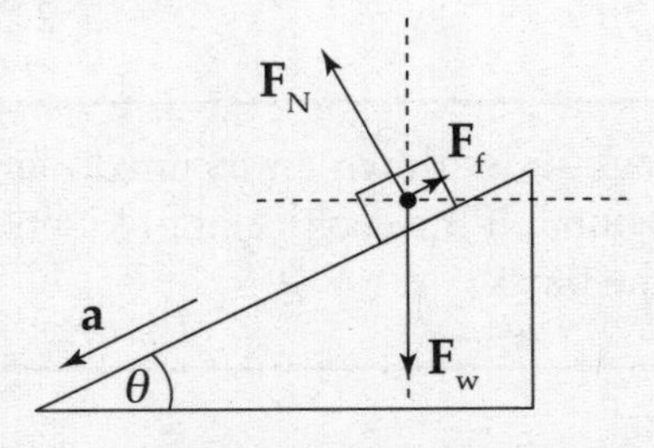

This is a non-rotated coordinate system—notice $\mathbf{F}_f$, $\mathbf{F}_N$, and **a** will each have to be broken into x and y components.

However, if you rotate the coordinate system so that x is parallel to the ramp, gravity becomes split into components, but the normal force acts only in the y direction, friction acts only in the x direction, and the acceleration acts only in the x direction. This makes the math much cleaner.

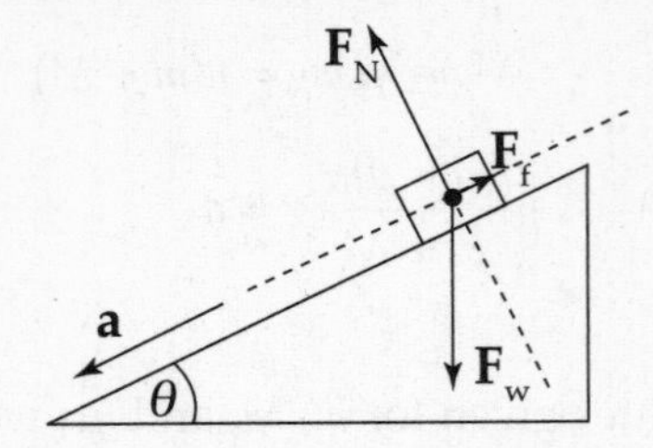

This is a rotated coordinate system—$\mathbf{F}_N$ acts only in the y direction, $\mathbf{F}_f$ and the acceleration act only in the x direction and only $\mathbf{F}_w$ must be broken into components. If an object of mass m is on the ramp, then the force of gravity on the object, $\mathbf{F}_w = m\mathbf{g}$, has two components: One that's parallel to the ramp ($mg \sin\theta$) and one that's normal to the ramp ($mg \cos\theta$), where θ is the incline angle. The force driving the block down the inclined plane is the component of the block's weight that's parallel to the ramp: $mg \sin\theta$.

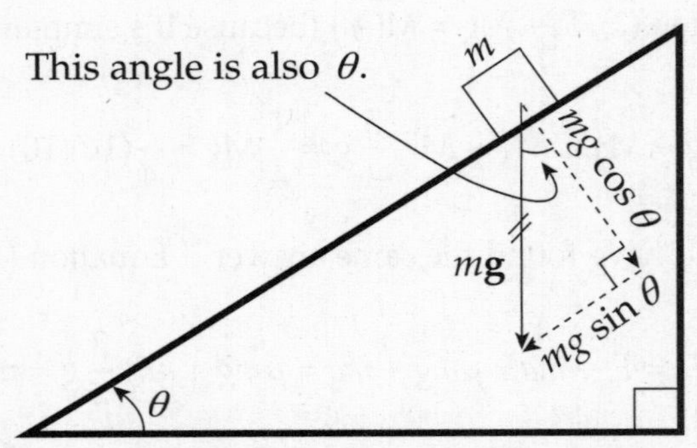

Example 3.17 A block slides down a frictionless, inclined plane that makes a 30° angle with the horizontal. Find the acceleration of this block.

Solution. Let m denote the mass of the block, so the force that pulls the block down the incline is $mg \sin\theta$, and the block's acceleration down the plane is

$$a = \frac{F}{m} = \frac{mg\sin\theta}{m} = g\sin\theta = g\sin 30^\circ = \frac{1}{2}g = 5 \text{ m/s}^2$$

Example 3.18 A block slides down an inclined plane that makes a 30° angle with the horizontal. If the coefficient of kinetic friction is 0.3, find the acceleration of the block.

Solution. First draw a free-body diagram. Notice that, in the diagram shown below, the weight of the block, F_w = mg, has been written in terms of its scalar components: $F_w \sin\theta$ parallel to the ramp and $F_w \cos\theta$ normal to the ramp:

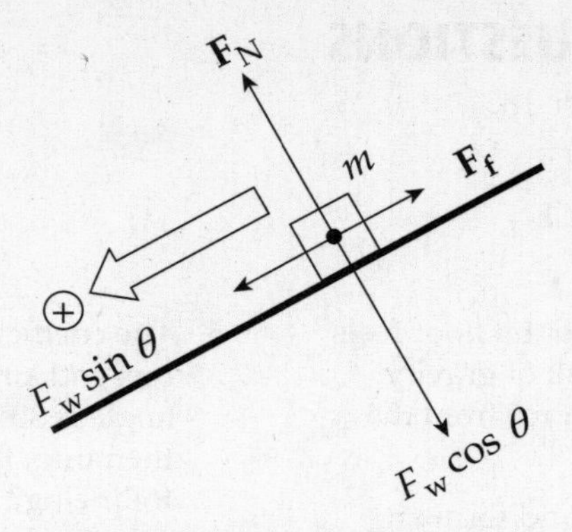

The force of friction, $\mathbf{F}_f$, that acts up the ramp (opposite to the direction in which the block slides) has magnitude $F_f = \mu F_N$. But the diagram shows that $F_N = F_w \cos\theta$, so $F_f = \mu(mg\cos\theta)$. Therefore the net force down the ramp is

$$F_w \sin\theta - F_f = mg\sin\theta - \mu mg\cos\theta = mg(\sin\theta - \mu\cos\theta)$$

Then, setting F_{net} equal to ma, we solve for a:

$$\begin{aligned} a = \frac{F_{net}}{m} &= \frac{mg(\sin\theta - \mu\cos\theta)}{m} \\ &= g(\sin\theta - \mu\cos\theta) \\ &= (10 \text{ m/s}^2)(\sin 30^\circ - 0.3\cos 30^\circ) \\ &= 2.4 \text{ m/s}^2 \end{aligned}$$

CHAPTER 3 REVIEW QUESTIONS

Solutions can be found in Chapter 18.

SECTION I: MULTIPLE CHOICE

1. A person standing on a horizontal floor feels two forces: the downward pull of gravity and the upward supporting force from the floor. These two forces

   (A) have equal magnitudes and form an action/reaction pair
   (B) have equal magnitudes but do not form an action/reaction pair
   (C) have unequal magnitudes and form an action/reaction pair
   (D) have unequal magnitudes and do not form an action/reaction pair
   (E) none of the above

2. A person who weighs 800 N steps onto a scale that is on the floor of an elevator car. If the elevator accelerates upward at a rate of 5 m/s^2, what will the scale read?

   (A) 400 N
   (B) 800 N
   (C) 1000 N
   (D) 1200 N
   (E) 1600 N

3. A frictionless inclined plane of length 20 m has a maximum vertical height of 5 m. If an object of mass 2 kg is placed on the plane, which of the following best approximates the net force it feels?

   (A) 5 N
   (B) 10 N
   (C) 15 N
   (D) 20 N
   (E) 30 N

4. A 20 N block is being pushed across a horizontal table by an 18 N force. If the coefficient of kinetic friction between the block and the table is 0.4, find the acceleration of the block.

   (A) 0.5 m/s^2
   (B) 1 m/s^2
   (C) 5 m/s^2
   (D) 7.5 m/s^2
   (E) 9 m/s^2

5. The coefficient of static friction between a box and a ramp is 0.5. The ramp's incline angle is 30°. If the box is placed at rest on the ramp, the box will do which of the following?

   (A) Accelerate down the ramp.
   (B) Accelerate briefly down the ramp but then slow down and stop.
   (C) Move with constant velocity down the ramp.
   (D) Not move.
   (E) It cannot be determined from the information given.

6.

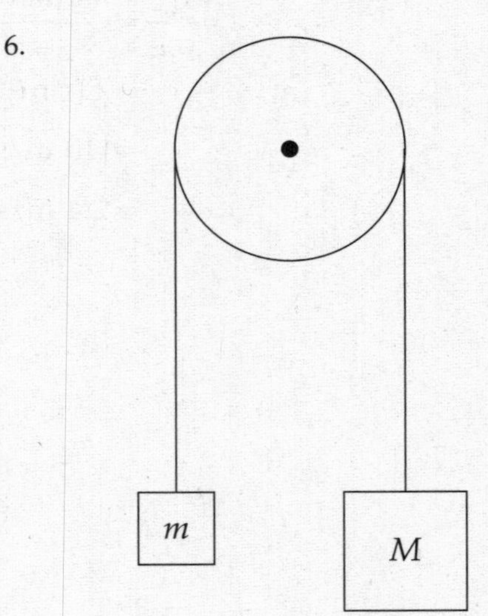

Assuming a frictionless, massless pulley, determine the acceleration of the blocks once they are released from rest.

   (A) $\frac{m}{M+m}g$
   (B) $\frac{M}{M+m}g$
   (C) $\frac{M}{m}g$
   (D) $\frac{M+m}{M-m}g$
   (E) $\frac{M-m}{M+m}g$

7. If all of the forces acting on an object balance so that the net force is zero, then

(A) the object must be at rest
(B) the object's speed will decrease
(C) the object will follow a parabolic trajectory
(D) the object's direction of motion can change, but not its speed
(E) None of the above will occur.

8. A block of mass m is at rest on a frictionless, horizontal table placed in a laboratory on the surface of the Earth. An identical block is at rest on a frictionless, horizontal table placed on the surface of the Moon. Let **F** be the net force necessary to give the Earth-bound block an acceleration of **a** across the table. Given that g_{Moon} is one-sixth of g_{Earth}, the force necessary to give the Moon-bound block the same acceleration **a** across the table is

(A) **F**/12
(B) **F**/6
(C) **F**/3
(D) **F**
(E) 6**F**

9. A crate of mass 100 kg is at rest on a horizontal floor. The coefficient of static friction between the crate and the floor is 0.4, and the coefficient of kinetic friction is 0.3. A force **F** of magnitude 344 N is then applied to the crate, parallel to the floor. Which of the following is true?

(A) The crate will accelerate across the floor at 0.5 m/s^2.
(B) The static friction force, which is the reaction force to **F** as guaranteed by Newton's Third Law, will also have a magnitude of 344 N.
(C) The crate will slide across the floor at a constant speed of 0.5 m/s.
(D) The crate will not move.
(E) None of the above

10. Two crates are stacked on top of each other on a horizontal floor; Crate #1 is on the bottom, and Crate #2 is on the top. Both crates have the same mass. Compared to the strength of the force $\mathbf{F}_1$ necessary to push only Crate #1 at a constant speed across the floor, the strength of the force $\mathbf{F}_2$ necessary to push the stack at the same constant speed across the floor is greater than F_1 because

(A) the normal force on Crate #1 is greater
(B) the coefficient of kinetic friction between Crate #1 and the floor is greater
(C) the force of kinetic friction, but not the normal force, on Crate #1 is greater
(D) the coefficient of static friction between Crate #1 and the floor is greater
(E) the weight of Crate #1 is greater

SECTION II: FREE RESPONSE

1. This question concerns the motion of a crate being pulled across a horizontal floor by a rope. In the diagram below, the mass of the crate is m, the coefficient of kinetic friction between the crate and the floor is μ, and the tension in the rope is $\mathbf{F}_T$.

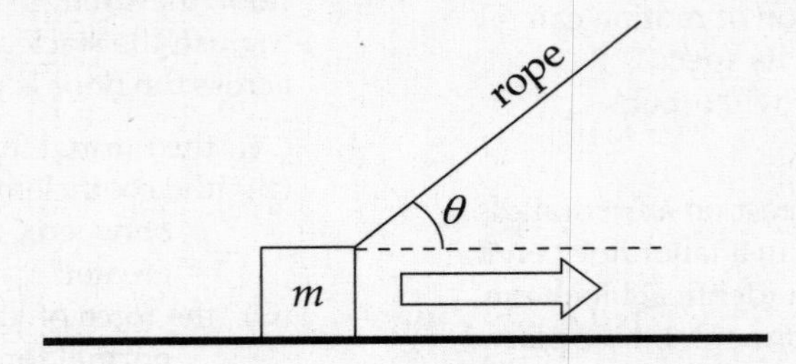

   (a) Draw and label all of the forces acting on the crate.

   (b) Compute the normal force acting on the crate in terms of m, F_T, θ, and g.

   (c) Compute the acceleration of the crate in terms of m, F_T, θ, μ, and g.

2. In the diagram below, a massless string connects two blocks—of masses m_1 and m_2, respectively—on a flat, frictionless tabletop. A force **F** pulls on Block #2, as shown:

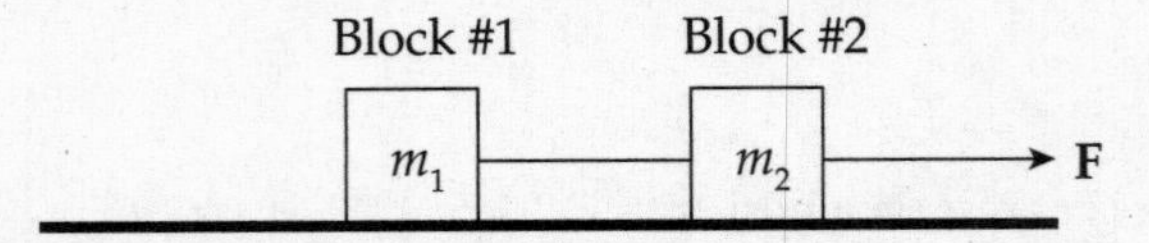

   (a) Draw and label all of the forces acting on Block #1.

   (b) Draw and label all of the forces acting on Block #2.

   (c) What is the acceleration of Block #1?

   (d) What is the tension in the string connecting the two blocks?

   (e) If the string connecting the blocks were not massless, but instead had a mass of m, figure out

   (i) the acceleration of Block #1, and

   (ii) the difference between the strength of the force that the connecting string exerts on Block #2 and the strength of the force that the connecting string exerts on Block #1.

3. In the figure shown, assume that the pulley is frictionless and massless.

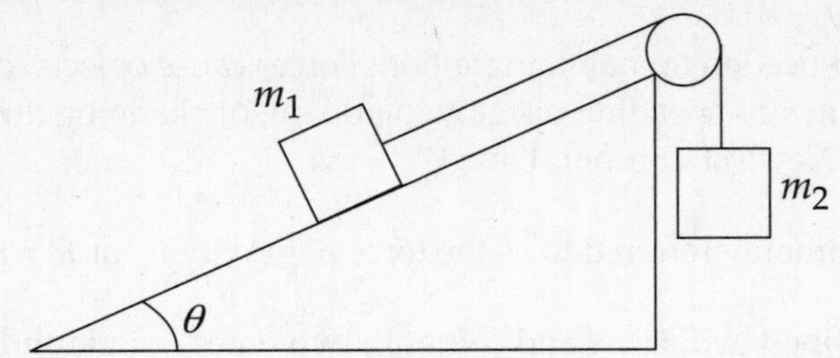

(a) If the surface of the inclined plane is frictionless, determine what value(s) of θ will cause the box of mass m_1 to

(i) accelerate up the ramp;

(ii) slide up the ramp at constant speed.

(b) If the coefficient of kinetic friction between the surface of the inclined plane and the box of mass m_1 is μ_k, derive (but do not solve) an equation satisfied by the value of θ which will cause the box of mass m_1 to slide up the ramp at constant speed.

4. A sky diver is falling with speed v_0 through the air. At that moment (time $t = 0$), she opens her parachute and experiences the force of air resistance whose strength is given by the equation $F = kv$, where k is a proportionality constant and v is her descent speed. The total mass of the sky diver and equipment is m. Assume that g is constant throughout her descent.

(a) Draw and label all the forces acting on the sky diver after her parachute opens.

(b) Determine the sky diver's acceleration in terms of m, v, k, and g.

(c) Determine the sky diver's terminal speed (that is, the eventual constant speed of descent).

(d) Sketch a graph of v as a function of time, being sure to label important values on the vertical axis.

SUMMARY

- Forces are not needed to maintain motion. Forces cause objects to change their motion, whether that means speeding up, slowing down, or changing direction. This idea is expressed by Newton's Second Law: $F_{net} = ma$.

- Weight is commonly referred to as the force of gravity F_w or $F_g = mg$ where g is $-10\ m/s^2$.

- Friction is defined by $F_f = \mu N$ and comes in two types—static and kinetic where $\mu_{kinetic} < \mu_{static}$.

- The normal force (N or sometimes F_N) is frequently given by N or $F_N = mg \cos \theta$, where θ is the angle between the horizontal axis and the surface on which the object rests.

- To solve almost any Newton's Second Law problem, use the following strategy:

I. You must be able to visualize what's going on. Make a sketch if it helps. Make sure to make a free-body diagram (or FBD).

II. Clearly define an appropriate coordinate system.

III. Write out Newton's Second Law in the form of $\sum F_x = ma_x$ and/or $\sum F_y = ma_y$ using the forces identified in the FBD to fill in the appropriate forces.

IV. Do the math.

4

Work, Energy, and Power

INTRODUCTION

It wasn't until more than one hundred years after Newton that the idea of energy became incorporated into physics, but today it permeates every branch of the subject.

It's difficult to give a precise definition of energy; there are different forms of energy because there are different kinds of forces. There's gravitational energy and kinetic energy (a meteor crashing into the earth), elastic energy (a stretched rubber band), thermal energy (an oven), radiant energy (sunlight), electrical energy (a lamp plugged into a wall socket), nuclear energy (nuclear power plants), and mass energy (the heart of Einstein's equation $E = mc^2$). Energy can come into a system or leave it via various interactions that produce changes. One of the best definitions we know reads as follows: **Force** is the agent of change, **energy** is the measure of change, and **work** is the way of transferring energy from one system to another. And one of the most important laws in physics (the **Law of Conservation of Energy**, also known as the **First Law of Thermodynamics**) says that if you account for all its various forms, the total amount of energy in a given process will stay constant; that is, it will be *conserved*. For example, electrical energy can be converted into light and heat (this is how a light bulb works), but the amount of electrical energy *coming in* to the light bulb equals the total amount of light and heat *given off*. Energy cannot be created or destroyed; it can only be transferred (from one system to another) or transformed (from one form to another).

WORK

When you lift a book from the floor, you exert a force on it, over a distance, and when you push a crate across a floor, you also exert a force on it, over a distance. The application of force over a distance, and the resulting change in energy of the system that the force acted on, give rise to the concept of **work**. When you hold a book in your hand, you exert a force on the book (normal force) but, since the book is at rest, the force does not act through a distance, so you do no work on the book. Although you did work on the book as you lifted it from the floor, once it's at rest in your hand, you are no longer doing work on it.

Definition. If a force **F** acts over a distance d, and **F** is parallel to **d**, then the work done by **F** is the product of force and distance: $W = Fd$.

Notice that, although work depends on two vectors (**F** and **d**), work itself is *not* a vector. *Work is a scalar quantity.*

Example 4.1 You slowly lift a book of mass 2 kg at constant velocity a distance of 3 m. How much work did you do on the book?

Solution. In this case, the force you exert must balance the weight of the book (otherwise the velocity of the book wouldn't be constant), so $F = mg = (2 \text{ kg})(10 \text{ m/s}^2) = 20$ N. Since this force is straight upward and the displacement of the book is also straight upward, F and d are parallel, so the work done by your lifting force is $W = Fd = (20 \text{ N})(3 \text{ m}) = 60$ N·m. The unit for work, the newton-meter (N·m) is renamed a joule, and abbreviated as J. So the work done here is 60 J.

The definition above takes care of cases in which **F** is parallel to the motion. If **F** is not parallel to the motion, then the definition needs to be generalized.

Definition. If a force **F** acts over a distance d, and θ is the angle between **F** and **d**, then the work done by **F** is the product of the component of force in the direction of the motion and the distance: $W = (F \cos \theta)d$.

Example 4.2 A 15 kg crate is moved along a horizontal floor by a warehouse worker who's pulling on it with a rope that makes a 30° angle with the horizontal. The tension in the rope is 69 N and the crate slides a distance of 10 m. How much work is done on the crate by the worker?

Solution. The figure below shows that $\mathbf{F}_T$ and $\mathbf{d}$ are not parallel. It's only the component of the force acting along the direction of motion, $\mathbf{F}_T \cos\theta$, that does work.

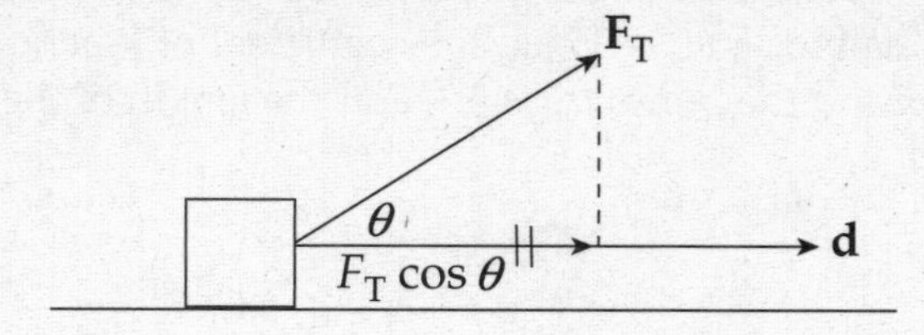

Therefore,

$$W = (F_T \cos\theta)d = (69\text{ N} \cdot \cos 30°)(10\text{ m}) = 600\text{ J}$$

> **Example 4.3** In the previous example, assume that the coefficient of kinetic friction between the crate and the floor is 0.4.
> (a) How much work is done by the normal force?
> (b) How much work is done by the friction force?

Solution.

(a) Clearly, the normal force is not parallel to the motion, so we use the general definition of work. Since the angle between $\mathbf{F}_N$ and $\mathbf{d}$ is 90° (by definition of *normal*) and $\cos 90° = 0$, the normal force does zero work.

(b) The friction force, $\mathbf{F}_f$, is also not parallel to the motion; it's *antiparallel*. That is, the angle between $\mathbf{F}_f$ and $\mathbf{d}$ is 180°. Since $\cos 180° = -1$, and since the strength of the normal force is $F_N = F_w = mg = (15\text{ kg})(10\text{ m/s}^2) = 150\text{ N}$, the work done by the friction force is:

$$W = -F_f d = -\mu_k F_N d = -(0.4)(150\text{ N})(10\text{ m}) = -600\text{ J}$$

Note that $F_T \cos\theta = 60\text{ N}$

$$\mathbf{F}_f = \mu F_N$$

$$= -(0.4)150\text{ N}$$

$$= -60\text{ N}$$

Therefore, $F_{net} = 0$, as we expect.

The two previous examples show that work, which is a scalar quantity, may be positive, negative, or zero. If the angle between **F** and **d** (θ) is less than 90°, then the work is positive (because $\cos\theta$ is positive in this case); if $\theta = 90°$, the work is zero (because $\cos 90° = 0$); and if $\theta > 90°$, then the work is negative (because $\cos\theta$ is negative). Intuitively, if a force helps the motion, the work done by the force is positive, but if the force opposes the motion, then the work done by the force is negative.

Example 4.4 A box slides down an inclined plane (incline angle = 37°). The mass of the block, m, is 35 kg, the coefficient of kinetic friction between the box and the ramp, μ_k, is 0.3, and the length of the ramp, d, is 8 m.

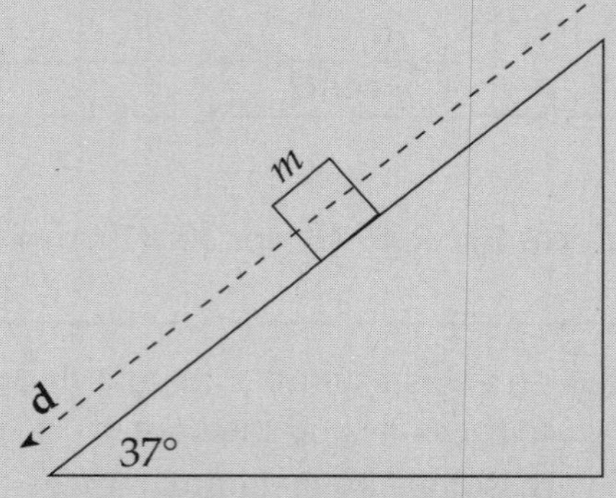

(a) How much work is done by gravity?
(b) How much work is done by the normal force?
(c) How much work is done by friction?
(d) What is the total work done?

Solution.

(a) Recall that the force that's directly responsible for pulling the box down the plane is the component of the gravitational force that's parallel to the ramp: $F_w \sin\theta = mg \sin\theta$ (where θ is the incline angle). This component is parallel to the motion, so the work done by gravity is

$$W_{\text{by gravity}} = (mg \sin\theta)d = (35\text{ kg})(10\text{ N/kg})(\sin 37°)(8\text{ m}) = 1690\text{ J}$$

Note that the work done by gravity is positive, as we would expect it to be, since gravity is helping the motion. Also, be careful with the angle θ. The general definition of work reads $W = (F\cos\theta)d$, where θ is the angle between **F** and **d**. However, the angle between $\mathbf{F}_w$ and **d** is *not* 37° here, so the work done by gravity is not $(mg \cos 37°)d$. The angle θ used in the calculation above is the incline angle.

(b) Since the normal force is perpendicular to the motion, the work done by this force is zero.

(c) The strength of the normal force is $F_w \cos\theta$ (where θ is the incline angle), so the strength of the friction force is $F_f = \mu_k F_N = \mu_k F_w \cos\theta = \mu_k mg \cos\theta$. Since $\mathbf{F}_f$ is antiparallel to **d**, the cosine of the angle between these vectors (180°) is –1, so the work done by friction is

$$W_{\text{by friction}} = -F_f d = -(\mu_k mg \cos\theta)(d) = -(0.3)(35\text{ kg})(10\text{ N/kg})(\cos 37°)(8\text{ m}) = -671\text{ J}$$

Note that the work done by friction is negative, as we expect it to be, since friction is opposing the motion.

(d) The total work done is found simply by adding the values of the work done by each of the forces acting on the box:

$$W_{total} = \Sigma W = W_{by\ gravity} + W_{by\ normal\ force} + W_{by\ friction} = 1690 + 0 + (-671) = 1019\ \text{J}$$

WORK DONE BY A VARIABLE FORCE

If a force remains constant over the distance through which it acts, then the work done by the force is simply the product of force and distance. However, if the force does not remain constant, then the work done by the force is given by the area under the curve of a force vs. displacement graph. In physics language, the term "under the curve" really means between the line itself and zero.

Example 4.5 A spring exerts a force as shown on the graph below. How much work is done as the spring stretches from 20 to 40 cm?

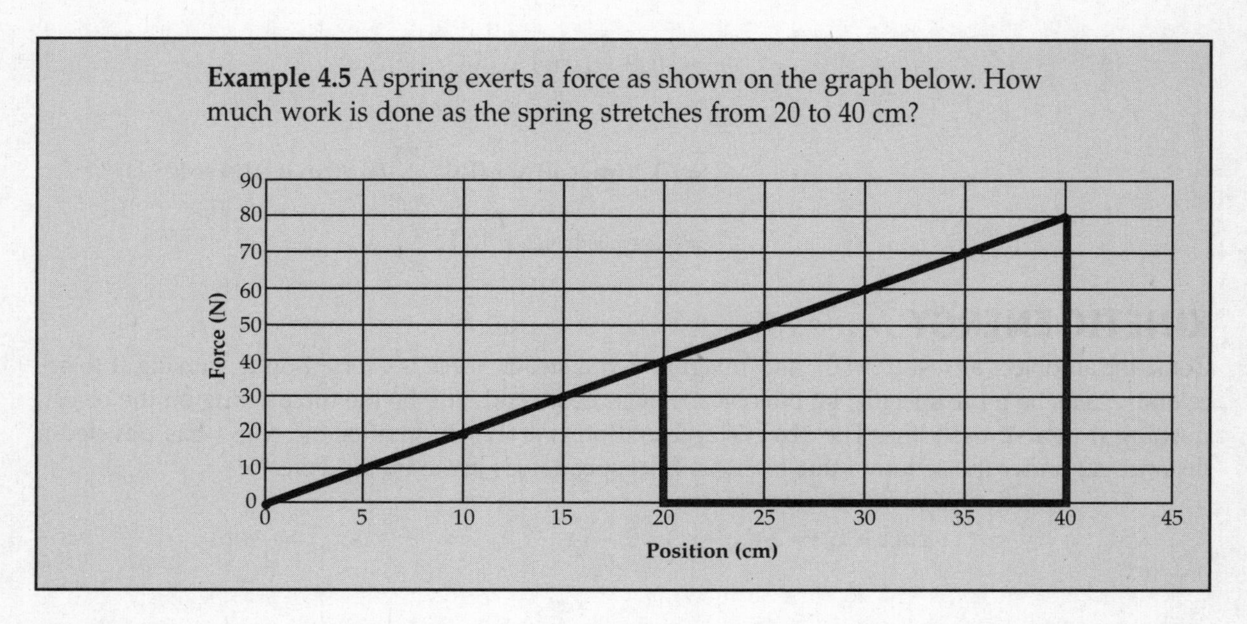

Solution. The area under the curve will be equal to the work done. In this case we have some choices. You may recognize this shape as a trapezoid (it might help to momentarily rotate your head, or this book, 90 degrees) to see this.

$$A = \frac{1}{2}(b_1 + b_2)h$$

$$A = \frac{1}{2}(40\text{N} + 80\text{N})(0.20\text{m})$$

$$= 12\ \text{Nm or } 12\ \text{J}$$

Similar to the previous chapter, units for work and energy should be confined to kg, m, and s, which is why we converted here.

An alternative choice is to recognize this shape as a triangle sitting on top of a rectangle. The total area is simply the area of the rectangle plus the area of the triangle.

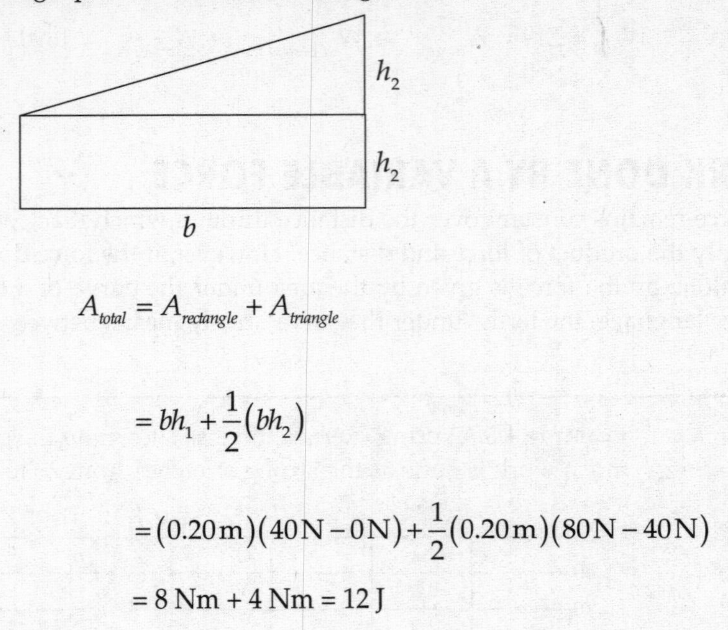

$$A_{total} = A_{rectangle} + A_{triangle}$$

$$= bh_1 + \frac{1}{2}(bh_2)$$

$$= (0.20\,\text{m})(40\,\text{N} - 0\,\text{N}) + \frac{1}{2}(0.20\,\text{m})(80\,\text{N} - 40\,\text{N})$$

$$= 8\text{ Nm} + 4\text{ Nm} = 12\text{ J}$$

KINETIC ENERGY

Consider an object at rest ($v_0 = 0$), and imagine that a steady force is exerted on it, causing it to accelerate. Let's be more specific; let the object's mass be m, and let **F** be the force acting on the object, pushing it in a straight line. The object's acceleration is $a = F/m$, so after the object has traveled a distance Δx under the action of this force, its final speed, v, is given by Big Five #5:

$$v^2 = v_0^2 + 2a(x - x_0) = 2a(x - x_0) = 2\frac{F}{m}(x - x_0) \quad \Rightarrow \quad F(x - x_0) = \frac{1}{2}mv^2$$

But the quantity $F(x - x_0)$ is the work done by the force, so $W = \frac{1}{2}mv^2$. The work done on the object has transferred energy to it, in the amount $\frac{1}{2}mv^2$. The energy an object possesses by virtue of its motion is therefore defined as $\frac{1}{2}mv^2$ and is called **kinetic energy**:

$$K = \frac{1}{2}mv^2$$

THE WORK–ENERGY THEOREM

Kinetic energy is expressed in joules, just like work, since in the case we just looked at, $W = K$. In fact, the derivation above can be extended to an object with a nonzero initial speed, and the same analysis will show that the total work done on an object—or, equivalently, the work done by the net force—will equal its change in kinetic energy; this is known as the **work–energy theorem**:

$$W_{total} = \Delta K$$

Note that kinetic energy, like work, is a scalar quantity.

Example 4.6 What is the kinetic energy of a ball (mass = 0.10 kg) moving with a speed of 30 m/s?

Solution. From the definition,

$$K = \frac{1}{2}mv^2 = \frac{1}{2}(0.10 \text{ kg})\,(30 \text{ m/s})^2 = 45 \text{ J}$$

Example 4.7 A tennis ball (mass = 0.06 kg) is hit straight upward with an initial speed of 50 m/s. How high would it go if air resistance were negligible?

Solution. This could be done using the Big Five, but let's try to solve it using the concepts of work and energy. As the ball travels upward, gravity acts on it by doing negative work. [The work is negative because gravity is opposing the upward motion. F_w and d are in opposite directions, so $\theta = 180°$, which tells us that $W = (F_w \cos \theta)d = -F_w d$.] At the moment the ball reaches its highest point, its speed is 0, so its kinetic energy is also 0. The work–energy theorem says

$$W = \Delta K \;\Rightarrow\; -F_w d = 0 - \frac{1}{2}mv_0^2 \;\Rightarrow\; d = \frac{\frac{1}{2}mv_0^2}{F_w} = \frac{\frac{1}{2}mv_0^2}{mg} = \frac{\frac{1}{2}v_0^2}{g} = \frac{\frac{1}{2}(50 \text{ m/s})^2}{10 \text{ m/s}^2} = 125 \text{ m}$$

Example 4.8 Consider the box sliding down the inclined plane in Example 4.4. If it starts from rest at the top of the ramp, with what speed does it reach the bottom?

Solution. It was calculated in Example 4.4 that W_{total} = 1019 J. According to the work–energy theorem,

$$W_{total} = \Delta K \;\Rightarrow\; W_{total} = K_f - K_i = K_f = \frac{1}{2}mv^2 \;\Rightarrow\; v = \sqrt{\frac{2W_{total}}{m}} = \sqrt{\frac{2(1019 \text{ J})}{35 \text{ kg}}} = 7.6 \text{ m/s}$$

Example 4.9 A pool cue striking a stationary billiard ball (mass = 0.25 kg) gives the ball a speed of 2 m/s. If the average force of the cue on the ball was 200 N, over what distance did this force act?

Solution. The kinetic energy of the ball as it leaves the cue is

$$K = \frac{1}{2}mv^2 = \frac{1}{2}(0.25 \text{ kg})(2 \text{ m/s})^2 = 0.50 \text{ J}$$

The work W done by the cue gave the ball this kinetic energy, so

$$W = \Delta K \;\Rightarrow\; W = K_f \;\Rightarrow\; Fd = K \;\Rightarrow\; d = \frac{K}{F} = \frac{0.50 \text{ J}}{200 \text{ N}} = 0.0025 \text{ m} = 0.25 \text{ cm}$$

POTENTIAL ENERGY

Kinetic energy is the energy an object has by virtue of its motion. Potential energy is independent of motion; it arises from the object's position (or the system's configuration). For example, a ball at the edge of a tabletop has energy that could be transformed into kinetic energy if it falls off. An arrow in an archer's pulled-back bow has energy that could be transformed into kinetic energy if the archer releases the arrow. Both of these examples illustrate the concept of **potential energy**, the energy an object or system has by virtue of its position or configuration. In each case, work was done on the object to put it in the given configuration (the ball was lifted to the tabletop, the bowstring was pulled back), and since work is the means of transferring energy, these things have *stored energy that can be retrieved*, as kinetic energy. This is **potential energy**, denoted by U.

Because there are different types of forces, there are different types of potential energy. The ball at the edge of the tabletop provides an example of **gravitational potential energy**, U_g, which is the energy stored by virtue of an object's position in a gravitational field. This energy would be converted to kinetic energy as gravity pulled the ball down to the floor. For now, let's concentrate on gravitational potential energy.

Assume the ball has a mass m of 2 kg, and that the tabletop is $h = 1.5$ m above the floor. How much work did gravity do as the ball was lifted from the floor to the table? The strength of the gravitational force on the ball is $F_w = mg = (2\text{ kg})(10\text{ N/kg}) = 20\text{ N}$. The force $\mathbf{F}_w$ points downward, and the ball's motion was upward, so the work done by gravity during the ball's ascent was

$$W_{\text{by gravity}} = -F_w h = -mgh = -(20\text{ N})(1.5\text{ m}) = -30\text{ J}$$

So someone performed +30 J of work to raise the ball from the floor to the tabletop. That energy is now stored and, if the ball was given a push to send it over the edge, by the time the ball reached the floor it would acquire a kinetic energy of 30 J. We therefore say that the change in the ball's gravitational potential energy in moving from the floor to the table was +30 J. That is,

$$\Delta U_g = -W_{\text{by gravity}}$$

Note that potential energy, like work (and kinetic energy), is expressed in joules.

In general, if an object of mass m is raised a height h (which is small enough that g stays essentially constant over this altitude change), then the increase in the object's gravitational potential energy is

$$\Delta U_g = mgh$$

An important fact that makes the above equation possible is that the work done by gravity as the object is raised does not depend on the path taken by the object. The ball could be lifted straight upward, or in some curvy path; it would make no difference. Gravity is said to be a **conservative** force because of this property.

If we decide on a reference level to call $h = 0$, then we can say that the gravitational potential energy of an object of mass m at a height h is $U_g = mgh$. In order to use this last equation, it's essential that we choose a reference level for height. For example, consider a passenger in an airplane reading a book. If the book is 1 m above the floor of the plane then, to the passenger, the gravitational potential energy of the book is mgh, where $h = 1$ m. However, to someone on the ground looking up, the floor of the plane may be, say, 9000 m above the ground. So, to this person, the gravitational potential energy of the book is mgH, where $H = 9001$ m. What both would agree on, though, is that the difference in potential energy between the floor of the plane and the position of the book is $mg \times (1\text{ m})$, since the airplane passenger would calculate the difference as $mg \times (1\text{ m} - 0\text{ m})$, while the person on the ground would calculate it as $mg \times (9001\text{ m} - 9000\text{ m})$. Differences, or changes, in potential energy are unambiguous, but values of potential energy are relative.

Example 4.10 A stuntwoman (mass = 60 kg) scales a 40-meter-tall rock face. What is her gravitational potential energy (relative to the ground)?

Solution. Calling the ground $h = 0$, we find

$$U_g = mgh = (60 \text{ kg})(10 \text{ m/s}^2)(40 \text{ m}) = 24{,}000 \text{ J}$$

Example 4.11 If the stuntwoman in the previous example were to jump off the cliff, what would be her final speed as she landed on a large, air-filled cushion lying on the ground?

Solution. The gravitational potential energy would be transformed into kinetic energy. So

$$U \to K \quad \Rightarrow \quad U \to \frac{1}{2}mv^2 \quad \Rightarrow \quad v = \sqrt{\frac{2 \cdot U}{m}} = \sqrt{\frac{2(24{,}000 \text{ J})}{60 \text{ kg}}} = 28 \text{ m/s}$$

CONSERVATION OF MECHANICAL ENERGY

We have seen energy in its two basic forms: Kinetic energy (K) and potential energy (U). The sum of an object's kinetic and potential energies is called its **mechanical energy**, E:

$$E = K + U$$

(Note that because U is relative, so is E.) Assuming that no nonconservative forces (friction, for example) act on an object or system while it undergoes some change, then mechanical energy is conserved. That is, the initial mechanical energy, E_i, is equal to the final mechanical energy, E_f, or

$$K_i + U_i = K_f + U_f$$

This is the simplest form of the Law of Conservation of Total Energy, which we mentioned at the beginning of this section.

Example 4.12 A ball of mass 2 kg is gently pushed off the edge of a tabletop that is 5.0 m above the floor. Find the speed of the ball as it strikes the floor.

Solution. Ignoring the friction due to the air, we can apply Conservation of Mechanical Energy. Calling the floor our $h = 0$ reference level, we write

$$\begin{aligned} K_i + U_i &= K_f + U_f \\ 0 + mgh &= \frac{1}{2}mv^2 + 0 \\ v &= \sqrt{2gh} \\ &= \sqrt{2(10 \text{ m/s}^2)(5.0 \text{ m})} \\ &= 10 \text{ m/s} \end{aligned}$$

Note that the ball's potential energy decreased, while its kinetic energy increased. This is the basic idea behind Conservation of Mechanical Energy: One form of energy decreases while the other increases.

> **Example 4.13** A box is projected up a long ramp (incline angle with the horizontal = 37°) with an initial speed of 10 m/s. If the surface of the ramp is very smooth (essentially frictionless), how high up the ramp will the box go? What distance along the ramp will it slide?

Solution. Because friction is negligible, we can apply Conservation of Mechanical Energy. Calling the bottom of the ramp our h = 0 reference level, we write

$$K_i + U_i = K_f + U_f$$

$$\frac{1}{2}mv_0^2 + 0 = 0 + mgh$$

$$h = \frac{\frac{1}{2}v_0^2}{g}$$

$$= \frac{\frac{1}{2}(10 \text{ m/s})^2}{10 \text{ m/s}^2}$$

$$= 5 \text{ m}$$

Since the incline angle is θ = 37°, the distance d it slides up the ramp is found in this way:

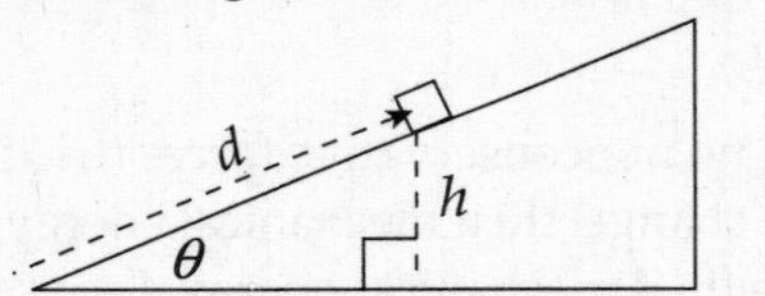

$$h = d \sin\theta$$

$$d = \frac{h}{\sin\theta} = \frac{5 \text{ m}}{\sin 37°} = \frac{25}{3} \text{ m} = 8.3 \text{ m}$$

> **Example 4.14** A skydiver jumps from a hovering helicopter that's 3000 meters above the ground. If air resistance can be ignored, how fast will he be falling when his altitude is 2000 m?

Solution. Ignoring air resistance, we can apply Conservation of Mechanical Energy. Calling the ground our h = 0 reference level, we write

$$K_i + U_i = K_f + U_f$$

$$0 + mgH = \frac{1}{2}mv^2 + mgh$$

$$v = \sqrt{2g(H - h)}$$

$$= \sqrt{2(10 \text{ m/s}^2)(3000 \text{ m} \quad 2000 \text{ m})}$$

$$= 140 \text{ m/s}$$

(That's over 300 mph! This shows that air resistance *does* play a role, even before the parachute is opened.)

The equation $K_i + U_i = K_f + U_f$ holds if no nonconservative forces are doing work. However, if work is done by such forces during the process under investigation, then the equation needs to be modified to account for this work as follows:

$$K_i + U_i + W_{other} = K_f + U_f$$

> **Example 4.15** Wile E. Coyote (mass = 40 kg) falls off a 50-meter-high cliff. On the way down, the force of air resistance has an average strength of 100 N. Find the speed with which he crashes into the ground.

Solution. The force of air resistance opposes the downward motion, so it does negative work on the coyote as he falls: $W_r = -F_r h$. Calling the ground $h = 0$, we find that

$$\begin{aligned} K_i + U_i + W_r &= K_f + U_f \\ 0 + mgh + (-F_r h) &= \frac{1}{2}mv^2 + 0 \\ v &= \sqrt{2h(g - F_r / m)} = \sqrt{2(50)(10 \quad 100/40)} = 27 \text{ m/s} \end{aligned}$$

> **Example 4.16** A skier starts from rest at the top of a 20° incline and skis in a straight line to the bottom of the slope, a distance d (measured along the slope) of 400 m. If the coefficient of kinetic friction between the skis and the snow is 0.2, calculate the skier's speed at the bottom of the run.

Solution. The strength of the friction force on the skier is $F_f = \mu_k F_N = \mu_k(mg \cos \theta)$, so the work done by friction is $-F_f d = \mu_k(mg \cos \theta) \cdot d$. The vertical height of the slope above the bottom of the run (which we designate the $h = 0$ level) is $h = d \sin \theta$. Therefore, Conservation of Mechanical Energy (including the negative work done by friction) gives

$$\begin{aligned} K_i + U_i + W_{friction} &= K_{ff} + U \\ 0 + mgh + (-\mu_k mg \cos\theta \cdot d) &= \tfrac{1}{2}mv^2 + 0 \\ mg(d \sin\theta) + (-\mu_k mg \cos\theta \cdot d) &= \tfrac{1}{2}mv^2 \\ gd(\sin\theta - \mu_k \cos\theta) &= \tfrac{1}{2}v^2 \\ v &= \sqrt{2gd(\sin\theta - \mu_k \cos\theta)} \\ &= \sqrt{2(10)(400)[\sin 20° - (0.2)\cos 20°]} \\ &= 35 \text{ m/s} \end{aligned}$$

So far, any of the problems we have solved this chapter could have been solved using the kinematics equations and Newton's Laws. The truly powerful thing about energy is that in a closed system, changes in energy are independent of the path you take. This allows you to solve many problems you would not otherwise be able to solve. With many energy problems you do not need to measure time with a stopwatch, you do not need to know the mass of the object, you do not need a constant acceleration (remember that is required for our Big Five equations from kinematics), and you do not need to know the path the object takes.

Example 4.17 A roller coaster at an amusement park is at rest on top of a 30 m hill (point A). The car starts to roll down the hill and reaches point B, which is 10 m above the ground, and then rolls up the track to point C, which is 20 m above the ground.

(a) A student assumes no energy is lost, and solves for how fast is the car moving at point C using energy arguments. What answer does he get?

(b) If the final speed at C is actually measured to be 2 m/s, what percentage of energy was "lost" and where did it go?

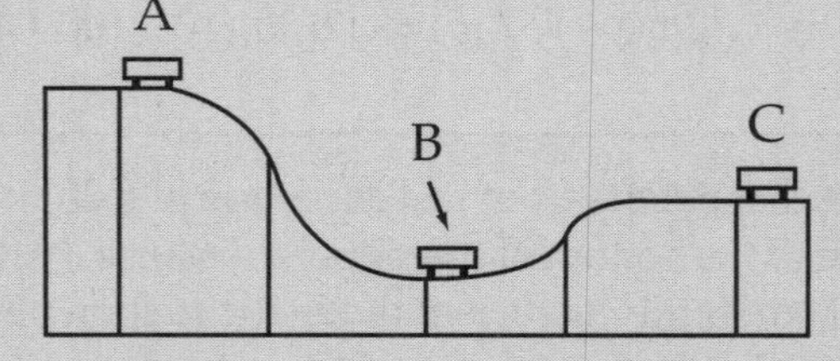

Solution.

(a) Our standard energy equation states

$$K_i + U_i = K_f + U_f$$

or

$$\frac{1}{2}mv_i^2 + mgh_i = \frac{1}{2}mv_f^2 + mgh_f$$

Canceling the mass, setting $v_i = 0$ m/s and rearranging terms we get

$$v_f = \sqrt{2g\Delta h}$$

$$v_f = \sqrt{2\left(9.8\ \frac{\text{m}}{\text{s}^2}\right)(30\ \text{m} - 10\ \text{m})}$$

$$v_f = 20\ \text{m/s}$$

(b) The initial energy of the system was all potential energy, that is,

$$E_i = mgh_i$$

The final energy of the system is given by

$$E_f = \frac{1}{2}mv_f^2 + mgh_f$$

The percent of energy lost is given by the ratio of final energy to initial energy

$$\frac{\frac{1}{2}mv_f^2 + mgh_f}{mgh_i}$$

Canceling mass and substituting values yields

$$\frac{\frac{1}{2}\left(2\ \frac{\text{m}}{\text{s}}\right)^2 + \left(9.8\ \frac{\text{m}}{\text{s}^2}\right)(20\ \text{m})}{\left(9.8\ \frac{\text{m}}{\text{s}^2}\right)(30\ \text{m})}$$

or 0.673 or 67.3%

The other 32.7% was likely lost as heat.

It is worthy to note that the simplification of the energy equation to $v_f = \sqrt{2g\Delta h}$ can be used in quite a number of problems. It can also be inappropriately overused. The equation is valid only if the initial velocity is zero and there is no additional work being done such as a loss of energy due to frictional force, a collision that is not perfectly elastic, or some other applied force. In the above problem there was probably frictional force acting on the cart (which created the heat). Also, be careful about units. If gravity is being measured in meters per second squared, you must measure the change in height in meters.

POWER

Simply put, **power** is the rate at which work gets done (or energy gets transferred, which is the same thing). Suppose you and I each do 1000 J of work, but I do the work in 2 minutes while you do it in 1 minute. We both did the same amount of work, but you did it more quickly; you were more powerful. Here's the definition of power:

$$\text{Power} = \frac{\text{Work}}{\text{time}} \qquad \text{in symbols} \rightarrow \qquad P = \frac{W}{t}$$

Because $W = Fd\cos\theta$, we could write $P = \frac{Fd\cos\theta}{t}$ or $P = F\left(\frac{d}{t}\right)\cos\theta$ or $P = Fv\cos\theta$.

The unit of power is the joule per second (J/s), which is renamed the **watt**, and symbolized W (not to be confused with the symbol for work, *W*). One watt is 1 joule per second: 1 W = 1 J/s. Here in the United States, which still uses older units like inches, feet, yards, miles, ounces, pounds, and so forth, you still hear of power ratings expressed in horsepower (particularly of engines). One horsepower is defined as, well, the power output of a large horse. Horses can pull a 150-pound weight at a speed of $2\frac{1}{2}$ mph for quite a while. Let's assume that **F** and **d** are parallel, so that

$$P = Fv \quad \Rightarrow \quad 1 \text{ horsepower (hp)} = (150 \text{ lb})(2\tfrac{1}{2} \text{ mph})$$

Now for some unit conversions:

$$150 \text{ lb} = 150 \text{ lb} \times \frac{4.45 \text{ N}}{\text{lb}} = 667.5 \text{ N}$$

$$2\frac{1}{2} \text{ mph} = \frac{2\frac{1}{2} \text{ mi}}{\text{hr}} \times \frac{1609 \text{ m}}{1 \text{ mi}} \times \frac{1 \text{ hr}}{3600 \text{ s}} = 1.117 \text{ m/s}$$

Therefore,

$$1 \text{ hp} = (667.5 \text{ N})(1.117 \text{ m/s}) = 746 \text{ W}$$

By contrast, a human in good physical condition can do work at a steady rate of about 75 W (about 1/10 that of a horse!) but can attain power levels as much as twice this much for short periods of time.

Example 4.18 A mover pushes a large crate (mass m = 75 kg) from the inside of the truck to the back end (a distance of 6 m), exerting a steady push of 300 N. If he moves the crate this distance in 20 s, what is his power output during this time?

Solution. The work done on the crate by the mover is $W = Fd$ = (300 N)(6 m) = 1800 J. If this much work is done in 20 s, then the power delivered is $P = W/t$ = (1800 J)/(20 s) = 90 W.

Example 4.19 What must be the power output of an elevator motor that can lift a total mass of 1000 kg and give the elevator a constant speed of 8.0 m/s?

Solution. The equation $P = Fv$, with $F = mg$, yields

$$P = mgv = (1000 \text{ kg})(10 \text{ N/kg})(8.0 \text{ m/s}) = 80{,}000 \text{ W} = 80 \text{ kW}$$

CHAPTER 4 REVIEW QUESTIONS

Solutions can be found in Chapter 18.

SECTION I: MULTIPLE CHOICE

1. A force **F** of strength 20 N acts on an object of mass 3 kg as it moves a distance of 4 m. If **F** is perpendicular to the 4 m displacement, the work it does is equal to

   (A) 0 J
   (B) 60 J
   (C) 80 J
   (D) 600 J
   (E) 2400 J

2. Under the influence of a force, an object of mass 4 kg accelerates from 3 m/s to 6 m/s in 8 s. How much work was done on the object during this time?

   (A) 27 J
   (B) 54 J
   (C) 72 J
   (D) 96 J
   (E) Cannot be determined from the information given

3. A box of mass m slides down a frictionless inclined plane of length L and vertical height h. What is the change in its gravitational potential energy?

   (A) $-mgL$
   (B) $-mgh$
   (C) $-mgL/h$
   (D) $-mgh/L$
   (E) $-mghL$

4. An object of mass m is traveling at constant speed v in a circular path of radius r. How much work is done by the centripetal force during one-half of a revolution?

   (A) πmv^2
   (B) $2\pi mv^2$
   (C) 0
   (D) πmv^2r
   (E) $2\pi mv^2r$

5. While a person lifts a book of mass 2 kg from the floor to a tabletop, 1.5 m above the floor, how much work does the gravitational force do on the book?

   (A) –30 J
   (B) –15 J
   (C) 0 J
   (D) 15 J
   (E) 30 J

6. A block of mass 3.5 kg slides down a frictionless inclined plane of length 6.4 m that makes an angle of 30° with the horizontal. If the block is released from rest at the top of the incline, what is its speed at the bottom?

   (A) 5.0 m/s
   (B) 5.7 m/s
   (C) 6.4 m/s
   (D) 8.0 m/s
   (E) 10 m/s

7. A block of mass m slides from rest down an inclined plane of length s and height h. If F is the magnitude of the force of kinetic friction acting on the block as it slides, then the kinetic energy of the block when it reaches the bottom of the incline will be equal to

   (A) mgh
   (B) $mgh - Fh$
   (C) $mgs - Fh$
   (D) $mgh - Fs$
   (E) $mgs - Fs$

8. As a rock of mass 4 kg drops from the edge of a 40-meter-high cliff, it experiences air resistance, whose average strength during the descent is 20 N. At what speed will the rock hit the ground?

   (A) 8 m/s
   (B) 10 m/s
   (C) 12 m/s
   (D) 16 m/s
   (E) 20 m/s

9. An astronaut drops a rock from the top of a crater on the Moon. When the rock is halfway down to the bottom of the crater, its speed is what fraction of its final impact speed?

(A) $\frac{1}{4\sqrt{2}}$

(B) $\frac{1}{4}$

(C) $\frac{1}{2\sqrt{2}}$

(D) $\frac{1}{2}$

(E) $\frac{1}{\sqrt{2}}$

10. A force of 200 N is required to keep an object sliding at a constant speed of 2 m/s across a rough floor. How much power is being expended to maintain this motion?

(A) 50 W
(B) 100 W
(C) 200 W
(D) 400 W
(E) Cannot be determined from the information given

SECTION II: FREE RESPONSE

1. A box of mass m is released from rest at Point A, the top of a long, frictionless slide. Point A is at height H above the level of Points B and C. Although the slide is frictionless, the horizontal surface from Point B to C is not. The coefficient of kinetic friction between the box and this surface is μ_k, and the horizontal distance between Point B and C is x.

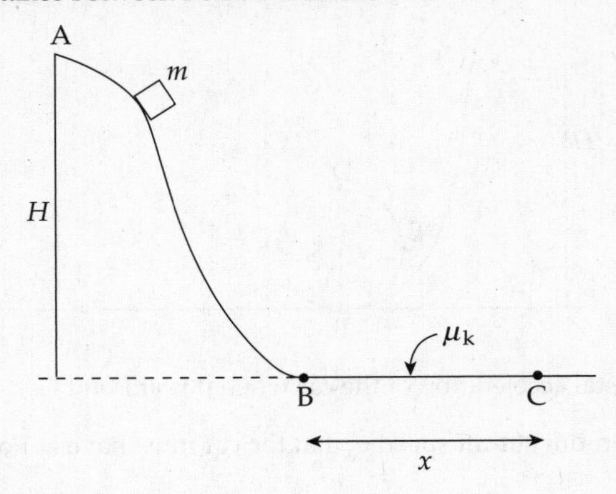

(a) Find the speed of the box when its height above Point B is $\frac{1}{2}H$.

(b) Find the speed of the box when it reaches Point B.

(c) Determine the value of μ_k so that the box comes to rest at Point C.

(d) Now assume that Points B and C were not on the same horizontal level. In particular, assume that the surface from B to C had a uniform upward slope so that Point C were still at a horizontal distance of x from B but now at a vertical height of y above B. Answer the question posed in part (c).

(e) If the slide were not frictionless, determine the work done by friction as the box moved from Point A to Point B if the speed of the box as it reached Point B were half the speed calculated in part (b).

2. The diagram below shows a roller-coaster ride, which contains a circular loop of radius r. A car (mass m) begins at rest from Point A and moves down the frictionless track from A to B where it then enters the vertical loop (also frictionless), traveling once around the circle from B to C to D to E and back to B, after which it travels along the flat portion of the track from B to F (which is not frictionless).

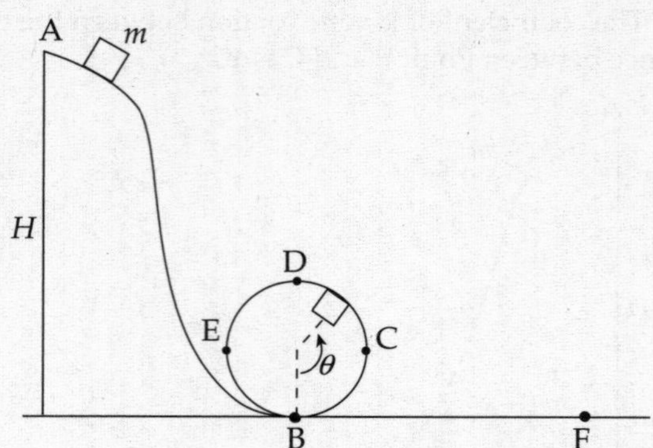

(a) Find the centripetal acceleration of the car when it is at Point C.

(b) What is the minimum cut-off speed v_c that the car must have at Point D to make it around the loop?

(c) What is the minimum height H necessary to ensure that the car makes it around the loop?

(d) If $H = 6r$ and the coefficient of friction between the car and the flat portion of the track from B to F is 0.5, how far along this flat portion of the track will the car travel before coming to rest at Point F?

3. A student uses a digital camera and computer to collect the following data about a ball as it slides down a curved frictionless track. The initial release point is 1.5 meters above the ground and the ball is released from rest. He prints up the following data and then tries to analyze it.

time (s)	velocity (m/s)
0.00	0.00
0.05	1.41
0.10	2.45
0.15	3.74
0.20	3.74
0.25	3.46
0.30	3.16
0.35	2.83
0.40	3.46
0.45	4.24
0.50	4.47

(a) Based on the data, what are the corresponding heights for each data point?

(b) What time segment experiences the greatest acceleration and what is the value of this acceleration?

(c) How would the values change if the mass of the ball were doubled?

SUMMARY

- Work is force applied across a displacement. Work can cause a change in energy. Positive work puts energy into a system, while negative work takes energy out of a system. Basic equations for work include:

  $W = Fd\cos\theta$ Work $= \Delta KE$ W = area under an F vs. d graph

- Energy is a conserved quantity. By that we mean the total initial energy is equal to the total final energy. Basic equations with energy include:

  $K = \frac{1}{2}mv^2$ $U_g = \text{mgh}$ $U_s = \frac{1}{2}kx^2$

- Often, we limit ourselves to mechanical energy with no heat lost or gained. In this case:

  $K_i + U_i \pm W = K_f + U_f$

- Power is the rate at which one does work and is given by:

  $P = \frac{W}{t}$ or $P = Fv$

5

Linear Momentum

INTRODUCTION

When Newton first expressed his Second Law, he didn't write $\mathbf{F}_{\text{net}} = m\mathbf{a}$. Instead, he expressed the law in the words, *The alteration of motion is . . . proportional to the . . . force impressed. . . .* By "motion," he meant the product of mass and velocity, a vector quantity known as **linear momentum** and denoted by **p**:

$$\mathbf{p} = m\mathbf{v}$$

So Newton's original formulation of the Second Law read $\Delta\mathbf{p} \propto \mathbf{F}$, or, equivalently, $\mathbf{F} \propto \Delta\mathbf{p}$. But a large force that acts for a short period of time can produce the same change in linear momentum as a small force acting for a greater period of time. Knowing this, we can turn the proportion above into an equation, if we take the average force that acts over the time interval Δt:

$$\bar{\mathbf{F}} = \frac{\Delta\mathbf{p}}{\Delta t}$$

This equation becomes $\mathbf{F} = m\mathbf{a}$, since $\Delta\mathbf{p}/\Delta t = \Delta(m\mathbf{v})/\Delta t = m\,(\Delta\mathbf{v}/\Delta t) = m\mathbf{a}$ (assuming that m remains constant).

Example 5.1 A golfer strikes a golf ball of mass 0.05 kg, and the time of impact between the golf club and the ball is 1 ms. If the ball acquires a velocity of magnitude 70 m/s, calculate the average force exerted on the ball.

Solution. Using Newton's Second Law, we find

$$\bar{F} = \frac{\Delta p}{\Delta t} = \frac{\Delta(mv)}{\Delta t} = m\frac{v-0}{\Delta t} = (0.05 \text{ kg})\frac{70 \text{ m/s}}{10^{-3} \text{ s}} = 3500 \text{ N} \quad [\approx 790 \text{ lb (!)}]$$

IMPULSE

The product of force and the time during which it acts is known as **impulse**; it's a vector quantity that's denoted by **J**:

$$\mathbf{J} = \bar{\mathbf{F}}\Delta t$$

In terms of impulse, Newton's Second Law can be written in yet another form:

$$\mathbf{J} = \Delta\mathbf{p}$$

Sometimes this is referred to as the **impulse–momentum theorem**, but it's just another way of writing Newton's Second Law. The impulse delivered to an object may be found by taking the area under a force-vs.-time graph. In physics, "the area under a graph," really means the area between the line and the horizontal axis.

You might notice some similarities between impulse causing a change in momentum (J = Δp) and work causing a change in kinetic energy (W = ΔK). Impulse talks about pushing or pulling for some amount of *time* and work deals with pushing or pulling for some *displacement*. Both can change the motion of an object. As a matter of fact, many times when you exert a force on an object, you change both the kinetic energy and the momentum of the object. Momentum and energy are both conserved. However, these two concepts are quite different. Whereas work can go into different forms of energy (heat, potential energy, etc.), impulse has one and only one place to go—momentum. Finally, energy is a scalar and has no direction whereas momentum is a vector and must follow all the rules of vectors as covered in Chapter 1.

Example 5.2 A football team's kicker punts the ball (mass = 0.4 kg) and gives it a launch speed of 30 m/s. Find the impulse delivered to the football by the kicker's foot and the average force exerted by the kicker on the ball, given that the impact time is 8 ms.

Solution. Impulse is equal to change in linear momentum, so

$$J = \Delta p = p_f - p_i = p_f = mv = (0.4 \text{ kg})(30 \text{ m/s}) = 12 \text{ kg·m/s}$$

Using the equation $\bar{F} = J / \Delta t$, we find that the average force exerted by the kicker is

$$\bar{F} = J / \Delta t = (12 \text{ kg} \cdot \text{ m/s}) / (8 \times 10^{-3} \text{ s}) = 1500 \text{ N} \quad [\approx 340 \text{ lb}]$$

Example 5.3 An 80 kg stuntman jumps out of a window that's 45 m above the ground.

(a) How fast is he falling when he reaches ground level?

(b) He lands on a large, air-filled target, coming to rest in 1.5 s. What average force does he feel while coming to rest?

(c) What if he had instead landed on the ground (impact time = 10 ms)?

Solution.

(a) His gravitational potential energy turns into kinetic energy: $mgh = \frac{1}{2}mv^2$, so

$$v = \sqrt{2gh} = \sqrt{2(10)(45)} = 30 \text{ m/s} \quad [\approx 70 \text{ mph}]$$

(You could also have answered this question using Big Five #5.)

(b) Using $\bar{\mathbf{F}} = \Delta \mathbf{p} / \Delta t$, we find that

$$\bar{\mathbf{F}} = \frac{\Delta \mathbf{p}}{\Delta t} = \frac{\mathbf{p}_f - \mathbf{p}_i}{\Delta t} = \frac{0 - m\mathbf{v}_i}{\Delta t} = \frac{-(80 \text{ kg})(30 \text{ m/s})}{1.5 \text{ s}} = -1600 \text{ N} \Rightarrow \bar{F} = 1600 \text{ N}$$

(c) In this case,

$$\bar{\mathbf{F}} = \frac{\Delta \mathbf{p}}{\Delta t} = \frac{\mathbf{p}_f - \mathbf{p}_i}{\Delta t} = \frac{0 - m\mathbf{v}_i}{\Delta t} = \frac{-(80 \text{ kg})(30 \text{ m/s})}{10 \times 10^{-3} \text{ s}} = -240{,}000 \text{ N} \Rightarrow \bar{F} = 240{,}000 \text{ N}$$

The negative signs in the vector answers to (b) and (c) simply tell you that the forces are acting in the opposite direction of motion and will cause the object to slow down. This force is equivalent to about 27 tons (!), more than enough to break bones and cause fatal brain damage. Notice how crucial impact time is: Increasing the slowing-down time reduces the acceleration and the force, ideally enough to prevent injury. This is the purpose of air bags in cars, for instance.

Example 5.4 A small block of mass $m = 0.07$ kg, initially at rest, is struck by an impulsive force **F** of duration 10 ms whose strength varies with time according to the following graph:

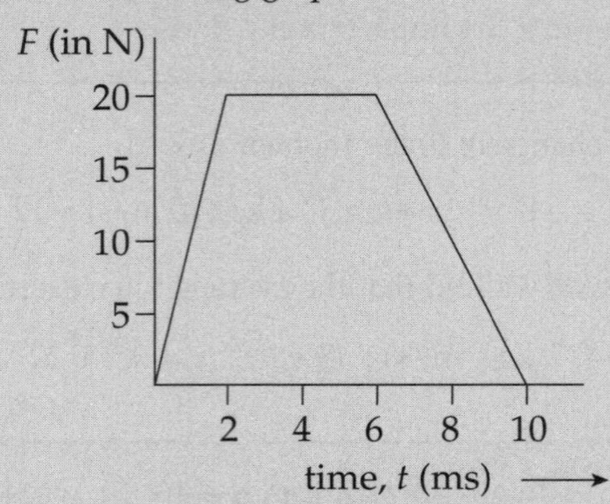

What is the resulting speed of the block?

Solution. The impulse delivered to the block is equal to the area under the *F*-vs.-*t* graph. The region is a trapezoid, so its area, $\frac{1}{2}(\text{base}_1 + \text{base}_2) \times \text{height}$, can be calculated as follows:

$$J = \frac{1}{2}[(10 \text{ ms} - 0) + (6 \text{ ms} - 2 \text{ ms})] \times (20 \text{ N}) = 0.14 \text{ N} \cdot \text{s}$$

Now, by the impulse–momentum theorem,

$$J = \Delta p = p_f - p_i = p_f = mv_f \quad \Rightarrow \quad v_f = \frac{J}{m} = \frac{0.14 \text{ N} \cdot \text{s}}{0.07 \text{ kg}} = 2 \text{ m/s}$$

CONSERVATION OF LINEAR MOMENTUM

Newton's Third Law says that when one object exerts a force on a second object, the second object exerts an equal but opposite force on the first. Since Newton's Second Law says that the impulse delivered to an object is equal to the resulting change in its linear momentum, $\mathbf{J} = \Delta\mathbf{p}$, the two interacting objects experience equal but opposite momentum changes (assuming that we have an isolated system in which there are no external forces), which implies that the total linear momentum of the system remains constant. In fact, given any number of interacting objects, each pair that comes in contact will undergo equal but opposite momentum changes, so the result described for two interacting objects will actually hold for any number of objects, given that the only forces they feel are from each other. This means that, in an isolated system, *the total linear momentum will remain constant*. This is the **Law of Conservation of Linear Momentum**.

Example 5.5 An astronaut is floating in space near her shuttle when she realizes that the cord that's supposed to attach her to the ship has become disconnected. Her total mass (body + suit + equipment) is 91 kg. She reaches into her pocket, finds a 1 kg metal tool, and throws it out into space with a velocity of 9 m/s directly away from the ship. If the ship is 10 m away, how long will it take her to reach it?

Solution. Here, the astronaut + tool are the system. Because of Conservation of Linear Momentum,

$$m_{\text{astronaut}}\mathbf{v}_{\text{astronaut}} + m_{\text{tool}}\mathbf{v}_{\text{tool}} = 0$$

$$m_{\text{astronaut}}\mathbf{v}_{\text{astronaut}} = -m_{\text{tool}}\mathbf{v}_{\text{tool}}$$

$$\mathbf{v}_{\text{astronaut}} = -\frac{m_{\text{tool}}}{m_{\text{astronaut}}}\mathbf{v}_{\text{tool}}$$

$$= -\frac{1\text{ kg}}{90\text{ kg}}(-9\text{ m/s}) = +0.1\text{ m/s}$$

Using *distance* = *rate* × *time*, we find

$$t = \frac{d}{v} = \frac{10\text{ m}}{0.1\text{ m/s}} = 100\text{ s} =$$

COLLISIONS

Conservation of Linear Momentum is routinely used to analyze **collisions**. The objects whose collision we will analyze form the *system*, and although the objects exert forces on each other during the impact, these forces are only *internal* (they occur within the system), and the system's total linear momentum is conserved.

Collisions are classified into two major categories: (1) **elastic** and (2) **inelastic**. A collision is said to be *elastic* if kinetic energy is conserved. Ordinary macroscopic collisions are never truly elastic, because there is always a change in energy due to energy transferred as heat, deformation of the objects, and the sound of the impact. However, if the objects do not deform very much (for example, two billiard balls or a hard glass marble bouncing off a steel plate), then the loss of initial kinetic energy is small enough to be ignored, and the collision can be treated as virtually elastic. *Inelastic* collisions, then, are ones in which the total kinetic energy is different after the collision. An extreme example of inelasticism is **completely** (or **perfectly** or **totally**) **inelastic**. In this case, the objects stick together after the collision and move as one afterward. In all cases of isolated collisions (elastic or not), Conservation of Linear Momentum states that

$$\text{total } \mathbf{p}_{\text{before collision}} = \text{total } \mathbf{p}_{\text{after colision}}$$

Example 5.6 Two balls roll toward each other. The red ball has a mass of 0.5 kg and a speed of 4 m/s just before impact. The green ball has a mass of 0.2 kg and a speed of 2 m/s. After the head-on collision, the red ball continues forward with a speed of 2 m/s. Find the speed of the green ball after the collision. Was the collision elastic?

Solution. First remember that momentum is a vector quantity, so the direction of the velocity is crucial. Since the balls roll toward each other, one ball has a positive velocity while the other has a negative velocity. Let's call the red ball's velocity before the collision positive; then v_{red} = +4 m/s, and v_{green} = –2 m/s. Using a prime to denote *after the collision*, Conservation of Linear Momentum gives us the following:

$$\text{total } \mathbf{p}_{\text{before}} = \text{total } \mathbf{p}_{\text{after}}$$
$$m_{\text{red}}\mathbf{v}_{\text{red}} + m_{\text{green}}\mathbf{v}_{\text{green}} = m_{\text{red}}\mathbf{v}'_{\text{red}} + m_{\text{green}}\mathbf{v}'_{\text{green}}$$
$$(0.5)(+4) + (0.2)(-2) = (0.5)(+2) + (0.2)\mathbf{v}'_{\text{green}}$$
$$\mathbf{v}'_{\text{green}} = +3.0 \text{ m/s}$$

Notice that the green ball's velocity was reversed as a result of the collision; this typically happens when a lighter object collides with a heavier object. To see whether the collision was elastic, we need to compare the total kinetic energies before and after the collision.

Initially

$$K_t = K_1 + K_2$$

$$= \frac{1}{2}m_1 v_{i1}^2 + \frac{1}{2}m_2 v_{i2}^2$$

$$= \frac{1}{2}(0.5)(+4)^2 + \frac{1}{2}(0.2)(-2)^2$$
$$= 4.4 \text{ J}$$

There are 4.4 joules at the beginning. At the end

$$K_t = K_1 + K_2$$

$$= \frac{1}{2}m_1 v_{f1}^2 + \frac{1}{2}m_2 v_{f2}^2$$

$$= \frac{1}{2}(0.5)(+2)^2 + \frac{1}{2}(0.2)(3)^2$$
$$= 1.9 \text{ J}$$

So, there is less kinetic energy at the end compared to the beginning. Kinetic energy was lost (so the collision was inelastic); this is usually the case with macroscopic collisions. Most of the lost energy was transferred as heat; the two objects are both slightly warmer as a result of the collision.

Example 5.7 Two balls roll toward each other. The red ball has a mass of 0.5 kg and a speed of 4 m/s just before impact. The green ball has a mass of 0.3 kg and a speed of 2 m/s. If the collision is completely inelastic, determine the velocity of the composite object after the collision.

Solution. If the collision is completely inelastic, then, by definition, the masses stick together after impact, moving with a velocity, v′. Applying Conservation of Linear Momentum, we find

$$\text{total } \mathbf{p}_{\text{before}} = \text{total } \mathbf{p}_{\text{after}}$$

$$m_{\text{red}}\mathbf{v}_{\text{red}} + m_{\text{green}}\mathbf{v}_{\text{green}} = (m_{\text{red}} + m_{\text{green}})\mathbf{v}'$$

$$(0.5)(+4) + (0.3)(-2) = (0.5 + 0.3)\mathbf{v}'$$

$$\mathbf{v}' = +1.8 \text{ m/s}$$

Example 5.8 A 500 kg car travels 20 m/s due north. It hits a 500 kg car traveling due west at 30 m/s. The cars lock bumpers and stick together. What is the velocity the instant after impact?

Solution. This problem illustrates the vector nature of numbers. First, look only at the x (east-west) direction. There is only one car moving west and its momentum is given by:

$$\mathbf{p} = m\mathbf{v} \Rightarrow (500\text{kg})(30\text{m/s}) = 15{,}000 \text{ kg} \cdot \text{m/s west}$$

Next, look only at the y (north-south) direction. There is only one car moving north and its momentum is given by:

$$\mathbf{p} = m\mathbf{v} \Rightarrow (500\text{kg})(20\text{m/s}) = 10{,}000 \text{ kg} \cdot \text{m/s north}$$

The total final momentum is the resultant of these two vectors. Use Pythagorean theorem:

$$(10{,}000)^2 + (15{,}000)^2 = \mathbf{p}_{\text{tf}}^{\ 2}$$

18,027 kg · m/s

10,000

p

15,000

To find the total velocity you need to solve for v using

$$\mathbf{p}_{\text{tf}} = mv \Rightarrow 18{,}027 \text{ kg} \cdot \text{m/s} = (500\text{kg}+500\text{kg})(v)$$

$$v = 18\text{m/s}$$

However, we are not done yet because velocity has both a magnitude (which we now know is 12 m/s) and a direction. The direction can be expressed using $\tan\theta = 10{,}000/15{,}000$ so $\theta = \tan^{-1}(10{,}000/15{,}000)$ or 33.7 degrees north of west. Again, most of the Physics B test is in degrees, not radians.

Example 5.9 An object of mass m moves with velocity **v** toward a stationary object of mass $2m$. After impact, the objects move off in the directions shown in the following diagram:

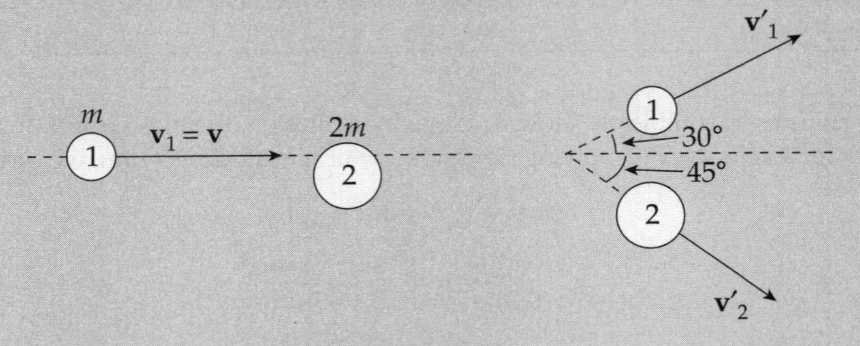

(a) Determine the magnitudes of the velocities after the collision (in terms of v).

(b) Is the collision elastic? Explain your answer.

Solution.

(a) Conservation of Linear Momentum is a principle that establishes the equality of two vectors: $\mathbf{p}_{\text{total}}$ before the collision and $\mathbf{p}_{\text{total}}$ after the collision. Writing this single vector equation as two equations, one for the x component and one for the y, we have

$$x \text{ component:}\quad mv = mv_1' \cos 30° + 2mv_2' \cos 45° \quad (1)$$

$$y \text{ component:}\quad 0 = mv_1' \sin 30° - 2mv_2' \sin 45° \quad (2)$$

Adding these equations eliminates v_2', because cos 45° = sin 45°.

$$mv = mv'_1(\cos 30° + \cos 30°)$$

and lets us determine v_1':

$$v_1' = \frac{v}{\cos 30° + \sin 30°} = \frac{2v}{1+\sqrt{3}}$$

Substituting this result into Equation (2) gives us

$$0 = m\frac{2v}{1+\sqrt{3}}\sin 30^\circ \; - 2mv_2'\sin 45^\circ$$

$$2mv_2'\sin 45^\circ \; = m\frac{2v}{1+\sqrt{3}}\sin 30^\circ$$

$$v_2' = \frac{\frac{2v}{1+\sqrt{3}}\sin 30^\circ}{2\sin 45^\circ} = \frac{v}{\sqrt{2}(1+\sqrt{3})}$$

(b) The collision is elastic only if kinetic energy is conserved. The total kinetic energy after the collision, K', is calculated as follows:

$$\begin{aligned} K' &= \frac{1}{2}\cdot\, mv_1'^2 + \frac{1}{2}\cdot\, 2mv_2'^2 \\ &= \frac{1}{2}m\left(\frac{2v}{1+\sqrt{3}}\right)^2 + m\left(\frac{v}{\sqrt{2}(1+\sqrt{3})}\right)^2 \\ &= mv^2\left[\frac{2}{(1+\sqrt{3})^2} + \frac{1}{2(1+\sqrt{3})^2}\right] \\ &= \frac{5}{2(1+\sqrt{3})^2}mv^2 \end{aligned}$$

However, the kinetic energy before the collision is just $K = \frac{1}{2}mv^2$, so the fact that

$$\frac{5}{2(1+\sqrt{3})^2} < \frac{1}{2}$$

tells us that K' is less than K, so some kinetic energy is lost; the collision is inelastic.

CHAPTER 5 REVIEW QUESTIONS

Solutions can be found in Chapter 18.

SECTION I: MULTIPLE CHOICE

1. An object of mass 2 kg has a linear momentum of magnitude 6 kg · m/s. What is this object's kinetic energy?

   (A) 3 J
   (B) 6 J
   (C) 9 J
   (D) 12 J
   (E) 18 J

2. A ball of mass 0.5 kg, initially at rest, acquires a speed of 4 m/s immediately after being kicked by a force of strength 20 N. For how long did this force act on the ball?

   (A) 0.01 s
   (B) 0.02 s
   (C) 0.1 s
   (D) 0.2 s
   (E) 1 s

3. A box with a mass of 2 kg accelerates in a straight line from 4 m/s to 8 m/s due to the application of a force whose duration is 0.5 s. Find the average strength of this force.

   (A) 2 N
   (B) 4 N
   (C) 8 N
   (D) 12 N
   (E) 16 N

4. A ball of mass m traveling horizontally with velocity **v** strikes a massive vertical wall and rebounds back along its original direction with no change in speed. What is the magnitude of the impulse delivered by the wall to the ball?

   (A) 0
   (B) $\frac{1}{2}mv$
   (C) mv
   (D) $2mv$
   (E) $4mv$

5. Two objects, one of mass 3 kg and moving with a speed of 2 m/s and the other of mass 5 kg and speed 2 m/s, move toward each other and collide head-on. If the collision is perfectly inelastic, find the speed of the objects after the collision.

   (A) 0.25 m/s
   (B) 0.5 m/s
   (C) 0.75 m/s
   (D) 1 m/s
   (E) 2 m/s

6. Object 1 moves toward Object 2, whose mass is twice that of Object 1 and which is initially at rest. After their impact, the objects lock together and move with what fraction of Object 1's initial kinetic energy?

   (A) 1/18
   (B) 1/9
   (C) 1/6
   (D) 1/3
   (E) None of the above

7. Two objects move toward each other, collide, and separate. If there was no net external force acting on the objects, but some kinetic energy was lost, then

   (A) the collision was elastic and total linear momentum was conserved
   (B) the collision was elastic and total linear momentum was not conserved
   (C) the collision was not elastic and total linear momentum was conserved
   (D) the collision was not elastic and total linear momentum was not conserved
   (E) None of the above

8. Two frictionless carts (mass = 500g each) are sitting at rest on a perfectly level table. Teacher taps the release so that one cart pushes off the other. If one of the carts has a speed of 2 m/s, then what is the final momentum of the system (in kg · m/s)?

   (A) Not enough information
   (B) 2000
   (C) 1000
   (D) 2
   (E) 0

9. A wooden block of mass M is moving at speed V in a straight line.

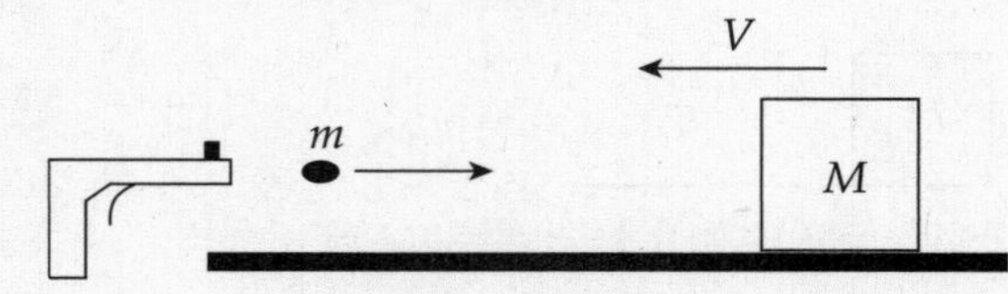

   How fast would the bullet of mass m need to travel to stop the block (assuming that the bullet became embedded inside)?

   (A) $mV/(m+M)$
   (B) $MV/(m+M)$
   (C) mV/M
   (D) MV/m
   (E) $(m+M)V/m$

10. Which of the following best describes a perfectly inelastic collision free of external forces?

   (A) Total linear momentum is never conserved.
   (B) Total linear momentum is sometimes conserved.
   (C) Kinetic energy is never conserved.
   (D) Kinetic energy is sometimes conserved.
   (E) Kinetic energy is always conserved.

SECTION II: FREE RESPONSE

1. A steel ball of mass m is fastened to a light cord of length L and released when the cord is horizontal. At the bottom of its path, the ball strikes a hard plastic block of mass $M = 4m$, initially at rest on a frictionless surface. The collision is elastic.

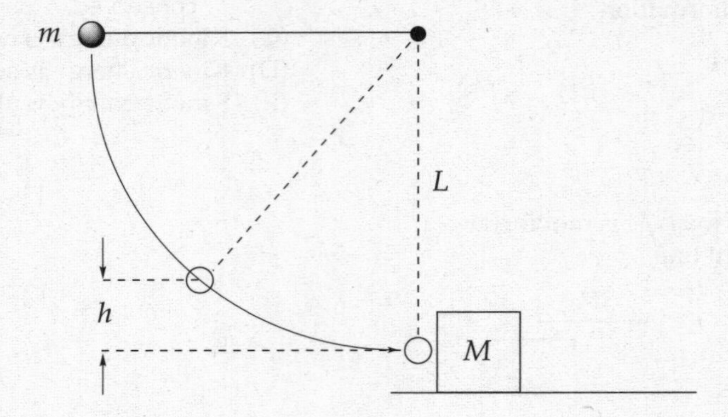

(a) Find the tension in the cord when the ball's height above its lowest position is $\frac{1}{2}L$. Write your answer in terms of m and g.

(b) Find the speed of the block immediately after the collision.

(c) To what height h will the ball rebound after the collision?

2. A *ballistic pendulum* is a device that may be used to measure the muzzle speed of a bullet. It is composed of a wooden block suspended from a horizontal support by cords attached at each end. A bullet is shot into the block, and as a result of the perfectly inelastic impact, the block swings upward. Consider a bullet (mass m) with velocity v as it enters the block (mass M). The length of the cords supporting the block each have length L. The maximum height to which the block swings upward after impact is denoted by y, and the maximum horizontal displacement is denoted by x.

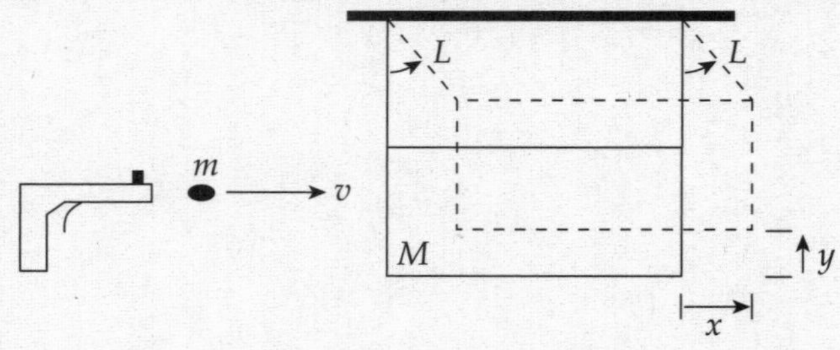

(a) In terms of m, M, g, and y, determine the speed v of the bullet.

(b) What fraction of the bullet's original kinetic energy is lost as a result of the collision? What happens to the lost kinetic energy?

(c) If y is very small (so that y^2 can be neglected), determine the speed of the bullet in terms of m, M, g, x, and L.

(d) Once the block begins to swing, does the momentum of the block remain constant? Why or why not?

SUMMARY

- Momentum is a vector quantity given by $p = mv$. If you push on an object for some amount of time we call that an impulse (J). Impulses cause a change in momentum. Impulse is also a vector quantity and these ideas are summed up in the equations $J = \bar{F}\Delta t$ or $J = \Delta p$.

- Momentum is a conserved quantity in a closed system (that is, a system with no external forces). That means:

  total p_i = total p_f.

- Overall strategy for conservation of momentum problems:

  1. Create a coordinate system.

  2. Break down each object's momentum into x and y components. That is $p_x = p \cos \theta$ and $p_y = p \sin \theta$ for any given object.

  3. $\sum p_{xi} = \sum p_{xf}$ and $\sum p_{yi} = \sum p_{yf}$

  4. Sometimes you end up rebuilding vectors in the end.
     Remember total $p = \sqrt{p_x^2 + p_y^2}$ and the angle is given by $\theta = \tan^{-1}\left(\frac{p_y}{p_x}\right)$.

6

Uniform Circular Motion and Newton's Law of Gravitation

INTRODUCTION

In Chapter 2, we considered two types of motion: straight-line motion and parabolic motion. We will now look at motion that follows a circular path, such as a rock on the end of a string, a horse on a merry-go-round, and (to a good approximation) the Moon around Earth and Earth around the Sun.

UNIFORM CIRCULAR MOTION

Let's simplify matters and consider the object's speed around its path to be constant. This is called **uniform circular motion**. You should remember that although the speed may be constant, the velocity is not, because the direction of the velocity is always changing. Since the velocity is changing, there must be acceleration. This acceleration does not change the speed of the object; it only changes the direction of the velocity to keep the object on its circular path. Also, in order to produce an acceleration, there must be a force; otherwise, the object would move off in a straight line (Newton's First Law).

The figure on the left on the next page shows an object moving along a circular trajectory, along with its velocity vectors at two nearby points. The vector $\mathbf{v}_1$ is the object's velocity at time $t = t_1$, and $\mathbf{v}_2$ is the object's velocity vector a short time later (at time $t = t_2$). The velocity vector is always tangential to the object's path (whatever the shape of the trajectory). Notice that since we are assuming constant speed, the lengths of $\mathbf{v}_1$ and $\mathbf{v}_2$ (their magnitudes) are the same.

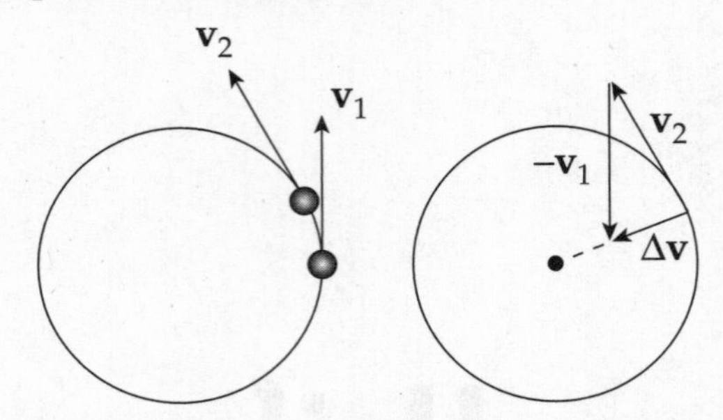

Since $\Delta\mathbf{v} = \mathbf{v}_2 - \mathbf{v}_1$ points toward the center of the circle (see the figure on the right), so does the acceleration, since $\mathbf{a} = \Delta\mathbf{v}/\Delta t$. Because the acceleration vector points toward the center of the circle, it's called **centripetal acceleration**, or $\mathbf{a}_c$. The centripetal acceleration is what turns the velocity vector to keep the object traveling in a circle. The magnitude of the centripetal acceleration depends on the object's speed, v, and the radius, r, of the circular path according to the equation

$$a_c = \frac{v^2}{r}$$

> **Example 6.1** An object of mass 5 kg moves at a constant speed of 6 m/s in a circular path of radius 2 m. Find the object's acceleration and the net force responsible for its motion.

Solution. By definition, an object moving at constant speed in a circular path is undergoing uniform circular motion. Therefore, it experiences a centripetal acceleration of magnitude v^2/r, always directed toward the center of the circle:

$$a_c = \frac{v^2}{r} = \frac{(6\text{ m/s})^2}{2\text{ m}} = 18\text{ m/s}^2$$

The force that produces the centripetal acceleration is given by Newton's Second Law, coupled with the equation for centripetal acceleration:

$$F_c = ma_c = m\frac{v^2}{r}$$

This equation gives the magnitude of the force. As for the direction, recall that because **F** = m**a**, the directions of **F** and **a** are always the same. Since centripetal acceleration points toward the center of the circular path, so does the force that produces it. Therefore, it's called **centripetal force**. The centripetal force acting on this object has a magnitude of $F_c = ma_c = (5 \text{ kg})(18 \text{ m/s}^2) = 90 \text{ N}$.

> **Example 6.2** A 10.0 kg mass is attached to a string that has a breaking strength of 200 N. If the mass is whirled in a horizontal circle of radius 80 cm, what maximum speed can it have?

Solution. The first thing to do in problems like this is to identify what force(s) provide the centripetal acceleration. Notice that this is a horizontal circle. We can limit our examination to the horizontal (x) direction. Because gravity exerts a force in the y direction, it can be ignored. If we were given a problem with a vertical circle, we would have to include the effects of gravity, as demonstrated in Example 6.4. In this example, the tension in the string provides the centripetal force:

$$\mathbf{F}_\text{T} \text{ provides } \mathbf{F}_\text{c} \Rightarrow F_\text{T} = \frac{mv^2}{r} \Rightarrow v = \sqrt{\frac{rF_\text{T}}{m}} \Rightarrow v_\text{max} = \sqrt{\frac{rF_\text{T, max}}{m}}$$

$$= \sqrt{\frac{(0.80 \text{ m})(200 \text{ N})}{10 \text{ kg}}}$$

$$= 4 \text{ m/s}$$

Notice the unit change from 80 cm to 0.80 m. As a general rule, stick to kg, m, and s because the Newton is composed of these units.

> **Example 6.3** An athlete who weighs 800 N is running around a curve at a speed of 5.0 m/s in an arc whose radius of curvature, r, is 5.0 m. Find the centripetal force acting on him. What provides the centripetal force? What could happen to him if r were smaller?

Solution. Using the equation for the strength of the centripetal force, we find that

$$F_\text{c} = m\frac{v^2}{r} = \frac{F_\text{w}}{g} \cdot \frac{v^2}{r} = \frac{800 \text{ N}}{10 \text{ N/kg}} \cdot \frac{(5.0 \text{ m/s})^2}{5.0 \text{ m}} = 400 \text{ N}$$

In this case, static friction provides the centripetal force. Since the coefficient of static friction between his shoes and the ground is most likely around 1, the maximum force that static friction can exert is $\mu_s F_\text{N} \approx F_\text{N} = F_\text{w} = 800 \text{ N}$. Fortunately, 800 N is greater than 400 N. But notice that if the radius of curvature of the arc were much smaller, then F_c would become greater than what static friction could handle, and he would slip.

Example 6.4 A roller-coaster car enters the circular-loop portion of the ride. At the very top of the circle (where the people in the car are upside down), the speed of the car is 15 m/s, and the acceleration points straight down. If the diameter of the loop is 40 m and the total mass of the car (plus passengers) is 1200 kg, find the magnitude of the normal force exerted by the track on the car at this point.

Solution. There are two forces acting on the car at its topmost point: The normal force exerted by the track and the gravitational force, both of which point downward. Assume that the positive direction is down, and $g = +10\ \text{m/s}^2$.

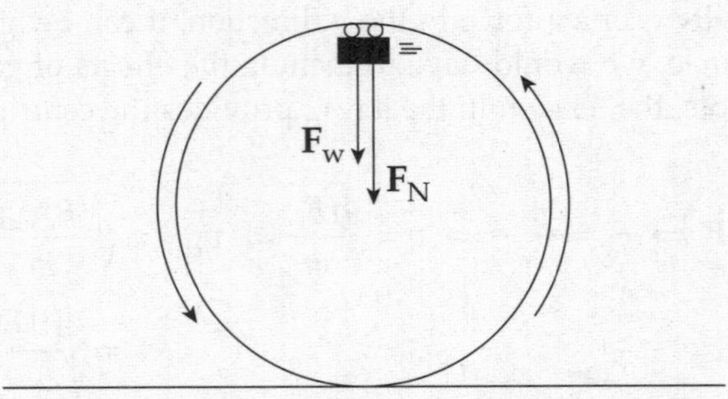

The combination of these two forces, $\mathbf{F}_N + \mathbf{F}_w$, provides the centripetal force:

$$F_N + F_w = \frac{mv^2}{r} \Rightarrow F_N = \frac{mv^2}{r} - F_w$$

$$= \frac{mv^2}{r} - mg$$

$$= m\left(\frac{v^2}{r} - g\right)$$

$$= (1200\ \text{kg})\left[\frac{(15\ \text{m/s})^2}{\frac{1}{2}(40\ \text{m})} - 10\ \text{m/s}^2\right]$$

$$= 1500\ \text{N}$$

Example 6.5 In the previous example, if the net force on the car at its topmost point is straight down, why doesn't the car fall straight down?

Solution. Remember that force tells an object how to accelerate. If the car had zero velocity at this point, then it would certainly fall straight down, but the car has a nonzero velocity (to the left) at this point. The fact that the acceleration is downward means that, at the next moment v will point down to the left at a slight angle, ensuring that the car remains on a circular path, in contact with the track.

Example 6.6 How would the normal force change in Example 6.4 if the car was at the bottom of the circle?

Solution. There are still two forces acting on the cart: The gravitational force still points downward but the normal force pushes ninety degrees to the surface (upward). These forces now oppose one another. The combination of these two forces still provides the centripetal force. Because the centripetal acceleration points inward, we will make anything that points toward the center of the circle positive and anything that points away from the circle negative. Therefore our equation becomes:

$$F_N - F_w = \frac{mv^2}{r} \Rightarrow F_N = \frac{mv^2}{r} + F_w$$

$$= \frac{mv^2}{r} + mg$$

$$= m\left(\frac{v^2}{r} + g\right)$$

$$= 1200 \text{ kg}\left[\frac{(15 \text{ m/s})^2}{\frac{1}{2}(40 \text{ m})} + 10 \text{ m/s}^2\right]$$

$$= 25{,}500 \text{ N}$$

Notice the big difference between this answer and the answer from Example 6.4. This is why you would feel very little force between you and the seat at the top of the loop, but you would feel a big slam at the bottom of the loop.

TORQUE

Intuitively, torque describes the effectiveness of a force in producing rotational acceleration. Consider a uniform rod that pivots around one of its ends, which is fixed. For simplicity, let's assume that the rod is at rest. What effect, if any, would each of the four forces in the figure below have on the potential rotation of the rod?

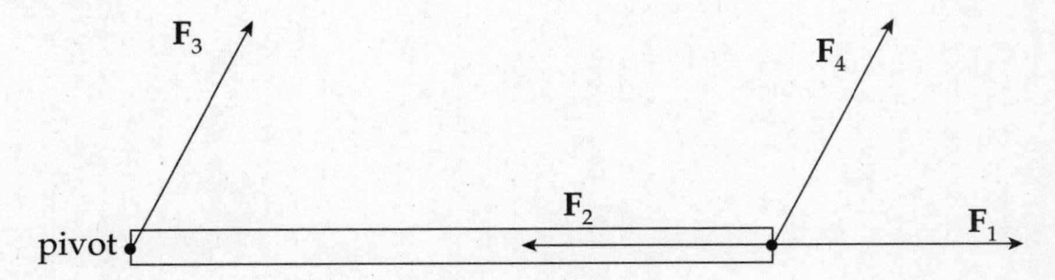

Our intuition tells us the $\mathbf{F}_1$, $\mathbf{F}_2$, and $\mathbf{F}_3$ would not cause the rod to rotate, but $\mathbf{F}_4$ would. What's different about $\mathbf{F}_4$? It has torque. Clearly, torque has something to do with rotation. Just like a force is a vector quantity that produces linear acceleration, a torque is a vector quantity that produces angular acceleration. Note that, just like for linear acceleration, an **angular acceleration** is something that either changes the direction of the angular velocity or changes the angular speed. A torque can be thought of as being positive if it produces counterclockwise rotation, or negative if it produces clockwise rotation.

The **torque** of a force can be defined as follows. Let r be the distance from the pivot (axis of rotation) to the point of application of the force **F**, and let θ be the angle between vectors **r** and **F**.

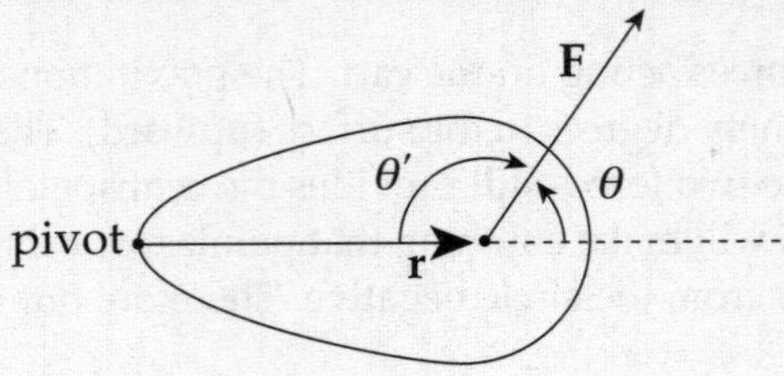

Then the torque of **F**, denoted by τ (*tau*), is defined as:

$$\tau = rF \sin \theta$$

In the figure above, the angle between the vectors **r** and **F** is θ. Imagine sliding **r** over so that its initial point is the same as that of **F**. *The angle between two vectors is the angle between them when they start at the same point*. However, the supplementary angle θ' can be used in place of θ in the definition of torque. This is because torque depends on $\sin \theta$, and the sine of an angle and the sine of its supplement are always equal. Therefore, when figuring out torque, use whichever of these angles is more convenient.

We will now see if this mathematical definition of torque supports our intuition about forces $\mathbf{F}_1$, $\mathbf{F}_2$, $\mathbf{F}_3$, and $\mathbf{F}_4$.

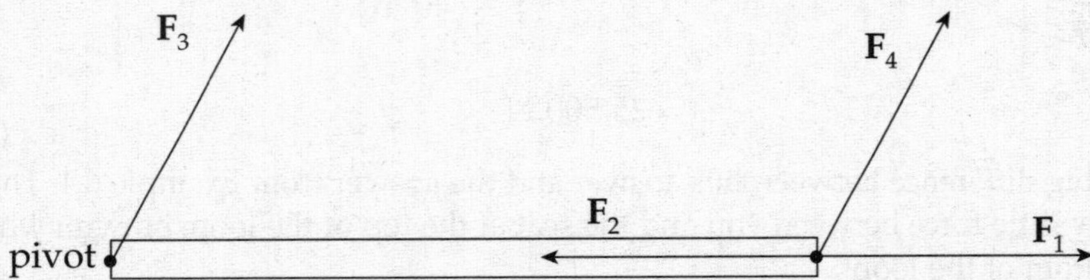

The angle between **r** and $\mathbf{F}_1$ is 0, and $\theta = 0$ implies $\sin \theta = 0$, so by the definition of torque, $\tau = 0$ as well. The angle between **r** and $\mathbf{F}_2$ is 180°, and $\theta = 180°$ gives us $\sin \theta = 0$, so $\tau = 0$. For $\mathbf{F}_3$, $r = 0$ (because $\mathbf{F}_3$ acts *at* the pivot, so the distance from the pivot to the point of application of $\mathbf{F}_3$ is zero); since $r = 0$, the torque is 0 as well. However, for $\mathbf{F}_4$, neither r nor $\sin \theta$ is zero, so $\mathbf{F}_4$ has a nonzero torque. Of the four forces shown in that figure, only $\mathbf{F}_4$ has torque and would produce rotation.

Example 6.7 A student pulls down with a force of 40 N on a rope that winds around a pulley of radius 5 cm.

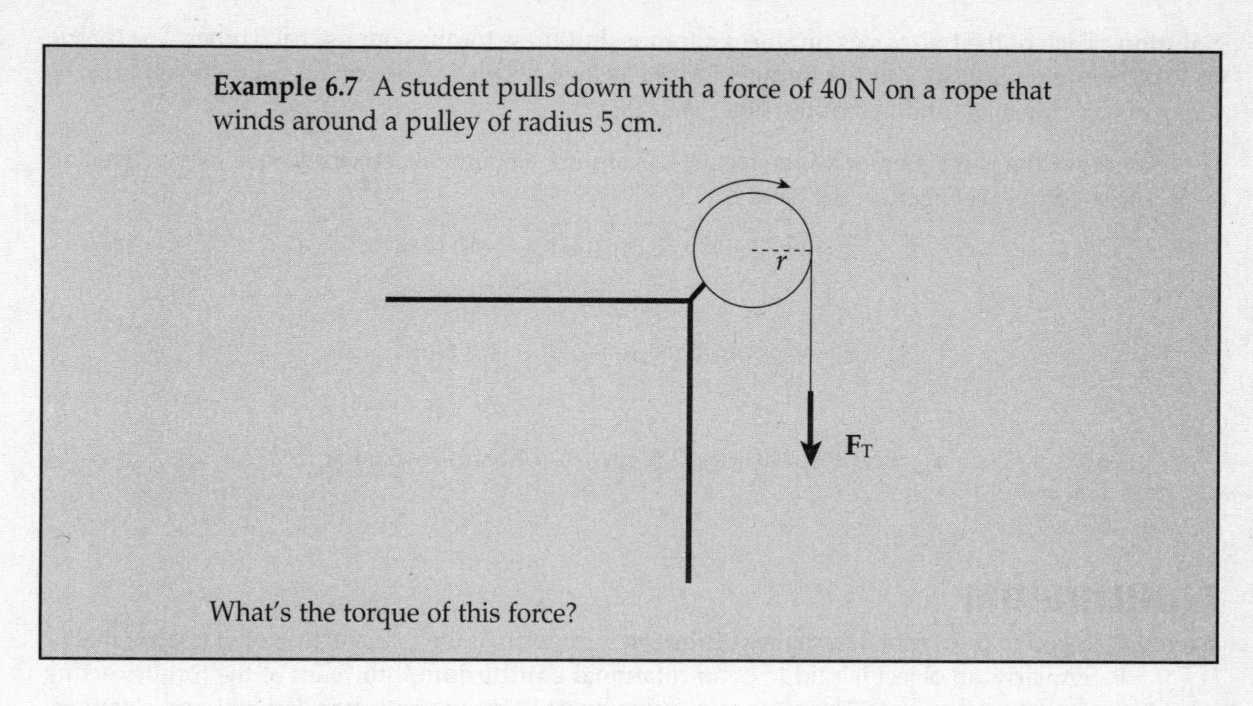

What's the torque of this force?

Solution. Since the tension force, $\mathbf{F}_T$, is tangent to the pulley, it is perpendicular to the radius vector **r** at the point of contact:

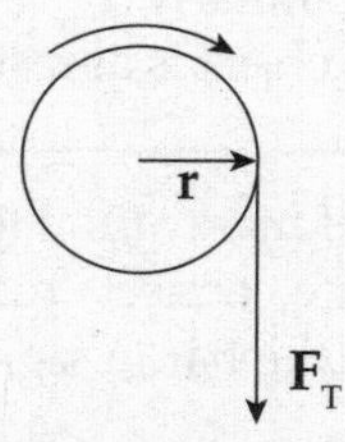

Therefore, the torque produced by this tension force is simply

$$\tau = rF_T = (0.05 \text{ m})(40 \text{ N}) = 2 \text{ N·m}$$

Example 6.8 What is the net torque on the cylinder shown below, which is pinned at the center?

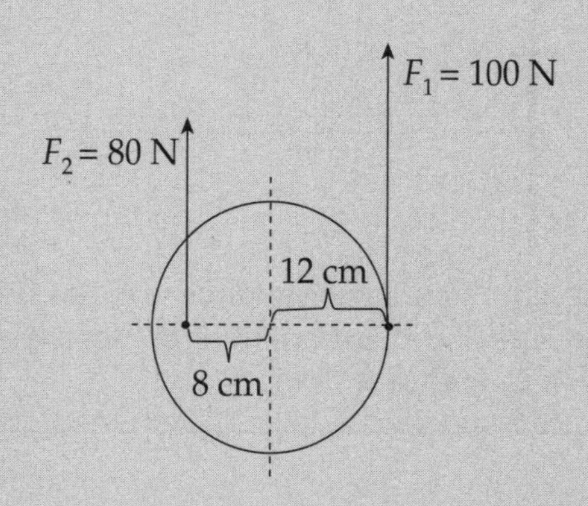

Solution. Each of the two forces produces a torque, but these torques oppose each other. The torque of F_1 is counterclockwise, and the torque of F_2 is clockwise. This can be visualized by imagining the effect of each force, assuming that the other was absent.

The **net torque** is the sum of all the torques. Counting a counterclockwise torque as positive and a clockwise torque as negative, we have

$$\tau_1 = +r_1F_1 = +(0.12 \text{ m})(100 \text{ N}) = +12 \text{ N·m}$$

and

$$\tau_2 = -r_2F_2 = -(0.08 \text{ m})(80 \text{ N}) = -6.4 \text{ N·m}$$

so

$$\tau_{net} = \Sigma\tau = \tau_1 + \tau_2 = (+12 \text{ N·m}) + (-6.4 \text{ N·m}) = +5.6 \text{ N·m}$$

EQUILIBRIUM

An object is said to be in **translational equilibrium** if the sum of the forces acting on it is zero; that is, if $F_{net} = 0$. Similarly, an object is said to be in **rotational equilibrium** if the sum of the torques acting on it is zero; that is, if $\tau_{net} = 0$. The term *equilibrium* by itself means both translational and rotational equilibrium. A body in equilibrium may be in motion; $F_{net} = 0$ does not mean that the velocity is zero; it only means that the velocity is constant. Similarly, $\tau_{net} = 0$ does not mean that the angular velocity is zero; it only means that it's constant. If an object is at rest, then it is said to be in **static equilibrium**.

Example 6.9 A uniform bar of mass m and length L extends horizontally from a wall. A supporting wire connects the wall to the bar's midpoint, making an angle of 55° with the bar. A sign of mass M hangs from the end of the bar.

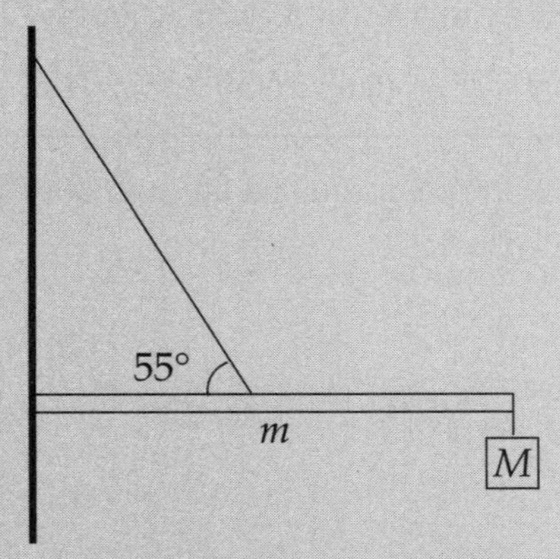

If the system is in static equilibrium and the wall has friction, determine the tension in the wire and the strength of the force exerted on the bar by the wall if $m = 8$ kg and $M = 12$ kg.

Solution. Let F_C denote the (contact) force exerted by the wall on the bar. In order to simplify our work, we can write F_C in terms of its horizontal component, F_{Cx}, and its vertical component, F_{Cy}. Also, if F_T is the tension in the wire, then $F_{Tx} = F_T \cos 55°$ and $F_{Ty} = F_T \sin 55°$ are its components. This gives us the following force diagram:

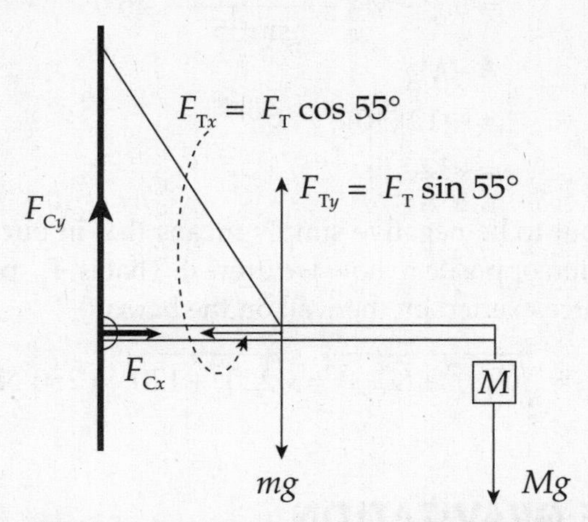

The first condition for equilibrium requires that the sum of the horizontal forces is zero and the sum of the vertical forces is zero:

$$\Sigma F_x = 0: \qquad F_{Cx} - F_T \cos 55° = 0 \qquad (1)$$
$$\Sigma F_y = 0: \qquad F_{Cy} + F_T \sin 55° - mg - Mg = 0 \qquad (2)$$

We notice immediately that we have more unknowns (F_{Cx}, F_{Cy}, F_T) than equations, so this system cannot be solved as is. The second condition for equilibrium requires that the sum of the torques about any point is equal to zero. Choosing the contact point between the bar and the wall as our pivot, only three of the forces in the diagram above produce torque: $\mathbf{F}_{Ty}$ produces a counterclockwise torque, and both mg and Mg produce clockwise torques, and the sum of the three torques must equal zero. From the definition $\tau = rF\sin 0$, and taking counterclockwise torque as positive and clockwise torque as negative, we have

$$\Sigma\tau = 0: \qquad \tfrac{L}{2}F_{Ty} - \tfrac{L}{2}mg - LMg = 0 \qquad (3)$$

This equation contains only one unknown and can be solved immediately:

$$\tfrac{L}{2}F_{Ty} = \tfrac{L}{2}mg + LMg$$
$$F_{Ty} = mg + 2Mg = (m + 2M)g$$

Since $F_{Ty} = F_T \sin 55°$, we can find that

$$F_T \sin 55° = (m+2M)g \;\Rightarrow\; F_T = \frac{(m+2M)g}{\sin 55°} = \frac{(8+2\cdot 12)(10)}{\sin 55°} = 390 \text{ N}$$

Substituting this result into Equation (1) gives us F_{Cx}:

$$F_{Cx} = F_T \cos 55° = \frac{(m+2M)g}{\sin 55°}\cos 55° = (8+2\cdot 12)(10)\cot 55° = 220 \text{ N}$$

And finally, from Equation (2), we get

$$\begin{aligned} F_{Cy} &= mg + Mg - F_T \sin 55° \\ &= mg + Mg - \frac{(m+2M)g}{\sin 55°} \sin 55° \\ &= -Mg \\ &= -(12)(10) \\ &= -120 \text{ N} \end{aligned}$$

The fact that F_{Cy} turned out to be negative simply means that in our original force diagram, the vector $\mathbf{F}_{Cy}$ points in the direction opposite to how we drew it. That is, $\mathbf{F}_{Cy}$ points downward. Therefore, the magnitude of the total force exerted by the wall on the bar is

$$F_C = \sqrt{(F_{Cx})^2 + (F_{Cy})^2} = \sqrt{220^2 + 120^2} = 250 \text{ N}$$

NEWTON'S LAW OF GRAVITATION

Newton eventually formulated a law of gravitation: Any two objects in the universe exert an attractive force on each other—called the **gravitational force**—whose strength is proportional to the product of the objects' masses and inversely proportional to the square of the distance between them as measured from center to center. If we let G be the **universal gravitational constant**, then the strength of the gravitational force is given by the equation:

$$F_G = \frac{Gm_1m_2}{r^2}$$

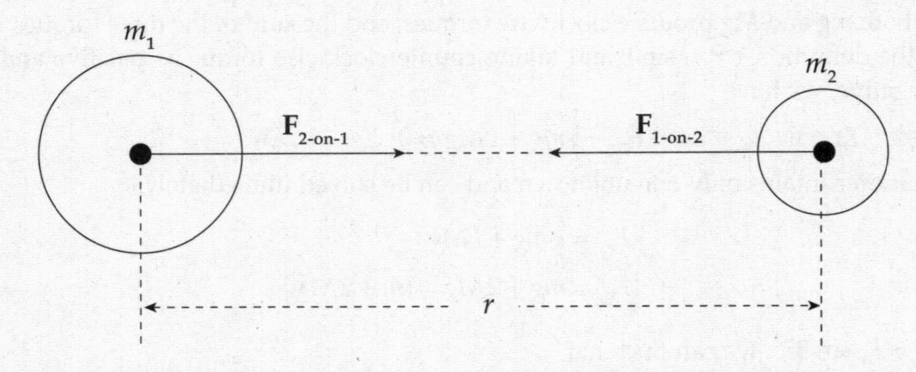

The forces $\mathbf{F}_{1\text{-on-}2}$ and $\mathbf{F}_{2\text{-on-}1}$ act along the line that joins the bodies and form an action/reaction pair.

The first reasonably accurate numerical value for G was determined by Cavendish more than one hundred years after Newton's Law was published. To three decimal places, the currently accepted value of G is

$$G = 6.67 \times 10^{-11} \text{ N} \cdot \text{m}^2/\text{kg}^2$$

THE GRAVITATIONAL ATTRACTION DUE TO AN EXTENDED BODY

Newton's Law of Gravitation is really a statement about the force between two point particles; objects that are very small in comparison to the distance between them. Newton also proved that a uniform sphere attracts another body as if all of the sphere's mass were concentrated at its center.

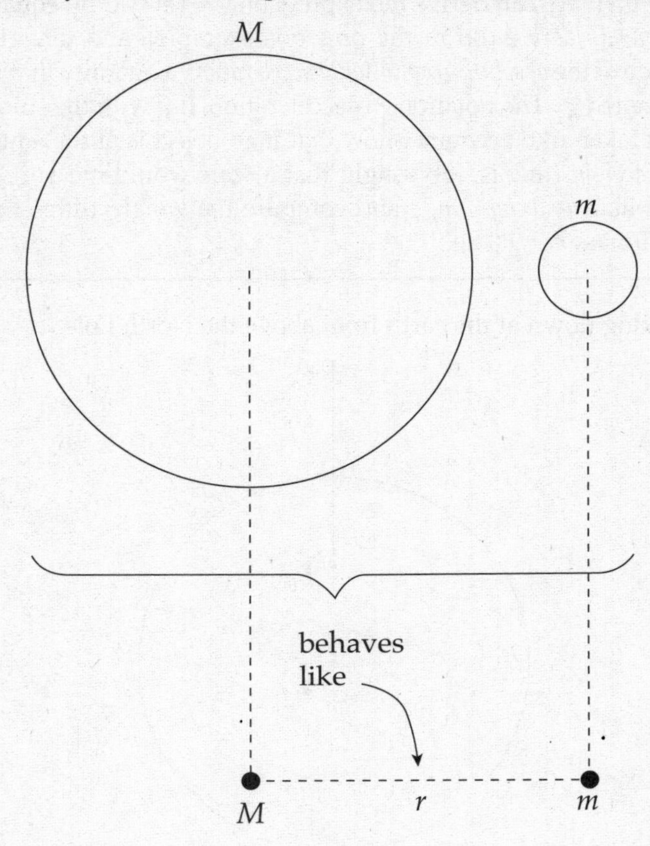

For this reason, we can apply Newton's Law of Gravitation to extended bodies, that is, to objects that are not small relative to the distance between them.

Additionally, a uniform shell of mass does not exert a gravitational force on a particle inside it. This means that if a spherical planet is uniform, then as we descend into it, only the mass of the sphere *underneath* us exerts a gravitational force; the shell *above* exerts no force because we're inside it.

Example 6.10 Given that the radius of the earth is 6.37×10^6 m, determine the mass of the earth.

Solution. Consider a small object of mass m near the surface of the earth (mass M). Its weight is mg, but its weight is just the gravitational force it feels due to the earth, which is GMm/R^2. Therefore,

$$mg = G\frac{Mm}{R^2} \quad \Rightarrow \quad M = \frac{gR^2}{G}$$

Since we know that $g = 10\ \text{m/s}^2$ and $G = 6.67 \times 10^{-11}\ \text{N·m}^2/\text{kg}^2$, we can substitute to find

$$M = \frac{gR^2}{G} = \frac{(10\ \text{m/s}^2)(6.37 \times 10^6\ \text{m})^2}{6.67 \times 10^{-11}\ \text{N} \cdot \text{m}^2/\text{kg}^2} = 6.1 \times 10^{24}\ \text{kg}$$

Example 6.11 We can derive the expression $g = GM/R^2$ by equating mg and GMm/R^2 (as we did in the previous example), and this gives the magnitude of the *absolute gravitational acceleration*, a quantity that's sometimes denoted g_0. The notation g is acceleration, but with the spinning of the earth taken into account. Show that if an object is at the equator, its *measured weight* (that is, the weight that a scale would measure), mg, is less than its *true weight*, mg_0, and compute the weight difference for a person of mass $m = 60$ kg.

Solution. Imagine looking down at the earth from above the North Pole.

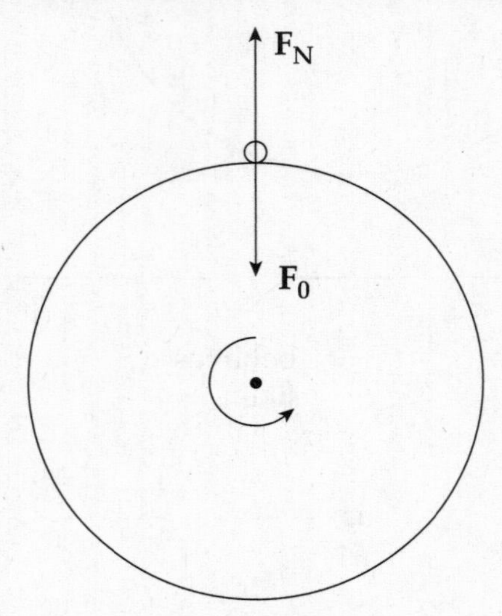

The net force toward the center of the earth is $\mathbf{F}_0 - \mathbf{F}_N$, which provides the centripetal force on the object. Therefore,

$$F_0 - F_N = \frac{mv^2}{R}$$

Since $v = 2\pi R/T$, where T is the earth's rotation period we have

$$F_0 - F_N = \frac{m}{R}\left(\frac{2\pi R}{T}\right)^2 = \frac{4\pi^2 mR}{T^2}$$

or, since $F_0 = mg_0$ and $F_N = mg$,

$$mg_0 - mg = \frac{4\pi^2 mR}{T^2}$$

Since the quantity $\frac{4\pi^2 mR}{T^2}$ is positive, mg must be less than mg_0. The difference between mg_0 and mg, for a person of mass m = 60 kg, is only:

$$\frac{4\pi^2 mR}{T^2} = \frac{4\pi^2(60 \text{ kg})(6.37\times10^6 \text{ m})}{\left(24 \text{ hr} \times \frac{60 \text{ min}}{\text{hr}} \times \frac{60 \text{ s}}{\text{min}}\right)^2} = 2.0 \text{ N}$$

and the difference between g_0 and g is

$$g_0 - g = \frac{mg_0 - mg}{m} = \frac{4\pi^2 R}{T^2} = \frac{4\pi^2(6.37\times10^6 \text{ m})}{\left(24 \text{ hr} \times \frac{60 \text{ min}}{\text{hr}} \times \frac{60 \text{ s}}{\text{min}}\right)^2} = 0.034 \text{ m/s}^2$$

Note that this difference is so small (< 0.3%) that it can usually be ignored.

Example 6.12 Communications satellites are often parked in geosynchronous orbits above Earth's surface. These satellites have orbit periods that are equal to Earth's rotation period, so they remain above the same position on Earth's surface. Determine the altitude that a satellite must have to be in a geosynchronous orbit above a fixed point on Earth's equator. (The mass of the earth is 5.98×10^{24} kg.)

Solution. Let m be the mass of the satellite, M the mass of Earth, and R the distance from the center of Earth to the position of the satellite. The gravitational pull of Earth provides the centripetal force on the satellite, so

$$G\frac{Mm}{R^2} = \frac{mv^2}{R} \quad \Rightarrow \quad G\frac{M}{R} = v^2$$

The orbit speed of the satellite is $2\pi R/T$, so

$$G\frac{M}{R} = \left(\frac{2\pi R}{T}\right)^2$$

which implies that

$$G\frac{M}{R} = \frac{4\pi^2 R^2}{T^2} \quad \Rightarrow \quad 4\pi^2 R^3 = GMT^2 \quad \Rightarrow \quad R = \sqrt[3]{\frac{GMT^2}{4\pi^2}}$$

Now the key feature of a geosynchronous orbit is that its period matches Earth's rotation period, T = 24 hr. Substituting the numerical values of G, M, and T into this expression, we find that

$$R = \sqrt[3]{\frac{GMT^2}{4\pi^2}} = \sqrt[3]{\frac{(6.67\times10^{-11})(5.98\times10^{24})(24\cdot60\cdot60)^2}{4\pi^2}}$$
$$= 4.23\times10^7 \text{ m}$$

Therefore, if r_E is the radius of Earth, then the satellite's altitude above Earth's surface must be

$$h = R - r_E = (4.23 \times 10^7 \text{ m}) - (6.37 \times 10^6 \text{ m}) = 3.59 \times 10^7 \text{ m}$$

Example 6.13 An artificial satellite of mass m travels at a constant speed in a circular orbit of radius R around the earth (mass M). What is the speed of the satellite?

Solution. The centripetal force on the satellite is provided by Earth's gravitational pull. Therefore,

$$\frac{mv^2}{R} = G\frac{Mm}{R^2}$$

Solving this equation for v yields

$$v = \sqrt{G\frac{M}{R}}$$

Notice that the satellite's speed doesn't depend on its mass; even if it were a baseball, if its orbit radius were R, then its orbit speed would still be $\sqrt{GM/R}$.

CHAPTER 6 REVIEW QUESTIONS

Solutions can be found in Chapter 18.

SECTION I: MULTIPLE CHOICE

1. An object moves at constant speed in a circular path. Which of the following statements is/are true?

    I. The velocity is constant.
    II. The acceleration is constant.
    III. The net force on the object is zero.

    (A) II only
    (B) I and III only
    (C) II and III only
    (D) I and II only
    (E) None of the above

Questions 2–3:

A 60 cm rope is tied to the handle of a bucket which is then whirled in a vertical circle. The mass of the bucket is 3 kg.

2. At the lowest point in its path, the tension in the rope is 50 N. What is the speed of the bucket?

    (A) 1 m/s
    (B) 2 m/s
    (C) 3 m/s
    (D) 4 m/s
    (E) 5 m/s

3. What is the critical speed below which the rope would become slack when the bucket reaches the highest point in the circle?

    (A) 0.6 m/s
    (B) 1.8 m/s
    (C) 2.4 m/s
    (D) 3.2 m/s
    (E) 4.8 m/s

4. An object moves at a constant speed in a circular path of radius r at a rate of 1 revolution per second. What is its acceleration?

    (A) 0
    (B) $2\pi^2 r$
    (C) $2\pi^2 r^2$
    (D) $4\pi^2 r$
    (E) $4\pi^2 r^2$

5.

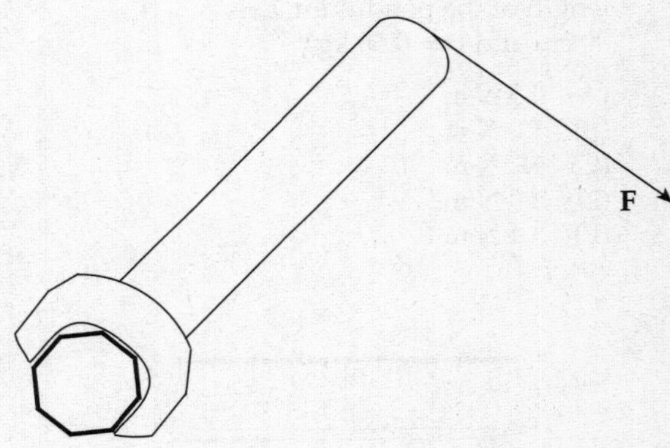

In an effort to tighten a bolt, a force **F** is applied as shown in the figure above. If the distance from the end of the wrench to the center of the bolt is 20 cm and $F = 20$ N, what is the magnitude of the torque produced by **F**?

(A) 0 N·m
(B) 1 N·m
(C) 2 N·m
(D) 4 N·m
(E) 10 N·m

6.

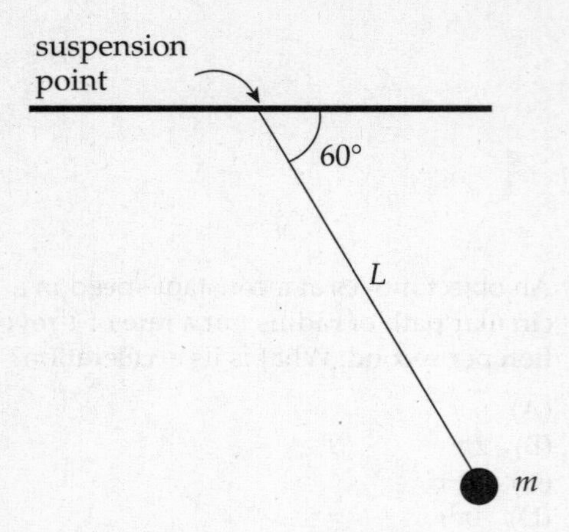

In the figure above, what is the torque about the pendulum's suspension point produced by the weight of the bob, given that the length of the pendulum, L, is 80 cm and $m = 0.50$ kg?

(A) 0.5 N·m
(B) 1.0 N·m
(C) 1.7 N·m
(D) 2.0 N·m
(E) 3.4 N·m

7.

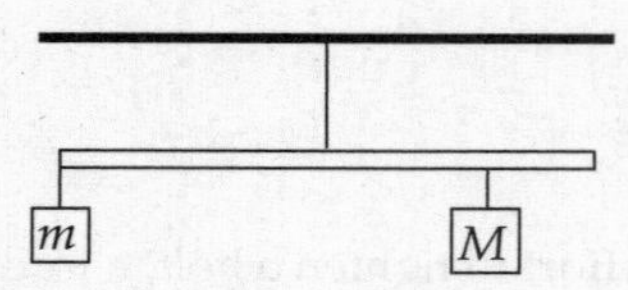

A uniform meter stick of mass 1 kg is hanging from a thread attached at the stick's midpoint. One block of mass $m = 3$ kg hangs from the left end of the stick, and another block, of unknown mass M, hangs below the 80 cm mark on the meter stick. If the stick remains at rest in the horizontal position shown above, what is M?

(A) 4 kg
(B) 5 kg
(C) 6 kg
(D) 8 kg
(E) 9 kg

8. If the distance between two point particles is doubled, then the gravitational force between them

(A) decreases by a factor of 4
(B) decreases by a factor of 2
(C) increases by a factor of 2
(D) increases by a factor of 4
(E) cannot be determined without knowing the masses

9. At the surface of the earth, an object of mass m has weight w. If this object is transported to an altitude that's twice the radius of the earth, then, at the new location,

(A) its mass is $m/2$ and its weight is $w/2$
(B) its mass is m and its weight is $w/2$
(C) its mass is $m/2$ and its weight is $w/4$
(D) its mass is m and its weight is $w/4$
(E) its mass is m and its weight is $w/9$

10. A moon of mass m orbits a planet of mass $100m$. Let the strength of the gravitational force exerted by the planet on the moon be denoted by F_1, and let the strength of the gravitational force exerted by the moon on the planet be F_2. Which of the following is true?

(A) $F_1 = 100F_2$
(B) $F_1 = 10F_2$
(C) $F_1 = F_2$
(D) $F_2 = 10F_1$
(E) $F_2 = 100F_1$

11. The dwarf planet Pluto has 1/500 the mass and 1/15 the radius of Earth. What is the value of g (in m/s^2) on the surface of Pluto?

(A) $\frac{50}{225}$

(B) $\frac{50}{15}$

(C) $\frac{15}{50}$

(D) $\frac{225}{50}$

(E) $\frac{225}{500}$

12. A satellite is currently orbiting Earth in a circular orbit of radius R; its kinetic energy is K_1. If the satellite is moved and enters a new circular orbit of radius $2R$, what will be its kinetic energy?

(A) $\frac{K_1}{4}$

(B) $\frac{K_1}{2}$

(C) K_1

(D) $2K_1$

(E) $4K_1$

13. A moon of Jupiter has a nearly circular orbit of radius R and an orbit period of T. Which of the following expressions gives the mass of Jupiter?

(A) $\frac{2\pi R}{T}$

(B) $\frac{4\pi^2 R}{T^2}$

(C) $\frac{2\pi R^3}{(GT^2)}$

(D) $\frac{4\pi R^2}{(GT^2)}$

(E) $\frac{4\pi^2 R^3}{(GT^2)}$

14. Two large bodies, Body A of mass m and Body B of mass $4m$, are separated by a distance R. At what distance from Body A, along the line joining the bodies, would the gravitational force on an object be equal to zero? (Ignore the presence of any other bodies.)

(A) $\frac{R}{16}$

(B) $\frac{R}{8}$

(C) $\frac{R}{5}$

(D) $\frac{R}{4}$

(E) $\frac{R}{3}$

15. You are looking at a top view of a planet orbiting the Sun in a clockwise direction. Which of the following would describe the velocity, acceleration, and force acting on the planet due to the Sun's pull at point P?

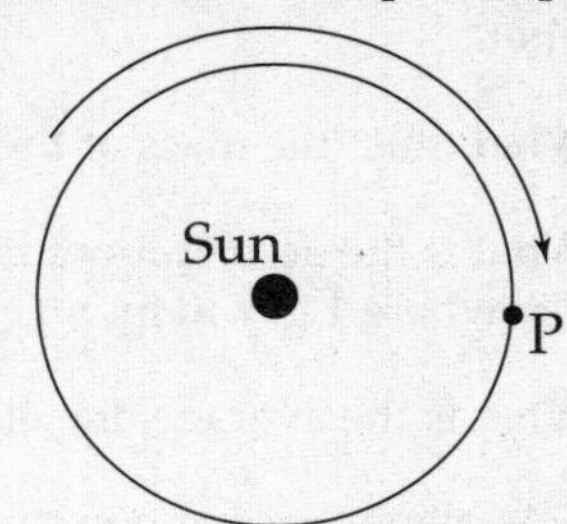

(A) v ↓ a ↑ F ↑
(B) v ↓ a ← F ←
(C) v ↓ a → F →
(D) v ← a ↓ F ↓
(E) v ↑ a ← F ←

16. Which of the following statements are true for a satellite in outer space orbiting the earth?

I. There are no forces acting on the satellite.
II. The force of gravity is the only force acting on the satellite.
III. The force of gravity is balanced by the outward force of the object.

(A) I only
(B) II only
(C) III only
(D) either I or III may be correct
(E) either II or III may be correct

SECTION II: FREE RESPONSE

1. A robot probe lands on a new, uncharted planet. It has determined the diameter of the planet to be 8×10^6 m. It weighs a standard 1 kg mass and determines that 1 kg weighs only 5 Newtons on this new planet.

   (a) What must the mass of the planet be?

   (b) What is the acceleration due to gravity on this planet? Express your answer in both m/s² and g's (where 1 g = 10 m/s²).

   (c) What is the average density of this planet?

   (d) This planet has a moon that orbits it at a distance 6×10^9 m from center to center. What is the moon's period?

2. The earth has a mass of 6×10^{24} kg and orbits the sun in 3.15×10^7 seconds at a constant circular distance of 1.5×10^{11} m.

   (a) What is the earth's centripetal acceleration around the Sun?

   (b) What is the gravitational force acting between the Sun and Earth?

   (c) What is the mass of the Sun?

3. An amusement park ride consists of a large cylinder that rotates around its central axis as the passengers stand against the inner wall of the cylinder. Once the passengers are moving at a certain speed v, the floor on which they were standing is lowered. Each passenger feels pinned against the wall of the cylinder as it rotates. Let r be the inner radius of the cylinder.

   (a) Draw and label all the forces acting on a passenger of mass m as the cylinder rotates with the floor lowered.

   (b) Describe what conditions must hold to keep the passengers from sliding down the wall of the cylinder.

   (c) Compare the conditions discussed in part (b) for an adult passenger of mass m and a child passenger of mass $m/2$.

4. A curved section of a highway has a radius of curvature of r. The coefficient of friction between standard automobile tires and the surface of the highway is μ_s.

    (a) Draw and label all the forces acting on a car of mass m traveling along this curved part of the highway.

    (b) Compute the maximum speed with which a car of mass m could make it around the turn without skidding in terms of μ_s, r, g, and m.

    City engineers are planning on banking this curved section of highway at an angle of θ to the horizontal.

    (c) Draw and label all of the forces acting on a car of mass m traveling along this banked turn. Do not include friction.

    (d) The engineers want to be sure that a car of mass m traveling at a constant speed v (the posted speed limit) could make it safely around the banked turn even if the road were covered with ice (that is, essentially frictionless). Compute this banking angle θ in terms of r, v, g, and m.

SUMMARY

- For objects undergoing uniform circular motion, the centripetal acceleration is given by $a_c = \frac{v^2}{r}$ and the centripetal force is given by $F_c = \frac{mv^2}{r}$.
- Torque is a turning force that makes an object rotate. The equation for torque is $\tau = rF\sin\theta$. Torques may be clockwise or counterclockwise. An object is in equilibrium if $\Sigma F_x = 0$, $\Sigma F_y = 0$, and $\Sigma\tau = 0$.
- For any two masses in the universe there is a gravitational attraction given by:

$$F_g = \frac{Gm_1m_2}{r^2} \text{ where } G = 6.67x10^{-11}\frac{Nm^2}{kg^2}.$$

- The acceleration due to gravity on any planet is given by:

$$g_{planet} = \frac{Gm_{planet}}{r^2}$$

- Many times universal gravitation is linked up with circular motion (because planetary orbits are very nearly circular). Therefore, it is useful to keep the following equations for circular motion mentally linked for those questions that include orbits:

$$v = \frac{2\pi r}{T} \qquad a_c = \frac{v^2}{r} \qquad a_c = \frac{4\pi^2 r}{T^2}$$

$$F = ma_c \qquad F_c = \frac{mv^2}{r} \qquad F_c = \frac{4\pi^2 mr}{T^2}$$

7 Oscillations

INTRODUCTION

In this chapter, we'll concentrate on a kind of periodic motion that's straightforward and that, fortunately, actually describes many real-life systems. This type of motion is called *simple harmonic motion*. The prototypical example of simple harmonic motion is a block that's oscillating on the end of a spring, and what we learn about this simple system, we can apply to many other oscillating systems.

SIMPLE HARMONIC MOTION (SHM): THE SPRING-BLOCK OSCILLATOR

When a spring is compressed or stretched from its natural length, a force is created. For most, but not all, springs, if the spring is displaced by **x** from its natural length, the force it exerts in response is given by the equation

$$\mathbf{F}_S = -k\mathbf{x}$$

This is known as **Hooke's Law**. The proportionality constant, k, is a positive number called the **spring** (or **force**) **constant** that indicates how stiff the spring is. The stiffer the spring, the greater the value of k. The minus sign in Hooke's Law tells us that $\mathbf{F}_S$ and **x** always point in opposite directions. For example, referring to the figure below, when the spring is stretched (**x** is to the right), the spring pulls back (**F** is to the left); when the spring is compressed (**x** is to the left), the spring pushes outward (**F** is to the right). In all cases, the spring wants to return to its original length. As a result, the spring tries to restore the attached block to the **equilibrium position**, which is the position at which the net force on the block is zero. For this reason, we say that the spring provides a **restoring force**.

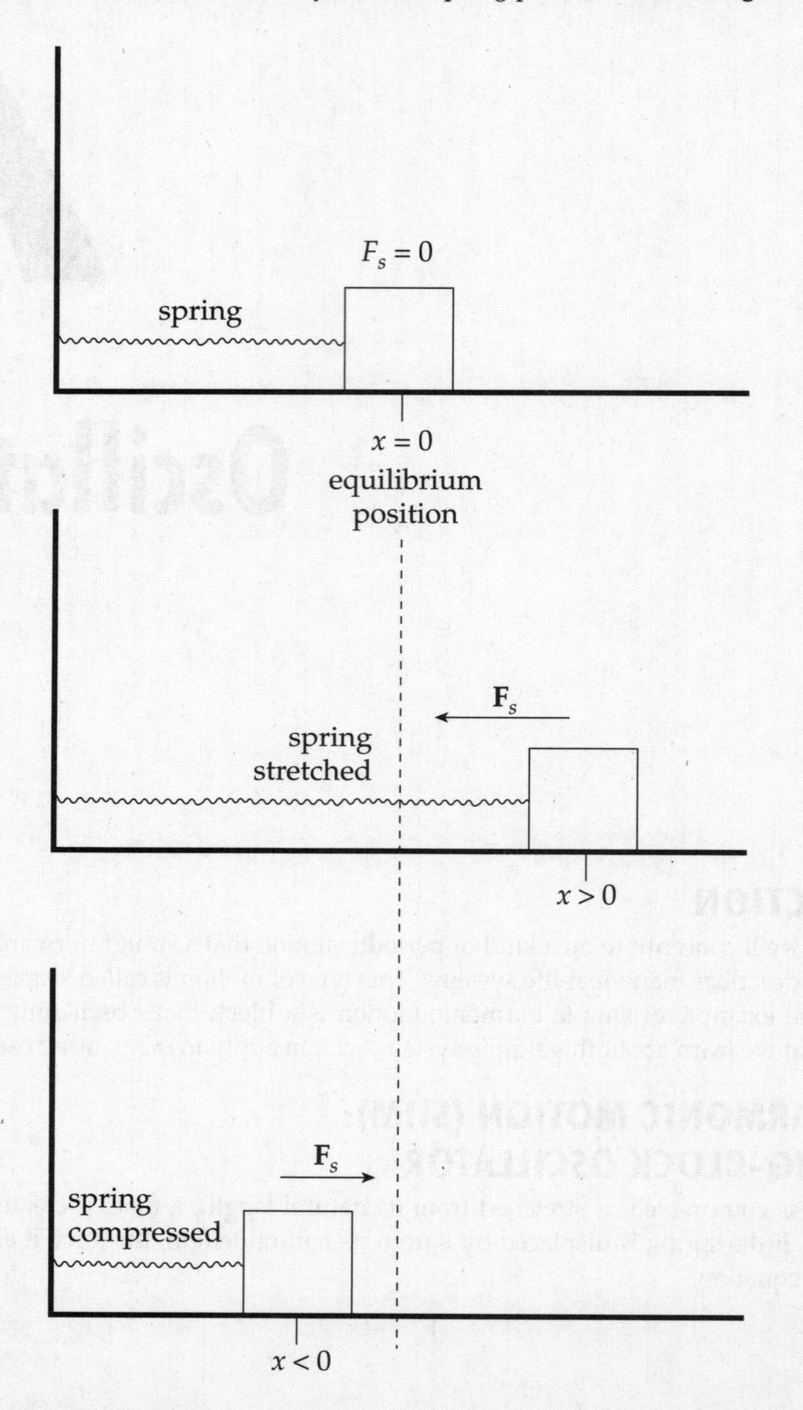

Example 7.1 A 12 cm-long spring has a force constant (k) of 400 N/m. How much force is required to stretch the spring to a length of 14 cm?

Solution. The displacement of the spring has a magnitude of 14 – 12 = 2 cm = 0.02 m so, according to Hooke's Law, the spring exerts a force of magnitude $F = kx$ = (400 N/m)(0.02 m) = 8 N. Therefore, we'd have to exert this much force to keep the spring in this stretched state.

Springs that obey Hooke's Law (called **ideal** or **linear** springs) provide an ideal mechanism for defining the most important kind of vibrational motion: simple harmonic motion.

Consider a spring with force constant k, attached to a vertical wall, with a block of mass m on a frictionless table attached to the other end.

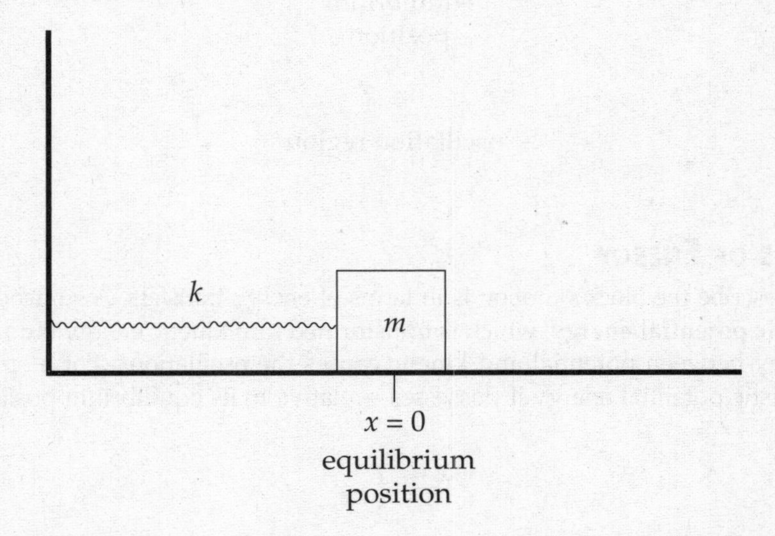

Grab the block, pull it some distance from its original position, and release it. The spring will pull the block back toward equilibrium. Of course, because of its momentum, the block will pass through the equilibrium position and compress the spring. At some point, the block will stop, and the compressed spring will push the block back. In other words, the block will oscillate.

During the oscillation, the force on the block is zero when the block is at equilibrium (the point we designate as $x = 0$). This is because Hooke's Law says that the strength of the spring's restoring force is given by the equation $F_S = kx$, so $F_S = 0$ at equilibrium. The acceleration of the block is also equal to zero at $x = 0$, since $F_S = 0$ at $x = 0$ and $a = F_S/m$. At the endpoints of the oscillation region, where the block's displacement, x, has the greatest magnitude, the restoring force and the magnitude of the acceleration are both at their maximum.

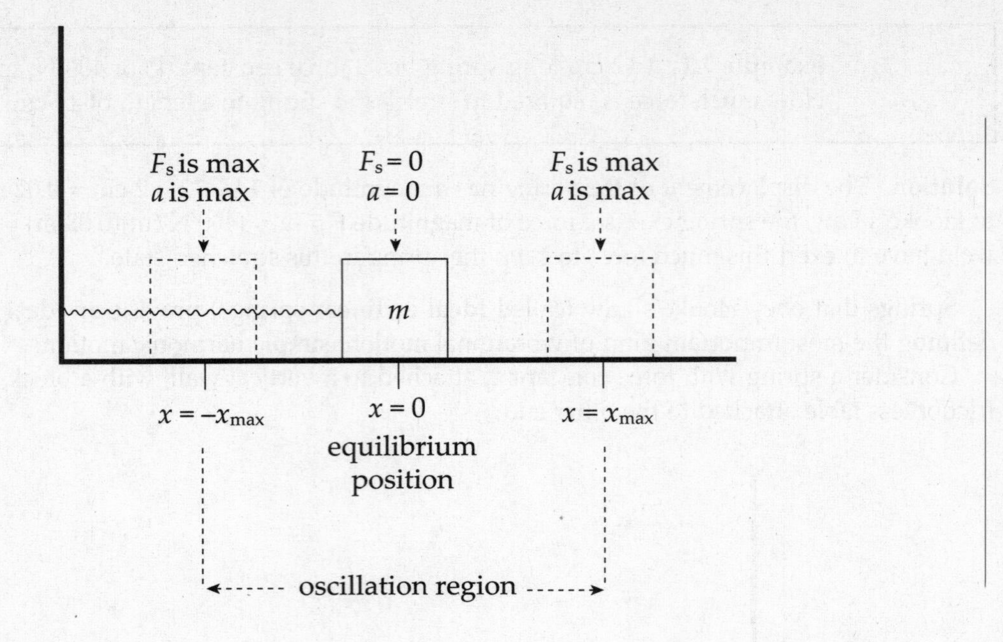

SHM IN TERMS OF ENERGY

Another way to describe the block's motion is in terms of energy transfers. A stretched or compressed spring stores **elastic potential energy**, which is transformed into kinetic energy (and back again); this shuttling of energy between potential and kinetic causes the oscillations. For a spring with spring constant k, the elastic potential energy it possesses—relative to its equilibrium position—is given by the equation

$$U_s = \frac{1}{2}kx^2$$

Notice that the farther you stretch or compress a spring, the more work you have to do, and, as a result, the more potential energy that's stored.

In terms of energy transfers, we can describe the block's oscillations as follows. When you initially pull the block out, you increase the elastic potential energy of the system. Upon releasing the block, this potential energy turns into kinetic energy, and the block moves. As it passes through equilibrium, $U_S = 0$, so all the energy is kinetic. Then, as the block continues through equilibrium, it compresses the spring and the kinetic energy is transformed back into elastic potential energy.

By Conservation of Mechanical Energy, the sum $K + U_S$ is a constant. Therefore, when the block reaches the endpoints of the oscillation region (that is, when $x = \pm x_{max}$), U_S is maximized, so K must be minimized; in fact, $K = 0$ at the endpoints. As the block is passing through equilibrium, $x = 0$, so $U_S = 0$ and K is maximized.

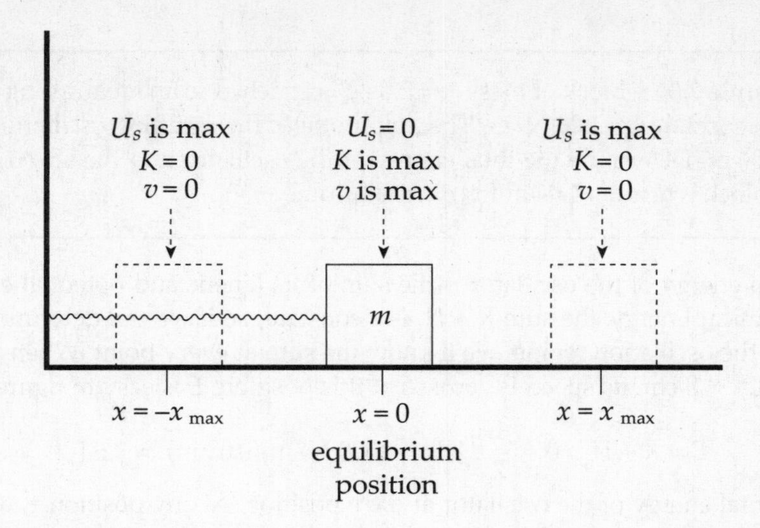

The maximum displacement from equilibrium is called the **amplitude** of oscillation, and is denoted by A. So instead of writing $x = x_{max}$, we write $x = A$ (and $x = -x_{max}$ will be written as $x = -A$).

> **Example 7.2** A block of mass $m = 0.05$ kg oscillates on a spring whose force constant k is 500 N/m. The amplitude of the oscillations is 4.0 cm. Calculate the maximum speed of the block.

Solution. First let's get an expression for the maximum elastic potential energy of the system:

$$U_S = \frac{1}{2}kx^2 \quad \Rightarrow \quad U_{S,\,max} = \frac{1}{2}kx_{max}^2 = \frac{1}{2}kA^2$$

When all this energy has been transformed into kinetic energy—which, as we discussed earlier, occurs just as the block is passing through equilibrium—the block will have a maximum kinetic energy and maximum speed of

$$U_{S,\,max} \rightarrow K_{max} \quad \Rightarrow \quad \frac{1}{2}kA^2 = \frac{1}{2}mv_{max}^2$$

$$v_{max} = \sqrt{\frac{kA^2}{m}}$$

$$= \sqrt{\frac{(500\ \text{N/m})(0.04\ \text{m})^2}{.05\ \text{kg}}}$$

$$= 4\ \text{m/s}$$

Example 7.3 A block of mass $m = 2.0$ kg is attached to an ideal spring of force constant $k = 500$ N/m. The amplitude of the resulting oscillations is 8.0 cm. Determine the total energy of the oscillator and the speed of the block when it's 4.0 cm from equilibrium.

Solution. The total energy of the oscillator is the sum of its kinetic and potential energies. By Conservation of Mechanical Energy, the sum $K + U_S$ is a constant, so if we can determine what this sum is at some point in the oscillation region, we'll know the sum at every point. When the block is at its amplitude position, $x = 8$ cm, its speed is zero; so at this position, E is easy to figure out:

$$E = K + U_S = 0 + \frac{1}{2}kA^2 = \frac{1}{2}(500\text{ N/m})(0.08\text{ m})^2 = 1.6\text{ J}$$

This gives the total energy of the oscillator at *every* position. At any position x, we have

$$\frac{1}{2}mv^2 + \frac{1}{2}kx^2 = E$$

$$v = \sqrt{\frac{E - \frac{1}{2}kx^2}{\frac{1}{2}m}}$$

so when we substitute in the numbers, we get

$$v = \sqrt{\frac{E - \frac{1}{2}kx^2}{\frac{1}{2}m}} = \sqrt{\frac{(1.6\text{ J}) - \frac{1}{2}(500\text{ N/m})(0.04\text{ m})^2}{\frac{1}{2}(2.0\text{ kg})}}$$

$$= 1.1\text{ m/s}$$

Example 7.4 A block of mass $m = 8.0$ kg is attached to an ideal spring of force constant $k = 500$ N/m. The block is at rest at its equilibrium position. An impulsive force acts on the block, giving it an initial speed of 2.0 m/s. Find the amplitude of the resulting oscillations.

Solution. The block will come to rest when all of its initial kinetic energy has been transformed into the spring's potential energy. At this point, the block is at its maximum displacement from equilibrium, that is, it's at one of its amplitude positions, and

$$K_i + U_i = K_f + U_f$$

$$\frac{1}{2}mv_i^2 + 0 = 0 + \frac{1}{2}kA^2$$

$$A = \sqrt{\frac{mv_i^2}{k}}$$

$$= \sqrt{\frac{(8.0\text{ kg})(2.0\text{ m/s})^2}{500\text{ N/m}}}$$

$$= 0.25\text{ m}$$

THE KINEMATICS OF SHM

Now that we've explored the dynamics of the block's oscillations in terms of force and energy, let's talk about motion—or kinematics. As you watch the block oscillate, you should notice that it repeats each **cycle** of oscillation in the same amount of time. A cycle is a *round-trip*: for example, from position $x = A$ over to $x = -A$ and back again to $x = A$. The amount of time it takes to complete a cycle is called the **period** of the oscillations, or T. If T is short, the block is oscillating rapidly, and if T is long, the block is oscillating slowly.

Another way of indicating the rapidity of the oscillations is to count the number of cycles that can be completed in a given time interval; the more completed cycles, the more rapid the oscillations. The number of cycles that can be completed per unit time is called the **frequency** of the oscillations, or f, and frequency is expressed in cycles per second. One cycle per second is one **hertz** (abbreviated **Hz**). Do not confuse lower case "f" (frequency) with upper case "F" (Force).

One of the most basic equations of oscillatory motion expresses the fact that the period and frequency are reciprocals of each other:

$$\text{period} = \frac{\#\text{ seconds}}{\text{cycle}} \qquad \text{while} \qquad \text{frequency} = \frac{\#\text{ cycles}}{\text{second}}$$

Therefore,

$$T = \frac{1}{f} \qquad \text{and} \qquad f = \frac{1}{T}$$

Example 7.5 A block oscillating on the end of a spring moves from its position of maximum spring stretch to maximum spring compression in 0.25 s. Determine the period and frequency of this motion.

Solution. The period is defined as the time required for one full cycle. Moving from one end of the oscillation region to the other is only half a cycle. Therefore, if the block moves from its position of maximum spring stretch to maximum spring compression in 0.25 s, the time required for a full cycle is twice as much; $T = 0.5$ s. Because frequency is the reciprocal of period, the frequency of the oscillations is $f = 1/T = 1/(0.5 \text{ s}) = 2$ Hz.

Example 7.6 A student observing an oscillating block counts 45.5 cycles of oscillation in one minute. Determine its frequency (in hertz) and period (in seconds).

Solution. The frequency of the oscillations, in hertz (which is the number of cycles per second), is

$$f = \frac{45.5 \text{ cycles}}{\text{min}} \times \frac{1 \text{ min}}{60 \text{ s}} = \frac{0.758 \text{ cycles}}{\text{s}} = 0.758 \text{ Hz}$$

Therefore,

$$T = \frac{1}{f} = \frac{1}{0.758 \text{ Hz}} = 1.32 \text{ s}$$

One of the defining properties of the spring-block oscillator is that the frequency and period can be determined from the mass of the block and the force constant of the spring. The equations are as follows:

$$f = \frac{1}{2\pi}\sqrt{\frac{k}{m}} \quad \text{and} \quad T = 2\pi\sqrt{\frac{m}{k}}$$

Let's analyze these equations. Suppose we had a small mass on a very stiff spring; then intuitively, we would expect that this strong spring would make the small mass oscillate rapidly, with high frequency and short period. Both of these predictions are substantiated by the equations above, because if m is small and k is large, then the ratio k/m is large (high frequency) and the ratio m/k is small (short period).

> **Example 7.7** A block of mass $m = 2.0$ kg is attached to a spring whose force constant, k, is 300 N/m. Calculate the frequency and period of the oscillations of this spring–block system.

Solution. According to the equations above,

$$f = \frac{1}{2\pi}\sqrt{\frac{k}{m}} = \frac{1}{2\pi}\sqrt{\frac{300 \text{ N/m}}{2.0 \text{ kg}}} = 1.9 \text{ Hz}$$

$$T = 2\pi\sqrt{\frac{m}{k}} = 2\pi\sqrt{\frac{2.0 \text{ kg}}{300 \text{ N/m}}} = 0.51 \text{ s}$$

Notice that $f \approx 2$ Hz and $T \approx 0.5$ s, and that these values satisfy the basic equation $T = 1/f$.

> **Example 7.8** A block is attached to a spring and set into oscillatory motion, and its frequency is measured. If this block were removed and replaced by a second block with 1/4 the mass of the first block, how would the frequency of the oscillations compare to that of the first block?

Solution. Since the same spring is used, k remains the same. According to the equation given above, f is inversely proportional to the square root of the mass of the block: $f \propto 1/\sqrt{m}$. Therefore, if m decreases by a factor of 4, then f increases by a factor of $\sqrt{4} = 2$.

The equations we saw above for the frequency and period of the spring-block oscillator do not contain A, the amplitude of the motion. In simple harmonic motion, *both the frequency and the period are independent of the amplitude.* The reason that the frequency and period of the spring-block oscillator are independent of amplitude is that F, the strength of the restoring force, is proportional to x, the displacement from equilibrium, as given by Hooke's Law: $F_S = -kx$.

> **Example 7.9** A student performs an experiment with a spring-block simple harmonic oscillator. In the first trial, the amplitude of the oscillations is 3.0 cm, while in the second trial (using the same spring and block), the amplitude of the oscillations is 6.0 cm. Compare the values of the period, frequency, and maximum speed of the block between these two trials.

Solution. If the system exhibits simple harmonic motion, then the period and frequency are independent of amplitude. This is because the same spring and block were used in the two trials, so the period and frequency will have the same values in the second trial as they had in the first. But the maximum speed of the block will be greater in the second trial than in the first. Since the amplitude is greater in the second trial, the system possesses more total energy ($E = \frac{1}{2}kA^2$). So when the block is passing through equilibrium (its position of greatest speed), the second system has more energy to convert to kinetic, meaning that the block will have a greater speed. In fact, from Example 7.2, we know that $v_{max} = A\sqrt{k/m}$ so, since A is twice as great in the second trial than in the first, v_{max} will be twice as great in the second trial than in the first.

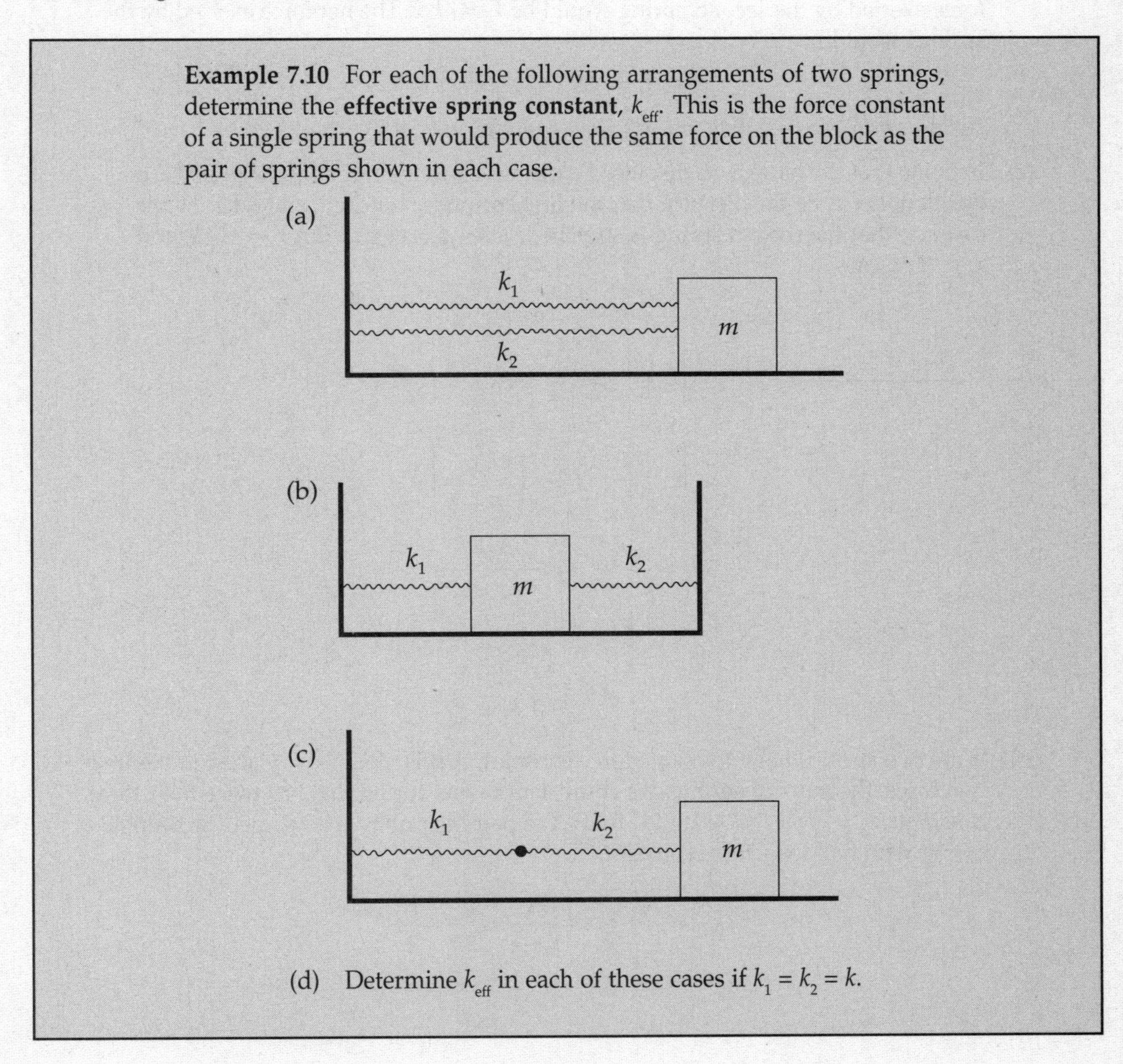

Example 7.10 For each of the following arrangements of two springs, determine the **effective spring constant,** k_{eff}. This is the force constant of a single spring that would produce the same force on the block as the pair of springs shown in each case.

(d) Determine k_{eff} in each of these cases if $k_1 = k_2 = k$.

Solution.

(a) Imagine that the block was displaced a distance x to the right of its equilibrium position. Then the force exerted by the first spring would be $F_1 = -k_1x$ and the force exerted by the second spring would be $F_2 = -k_2x$. The net force exerted by the springs would be

$$F_1 + F_2 = -k_1x + -k_2x = -(k_1 + k_2)x$$

Since $F_{eff} = -(k_1 + k_2)x$, we see that $k_{eff} = k_1 + k_2$.

(b) Imagine that the block was displaced a distance x to the right of its equilibrium position. Then the force exerted by the first spring would be $F_1 = -k_1x$ and the force exerted by the second spring would be $F_2 = -k_2x$. The net force exerted by the springs would be

$$F_1 + F_2 = -k_1x + -k_2x = -(k_1 + k_2)x$$

As in part (a), we see that, since $F_{eff} = -(k_1 + k_2)x$, we get $k_{eff} = k_1 + k_2$.

(c) Imagine that the block was displaced a distance x to the right of its equilibrium position. Let x_1 be the distance that the first spring is stretched, and let x_2 be the distance that the second spring is stretched. Then $x = x_1 + x_2$. But $x_1 = -F/k_1$ and $x_2 = -F/k_2$, so

$$\frac{-F}{k_1} + \frac{-F}{k_2} = x$$

$$-F\left(\frac{1}{k_1} + \frac{1}{k_2}\right) = x$$

$$F = -\left(\frac{1}{\frac{1}{k_1} + \frac{1}{k_2}}\right)x$$

$$F = -\frac{k_1k_2}{k_1 + k_2}x$$

Therefore,

$$k_{eff} = \frac{k_1k_2}{k_1 + k_2}$$

(d) If the two springs have the same force constant, that is, if $k_1 = k_2 = k$, then in the first two cases, the pairs of springs are equivalent to one spring that has twice their force constant: $k_{eff} = k_1 + k_2 = k + k = 2k$. In (c), the pair of springs is equivalent to a single spring with half their force constant:

$$k_{eff} = \frac{k_1k_2}{k_1 + k_2} = \frac{kk}{k + k} = \frac{k^2}{2k} = \frac{k}{2}$$

THE SPRING-BLOCK OSCILLATOR: VERTICAL MOTION

So far we've looked at a block sliding back and forth on a horizontal table, but the block could also oscillate vertically. The only difference would be that gravity would cause the block to move downward, to an equilibrium position at which, in contrast with the horizontal SHM we've examined, the spring would not be at its natural length. Of course, in calculating energy, the gravitational potential energy (mgh) must be included.

Consider a spring of negligible mass hanging from a stationary support. A block of mass m is attached to its end and allowed to come to rest, stretching the spring a distance d. At this point, the block is in equilibrium; the upward force of the spring is balanced by the downward force of gravity. Therefore,

$$kd = mg \quad \Rightarrow \quad d = \frac{mg}{k}$$

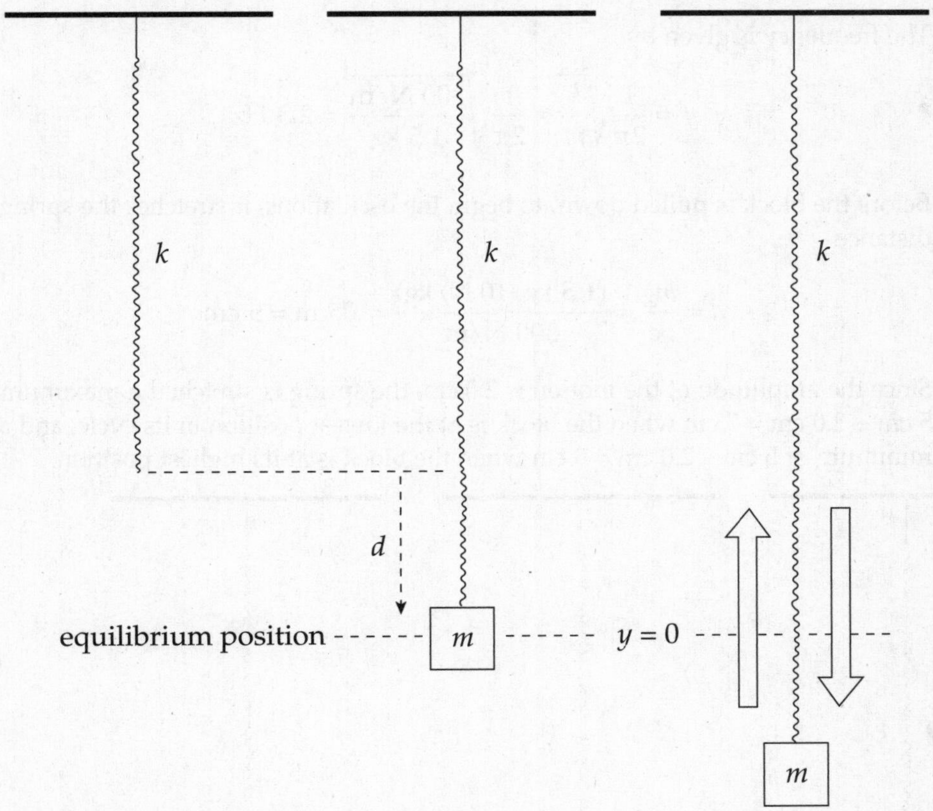

k
k
k
d
equilibrium position
m
y = 0
m

Next, imagine that the block is pulled down a distance A and released. The spring force increases (because the spring was stretched farther); it's stronger than the block's weight, and, as a result, the block accelerates upward. As the block's momentum carries it up, through the equilibrium position, the spring becomes less stretched than it was at equilibrium, so F_S is less than the block's weight. As a result, the block decelerates, stops, and accelerates downward again, and the up-and-down motion repeats.

When the block is at a distance y below its equilibrium position, the spring is stretched a total distance of $d + y$, so the upward spring force is equal to $k(d + y)$, while the downward force stays the same, mg. The net force on the block is

$$F = k(d + y) - mg$$

but this equation becomes $F = ky$, because $kd = mg$ (as we saw above). Since the resulting force on the block, $F = ky$, has the form of Hooke's Law, we know that the vertical simple harmonic oscillations of the block have the same characteristics as do horizontal oscillations, with the equilibrium position, $y = 0$, not at the spring's natural length, but at the point where the hanging block is in equilibrium.

> **Example 7.11** A block of mass $m = 1.5$ kg is attached to the end of a vertical spring of force constant $k = 300$ N/m. After the block comes to rest, it is pulled down a distance of 2.0 cm and released.
> (a) What is the frequency of the resulting oscillations?
> (b) What are the minimum and maximum amounts of stretch of the spring during the oscillations of the block?

Solution.

(a) The frequency is given by

$$f = \frac{1}{2\pi}\sqrt{\frac{k}{m}} = \frac{1}{2\pi}\sqrt{\frac{300 \text{ N/m}}{1.5 \text{ kg}}} = 2.3 \text{ Hz}$$

(b) Before the block is pulled down, to begin the oscillations, it stretches the spring by a distance

$$d = \frac{mg}{k} = \frac{(1.5 \text{ kg})(10 \text{ N/kg})}{300 \text{ N/m}} = .05 \text{ m} = 5 \text{ cm}$$

Since the amplitude of the motion is 2.0 cm, the spring is stretched a maximum of 5 cm + 2.0 cm = 7 cm when the block is at the lowest position in its cycle, and a minimum of 5 cm – 2.0 cm = 3 cm when the block is at its highest position.

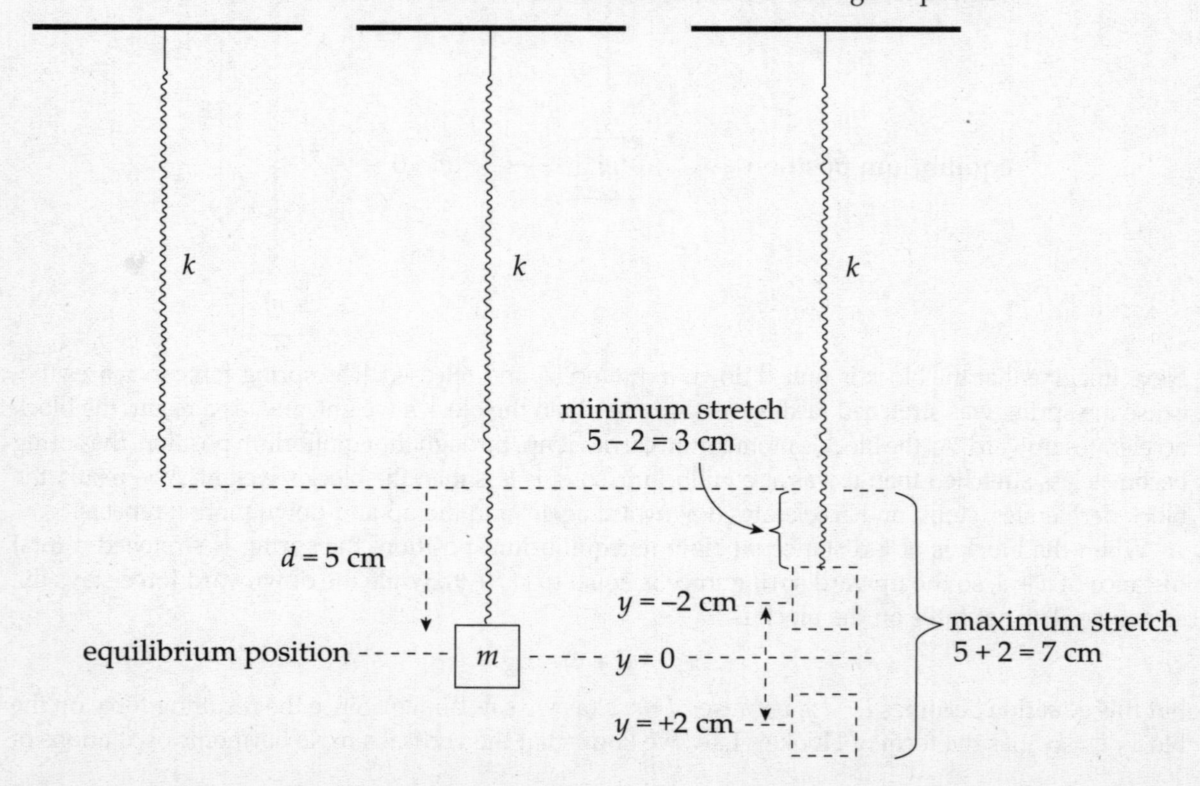

PENDULUMS

A **simple pendulum** consists of a weight of mass m attached to a string or a massless rod that swings, without friction, about the vertical equilibrium position. The restoring force is provided by gravity and, as the figure below shows, the magnitude of the restoring force when the bob is θ to an angle to the vertical is given by the equation:

$$F_{\text{restoring}} = mg \sin \theta$$

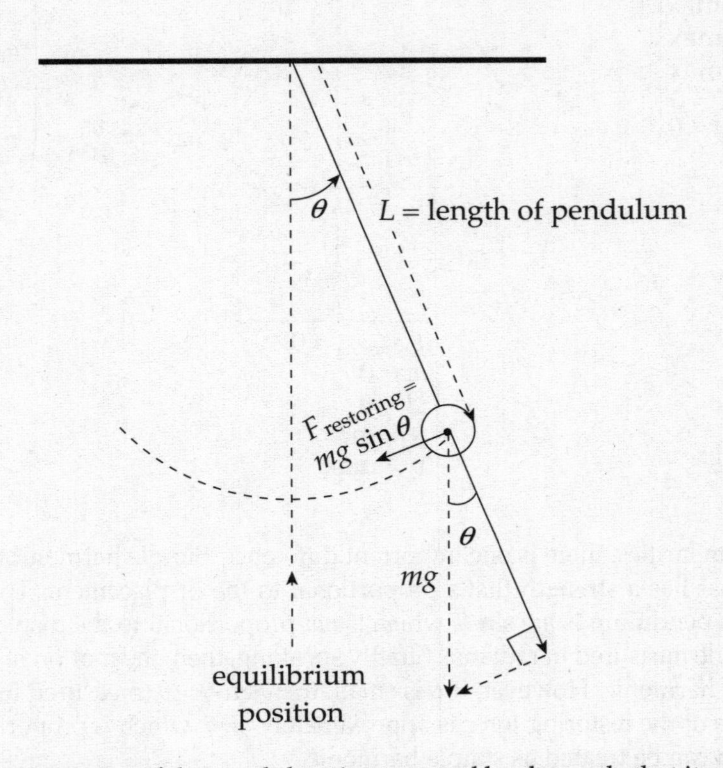

Although the displacement of the pendulum is measured by the angle that it makes with the vertical, rather than by its linear distance from the equilibrium position (as was the case for the spring–block oscillator), the simple pendulum shares many of the important features of the spring–block oscillator. For example,

- Displacement is zero at the equilibrium position.
- At the endpoints of the oscillation region (where $\theta = \pm\theta_{max}$), the restoring force and the tangential acceleration (a_t) have their greatest magnitudes, the speed of the pendulum is zero, and the potential energy is maximized.
- As the pendulum passes through the equilibrium position, its kinetic energy and speed are maximized.

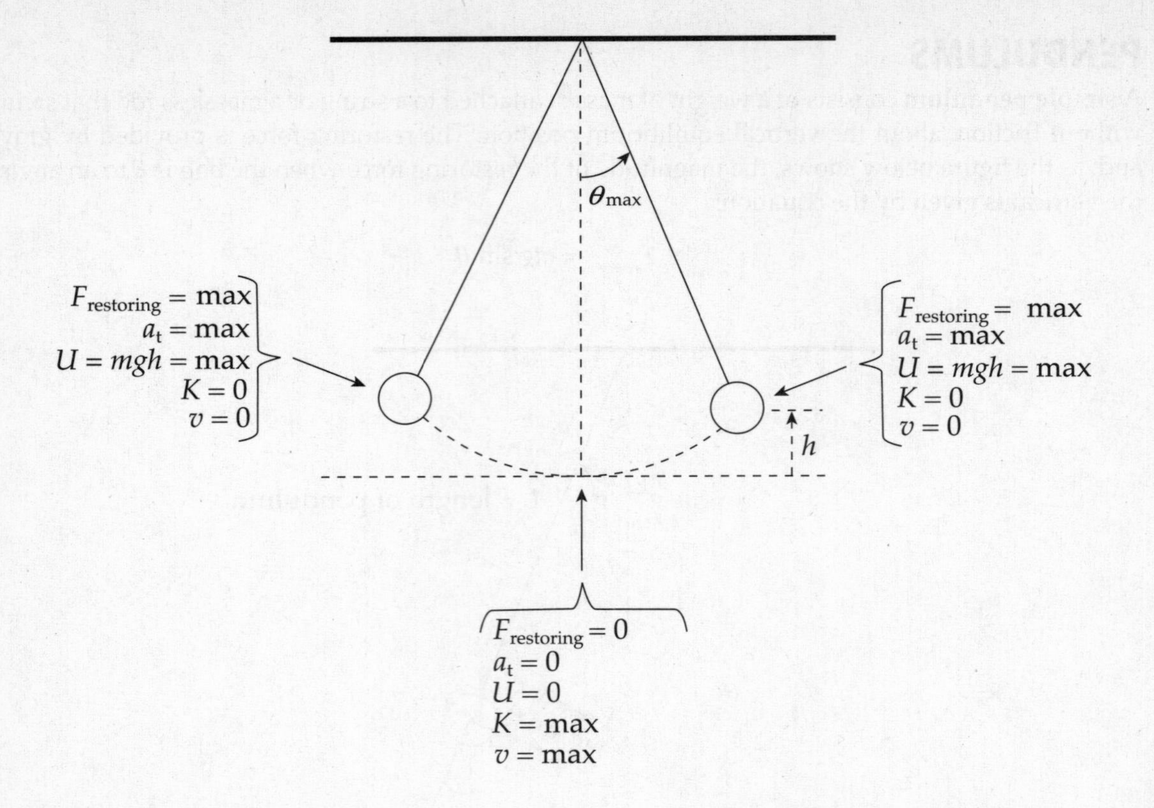

Despite these similarities, there is one important difference. Simple harmonic motion results from a restoring force that has a strength that's proportional to the displacement. The magnitude of the restoring force on a pendulum is $mg \sin \theta$, which is *not* proportional to the displacement (θL, the arc length, with the angle measured in radians). Strictly speaking, then, the motion of a simple pendulum is not really simple harmonic. However, if θ is small, then $\sin \theta \approx \theta$ (measured in radians) so, in this case, the magnitude of the restoring force is approximately $mg\theta$, which *is* proportional to θ. So if θ_{max} is small, the motion can be treated as simple harmonic.

If the restoring force is given by $mg\theta$, rather than $mg \sin \theta$, then the frequency and period of the oscillations depend only on the length of the pendulum and the value of the gravitational acceleration, according to the following equations:

$$f = \frac{1}{2\pi}\sqrt{\frac{g}{L}} \quad \text{and} \quad T = 2\pi\sqrt{\frac{L}{g}}$$

Note that neither frequency nor period depends on the amplitude (the maximum angular displacement, θ_{max}); this is a characteristic feature of simple harmonic motion. Also notice that neither depends on the mass of the weight.

Example 7.12 A simple pendulum has a period of 1 s on Earth. What would its period be on the Moon (where g is one-sixth of its value here)?

Solution. The equation $T = 2\pi\sqrt{L/g}$ shows that T is inversely proportional to $\sqrt{g}$, so if g decreases by a factor of 6, then T increases by factor of $\sqrt{6}$. That is,

$$T_{\text{on Moon}} = \sqrt{6} \times T_{\text{on Earth}} = (1\text{ s})\sqrt{6} = 2.4\text{ s}$$

CHAPTER 7 REVIEW QUESTIONS

Solutions can be found in Chapter 18.

SECTION I: MULTIPLE CHOICE

1. Which of the following is/are characteristics of simple harmonic motion?

   I. The acceleration is constant.
   II. The restoring force is proportional to the displacement.
   III. The frequency is independent of the amplitude.

   (A) II only
   (B) I and II only
   (C) I and III only
   (D) II and III only
   (E) I, II, and III

2. A block attached to an ideal spring undergoes simple harmonic motion. The acceleration of the block has its maximum magnitude at the point where

   (A) the speed is the maximum
   (B) the potential energy is the minimum
   (C) the speed is the minimum
   (D) the restoring force is the minimum
   (E) the kinetic energy is the maximum

3. A block attached to an ideal spring undergoes simple harmonic motion about its equilibrium position ($x = 0$) with amplitude A. What fraction of the total energy is in the form of kinetic energy when the block is at position $x = \frac{1}{2}A$?

   (A) $\frac{1}{3}$
   (B) $\frac{3}{8}$
   (C) $\frac{1}{2}$
   (D) $\frac{2}{3}$
   (E) $\frac{3}{4}$

4. A student measures the maximum speed of a block undergoing simple harmonic oscillations of amplitude A on the end of an ideal spring. If the block is replaced by one with twice the mass but the amplitude of its oscillations remains the same, then the maximum speed of the block will

   (A) decrease by a factor of 4
   (B) decrease by a factor of 2
   (C) decrease by a factor of $\sqrt{2}$
   (D) remain the same
   (E) increase by a factor of 2

5. A spring–block simple harmonic oscillator is set up so that the oscillations are vertical. The period of the motion is T. If the spring and block are taken to the surface of the Moon, where the gravitational acceleration is 1/6 of its value here, then the vertical oscillations will have a period of

   (A) $\frac{T}{6}$
   (B) $\frac{T}{3}$
   (C) $\frac{T}{\sqrt{6}}$
   (D) T
   (E) $T\sqrt{6}$

6. A linear spring of force constant k is used in a physics lab experiment. A block of mass m is attached to the spring and the resulting frequency, f, of the simple harmonic oscillations is measured. Blocks of various masses are used in different trials, and in each case, the corresponding frequency is measured and recorded. If f^2 is plotted versus $1/m$, the graph will be a straight line with slope

(A) $\frac{4\pi^2}{k^2}$

(B) $\frac{4\pi^2}{k}$

(C) $4\pi^2 k$

(D) $\frac{k}{4\pi^2}$

(E) $\frac{k^2}{4\pi^2}$

7. A simple pendulum swings about the vertical equilibrium position with a maximum angular displacement of 5° and period T. If the same pendulum is given a maximum angular displacement of 10°, then which of the following best gives the period of the oscillations?

(A) $\frac{T}{2}$

(B) $\frac{T}{\sqrt{2}}$

(C) T

(D) $T\sqrt{2}$

(E) $2T$

SECTION II: FREE RESPONSE

1. The figure below shows a block of mass m (Block 1) that's attached to one end of an ideal spring of force constant k and natural length L. The block is pushed so that it compresses the spring to 3/4 of its natural length and then released from rest. Just as the spring has extended to its natural length L, the attached block collides with another block (also of mass m) at rest on the edge of the frictionless table. When Block 1 collides with Block 2, half of its kinetic energy is lost to heat; the other half of Block 1's kinetic energy at impact is divided between Block 1 and Block 2. The collision sends Block 2 over the edge of the table, where it falls a vertical distance H, landing at a horizontal distance R from the edge.

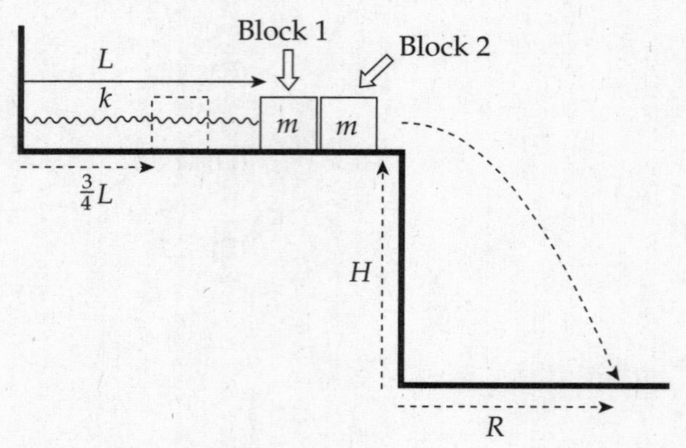

(a) What is the acceleration of Block 1 at the moment it's released from rest from its initial position? Write your answer in terms of k, L, and m.

(b) If v_1 is the velocity of Block 1 just before impact, show that the velocity of Block 1 just after impact is $\frac{1}{2}v_1$.

(c) Determine the amplitude of the oscillations of Block 1 after Block 2 has left the table. Write your answer in terms of L only.

(d) Determine the period of the oscillations of Block 1 after the collision, writing your answer in terms of T_0, the period of the oscillations that Block 1 would have had if it did not collide with Block 2.

(e) Find an expression for R in terms of H, k, L, m, and g.

2. A bullet of mass m is fired horizontally with speed v into a block of mass M initially at rest, at the end of an ideal spring on a frictionless table. At the moment the bullet hits, the spring is at its natural length, L. The bullet becomes embedded in the block, and simple harmonic oscillations result.

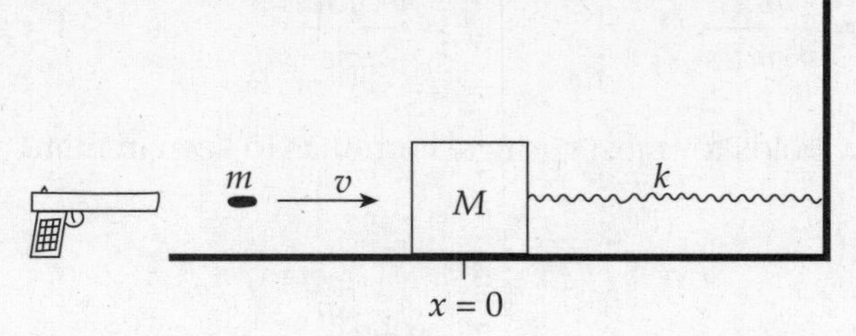

(a) Determine the speed of the block immediately after the impact by the bullet.

(b) Determine the amplitude of the resulting oscillations of the block.

(c) Compute the frequency of the resulting oscillations.

(d) Derive an equation which gives the position of the block as a function of time (relative to $x = 0$ at time $t = 0$).

3. A block of mass M oscillates with amplitude A on a frictionless horizontal table, connected to an ideal spring of force constant k. The period of its oscillations is T. At the moment when the block is at position $x = \frac{1}{2}A$ and moving to the right, a ball of clay of mass m dropped from above lands on the block.

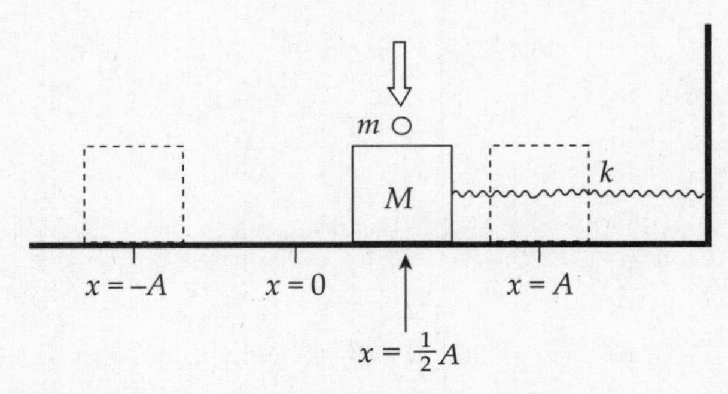

(a) What is the velocity of the block just before the clay hits?

(b) What is the velocity of the block just after the clay hits?

(c) What is the new period of the oscillations of the block?

(d) What is the new amplitude of the oscillations? Write your answer in terms of A, k, M, and m.

(e) Would the answer to part (c) be different if the clay had landed on the block when it was at a different position? Support your answer briefly.

(f) Would the answer to part (d) be different if the clay had landed on the block when it was at a different position? Support your answer briefly.

SUMMARY

$$T = \frac{time}{\#\,cycles} \qquad f = \frac{\#\,cycles}{time} \qquad T = \frac{1}{f}$$

- Hooke's Law holds for most springs. Formulas to keep in mind are:

$$F_s = -kx$$

$$T = 2\pi\sqrt{\frac{m}{k}}$$

$$U_s = \frac{1}{2}kx^2$$

- For small angle of a pendulum swing:

$$T = 2\pi\sqrt{\frac{L}{g}}$$

8

Fluid Mechanics

INTRODUCTION

In this chapter, we'll discuss some of the fundamental concepts dealing with substances that can flow, which are known as **fluids**. Although both liquids and gases are fluids, there are distinctions between them. At the molecular level, a substance in the liquid phase is similar to one in the solid phase in that the molecules are close to—and interact with—one another. The molecules in a liquid are able to move around a little more freely than those in a solid, where the molecules typically only vibrate around relatively fixed positions. By contrast, the molecules of a gas are not constrained and fly around in a chaotic swarm, with hardly any interaction. On a macroscopic level, there is also a distinction between liquids and gases. If you pour a certain volume of a liquid into a container, the liquid will occupy that same volume, whatever the shape of the container. However, if you introduce a sample of gas into a container, the molecules will fly around and fill the *entire* container.

DENSITY

Although the concept of *mass* was central to our study of mechanics in the preceding chapters (because of the all-important equation $F_{net} = ma$), it is the substance's *density* that turns out to be more useful in fluid mechanics.

By definition, the density of a substance is its mass per unit volume, and it's typically denoted by the letter ρ (the Greek letter *rho*):

$$\text{density} = \frac{\text{mass}}{\text{volume}}$$

$$\rho = \frac{m}{V}$$

Note that this equation immediately implies that m is equal to ρV.

For example, if 10^{-3} m^3 of oil has a mass of 0.8 kg, then the density of this oil is

$$\rho = \frac{m}{V} = \frac{0.8 \text{ kg}}{10^{-3} \text{ m}^3} = 800 \frac{\text{kg}}{\text{m}^3}$$

An old-fashioned but still very common way of expressing the density of a substance is to relate it to the density of water. The ratio of the density of any substance to the density of water is known as the **specific gravity** of the substance, abbreviated sp. gr.:

$$\text{sp. gr.} = \frac{\rho_{\text{substance}}}{\rho_{\text{water}}}$$

For example, if a substance has twice the density of water, we'd say that its specific gravity is 2. If the specific gravity is less than 1, this means the substance is less dense than water; if the specific gravity is greater than 1, then the substance is denser than water.

The density of liquid water is taken to be a constant, equal to 1000 kg/m^3. So, for example, if a certain type of oil has a density of 800 kg/m^3, then its specific gravity is

$$\text{sp. gr. of oil} = \frac{\rho_{\text{oil}}}{\rho_{\text{water}}} = \frac{800 \frac{\text{kg}}{\text{m}^3}}{1000 \frac{\text{kg}}{\text{m}^3}} = 0.8$$

Note: Because it is a ratio of densities, sp. gr. has no units.

Example 8.1 Turpentine has a specific gravity of 0.9. What is its density?

Solution. We can use the equation for specific gravity to find that

$$\rho_{\text{substance}} = (\text{sp. gr.})\,\rho_{\text{water}} \quad \Rightarrow \quad \rho_{\text{turpentine}} = (0.9)(1000 \tfrac{\text{kg}}{\text{m}^3}) = 900 \tfrac{\text{kg}}{\text{m}^3}$$

Example 8.2 A cork has a volume of 4 cm^3 and weighs 10^{-2} N. What is the specific gravity of cork?

Solution. Because the cork weighs 10^{-2} N, it has a mass of

$$m = \frac{F_g}{g} = \frac{10^{-2}\ \text{N}}{10\ \frac{\text{N}}{\text{kg}}} = 10^{-3}\ \text{kg}$$

Its density is therefore

$$\rho_{\text{cork}} = \frac{m}{V} = \frac{10^{-3}\ \text{kg}}{4\ \text{cm}^3} \times \left(\frac{10^2\ \text{cm}}{1\ \text{m}}\right)^3 = 250\ \frac{\text{kg}}{\text{m}^3}$$

This means that the specific gravity of the cork is

$$\text{sp. gr.} = \frac{\rho_{\text{cork}}}{\rho_{\text{water}}} = \frac{250\ \frac{\text{kg}}{\text{m}^3}}{1000\ \frac{\text{kg}}{\text{m}^3}} = 0.25$$

PRESSURE

If we place an object in a fluid, the fluid exerts a contact force on the object. How that force is distributed over any small area of the object's surface defines the **pressure**:

$$\text{pressure} = \frac{\text{force}_{\perp}}{\text{area}}$$

$$P = \frac{F_{\perp}}{A}$$

The subscript $\perp$ (which means *perpendicular*) is meant to emphasize that the pressure is defined to be the magnitude of the force that acts perpendicular to the surface, divided by the area. Because force is measured in newtons (N) and area is expressed in square meters (m^2), the SI unit for pressure is the newton per square meter. This unit is given its own name: the **pascal**, abbreviated Pa:

$$\text{SI unit of pressure: } 1\ \text{pascal} = 1\ \text{Pa} = 1\ \frac{\text{N}}{\text{m}^2}$$

One pascal is a very tiny amount of pressure; for example, a nickel on a table exerts about 140 Pa of pressure, just due to its weight alone. For this reason, you'll often see pressures expressed in kPa (kilopascals, where 1 kPa = 10^3 Pa) or even in MPa (megapascals, where 1 MPa = 10^6 Pa). Another common unit for pressure is the atmosphere (atm). At sea level, atmospheric pressure, P_{atm}, is about 101,300 Pa; this is 1 **atmosphere**.

Example 8.3 A vertical column made of cement has a base area of 0.5 m^2. If its height is 2 m, and the specific gravity of cement is 3, how much pressure does this column exert on the ground?

Solution. The force the column exerts on the ground is equal to its weight, mg, so we'll find the pressure it exerts by dividing this by the base area, A. The mass of the column is equal to ρV, which we calculate as follows:

$$\rho = \text{sp.gr.} \times \rho_{\text{water}} = 3 \times 1000 \tfrac{\text{kg}}{\text{m}^3} = 3000 \tfrac{\text{kg}}{\text{m}^3}$$

$$m = \rho V = \rho A h = (3 \times 10^3 \tfrac{\text{kg}}{\text{m}^3})(0.5\ \text{m}^2)(2\ \text{m}) = 3 \times 10^3\ \text{kg}$$

Therefore,

$$P = \frac{F}{A} = \frac{mg}{A} = \frac{(3 \times 10^3\ \text{kg})(10\ \tfrac{\text{N}}{\text{kg}})}{0.5\ \text{m}^2} = 6 \times 10^4\ \text{Pa} = 60\ \text{kPa}$$

HYDROSTATIC PRESSURE

Imagine that we have a tank with a lid on top, filled with some liquid. Suspended from this lid is a string, attached to a thin sheet of metal. The figures below show two views of this:

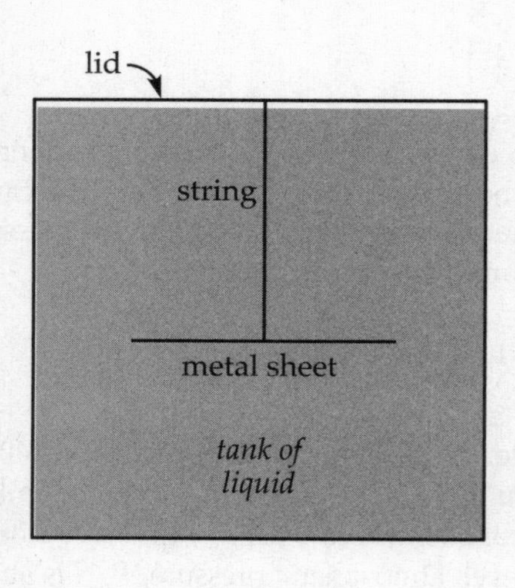

front view

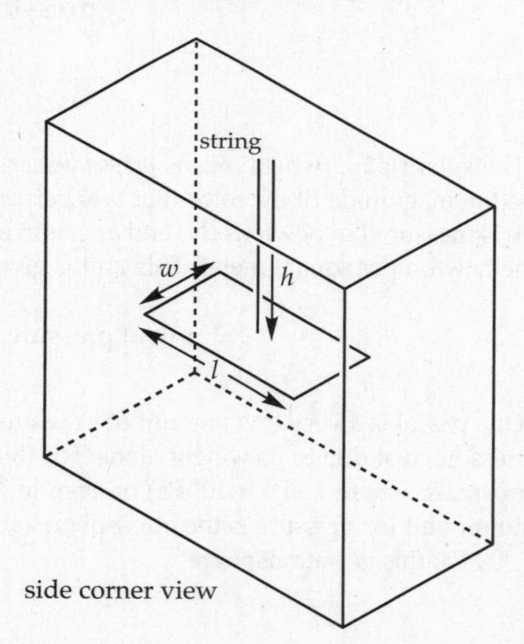

side corner view

The weight of the liquid produces a force that pushes down on the metal sheet. If the sheet has length l and width w, and is at depth h below the surface of the liquid, then the weight of the liquid on top of the sheet is

$$F_g = mg = \rho Vg = \rho(lwh)g$$

where ρ is the liquid's density. If we divide this weight by the area of the sheet ($A = lw$), we get the pressure due to the liquid:

$$P_{\text{liquid}} = \frac{\text{force}}{\text{area}} = \frac{F_{g\,\text{liquid}}}{A} = \frac{\rho(lwh)g}{lw} = \rho gh$$

Since the liquid is at rest, this is known as **hydrostatic pressure.**

Note that the hydrostatic pressure due to the liquid, $P_{\text{liquid}} = \rho gh$, depends only on the density of the liquid and the depth below the surface; in fact, it's proportional to both of these quantities. One important consequence of this is that the shape of the container doesn't matter. For example, if all the containers in the figure below are filled with the same liquid, then the pressure is the same at every point along the horizontal dashed line (and within a container), simply because every point on this line is at the same depth, h, below the surface of the liquid.

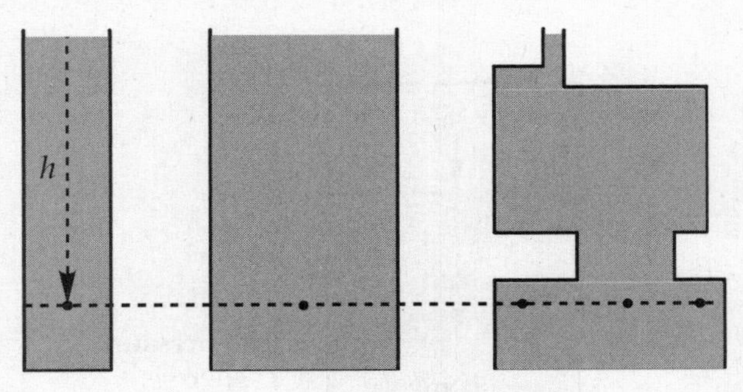

If the liquid in the tank were open to the atmosphere, then the total (or absolute) pressure at depth h would be equal to the pressure pushing down on the surface—the atmospheric pressure, P_{atm}—plus the pressure due to the liquid alone:

$$\text{total (absolute) pressure: } P_{\text{total}} = P_{\text{atm}} + P_{\text{liquid}} = P_{\text{atm}} + \rho gh$$

The difference between total pressure and atmospheric pressure is known as the **gauge pressure:**

$$P_{\text{gauge}} = P - P_{\text{atm}}$$

So, another way of writing the equation above is to say:

$$P_{\text{gauge}} = \rho gh$$

The following graphs show how gauge pressure and total pressure vary with depth. Note that although both graphs are straight lines, only gauge pressure is proportional to depth. In order for a graph to represent a direct proportion, it must be a straight line *through the origin.*

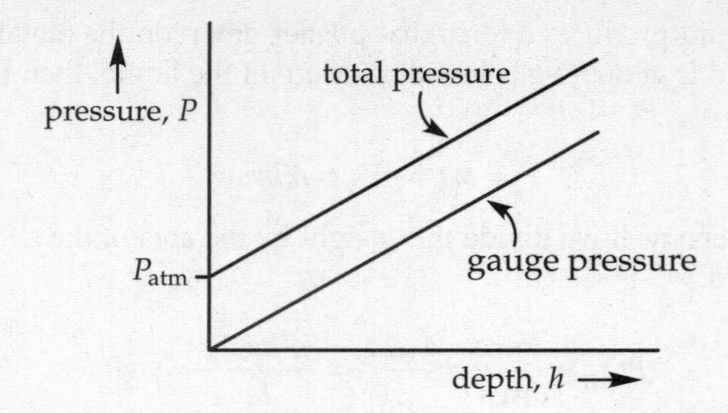

The lines will be straight as long as the density of the liquid remains constant as the depth increases. Actually, ρ increases as the depth increases, but the effect is small enough that we generally consider liquids to be incompressible—that is, that the density of a liquid remains constant and does not increase with depth.

If we have a container of liquid where the pressure above the surface of the liquid is P_0, then the total pressure at depth h below the surface would be P_0 plus the pressure due just to the liquid:

$$P = P_0 + \rho g h$$

lid
P_0
gas
h
liquid
(density = ρ)
The pressure
here is
$P = P_0 + \rho g h$

If there's no lid, then $P_0 = P_{atm}$, and we get the equation $P = P_{atm} + \rho g h$ again.

Because pressure is the *magnitude* of the force per area, pressure is a scalar. It has no direction. The direction of the force due to the pressure on any small surface is perpendicular to that surface. For example, in the figure below, the pressure at Point A is the same as the pressure at Point B, because they're at the same depth.

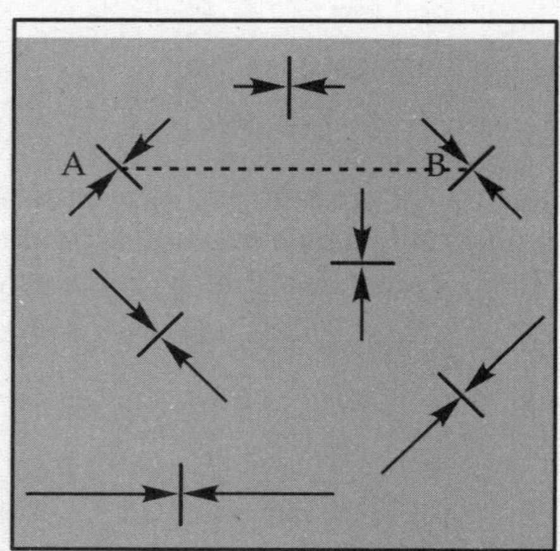

But, as you can see, the direction of the force due to the pressure varies depending on the orientation of the surface—and even which side of the surface—the force is pushing on.

Example 8.4 What is the hydrostatic gauge pressure at a point 10 m below the surface of the ocean? (Note: The specific gravity of seawater is 1.025.)

Solution. Using $\rho_{seawater}$ = sp. gr. × ρ_{water} = 1025 kg/m^3, we find that

$$P_{gauge} = \rho gh = (1025\,\tfrac{kg}{m^3})(10\,\tfrac{N}{kg})(10\text{ m}) = 1.025 \times 10^5\text{ Pa}$$

Note that this is just about equal to one atmosphere (1 atm = 1.013×10^5 Pa). In fact, a rule of thumb used by scuba and deep-sea divers is that the pressure increases by about 1 atmosphere for every 10 m (or 30.5 feet) of depth.

Example 8.5 A swimming pool has a depth of 4 m. What is the hydrostatic gauge pressure at a point 1 m below the surface?

Solution. Using $\rho = \rho_{water}$ = 1000 kg/m^3 for the density of the pool water, we find that

$$P_{gauge} = \rho gh = (1000\,\tfrac{kg}{m^3})(10\,\tfrac{N}{kg})(1\text{ m}) = 1 \times 10^4\text{ Pa}$$

Note that the total depth of the swimming pool is irrelevant; all that matters is the depth below the surface.

Example 8.6 What happens to the gauge pressure if we double our depth below the surface of a liquid? What happens to the total pressure?

Solution. Since $P_{gauge} = \rho gh$, we see that P_{gauge} is proportional to the depth, h. Therefore, if we double h, then P_{gauge} will double. However, the total pressure, $P_{atm} + \rho gh$, is not proportional to h. The term ρgh will double, but the term P_{atm}, being a constant, will not. So, the total pressure will increase, but it will not double; in fact, it will increase by less than a factor of 2.

Example 8.7 A flat piece of wood, of area 0.5 m^2, is lying at the bottom of a lake. If the depth of the lake is 30 m, what is the force on the wood due to the pressure? (Use $P_{atm} = 1 \times 10^5$ Pa.)

Solution. Using $\rho = \rho_{water}$ = 1000 kg/m^3 for the density of the water in the lake, we find that

$$P_{gauge} = \rho gh = (1000\,\tfrac{kg}{m^3})(10\,\tfrac{N}{kg})(30\text{ m}) = 3 \times 10^5\text{ Pa}$$

Therefore, the total pressure on the wood is $P = P_{atm} + P_{gauge} = 4 \times 10^5$ Pa. Now, by definition, we have $P = F/A$, so $F = PA$; this gives

$$F = PA = (4 \times 10^5\text{ Pa})(0.5\text{ m}^2) = 2 \times 10^5\text{ N}$$

Example 8.8 Consider a closed container, partially filled with a liquid of density $\rho = 1200$ kg/m^3, and a Point X that's 0.5 m below the surface of the liquid.

(a) If the space above the surface of the liquid is vacuum, what's the absolute pressure at Point X?

(b) If the space above the surface of the liquid is occupied by a gas whose pressure is 2.4×10^4 Pa, what's the absolute pressure at Point X?

Solution.

(a) The absolute pressure at X is equal to the pressure pushing down on the surface of the liquid plus the pressure due to the liquid alone: $P = P_0 + \rho gh$. However, since the space above the liquid is vacuum, we have $P_0 = 0$, so

$$P = \rho gh = (1200 \tfrac{\text{kg}}{\text{m}^3})(10 \tfrac{\text{N}}{\text{kg}})(0.5 \text{ m}) = 6 \times 10^3 \text{ Pa}$$

(b) In this case, $P_0 = 2.4 \times 10^4$ Pa, so

$$P = P_0 + \rho gh = (2.4 \times 10^4 \text{ Pa}) + (6 \times 10^3 \text{ Pa}) = 3 \times 10^4 \text{ Pa}$$

BUOYANCY

Let's place a block in our tank of fluid. Because the pressure on each side of the block depends on its average depth, we see that there's more pressure on the bottom of the block than there is on the top. Therefore, there's a greater force pushing up on the block than there is pushing down on it. The forces due to the pressure on the other four sides (left and right, front and back) cancel out, so the net force on the block is upward.

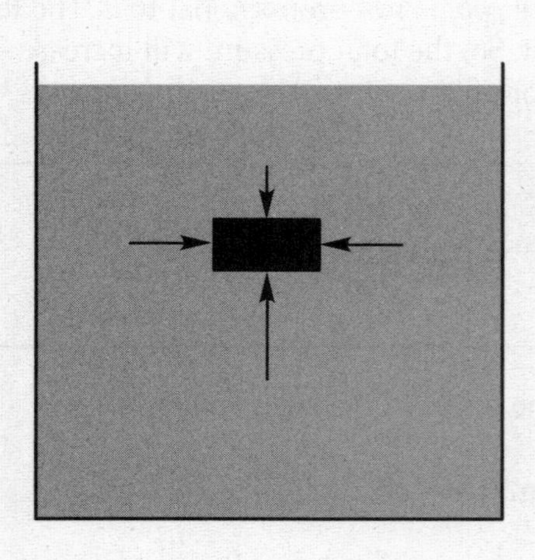

front view

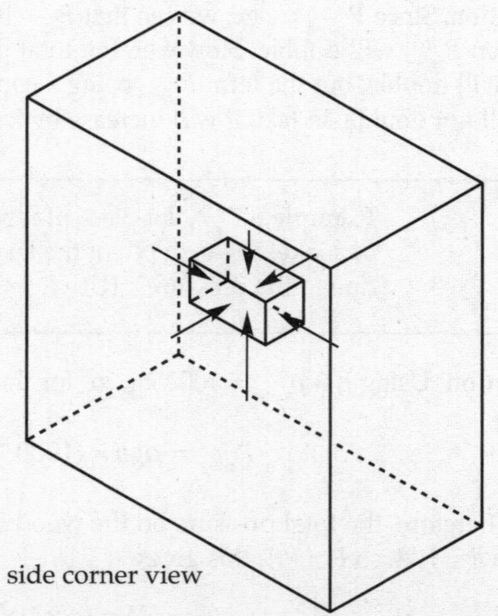

side corner view

This net upward force is called the **buoyant force** (or just **buoyancy** for short), denoted $\mathbf{F}_{buoy}$. We calculate the magnitude of the buoyant force using **Archimedes' principle**; in words, Archimedes' principle says

The strength of the buoyant force is equal to the weight of the fluid displaced by the object.

When an object is partially or completely submerged in a fluid, the volume of the object that's submerged, which we call V_{sub}, is the volume of the fluid displaced. By multiplying this volume by the density of the fluid, we get the mass of the fluid displaced; then, multiplying this mass by g gives us the weight of the fluid displaced. So, here's Archimedes' principle as a mathematical equation:

$$\text{Buoyant force: } F_{buoy} = \rho_{fluid} V_{sub} g$$

When an object floats, its submerged volume is just enough to make the buoyant force it feels balance its weight. So, if an object's density is ρ_{object} and its (total) volume is V, its weight will be $mg = \rho V g$. The buoyant force it feels is $\rho_{fluid} V_{sub} g$. Setting these equal to each other, we find that

$$\frac{V_{sub}}{V} = \frac{\rho_{object}}{\rho_{fluid}}$$

So, if $\rho_{object} < \rho_{fluid}$, then the object will float; and the fraction of its volume that's submerged is the same as the ratio of its density to the fluid's density. For example, if the object's density is 2/3 the density of the fluid, then the object will float, and 2/3 of the object will be submerged.

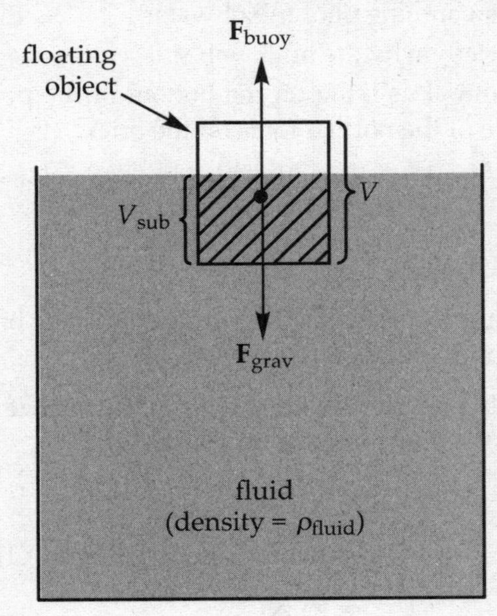

If an object is denser than the fluid, it will sink. In this case, even if the entire object is submerged (in an attempt to maximize V_{sub} and the buoyant force), its weight is still greater than the buoyant force, and down it goes. And if an object just happens to have the same density as the fluid, it will be happy hovering (in static equilibrium) anywhere underneath the fluid.

Example 8.9 An object with a mass of 150 kg and a volume of 0.75 m^3 is floating in ethyl alcohol, whose specific gravity is 0.8. What fraction of the object's volume is above the surface of the fluid?

Solution. The density of the object is

$$\rho_{\text{object}} = \frac{m}{V} = \frac{150 \text{ kg}}{0.75 \text{ m}^3} = \frac{150 \text{ kg}}{\frac{3}{4} \text{ m}^3} = 200 \frac{\text{kg}}{\text{m}^3}$$

Because the density of the fluid is (sp. gr.)(ρ_{water}) = (0.8)(1000 $\frac{\text{kg}}{m^3}$) = 800 $\frac{\text{kg}}{m^3}$, the ratio of the object's density to the fluid's density is

$$\frac{\rho_{\text{object}}}{\rho_{\text{fluid}}} = \frac{200 \frac{\text{kg}}{\text{m}^3}}{800 \frac{\text{kg}}{\text{m}^3}} = \frac{1}{4}$$

This means that 1/4 of the object's volume is *below* the surface of the fluid; therefore, the fraction that's *above* the surface is 1 – (1/4) = 3/4.

Example 8.10 A brick, of specific gravity 2 and volume 1.5×10^{-3} m^3, is dropped into a swimming pool full of water.

(a) Explain briefly why the brick will sink.

(b) When the brick is lying on the bottom of the pool, what is the magnitude of the normal force on the brick?

Solution.

(a) Since the specific gravity of the brick is greater than 1, the brick has a greater density than the surrounding fluid (water), so it will sink.

(b) When the brick is lying on the bottom surface of the pool, it is totally submerged, so $V_{\text{sub}} = V$; this means the buoyant force on the brick is

$$\begin{aligned} F_{\text{buoy}} &= \rho_{\text{fluid}} V_{\text{sub}} g = \rho_{\text{water}} V g \\ &= (1000 \tfrac{\text{kg}}{\text{m}^3})(1.5 \times 10^{-3} \text{ m}^3)(10 \tfrac{\text{N}}{\text{kg}}) \\ &= 15 \text{ N} \end{aligned}$$

The weight of the brick is

$$\begin{aligned} F_g &= mg = \rho_{\text{brick}} V g \\ &= (2000 \tfrac{\text{kg}}{\text{m}^3})(1.5 \times 10^{-3} \text{ m}^3)(10 \tfrac{\text{N}}{\text{kg}}) \\ &= 30 \text{ N} \end{aligned}$$

When the brick is lying on the bottom of the pool, the net force it feels is zero.

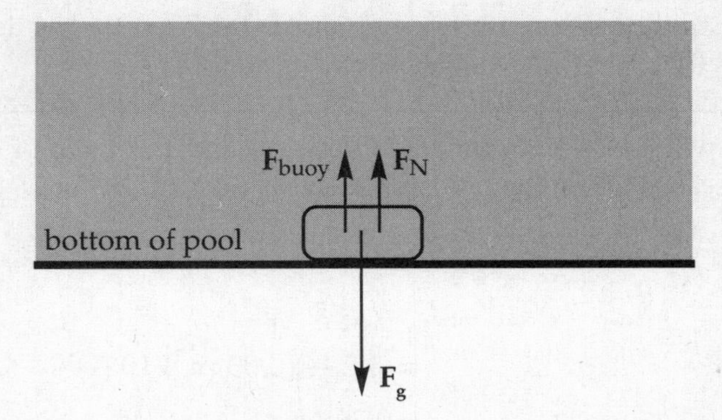

Therefore, we must have $F_{buoy} + F_N = F_g$, so $F_N = F_g - F_{buoy} = 30\text{ N} - 15\text{ N} = 15\text{ N}$.

Example 8.11 A glass sphere of specific gravity 2.5 and volume 10^{-3} m³ is completely submerged in a large container of water. What is the apparent weight of the sphere while immersed?

Solution. Because the buoyant force pushes up on the object, the object's *apparent weight*, $F_{g\text{ apparent}} = F_g - F_{buoy}$, is less than its true weight, F. Since the sphere is completely submerged, we have $V_{sub} = V$, so the buoyant force on the sphere is

$$\begin{aligned} F_{buoy} &= \rho_{fluid} V_{sub} g = \rho_{water} V g \\ &= (1000 \tfrac{\text{kg}}{\text{m}^3})(10^{-3}\text{ m}^3)(10 \tfrac{\text{N}}{\text{kg}}) \\ &= 10\text{ N} \end{aligned}$$

The true weight of the glass sphere is

$$\begin{aligned} F_g = mg = \rho_{glass} V g &= (\text{sp. gr.} \times \rho_{water}) V g \\ &= (2.5 \times 1000 \tfrac{\text{kg}}{\text{m}^3})(10^{-3}\text{ m}^3)(10 \tfrac{\text{N}}{\text{kg}}) \\ &= 25\text{ N} \end{aligned}$$

Therefore, the apparent weight of the sphere while immersed is

$$F_{g\text{ apparent}} = F_g - F_{buoy} = 25\text{ N} - 10\text{ N} = 15\text{ N}$$

> **Example 8.12** A helium balloon has a volume of 0.03 m^3. What is the net force on the balloon if it's surrounded by air? (Note: The density of helium is 0.2 kg/m^3, and the density of air is 1.2 kg/m^3.)

Solution. The balloon will feel a buoyant force upward, and the force of gravity—the balloon's weight—downward. Because the balloon is completely surrounded by air, $V_{sub} = V$, and the buoyant force is

$$\begin{aligned} F_{buoy} = \rho_{fluid} V_{sub} g = \rho_{air} Vg \\ = (1.2 \tfrac{kg}{m^3})(0.03\ m^3)(10\ \tfrac{N}{kg}) \\ = 0.36\ N \end{aligned}$$

The weight of the balloon is

$$\begin{aligned} F_g = mg = \rho_{helium} Vg = (0.2 \tfrac{kg}{m^3})(0.03\ m^3)(10\ \tfrac{N}{kg}) \\ = 0.06\ N \end{aligned}$$

Because $F_{buoy} > w$, the net force on the balloon is upward and has magnitude

$$F_{net} = F_{buoy} - F_g = 0.36\ N - 0.06\ N = 0.3\ N$$

FLOW RATE AND THE CONTINUITY EQUATION

Consider a pipe through which fluid is flowing. The **flow rate,** f, is the volume of fluid that passes a particular point per unit time; for example, the number of liters of water per minute that are coming out of a faucet. In SI units, flow rate is expressed in m^3/s. To find the flow rate, all we need to do is multiply the cross-sectional area of the pipe at any point by the average speed of the flow at that point.

$$\text{flow rate: } f = Av$$

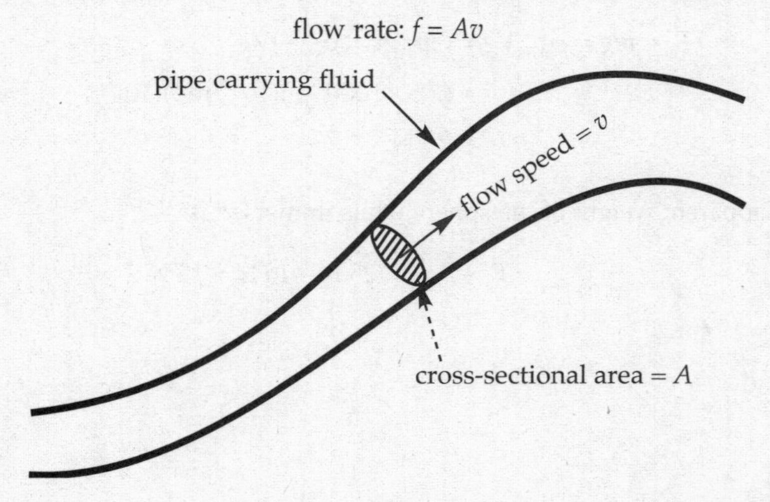

Be careful not to confuse *flow rate* with *flow speed;* flow rate tells us how *much* fluid flows per unit time; flow speed tells us how *fast* it's moving.

If the pipe is carrying a liquid—which we assume is incompressible (that is, its density remains constant)—then the flow rate must be the same everywhere along the pipe. Choose any two points, Point 1 and Point 2, in a pipe carrying a liquid. If there aren't any sources or sinks between these points, all the liquid that flows by Point 1 must also flow by Point 2. In other words, the flow rate at Point 1 must be the same as the flow rate at Point 2: $f_1 = f_2$. Rewriting this using $f = Av$, we get the **Continuity Equation:**

$$A_1v_1 = A_2v_2$$

Because the product Av is a constant, the flow speed will increase where the pipe narrows, and decrease where the pipe widens. In fact, we can say that the flow speed is inversely proportional to the cross-sectional area—or to the square of the radius—of the pipe.

Example 8.13 A pipe of non-uniform diameter carries water. At one point in the pipe, the radius is 2 cm and the flow speed is 6 m/s.

(a) What is the flow rate?

(b) What is the flow speed at a point where the pipe constricts to a radius of 1 cm?

Solution.

(a) At any point, the flow rate, f, is equal to the cross-sectional area of the pipe multiplied by the flow speed

$$f = Av = \pi r^2 v = \pi(2\times10^{-2}\text{ m})^2(6\ \tfrac{\text{m}}{\text{s}}) \approx 75\times10^{-4}\ \tfrac{\text{m}^3}{\text{s}} = 7.5\times10^{-3}\ \tfrac{\text{m}^3}{\text{s}}$$

(b) By the Continuity Equation, we know that v, the flow speed, is inversely proportional to A, the cross-sectional area of the pipe. If the pipe's radius decreases by a factor of 2 (from 2 cm to 1 cm), then A decreases by a factor of 4, because A is proportional to r^2. If A decreases by a factor of 4, then v will increase by a factor of 4. Therefore, the flow speed at a point where the pipe's radius is 1 cm will be $4 \times (6\text{ m/s}) = 24\text{ m/s}$.

Example 8.14 If the diameter of a pipe increases from 4 cm to 12 cm, what will happen to the flow speed?

Solution. The cross-sectional area of the pipe is proportional not only to the square of the radius (r), but also to the square of the diameter (d), since $A = \pi r^2 = \pi(\frac{1}{2}d)^2 = \frac{1}{4}\pi d^2$. So, if d increases by a factor of 3, you can say the flow speed will decrease by a factor of $3^2 = 9$ or you will have $\frac{1}{9}$ of the flow rate.

BERNOULLI'S EQUATION

The most important equation in fluid mechanics is **Bernoulli's Equation**, which is the statement of conservation of energy for ideal fluid flow. First, let's describe the conditions that make fluid flow *ideal*.

- *The fluid is incompressible.*

  This works very well for liquids and also applies to gases if the pressure changes are small.

- *The fluid's viscosity is negligible.*

  Viscosity is the force of cohesion between molecules in a fluid; think of viscosity as internal friction for fluids. For example, maple syrup is sticky and has a greater viscosity than water: there's more resistance to a flow of maple syrup than to a flow of water. While Bernoulli's Equation would give good results when applied to a flow of water, it would not give good results if it were applied to a flow of maple syrup.

- *The flow is streamline.*

  In a tube carrying a flowing fluid, a **streamline** is just what it sounds like: it's a *line* in the *stream*. If we were to inject a drop of dye into a clear glass pipe carrying, say, water, we'd see a streak of dye in the pipe, indicating a streamline.

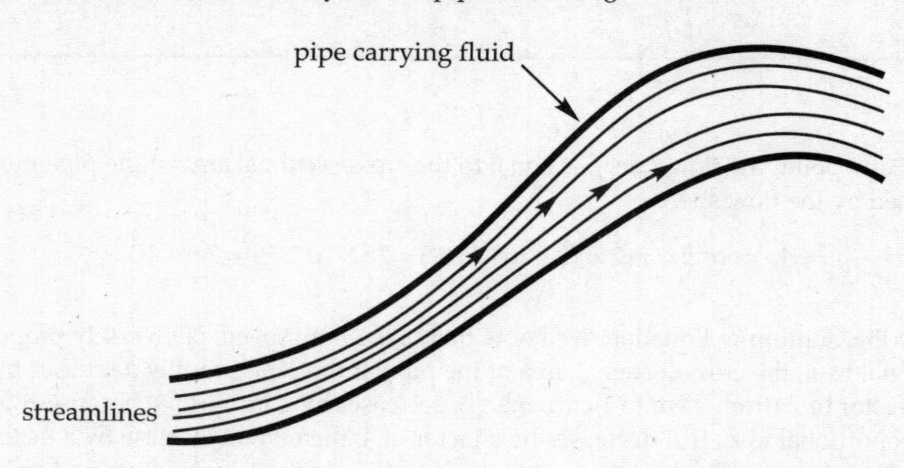

The entire flow is said to be **streamline** (as an adjective) or **laminar** if the individual streamlines don't curl up into vortices but instead remain steady and smooth. *When the flow is streamline, the fluid moves smoothly through the tube.* (The opposite of streamline flow is **turbulent** flow, which is characterized by rapidly swirling whirlpools; such chaotic flow is unpredictable.)

If the three conditions described above hold, and the flow rate, *f*, is steady, Bernoulli's Equation can be applied to any pair of points along a streamline within the flow. Let ρ be the density of the fluid that's flowing. Label the points we want to compare as Point 1 and Point 2. Choose a horizontal reference level, and let y_1 and y_2 be the heights of these points above this level. If the pressures at Points 1 and 2 are P_1 and P_2, and if the flow speeds at these points are v_1 and v_2, then Bernoulli's Equation says

$$P_1 + \rho g y_1 + \tfrac{1}{2}\rho v_1^2 = P_2 + \rho g y_2 + \tfrac{1}{2}\rho v_2^2$$

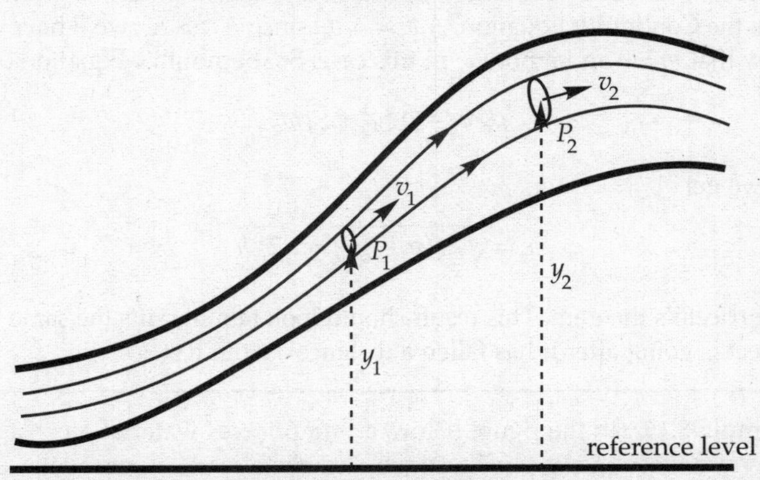

An alternative, but equivalent, way of stating Bernoulli's Equation is to say that the quantity

$$P + \rho g y + \tfrac{1}{2}\rho v^2 = \text{constant}$$

is constant along a streamline. We mentioned earlier that Bernoulli's Equation is a statement of conversation of energy. You should notice the similarity between $\rho g y$ and mgh (gravitational potential energy) as well as between $\frac{1}{2}\rho v^2$ and $\frac{1}{2}mv^2$ (kinetic energy).

Torricelli's Theorem

Imagine that we punch a small hole in the side of a tank of liquid. We can use Bernoulli's Equation to figure out the efflux speed: how fast the liquid will flow out of the hole.

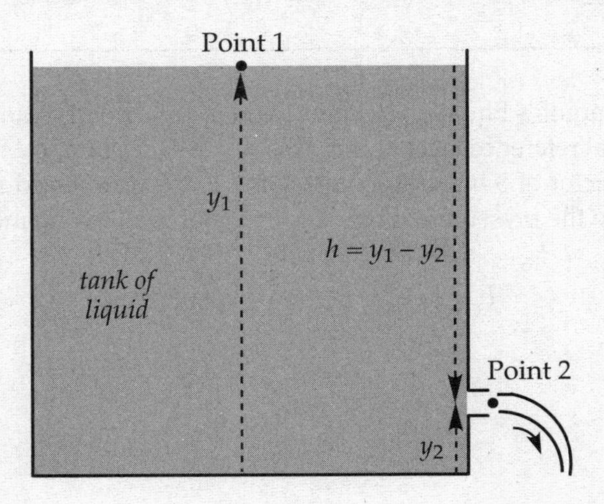

Let the bottom of the tank be our horizontal reference level, and choose Point 1 to be at the surface of the liquid and Point 2 to be at the hole where the water shoots out. First, the pressure at Point 1 is atmospheric pressure; and the emerging stream at Point 2 is open to the air, so it's at atmospheric pressure too. Therefore, $P_1 = P_2$, and these terms cancel out of Bernoulli's Equation. Next, since the area at Point 1 is so much greater than at Point 2, we can assume that v_1, the speed at which the water level in the tank drops, is much lower than v_2, the speed at which the water shoots out of the hole. (Remember that by the Continuity Equation, $A_1v_1 = A_2v_2$; since $A_1 >> A_2$, we'll have $v_1 << v_2$.) Because $v_1 << v_2$, we can say that $v_1 \approx 0$ and ignore v_1 in this case. So, Bernoulli's Equation becomes

$$\rho g y_1 = \rho g y_2 + \tfrac{1}{2}\rho v_2^2$$

Solving for v_2, we get

$$v_2 = \sqrt{2g(y_1 - y_2)} = \sqrt{2gh}$$

This is called Torricelli's theorem. This result should look familiar; it's the same formula that tells us how fast an object is going after it has fallen a distance h from rest.

Example 8.15 In the figure below, a pump forces water at a constant flow rate through a pipe whose cross-sectional area, A, gradually decreases: at the exit point, A has decreased to 1/3 its value at the beginning of the pipe. If y = 60 cm and the flow speed of the water just after it leaves the pump (Point 1 in the figure) is 1 m/s, what is the gauge pressure at Point 1?

Solution. We'll apply Bernoulli's Equation to Point 1 and the exit point, Point 2. We'll choose the level of Point 1 as the horizontal reference level; this makes $y_1 = 0$. Now, because the cross-sectional area of the pipe decreases by a factor of 3 between Points 1 and 2, the flow speed must increase by a factor of 3; that is, $v_2 = 3v_1$. Since the pressure at Point 2 is p_{atm}, Bernoulli's Equation becomes

$$P_1 + \tfrac{1}{2}\rho v_1^2 = P_{atm} + \rho g y_2 + \tfrac{1}{2}\rho v_2^2$$

Therefore,

$$\begin{aligned} P_1 - P_{\text{atm}} &= \rho g y_2 + \tfrac{1}{2}\rho v_2^2 - \tfrac{1}{2}\rho v_1^2 \\ &= \rho g y_2 + \tfrac{1}{2}\rho(3v_1)^2 - \tfrac{1}{2}\rho v_1^2 \\ &= \rho(g y_2 + 4v_1^2) \\ &= (1000\ \tfrac{\text{kg}}{\text{m}^3})[(10\ \tfrac{\text{m}}{\text{s}^2})(0.6\ \text{m}) + 4(1\ \tfrac{\text{m}}{\text{s}})^2] \\ &= 10^4\ \text{Pa} \end{aligned}$$

Example 8.16 What does Bernoulli's Equation tell us about a fluid at rest in a container open to the atmosphere?

Solution. Consider the figure below:

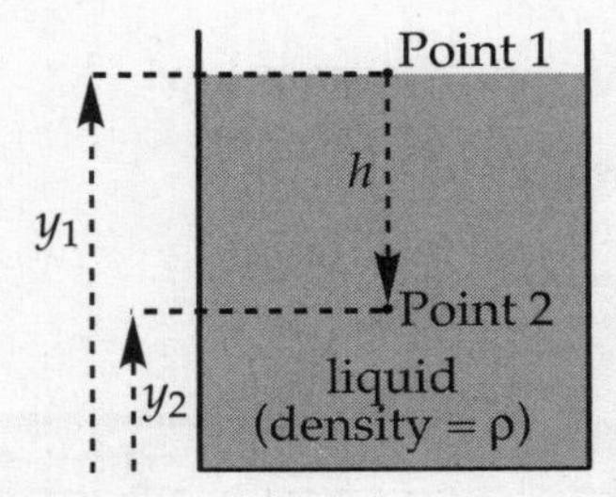

Because the fluid in the tank is at rest, both v_1 and v_2 are zero, and Bernoulli's Equation becomes

$$P_1 + \rho g y_1 = P_2 + \rho g y_2$$

Since $P_1 = P_{\text{atm}}$, if we solve this equation for P_2, we get

$$P_2 = P_{\text{atm}} + \rho g(y_1 - y_2) = P_{\text{atm}} + \rho g h$$

which is the same formula we found earlier for hydrostatic pressure.

> **Example 8.17** The side of an above-ground pool is punctured, and water gushes out through the hole. If the total depth of the pool is 2.5 m, and the puncture is 1 m above ground level, what is the efflux speed of the water?

Solution. Torricelli's Theorem is $v = \sqrt{2gh}$, where ***h*** is the distance from the surface of the pool down to the hole. If the puncture is 1 m above ground level, then it's 2.5 – 1 = 1.5 m below the surface of the water (because the pool is 2.5 m deep). Therefore, the efflux speed will be

$$v = \sqrt{2gh} = \sqrt{2(10\ \tfrac{\text{m}}{\text{s}^2})(1.5\ \text{m})} = \sqrt{30\ \tfrac{\text{m}^2}{\text{s}^2}} \approx 5.5\ \tfrac{\text{m}}{\text{s}}$$

THE BERNOULLI EFFECT

Consider the two points labeled in the pipe shown below:

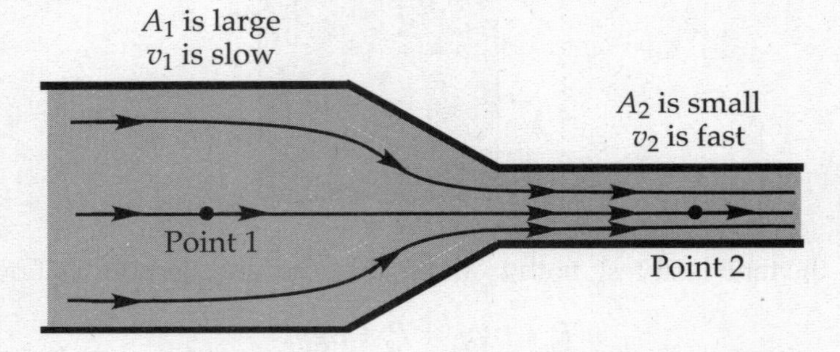

Since the heights y_1 and y_2 are equal in this case, the terms in Bernoulli's Equation that involve the heights will cancel, leaving us with

$$P_1 + \tfrac{1}{2}\rho v_1^2 = P_2 + \tfrac{1}{2}\rho v_2^2$$

We already know from the Continuity Equation ($f = Av$) that the speed increases as the cross-sectional area of the pipe decreases; that is, since $A_2 < A_1$, we know that $v_2 > v_1$, so the equation above tells us that $P_2 < P_1$. This shows that

The pressure is lower where the flow speed is greater.

This is known as the **Bernoulli** (or **Venturi**) **effect**, and is illustrated in the figure below.

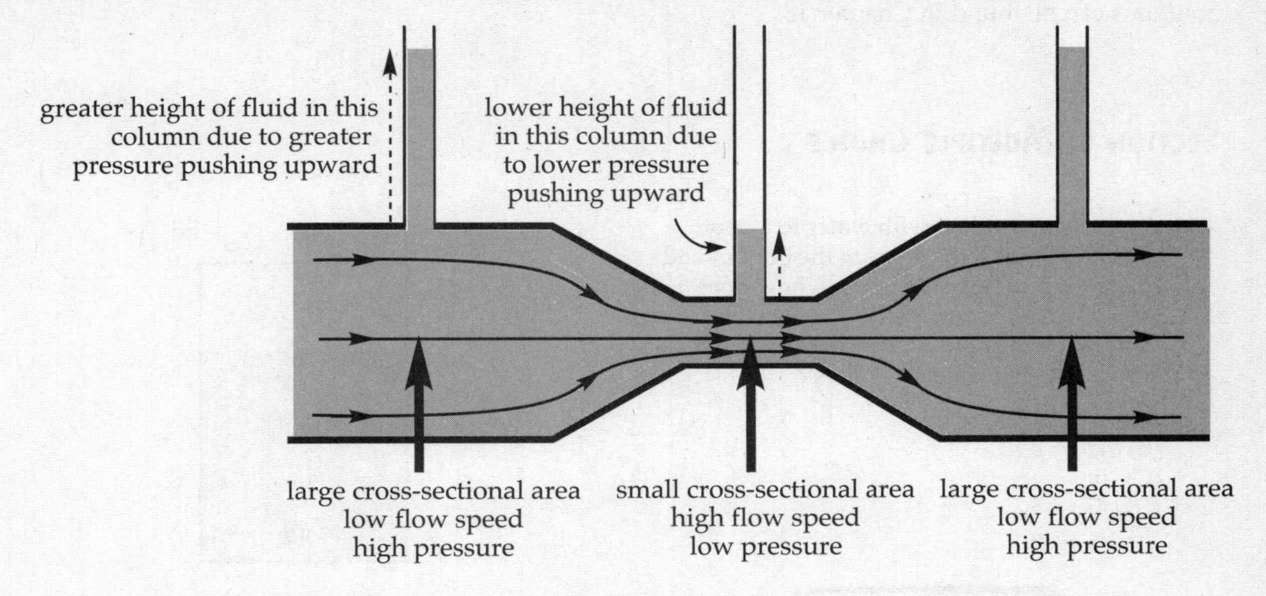

The height of the liquid column above Point 2 is less than the height of the liquid column above Point 1, because the pressure at Point 2 is lower than at Point 1, due to the fact that the flow speed at Point 2 is greater than at Point 1.

The Bernoulli Effect also accounts for many everyday phenomena. It's what allows airplanes to fly, curve balls to curve, and tennis balls hit with top spin to drop quickly. You may have seen sky divers or motorcycle riders wearing a jacket that seems to puff out as they move rapidly through the air. The essentially stagnant air trapped inside the jacket is at a much higher pressure than the air whizzing by outside, and as a result, the jacket expands outward. The drastic drop in air pressure that accompanies the high winds in a hurricane or tornado is yet another example. In fact, if high winds streak across the roof of a house whose windows are closed, the outside air pressure can be reduced so much that the air pressure inside the house (where the air speed is essentially zero) can be great enough to blow the roof off.

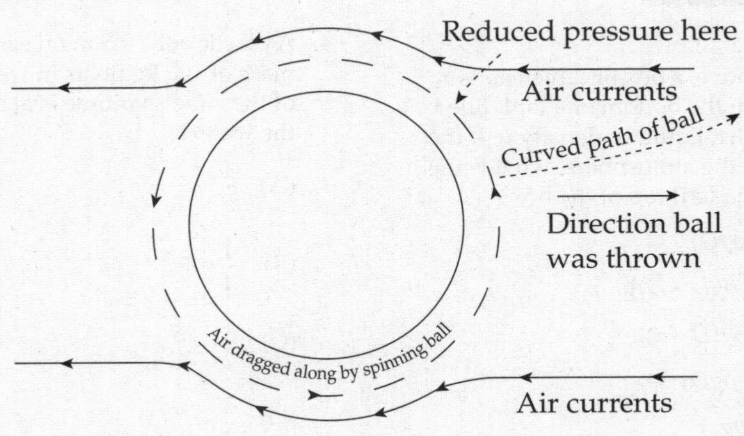

CHAPTER 8 REVIEW QUESTIONS

Solutions can be found in Chapter 18.

SECTION I: MULTIPLE CHOICE

1. A large tank is filled with water to a depth of 6 m. If Point X is 1 m from the bottom and Point Y is 2 m from the bottom, how does p_X, the hydrostatic pressure due to the water at Point X, compare to p_Y, the hydrostatic pressure due to the water at Point Y?

(A) $p_X = 2p_Y$
(B) $2p_X = p_Y$
(C) $5p_X = 4p_Y$
(D) $4p_X = 5p_Y$
(E) $p_X = 4p_Y$

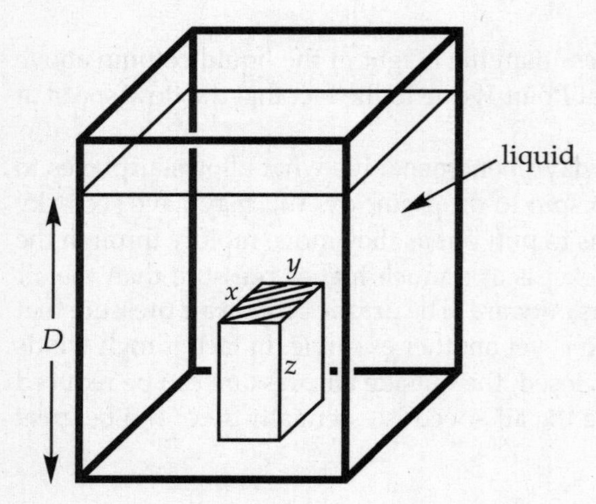

2. In the figure above, a box of dimensions x, y, and z rests on the bottom of a tank filled to depth D with a liquid of density ρ. If the tank is open to the atmosphere, what is the force on the (shaded) top of the box?

(A) $xy(p_{atm} + \rho gD)$
(B) $xyz[p_{atm} - \rho g(z - D)]$
(C) $xy[p_{atm} + \rho g(D - z)]$
(D) $xyz[p_{atm} + \rho gD]$
(E) $xy(p_{atm} + \rho gz)$

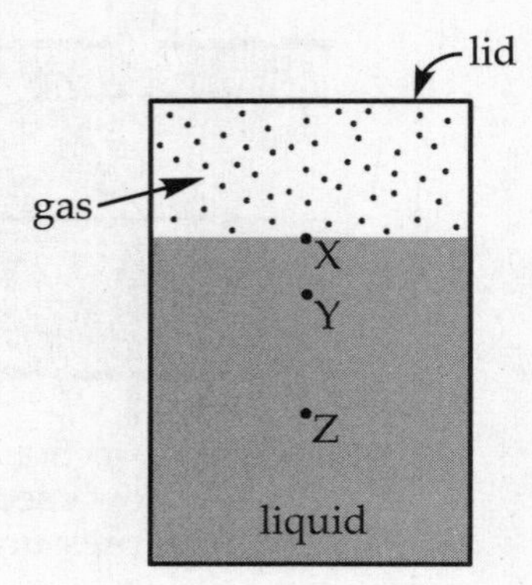

3. The figure above shows a closed container partially filled with liquid. Point Y is at a depth of 1 m, and Point Z is at a depth of 3 m. If the absolute pressure at Point Y is 13,000 Pa, and the absolute pressure at Point Z is 29,000 Pa, what is the pressure at the surface, Point X?

(A) 3,000 Pa
(B) 4,000 Pa
(C) 5,000 Pa
(D) 7,000 Pa
(E) 8,000 Pa

4. A plastic cube 0.5 m on each side and with a mass of 100 kg floats in water. What fraction of the cube's volume is above the surface of the water?

(A) $\frac{1}{5}$

(B) $\frac{1}{4}$

(C) $\frac{1}{2}$

(D) $\frac{3}{4}$

(E) $\frac{4}{5}$

5. A block of Styrofoam, with a density of ρ_S and volume V, is pushed completely beneath the surface of a liquid whose density is ρ_L, and released from rest. Given that $\rho_L > \rho_S$, which of the following expressions gives the magnitude of the block's initial upward acceleration?

(A) $(\rho_L - \rho_S)g$

(B) $\left(\frac{\rho_L}{\rho_S} - 1\right)g$

(C) $\left(\frac{\rho_L}{\rho_S} + 1\right)g$

(D) $[(\frac{\rho_L}{\rho_S})^2 - 1]g$

(E) $[(\frac{\rho_L}{\rho_S})^2 + 1]g$

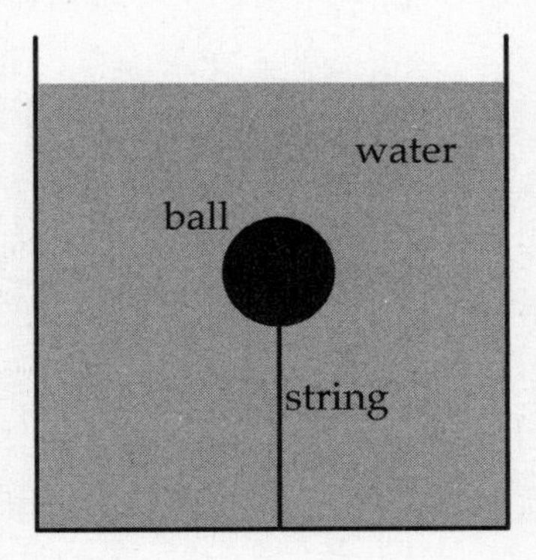

6. In the figure above, a ball of specific gravity 0.4 and volume 5×10^{-3} m^3 is attached to a string, the other end of which is fastened to the bottom of the tank. If the tank is filled with water, what is the tension in the string?

(A) 20 N
(B) 30 N
(C) 40 N
(D) 50 N
(E) 70 N

7. An object of specific gravity 2 weighs 100 N less when it's weighed while completely submerged in water than when it's weighed in air. What is the actual weight of this object?

(A) 200 N
(B) 300 N
(C) 400 N
(D) 600 N
(E) 800 N

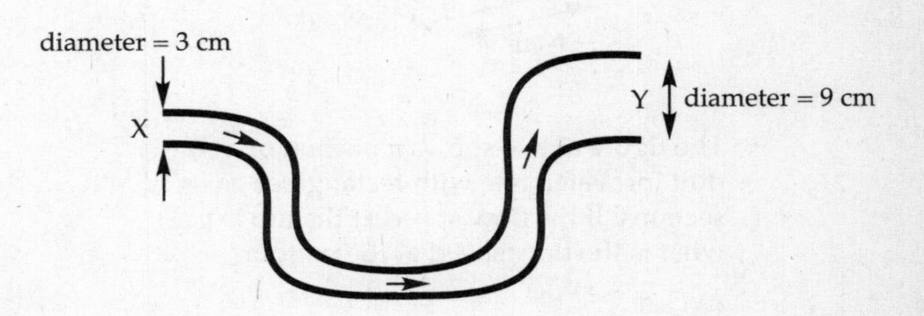

8. In the pipe shown above, which carries water, the flow speed at Point X is 6 m/s. What is the flow speed at Point Y?

(A) $\frac{2}{3}$ m/s

(B) 2 m/s

(C) 4 m/s

(D) 18 m/s

(E) 54 m/s

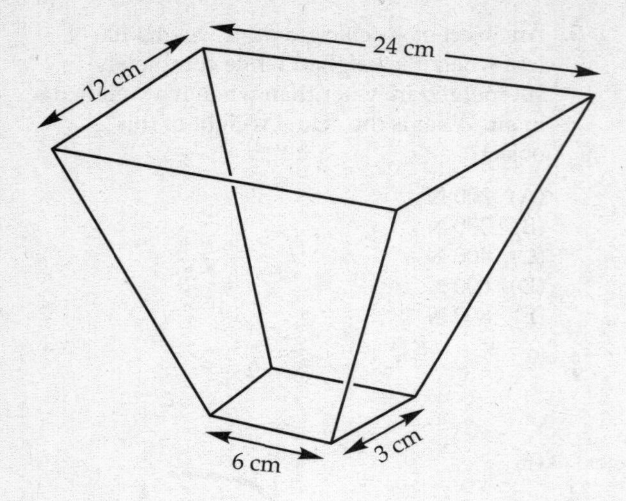

9. The figure above shows a portion of a conduit for water, one with rectangular cross sections. If the flow speed at the top is v, what is the flow speed at the bottom?

   (A) $2v$
   (B) $4v$
   (C) $8v$
   (D) $12v$
   (E) $16v$

10. A pump is used to send water through a hose, the diameter of which is 10 times that of the nozzle through which the water exits. If the nozzle is 1 m higher than the pump, and the water flows through the hose at 0.4 m/s, what is the gauge pressure of the water at the pump?

    (A) 108 kPa
    (B) 260 kPa
    (C) 400 kPa
    (D) 810 kPa
    (E) 1080 kPa

SECTION II: FREE RESPONSE

1. The figure below shows a tank open to the atmosphere and filled to depth D with a liquid of density ρ_L. Suspended from a string is a block of density ρ_B (which is greater than ρ_L), whose dimensions are x, y, and z (meters). The top of the block is at depth h meters below the surface of the liquid.

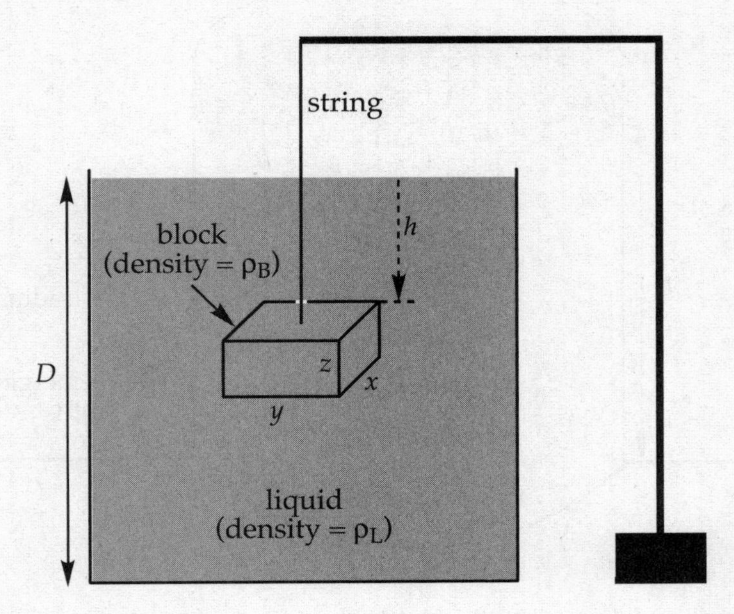

In each of the following, write your answer in simplest form in terms of ρ_L, ρ_B, x, y, z, h, D, and g.

(a) Find the force due to the pressure on the top surface of the block and on the bottom surface. Sketch these forces in the diagram below:

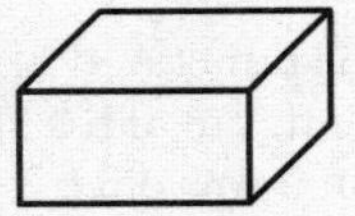

(b) What are the average forces due to the pressure on the other four sides of the block? Sketch these forces in the diagram above.

(c) What is the total force on the block due to the pressure?

(d) Find an expression for the buoyant force on the block. How does your answer here compare to your answer to part (c)?

(e) What is the tension in the string?

2. The figure below shows a large, cylindrical tank of water, open to the atmosphere, filled with water to depth D. The radius of the tank is R. At a depth h below the surface, a small circular hole of radius r is punctured in the side of the tank, and the point where the emerging stream strikes the level ground is labeled X.

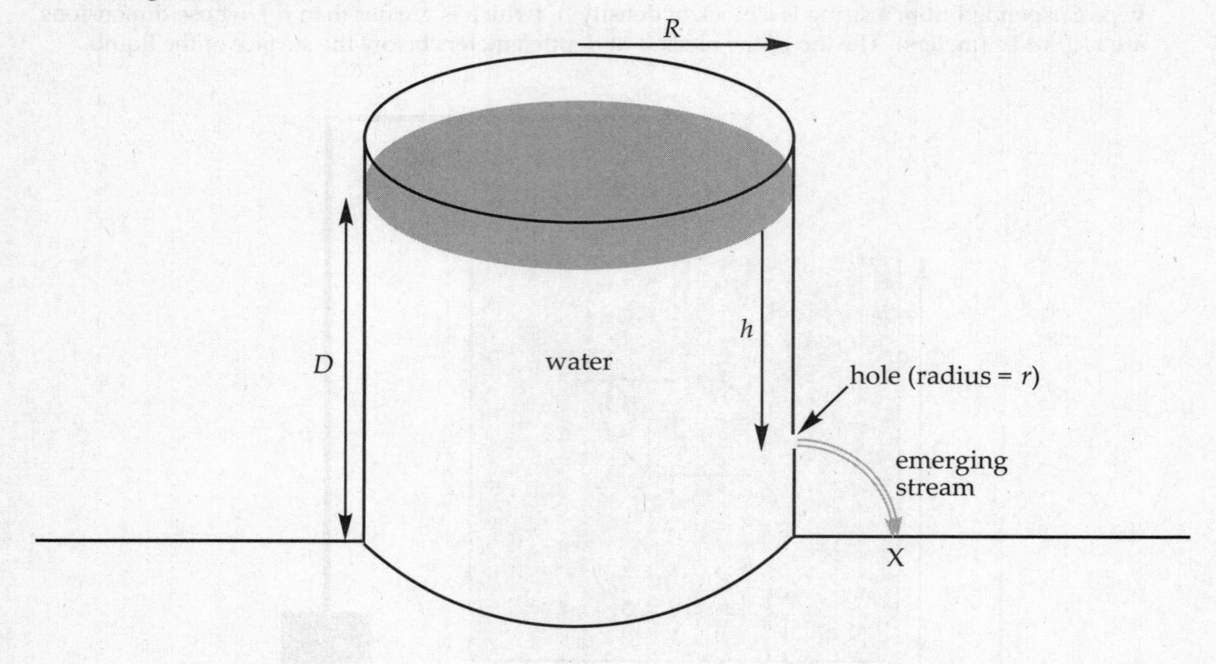

In parts (a) through (c), assume that the speed with which the water level in the tank drops is negligible.

(a) At what speed does the water emerge from the hole?

(b) How far is point X from the edge of the tank?

(c) Assume that a second small hole is punctured in the side of the tank, a distance of $h/2$ directly above the hole shown in the figure. If the stream of water emerging from this second hole also lands at Point X, find h in terms of D.

(d) For this part, do *not* assume that the speed with which the water level in the tank drops is negligible, and derive an expression for the speed of efflux from the hole punctured at depth h below the surface of the water. Write your answer in terms of r, R, h, and g.

3. The figure below shows a pipe fitted with a Venturi U-tube. Fluid of density ρ_F flows at a constant flow rate and with negligible viscosity through the pipe, which constricts from a cross-sectional area A_1 at Point 1 to a smaller cross-sectional area A_2 at Point 2. The upper portion of both sides of the Venturi U-tube contain the same fluid that's flowing through the pipe, while the lower portion is filled with a fluid of density ρ_V (which is greater than ρ_F). At Point 1 in the pipe, the pressure is P_1 and the flow speed is v_1; at Point 2 in the pipe, the pressure is P_2 and the flow speed is v_2. All the fluid within the Venturi U-tube is stationary.

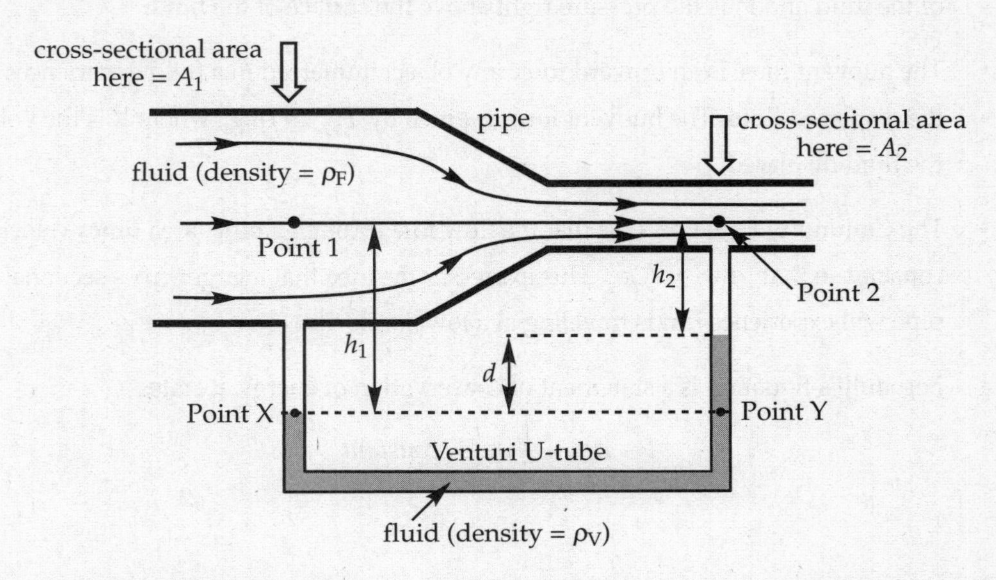

(a) What is P_X, the hydrostatic pressure at Point X? Write your answer in terms of P_1, ρ_F, h_1, and g.

(b) What is P_Y, the hydrostatic pressure at Point Y? Write your answer in terms of P_2, ρ_F, ρ_V, h_2, d, and g.

(c) Write down the result of Bernoulli's Equation applied to Points 1 and 2 in the pipe, and solve for $P_1 - P_2$.

(d) Since $P_X = P_Y$, set the expressions you derived in parts (a) and (b) equal to each other, and use this equation to find $P_1 - P_2$.

(e) Derive an expression for the flow speed, v_2, and the flow rate, f, in terms of A_1, A_2, d, ρ_F, ρ_V, and g. Show that v_2 and f are proportional to $\sqrt{d}$.

SUMMARY

- Density is given by $\rho = \frac{m}{V}$. Pressure is given by $P = \frac{F}{A}$. Hydrostatic pressure can be found using $P = P_0 + \rho gh$ where ρgh is the pressure at a given depth below the surface of the fluid and P_0 is the pressure right above the surface of the fluid.
- The buoyant force is an upward force any object immersed in a fluid experiences due to the displaced fluid. The buoyant force is given by $F_{buoy} = \rho Vg$, where V is the volume of the fluid displaced.
- The Continuity Equation says that the flow rate through a pipe (area times velocity) is constant so that $A_1 v_1 = A_2 v_2$. This expresses the idea that a larger cross sectional area of pipe will experience fluids traveling at a lower velocity.
- Bernoulli's Equation is a statement of conservation of energy. It states

$$P + \rho gy + \frac{1}{2}\rho v^2 = constant$$

9

Thermal Physics

INTRODUCTION

This chapter looks at heat and temperature, concepts that seem familiar from our everyday experience. Technically, **heat** is defined as thermal energy that's transmitted from one body to another. While an object can contain **thermal energy** (due to the random motion of its molecules), an object doesn't *contain* heat; heat is energy that's *in transit*. **Temperature**, on the other hand, is a measure of the concentration of an object's internal thermal energy, and is one of the basic SI units.

TEMPERATURE SCALES

In the United States, temperatures are still often expressed in **degrees Fahrenheit** (°F). On this scale, water freezes at 32°F and boils at 212°F. In other countries, temperature is expressed in **degrees Celsius** (°C); water freezes at 0°C and boils at 100°C. The size of a Fahrenheit degree is smaller than a Celsius degree, and the conversion between these two scales is given by the formula

$$T\ (°F) = \frac{9}{5}T\ (°C) + 32$$

The Celsius scale is sometimes used in scientific work, but it's giving way to the **absolute temperature scale**, in which temperatures are expressed in **kelvins** (K). On the Kelvin scale, water freezes at 273.15 K and boils at 373.15 K. Notice that the degree sign is *not* used for absolute temperature. The Kelvin scale assigns a value of 0 K to the lowest theoretically possible temperature and a value of 273.16 K to the **triple point of water** (the temperature at which the three phases of water—liquid water, ice, and vapor—can coexist when the pressure is one atmosphere). A kelvin is equal in size to a Celsius degree, and the conversion between kelvins and degrees Celsius is

$$T\text{ (K)} = T\text{ (°C)} + 273.15$$

For most purposes, you can ignore the .15 and use the simpler conversion equation $T = T\text{ (°C)} + 273$.

Example 9.1 Room temperature is 68°F. What's this temperature in kelvins?

Solution. Let's convert this to Celsius. Since $T\text{ (°F)} = \frac{9}{5}T\text{ (°C)} + 32$, it's also true that $T\text{ (°C)} = \frac{5}{9}(T\text{ (°F)} - 32)$, so 68°F is equal to $\frac{5}{9}(68 - 32) = 20$°C. Converting this to kelvins, we add 273; room temperature is 293 K.

PHYSICAL CHANGES DUE TO HEAT TRANSFER

When a substance absorbs or gives off heat, one of two things can happen:

(1) the temperature of the substance can change

(2) the substance can undergo a phase change

There are three phases of matter: **solid**, **liquid**, and **vapor** (or **gas**). When a solid **melts** (or **liquefies**), it becomes a liquid; the reverse process occurs when a liquid **freezes** (or **solidifies**) to become a solid. A liquid can **evaporate** (to become vapor), and vapor can **condense** to become liquid. These are the most common phase changes, but others exist: A solid can **sublimate**, going directly to vapor form, and vapor can experience **deposition**, going directly to solid.

Since either change (1) or change (2)—but *not both*—takes place upon heat transfer, let's study these changes separately.

CALORIMETRY I: HEAT TRANSFER AND TEMPERATURE CHANGE

The change in temperature that a substance experiences upon a transfer of heat depends upon two things: the identity and the amount of the substance present. For example, we could transfer 200 J of heat to a gold nugget and a piece of wood of equal mass and, even though they were infused with the same amount of thermal energy, the temperature of the gold would rise much more than the temperature of the wood. Also, if this heat were transferred to two nuggets of gold of unequal mass, the temperature of the smaller nugget would rise more than that of the bigger one. The equation that connects the amount of heat, Q, and the resulting temperature change, ΔT, is

$$Q = mc\Delta T$$

where m is the mass of the sample and c is an intrinsic property of the substance called its **specific heat**. Notice that positive Q is interpreted as heat coming *in* (ΔT is positive, so T increases), while negative Q corresponds to heat going *out* (ΔT is negative, so T decreases).

CALORIMETRY II: HEAT TRANSFER AND PHASE CHANGES

Consider an ice cube. Since water freezes at 0°C, the temperature of the ice cube is 0°C. If we add heat to the ice, its temperature does not rise. Instead, the thermal energy absorbed by the ice goes into loosening the intermolecular bonds of the solid, thereby transforming it into liquid. The temperature remains at 0°C throughout this melting process. Once the ice cube is completely melted, additional heat will increase the temperature of the liquid, until it reaches its boiling point temperature (100°C). At this point, additional heat does not increase the temperature; it breaks the intermolecular bonds, transforming it into steam. In each of the phase changes (solid to liquid, liquid to vapor), absorbed heat causes no temperature change, so $Q = mc\Delta T$ does not apply at these transitions. The equation that *does* apply is

$$Q = mL$$

where L is the **latent heat of transformation**. This equation tells us how much heat must be transferred in order to cause a sample of mass m to completely undergo a phase change. In the case of a solid to liquid (or vice versa) phase change, L is called the **latent heat of fusion**. For a phase change between liquid and vapor, L is called the **latent heat of vaporization**.

THERMAL EXPANSION

When a substance undergoes a temperature change, it changes in size. Steel beams that form railroad tracks or bridges expand when they get warmer; a balloon filled with air shrinks when it's placed in a freezer. The change in size of a substance due to a temperature change depends on the amount of the temperature change and the identity of the substance.

Let's first consider changes in length (of the steel beam, for example). When its temperature is T_i, its length is ℓ_i. Then, if its temperature changes to T_f, the length changes to ℓ_f, such that

$$\ell_f - \ell_i = \alpha L_i(T_f - T_i)$$

where α is the **coefficient of linear expansion** of the material. This equation is usually used in the simpler form

$$\Delta \ell = \alpha \ell_o \Delta T$$

Nearly all substances have a positive value of α, which means they expand upon heating.

Example 9.2 A brass rod 5 m long and 0.01 m in diameter increases in length by 0.05 m when its temperature is increased by 500°C. A similar brass rod of length 10 m has a diameter of 0.02 m. By how much will this rod's diameter increase if its temperature is increased by 1000°C?

Solution. First, let's use the equation $\Delta \ell = \alpha \ell_i \Delta T$ to determine α, the coefficient of linear expansion of the brass rod

$$\alpha = \frac{\Delta \ell}{\ell_i \Delta T} = \frac{0.05\text{ m}}{(5\text{ m})(500°\text{C})} = \frac{2\times 10^{-5}}{°\text{C}}$$

Like length, the diameter is also a linear dimension, so we now use this value of α in the same equation to determine the increase in diameter of the second brass rod. With ℓ_i now denoting the initial diameter of the second brass rod, we find that

$$\Delta\ell = \alpha\ell_i\Delta T = \frac{2\times10^{-5}}{^\circ\mathrm{C}}(0.02\ \mathrm{m})(1{,}000^\circ\mathrm{C}) = 4\times10^{-4}\ \mathrm{m}$$

As we've mentioned, substances also undergo volume changes when heat is lost or absorbed. The change in volume, ΔV, corresponding to a temperature change ΔT is given by the equation

$$\Delta V = \beta V_i \Delta T$$

where V_i is the sample's initial volume and β is the **coefficient of volume expansion** of the substance. Since we're now looking at the change in a three-dimensional quantity (volume) rather than a one-dimensional quantity (length), for most solids, $\beta \approx 3\alpha$. Nearly all substances have a positive value of β, which means that they expand upon heating. An extremely important example of a substance with a *negative* value of β is liquid water between 0°C and 4°C. Unlike the vast majority of substances, liquid water *expands* as it nears its freezing point and solidifies (which is why ice has a lower density and floats in water).

HEAT TRANSFER

Heat conducts from one point to another only if there is a temperature difference between the two objects. The rate at which heat is transferred is given by $H = \frac{Q}{t}$ or $H = \frac{kA\Delta T}{L}$ where k is the thermal conductivity (a property of the material), A is the cross-sectional area, ΔT is the temperature difference between the two sides and L is the thickness or distance between the two ends of the material.

Example 9.3 An aluminum rod (ρ = 2.7 × 10³ kg/m³) has a radius of 0.01 m and an initial length of 2 m at a temperature of 20°C. Heat is added to raise its temperature to 90°C. Its coefficient of linear expansion is α = 25 × 10⁻⁶/°C, the specific heat is cal = 900 J/kg°C, and a thermal conductivity of k = 200 J/s·m°C.

(a) What is the mass of the aluminum rod?
(b) What is the amount of heat added to the rod?
(c) What is the new length of the rod?
(d) If we were to use this rod to transfer heat between two objects one side being at 20°C and the other side at 90°C, what would the rate of heat transfer be?

Solution.

(a) Because $\rho = m/V$ we know $m = \rho V$. The term rod implies a long cylinder and the volume of a cylinder is given by $V = \pi r^2 h$. Therefore:

$$m = \rho\pi r^2 h.$$

This becomes

$$m = (2.7 \times 10^3\ \text{kg/m}^3)\ (3.14) \times (0.01\ \text{m})^2\ (2\ \text{m})$$

or

$$m = 1.7\text{kg}$$

(b) Given $\Delta Q = mc\Delta T$ we can now substitute in

$$\Delta Q = (1.7\ \text{kg})\ (900\ \text{J/kg°C})\ (90°\text{C} - 20°\text{C})$$

$$= 107{,}100\ \text{J}$$

(c) The new length of the rod can be given by

$$\Delta\ell = \alpha\ell_0\Delta T$$

$$= (25 \times 10^{-6}/°\text{C})(2\ \text{m})(90°\text{C} - 20°\text{C})$$

$$= 0.0035\ \text{m or } 3.5\ \text{mm}$$

Therefore the new length is given by

$$\ell = \ell_0 + \Delta\ell$$

Or $\ell = 2.0035$ m

(d) We know that $H = \dfrac{kA\Delta T}{L}$ and the cross-sectional area of a cylinder is $A = \pi r^2$ so

$$H = \frac{(200\ \text{J/s m °C})(3.14)(0.01\ \text{m})^2(90°\text{C} - 20°\text{C})}{2\ \text{m}} = 2.2\ \text{J/s}$$

THE KINETIC THEORY OF GASES

Unlike the condensed phases of matter—solid and liquid—the atoms or molecules that make up a gas do not move around relatively fixed positions. Rather, the molecules of a gas move freely and rapidly, in a chaotic swarm.

A confined gas exerts a force on the walls of its container, because the molecules are zipping around inside the container, striking the walls and rebounding. The magnitude of the force per unit area is called **pressure**, and is denoted by P

$$P = \frac{F}{A}$$

The SI unit for pressure is the N/m^2, the **pascal** (abbreviated Pa). As we'll see, the faster the gas molecules are moving, the more pressure they exert.

We also need a way to talk about the typically vast numbers of molecules in a given sample of gas. We will say $N = nN_A$ where N is the number of molecules, n is a **mole**, and

$$N_A = 6.022 \times 10^{23} \text{ molecules/mole}$$

The number N_A is known as **Avogadro's constant**, and one mole of any substance has a mass in grams that defines its atomic or molecular mass (these values are given in the Periodic Table of Elements). For example, the mass of a carbon-12 atom (the most abundant isotope of carbon) has a mass of exactly 12 atomic mass units, and a mole of these atoms has a mass of 12 grams. Oxygen has an atomic mass of 16 g, so a mole of carbon dioxide (CO_2), each molecule of which is composed of one carbon atom and two oxygen atoms, has a mass of 12 g + 2(16 g) = 44 g.

THE IDEAL GAS LAW

Three physical properties—pressure (P), volume (V), and temperature (T)—describe a gas. At low densities, all gases approach *ideal* behavior; this means that these three variables are related by the equation

$$PV = nRT$$

where n is the number of moles of gas and R is a constant (8.31 J/mol·K) called the **universal gas constant**. This equation is known as the **Ideal Gas Law**. It can also be written as $PV = Nk_BT$, where k_B is Boltzmann's constant ($K_B = 1.38 \times 10^{-23}$ J/K).

An important consequence of this equation is that, for a fixed volume of gas, an increase in P gives a proportional increase in T. The pressure increases when the gas molecules strike the walls of their container with more force, which occurs if they move more rapidly. Using Newton's Second Law (*rate of change of momentum = force*) we can find that the pressure exerted by N molecules of gas in a container of volume V is related to the average kinetic energy of the molecules by the equation $PV = NK_{avg}$. Comparing this to the Ideal Gas Law we see that $\frac{2}{3}NK_{avg} = nRT$. We can rewrite this equation in the form $\frac{2}{3}N_AK_{avg} = RT$, since, by definition, $N = nN_A$. The ratio R/N_A is a fundamental constant of nature called **Boltzmann's constant** ($k_B = 1.38 \times 10^{-23}$ J/K), so our equation becomes

$$K_{avg} = \frac{3}{2}k_BT$$

This tells us that the average translational kinetic energy of the gas molecules is directly proportional to the absolute temperature of the sample. Remember, this means you must use kelvins as your temperature unit.

Since the average kinetic energy of the gas molecules is $K_{avg} = \frac{1}{2}\mu(v^2)_{avg}$, where μ is the mass of each molecule, the equation above becomes $\frac{1}{2}\mu(v^2)_{avg} = \frac{3}{2}k_B T$, so

$$\sqrt{(v^2)_{avg}} = \sqrt{\frac{3k_B T}{\mu}}$$

The quantity on the left-hand side of this equation, the square root of the average of the square of v, is called the **root-mean-square speed**, v_{rms}, so

$$v_{rms} = \sqrt{\frac{3k_B T}{\mu}}$$

Because $k_B = R/N_A$ and $\mu N_A = M$ (the mass of one mole of the molecules—the **molar mass**), the equation for v_{rms} is also commonly written in the form

$$v_{rms} = \sqrt{\frac{3RT}{M}}$$

Note that these last two equations can only determine v_{rms}. It's important to realize that the molecules in the container have a wide range of speeds; some are much slower and others are much faster than v_{rms}. The root-mean-square speed is important because it gives us a type of average speed that's easy to calculate from the temperature of the gas.

Example 9.4 In order for the rms speed of the molecules in a given sample of gas to double, what must happen to the temperature?

Solution. Temperature is a measure of the average kinetic energy. The velocity is determined from the following equation:

$$v_{rms} = \sqrt{\frac{3RT}{M}}$$

Since v_{rms} is proportional to the square root of T, the temperature must quadruple, again, assuming the temperature is given in kelvins.

Example 9.5 Standard atmospheric pressure is 1.013×10^5 Pa (this is one **atmosphere**). At 0°C = 273.15 K, one mole of any ideal gas occupies the same volume. What is this volume?

Solution. Use the Ideal Gas Law:

$$V = \frac{nRT}{P} = \frac{(1.00 \text{ mol})(8.31 \text{ J/mol}\cdot\text{K})(273.15 \text{ K})}{1.013\times10^{5} \text{ Pa}} = 22.4\times10^{-3} \text{ m}^3$$

The quantity 10^{-3} m^3 is the definition of a **liter** (L), so the volume occupied by any ideal gas at **STP** (**s**tandard **t**emperature and **p**ressure, 0°C and 1 atm) is 22.4 L.

> **Example 9.6** A 0.1-mole sample of gas is confined to a jar whose volume is 4.0 L. To what temperature must the gas be raised to cause the pressure of the gas to increase to 2 atm?

Solution. Use the Ideal Gas Law:

$$T = \frac{PV}{nR} = \frac{[2(1.013\times10^{5} \text{ Pa})](4.0\times10^{-3} \text{ m}^3)}{(0.1 \text{ mol})(8.31 \text{ J/mol}\cdot\text{K})} = 975 \text{ K} = 702° \text{ C}$$

> **Example 9.7** A cylindrical container of radius 15 cm and height 30 cm contains 0.6 mole of gas at 433 K. How much force does the confined gas exert on the lid of the container?

Solution. The volume of the cylinder is $\pi r^2 h$, where r is the radius and h is the height. Since we know V and T, we can use the Ideal Gas Law to find P. Because pressure is force per unit area, we can find the force on the lid by multiplying the gas pressure times the area of the lid.

$$P = \frac{nRT}{V} = \frac{(0.6 \text{ mol})(8.31 \text{ J/mol}\cdot\text{K})(433 \text{ K})}{\pi(0.15 \text{ m})^2(0.30 \text{ m})} = 1.018\times10^{5} \text{ Pa}$$

So, since the area of the lid is πr^2, the force exerted by the confined gas on the lid is:

$$F = PA = (1.018\times10^{5} \text{ Pa})\cdot\pi(0.15 \text{ m})^2 = 7200 \text{ N}$$

This is about 1600 pounds of force, which seems like a lot. Why doesn't this pressure pop the lid off? Because, while the bottom of the lid is feeling a pressure (due to the confined gas) of 1.018×10^5 Pa that exerts a force upward, the top of the lid feels a pressure of 1.013×10^5 Pa (due to the atmosphere) that exerts a force downward. The net force on the lid is

$$F_{net} = (\Delta P)A = (0.005\times10^{5} \text{ Pa})\cdot\pi(0.15 \text{ m})^2 = 35 \text{ N}$$

which is only about 8 pounds.

THE LAWS OF THERMODYNAMICS

We've learned about two ways in which energy may be transferred between a system and its environment. One is work, which takes place when a force acts over a distance. The other is heat, which takes place when energy is transferred due to a difference in temperature. The study of the energy transfers involving work and heat, and the resulting changes in internal energy, temperature, volume, and pressure is called **thermodynamics**.

THE ZEROTH LAW OF THERMODYNAMICS

When two objects are brought into contact, heat will flow from the warmer object to the cooler one until they reach thermal equilibrium. This property of temperature is expressed by the Zeroth Law of Thermodynamics: *If Objects 1 and 2 are each in thermal equilibrium with Object 3, then Objects 1 and 2 are in thermal equilibrium with each other.*

> **Example 9.8** What is the final temperature of a mixture of 200 g of water that has an initial temperature of 20°C and 800 g of water that is initially at 60°C? Assume no heat lost to the environment.

Solution. For a closed system, $\Sigma Q = 0$. This means

$$Q_{\text{w1 gains}} + Q_{\text{w2 loses}} = 0$$

$$m_{w1}c_w\Delta T_{w2} + m_{w2}c_w\Delta T_{w2} = 0$$

The specific heat of water is the same for both samples so this can be eliminated

$$m_w(T_f - T_{iw1}) + m_{w2}(T_f - T_{iw2}) = 0$$

$$m_{w1}T_f - m_{w1}T_{iw1} + m_{w2}T_f - m_{w2}T_{iw2} = 0$$

$$m_{w1}T_f + m_{w2}T_f = m_{w1}T_{iw1} + m_{w2}T_{iw2}$$

$$(m_{w1} + m_{w2})T_f = m_{w1}T_{iw1} + m_{w2}T_{iw2}$$

$$T_f = \left(\frac{m_{w1}T_{iw1} + m_{w2}T_{iw2}}{m_{w1} + m_{w2}}\right)$$

$$T_f = \left(\frac{(200)(20)+(800)(60)}{(200)+(800)}\right)$$

$$= 52\ ^\circ\text{C}$$

> **Example 9.9** 500 g of water has an initial temperature of 20°C. 200 g of steel (c = 0.11 J/g°C) is initially at 80°C and is placed in the water. Assuming no heat is lost to the environment, what is the final temperature of the mixture?

Solution. For a closed system, $\Sigma Q = 0$. This means

$$Q_{\text{water gains}} + Q_{\text{steel loses}} = 0$$

$$m_w c_w \Delta T_w + m_s c_s \Delta T_s = 0$$

$$m_w c_w (T_f - T_{iw}) + m_s c_s (T_f - T_{is}) = 0$$

$$m_w c_w T_f - m_w c_w T_{iw} + m_s c_s T_f - m_s c_s T_{is} = 0$$

$$m_w c_w T_f + m_s c_s T_f = m_w c_w T_{iw} + m_s c_s T_{is}$$

$$(m_w c_w + m_s c_s) T_f = m_w c_w T_{iw} + m_s c_s T_{is}$$

$$T_f = \left(\frac{m_w c_w T_{iw} + m_s c_s T_{is}}{m_w c_w + m_s c_s} \right)$$

$$T_f = \left(\frac{(500)(1)(20) + (200)(0.11)(80)}{(500)(1) + (200)(0.11)} \right)$$

$$T_f = 22.5\ ^\circ\text{C}$$

The First Law of Thermodynamics

Simply put, the First Law of Thermodynamics is a statement of the Conservation of Energy that includes heat. Consider the following example, which is the prototype that's studied extensively in thermodynamics.

An insulated container filled with an ideal gas rests on a heat reservoir (that is, something that can act as a heat source or a heat sink). The container is fitted with a snug, but frictionless, weighted piston that can be raised or lowered. The confined gas is the *system*, and the piston and heat reservoir are the *surroundings*.

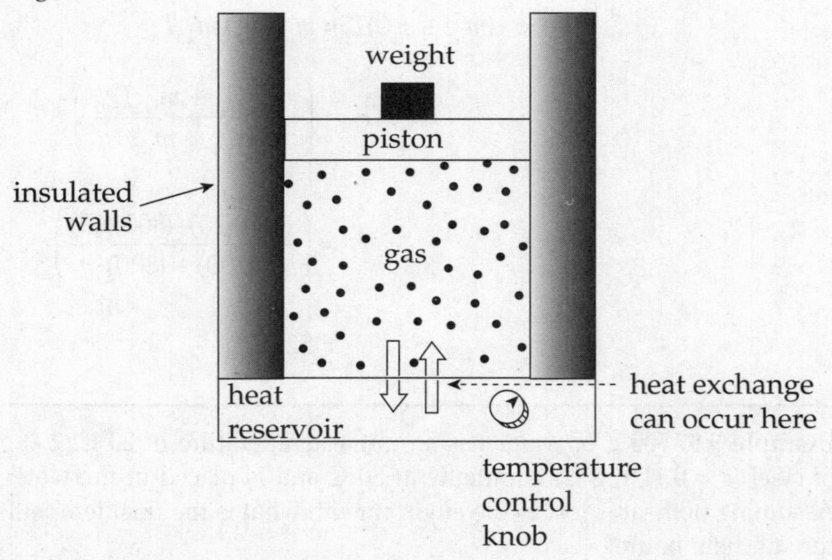

The **state** of the gas is given once its pressure, volume, and temperature are known, and the equation that connects these state variables is the Ideal Gas Law, $PV = nRT$. We'll imagine performing different experiments with the gas, such as heating it or allowing it to cool, increasing or decreasing the weight on the piston, etc., and study the energy transfers (work and heat) and the changes in the state variables. If each process is carried out such that, at each moment, the system and its surroundings are in thermal equilibrium, we can plot the pressure (P) vs. the volume (V) on a diagram. By following the path of this ***P-V* diagram**, we can study how the system is affected as it moves from one state to another.

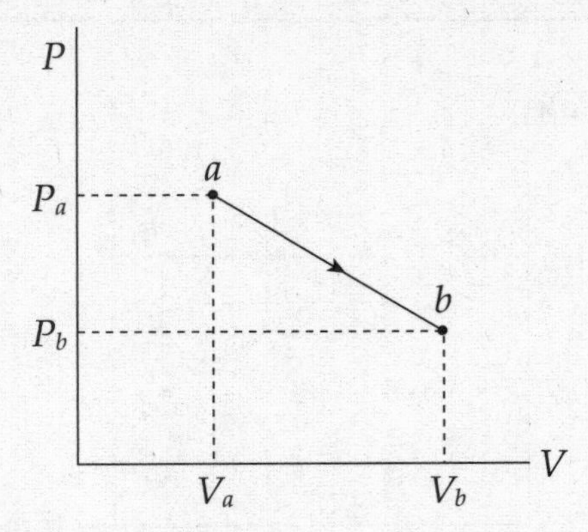

Work is done on or by the system when the piston is moved and the volume of the gas changes. For example, imagine that the weight pushes the piston downward a distance d, causing a decrease in volume. Assume that the pressure stays constant at P. (Heat must be removed via the reservoir to accomplish this.) We can calculate the work done on the gas during this compression as $W = -Fd$, but since $F = PA$, we have $W = -PAd$, and because $Ad = \Delta V$, we have

$$W = -P\Delta V$$

Textbooks differ about the circumstances under which work in thermodynamics is defined to be positive or negative. The negative signs we have included in the equations above are consistent with those used in the AP Physics Exam (appearing in the equation sheet given for Section II). For the exam, work in thermodynamics is considered to be positive when the work is being done *on the system.* This means that the volume of the system is *decreasing,* ΔV is *negative*, and, in agreement with intuition, energy is being *added* to the system. In other words, when work is done in compressing a system, ΔV is *negative* and the work done, $W = -P\Delta V$, is *positive*. This also means, conversely, that when the system is doing work *on the surroundings* (volume *increasing*, ΔV *positive*), the work is *negative*, in agreement with intuition that energy is leaving the system.

The equation $W = -P\Delta V$ assumes that the pressure P does not change during the process. If P *does* change, then the work is equal to the area under the curve in the P–V diagram; moving left to right gives a negative area (and negative work), while moving right to left gives a positive area (and positive work).

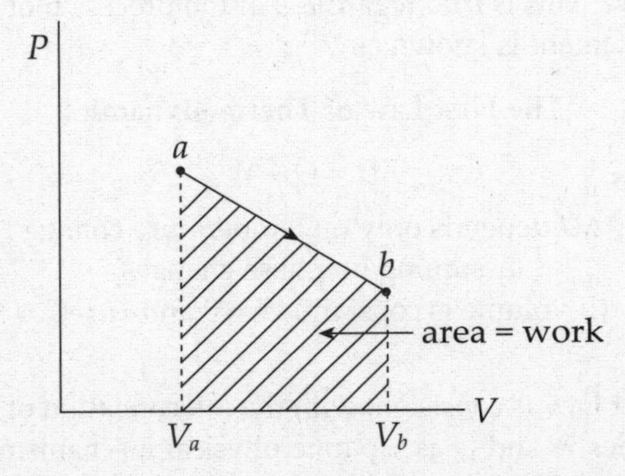

Example 9.10 What's the value of W for the process ab following path 1 and for the same process following path 2 (from a to d to b), shown in the P–V diagram below?

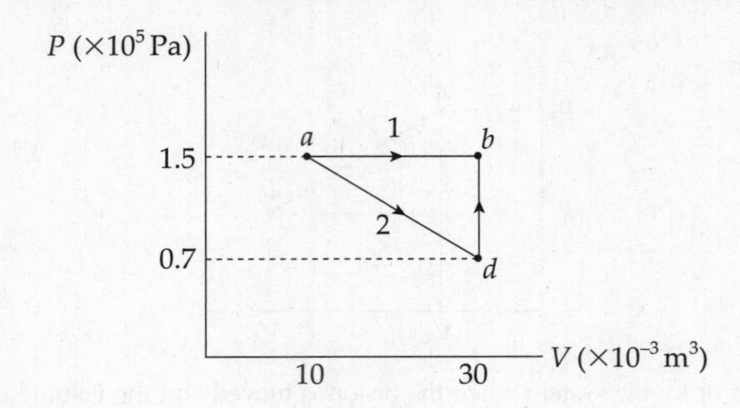

Solution.

Path 1. Since, in path 1, P remains constant, the work done is just $-P\,\Delta V$:

$$W = -P\Delta V = -(1.5 \times 10^5 \text{ Pa})[(30 \times 10^{-3} \text{ m}^3) - (10 \times 10^{-3} \text{ m}^3)] = -3{,}000 \text{ J}$$

Path 2. If the gas is brought from state a to state b, along path 2, then work is done only along the part from a to d. From d to b, the volume of the gas does not change, so no work can be performed. The area under the graph from a to d is

$$\begin{aligned} W &= -\frac{1}{2}h(b_1 + b_2) = -\frac{1}{2}(\Delta V)(P_a + P_d) \\ &= -\frac{1}{2}(20\times10^{-3}\ \text{m}^3)[(1.5\times10^5\ \text{Pa}) + (0.7\times10^5\ \text{Pa})] \\ &= -2{,}200 \text{ J} \end{aligned}$$

As this example shows, the value of W depends not only on the initial and final states of the system, but also on the path between the two. In general, different paths give different values for W.

Experiments have shown that the value of $Q + W$ is *not* path dependent; it depends only on the initial and the final state of the system, so it describes a change in a fundamental property. This property is called the system's **internal energy**, denoted U, and the change in the system's internal energy, ΔU, is equal to $Q + W$. This is true regardless of the process that brought the system from its initial to final state. This statement is known as

The First Law of Thermodynamics

$$\Delta U = Q + W$$

ΔU depends only on temperature change
(assuming no phase changes).
(If volume is constant, $W = 0$ and $U = Q = nC_v\Delta T$.)

This statement of the First Law is consistent with the interpretation of work ($W = -P\Delta V$) explained above. The First Law identifies W and Q as separate physical mechanisms for adding to or removing energy from the system, and the signs of both W and Q are defined consistently: Both are positive when they are adding energy to the system and negative when they are removing energy from the system.

Example 9.11 A 0.5 mol sample of an ideal gas is brought from state *a* to state *b* when 7500 J of heat is added along the path shown in the following *P–V* diagram:

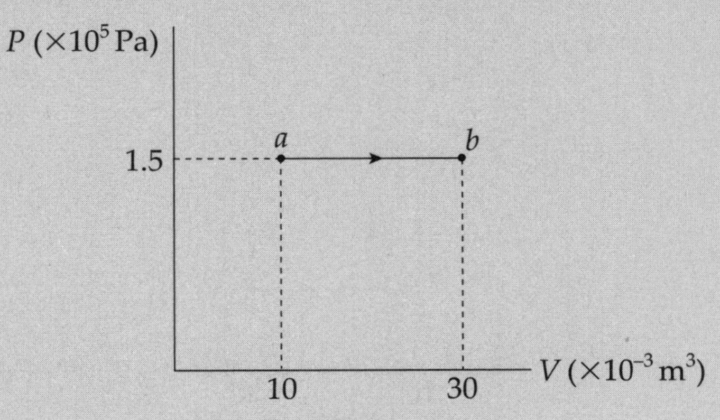

What are the values of each of the following?

(a) the temperature at *a*

(b) the temperature at *b*

(c) the work done by the gas during process *ab*

(d) the change in the internal energy of the gas

Solution.

(a, b) Both of these questions can be answered using the Ideal Gas Law, $T = PV/(nR)$:

$$T_a = \frac{P_a V_a}{nR} = \frac{(1.5 \times 10^5 \text{ Pa})(10 \times 10^{-3} \text{ m}^3)}{(0.5 \text{ mol})(8.31 \text{ J/mol} \cdot \text{K})} = 360 \text{ K}$$

$$T_b = \frac{P_b V_b}{nR} = \frac{(1.5 \times 10^5 \text{ Pa})(30 \times 10^{-3} \text{ m}^3)}{(0.5 \text{ mol})(8.31 \text{ J/mol} \cdot \text{K})} = 1{,}080 \text{ K}$$

(c) Since the pressure remains constant during the process, we can use the equation $W = -P\Delta V$. Because $\Delta V = (30 - 10) \times 10^{-3}$ m³ $= 20 \times 10^{-3}$ m³, we find that

$$W = -P\Delta V = (1.5 \times 10^5 \text{ Pa})(20 \times 10^{-3} \text{ m}^3) = -3000 \text{ J}$$

The expanding gas did negative work against its surroundings, pushing the piston upward. [Important note: If the pressure remains constant (which is designated by a horizontal line in the *P–V* diagram), the process is called **isobaric**.]

(d) By the First Law of Thermodynamics,

$$\Delta U = Q + W = 7{,}500 - 3{,}000 \text{ J} = 4{,}500 \text{ J}$$

Example 9.12 A 0.5 mol sample of an ideal monatomic gas is brought from state *a* to state *b* along the path shown in the following *P*–*V* diagram:

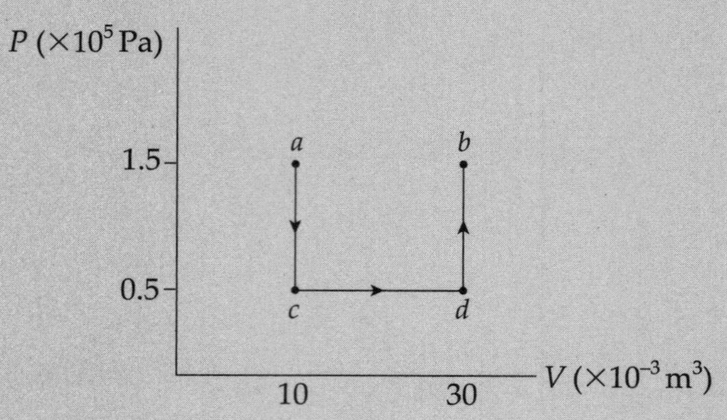

What are the values of each of the following?

(a) the work done by the gas during process *ab*
(b) the change in the internal energy of the gas
(c) the heat added to the gas during process *ab*

Solution. Note that the initial and final states of the gas are the same as in the preceding example, but the path is different.

(a) Let's break the path into 3 pieces:

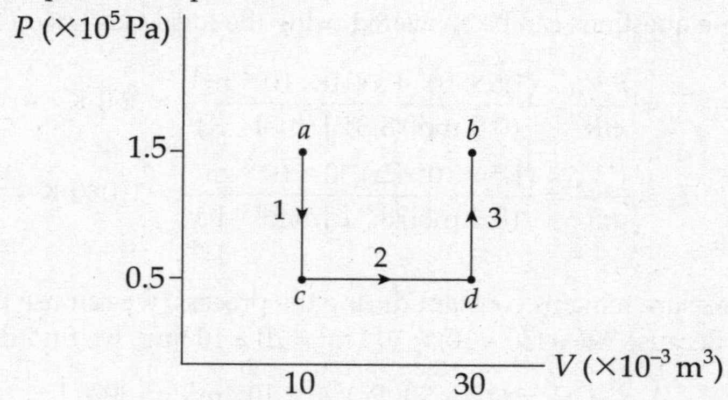

Over paths 1 and 3, the volume does not change, so no work is done. Work is done only over path 2:

$$W = -P\Delta V = -(0.5 \times 10^5 \text{ Pa})(20 \times 10^{-3} \text{ m}^3) = -1{,}000 \text{ J}$$

Once again, the expanding gas does negative work against its surroundings, pushing the piston upward.

(b) Because the initial and final states of the gas are the same here as they were in the preceding example, the change in internal energy, ΔU, *must* be the same. Therefore, $\Delta U = 4500$ J.

(c) By the First Law of Thermodynamics, $\Delta U = Q + W$, so

$$Q = \Delta U - W = 4{,}500\text{J} - (-1{,}000\text{ J}) = 5{,}500\text{ J}$$

> **Example 9.13** As **isochoric** process is one that takes place with no change in volume. What can you say about the change in the internal energy of a gas if it undergoes an isochoric change of state?

Solution. An isochoric process is illustrated by a vertical line in a P–V diagram and, since no change in volume occurs, $W = 0$. By the First Law of Thermodynamics, $\Delta U = Q + W = Q$. Therefore, the change in internal energy is entirely due to (and equal to) the heat transferred. If heat is transferred into the system (positive Q), then ΔU is positive; if heat is transferred out of the system (negative Q), then ΔU is negative.

> **Example 9.14** A 0.5 mol sample of an ideal gas is brought from state a back to state a along the path shown in the following P–V diagram:

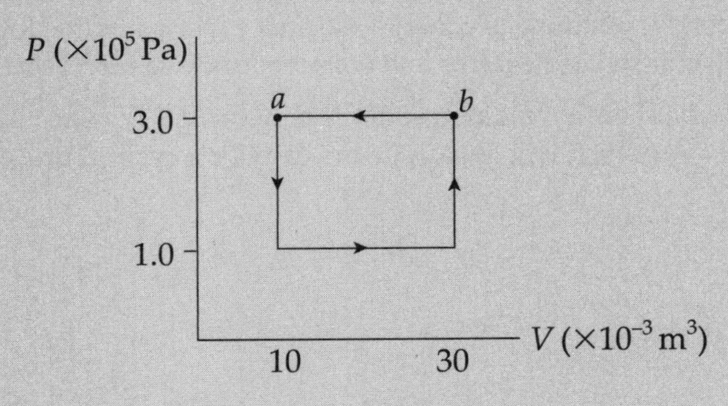

> What are the values of each of the following?
> (a) the change in the internal energy of the gas
> (b) the work done on the gas during the process
> (c) the heat added to the gas during the process

Solution. A process such as this, which begins and ends at the same state, is said to be **cyclical**.

(a) Because the final state is the same as the initial state, the internal energy of the system cannot have changed, so $\Delta U = 0$.

(b) The total work involved in the process is equal to the work done from c to d plus the work done from b to a,

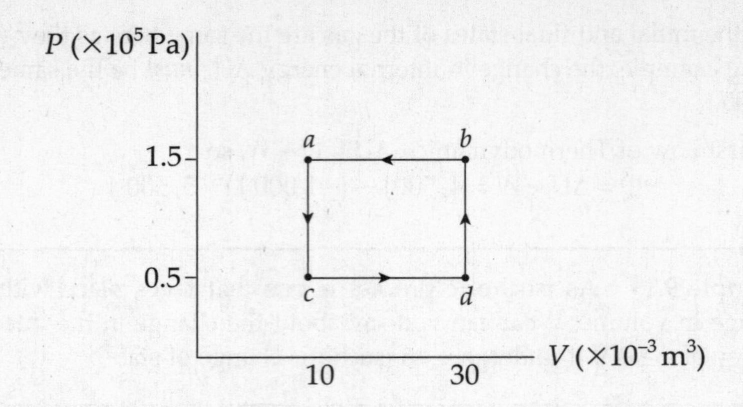

because only along these paths does the volume change. Along these portions, we find that

$$W_{cd} = -P\Delta V_{cd} = (0.5 \times 10^5 \text{ Pa})(+20 \times 10^{-3} \text{ m}^3) = -1{,}000 \text{ J}$$

$$W_{cd} = -P\Delta V_{ba} = (1.5 \times 10^5 \text{ Pa})(-20 \times 10^{-3} \text{ m}^3) = +3{,}000 \text{ J}$$

so the total work done is $W = +2{,}000$ J. The fact that W is positive means that, overall, work was done *on* the gas by the surroundings. Notice that for a cyclical process, the total work done is equal to the area enclosed by the loop, with clockwise travel taken as negative and counterclockwise travel taken as positive.

(c) The First Law of Thermodynamics states that $\Delta U = Q + W$. Since $\Delta U = 0$, it must be true that $Q = -W$ (which will always be the case for a cyclical process, so $Q = -2{,}000$ J.

Example 9.15 A 0.5 mol sample of an ideal gas is brought from state *a* to state *d* along an **isotherm**, then isobarically to state *c* and isochorically back to state *a*, as shown in the following *P–V* diagram:

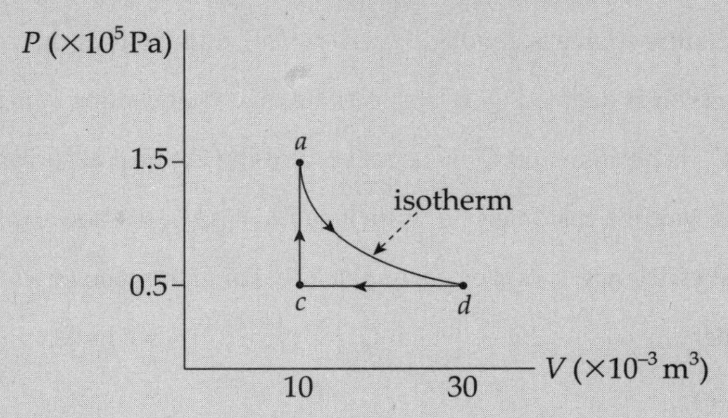

A process that takes place with no variation in temperature is said to be **isothermal**. Given that the work done during the isothermal part of the cycle is –1,650 J, how much heat is transferred during the isothermal process, from *a* to *d*?

Solution. Be careful that you don't confuse *isothermal* with *adiabatic*. A process is **isothermal** if the *temperature* remains constant; a process is **adiabatic** if $Q = 0$. You might ask, *How could a process be isothermal without also being adiabatic at the same time?* Remember that the temperature is determined by the internal energy of the gas, which is affected by changes in *Q*, *W*, or both. Therefore, it's possible for *U* to remain unchanged even if *Q* is not 0 (because there can be an equal but opposite *W* to cancel it out). In fact, this is the key to this problem. Since *T* doesn't change from *a* to *d*, neither can the internal energy, which depends entirely on *T*. Because $\Delta U_{ad} = 0$, it must be true that $Q_{ad} = -W_{ad}$. Since W_{ad} equals –1,650 J, Q_{ad} must be +1650 J. The gas absorbs heat from the reservoir and uses all this energy to do negative work as it expands, pushing the piston upward.

THE SECOND LAW OF THERMODYNAMICS

The form of the Second Law of Thermodynamics that we'll study for the AP Physics Exam deals with heat engines.

Converting work to heat is easy—rubbing your hands together in order to warm them up shows that work can be completely converted to heat. What we'll look at it is the reverse process: How efficiently can heat be converted into work? A device that uses heat to produce useful work is called a **heat engine**. The internal-combustion engine in a car is an example. In particular, we're only interested in engines that take their working substance (a mixture of air and fuel in this case) through a cyclic process, so that the cycle can be repeated. The basic components of any cyclic heat engine are simple: Energy in the form of heat comes into the engine from a high-temperature source, some of this energy is converted into useful work, the remainder is ejected as exhaust heat into a low-temperature sink, and the system returns to its original state to run through the cycle again.

Since we're looking at cyclic engines only, the system returns to its original state at the end of each cycle, so ΔU must be 0. Therefore, by the First Law of Thermodynamics, $Q_{net} = -W$. That is, the net heat absorbed by the system is equal to the work performed by the system. The heat that's absorbed from the high-temperature source is denoted Q_H (H for *hot*), and the heat that is discharged into the low-temperature reservoir is denoted Q_C (C for *cold*). Because heat coming *in* is positive and heat going *out* is negative, Q_H is positive and Q_C is negative, and the net heat absorbed is $Q_H + Q_C$. Instead of writing Q_{net} in this way, it's customary to write it as $Q_H - |Q_C|$, to show explicitly that Q_{net} is less than Q_H. The **thermal efficiency**, *e*, of the heat engine is equal to the ratio of what we get out to what we have to put in; that is, $e = \left|\frac{W}{Q_H}\right|$. Since $|W| = Q_{net} = Q_H - |Q_C|$, we have

$$e = \frac{Q_H - |Q_C|}{Q_H} = 1 - \frac{|Q_C|}{Q_H}$$

Notice that unless $Q_C = 0$, the engine's efficiency is *always* less than 1. This is one of the forms of

The Second Law of Thermodynamics

For any cyclic heat engine, some exhaust heat is always produced. Because $Q_C \neq 0$, no cyclic heat engine can operate at 100% efficiency; it's impossible to completely convert heat into useful work.

Example 9.16 A heat engine draws 800 J of heat from its high-temperature source and discards 450 J of exhaust heat into its cold-temperature reservoir during each cycle. How much work does this engine perform per cycle, and what is its thermal efficiency?

Solution. The absolute value of the work output per cycle is equal to the difference between the heat energy drawn in and the heat energy discarded:

$$|W| = Q_H - |Q_C| = 800 \text{ J} - 450 \text{ J} = 350 \text{ J}$$

The efficiency of this engine is

$$e = \left|\frac{W}{Q_H}\right| = \frac{350 \text{ J}}{800 \text{ J}} = 0.44 = 44\%$$

THE CARNOT CYCLE

The Second Law of Thermodynamics tells us that there are no perfect heat engines. But how can we construct, in principle, the best possible heat engine, meaning one with the maximum possible efficiency (that doesn't violate the Second Law)? Such an engine is called a **Carnot engine**, and the cycle through which its working substance (gas) is carried is called the **Carnot cycle**. It can be shown that a heat engine utilizing the Carnot cycle has the highest possible efficiency consistent with the Second Law of Thermodynamics. The cycle has four steps:

Step 1: Isothermal expansion

The system absorbs heat from the high-temperature source and expands isothermally. Because the temperature doesn't increase, all the heat drawn from the source goes into work performed by the expansion of the gas.

Step 2: Adiabatic expansion

The system is allowed to expand without exchanging heat with its surroundings. Because the system expands (that is, does negative work), its internal energy and temperature decrease because it receives no influx of heat from the surroundings.

Step 3: Isothermal compression

Work is done on the gas by the environment and, rather than increasing its internal energy, the system discards heat to the low-temperature reservoir.

Step 4: Adiabatic compression

The system is compressed to its initial state without exchanging heat with its surroundings. Because the system is compressed, its internal energy and temperature increase, because no heat is discarded.

These four steps can be illustrated on a *P–V* diagram as follows:

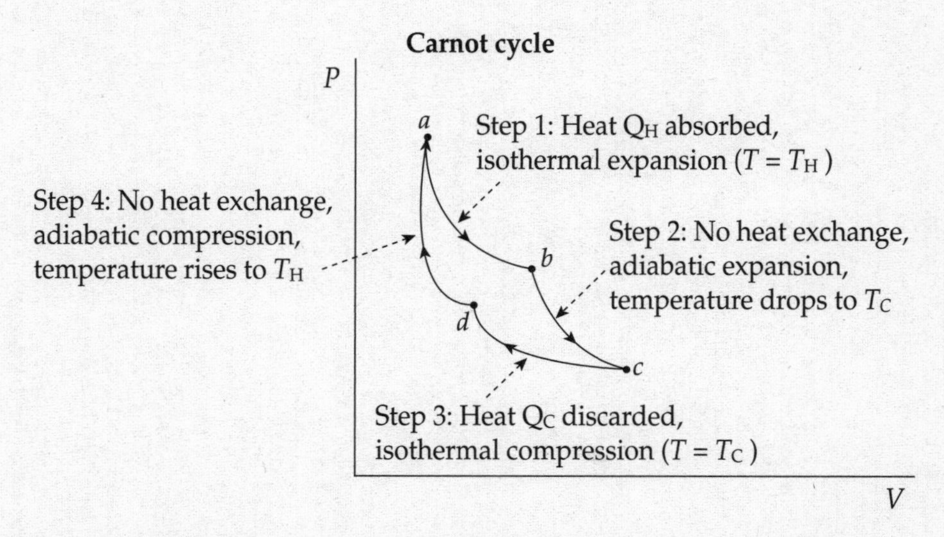

What is the efficiency of a Carnot engine? The answer is given by the equation

$$e_C = \frac{T_H - T_C}{T_H} = 1 - \frac{T_C}{T_H}$$

Note that the **Carnot efficiency** depends only on the absolute temperatures of the heat source and heat sink. The expression $1 - (T_C/T_H)$ gives the maximum theoretical efficiency of any cyclic heat engine. This cannot equal 1 unless $T_C = 0$, that is, unless the temperature of the cold reservoir is **absolute zero**. But the Third Law of Thermodynamics says, in part, that absolute zero can never be reached.

> **Example 9.17** An inventor proposes a design for a heat engine that operates between a heat source at 500°C and a cold reservoir at 25°C with an efficiency of 2/3. What's your reaction to the inventor's claim?

Solution. The highest possible efficiency for any heat engine is the Carnot efficiency, and a Carnot engine operating between the temperatures $T_H = 500°C = 773$ K and $T_C = 25°C = 298$ has an efficiency of

$$e_C = 1 - \frac{T_C}{T_H} = 1 - \frac{298\text{ K}}{773\text{ K}} = 0.61 = 61\%$$

The inventor's claim is that his engine has an efficiency of 2/3 = 67%, which is higher than e_C. Your reaction should therefore be one of extreme skepticism!

CHAPTER 9 REVIEW QUESTIONS

Solutions can be found in Chapter 18.

SECTION I: MULTIPLE CHOICE

1. A container holds a mixture of two gases, CO_2 and H_2, in thermal equilibrium. Let K_C and K_H denote the average kinetic energy of a CO_2 molecule and an H_2 molecule, respectively. Given that a molecule of CO_2 has 22 times the mass of a molecule of H_2, the ratio K_C/K_H is equal to

   (A) 1/22
   (B) $1/\sqrt{22}$
   (C) 1
   (D) $\sqrt{22}$
   (E) 22

2. If the temperature and volume of a sample of an ideal gas are both doubled, then the pressure

   (A) decreases by a factor of 4
   (B) decreases by a factor of 2
   (C) increases by a factor of 2
   (D) increases by a factor of 4
   (E) remains unchanged

3. In three separate experiments, a gas is transformed from state P_i, V_i to state P_f, V_f along the paths (1, 2, and 3) illustrated in the figure below:

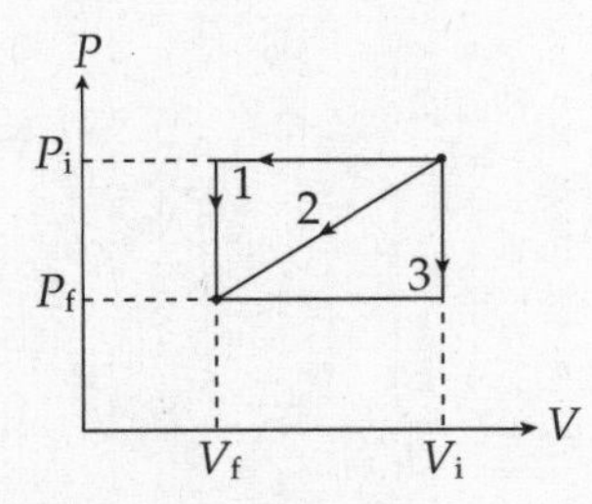

   The work done on the gas is

   (A) greatest for path 1
   (B) least for path 2
   (C) the same for paths 1 and 3
   (D) greatest for path 2
   (E) the same for all three paths

4. An ideal gas is compressed isothermally from 20 L to 10 L. During this process, 5 J of work is done to compress the gas. What is the change of internal energy for this gas?

   (A) –10 J
   (B) –5 J
   (C) 0 J
   (D) 5 J
   (E) 10 J

5. An ideal gas is confined to a container whose volume is fixed. If the container holds n moles of gas, by what factor will the pressure increase if the absolute temperature is increased by a factor of 2?

   (A) $2/(nR)$
   (B) 2
   (C) $2nR$
   (D) $2/n$
   (E) $2/R$

6. Two large glass containers of equal volume each hold 1 mole of gas. Container 1 is filled with hydrogen gas (2 g/mol), and Container 2 holds helium (4 g/mol). If the pressure of the gas in Container 1 equals the pressure of the gas in Container 2, which of the following is true?

   (A) The temperature of the gas in Container 1 is lower than the temperature of the gas in Container 2.
   (B) The temperature of the gas in Container 1 is greater than the temperature of the gas in Container 2.
   (C) The value of R for the gas in Container 1 is $\frac{1}{2}$ the value of R for the gas in Container 2.
   (D) The rms speed of the gas molecules in Container 1 is lower than the rms speed of the gas molecules in Container 2.
   (E) The rms speed of the gas molecules in Container 1 is greater than the rms speed of the gas molecules in Container 2.

7. Through a series of thermodynamic processes, the internal energy of a sample of confined gas is increased by 560 J. If the net amount of work done on the sample by its surroundings is 320 J, how much heat was transferred between the gas and its environment?

    (A) 240 J absorbed
    (B) 240 J dissipated
    (C) 880 J absorbed
    (D) 880 J dissipated
    (E) None of the above

8. What's the total work performed on the gas as it's transformed from state a to state c, along the path indicated?

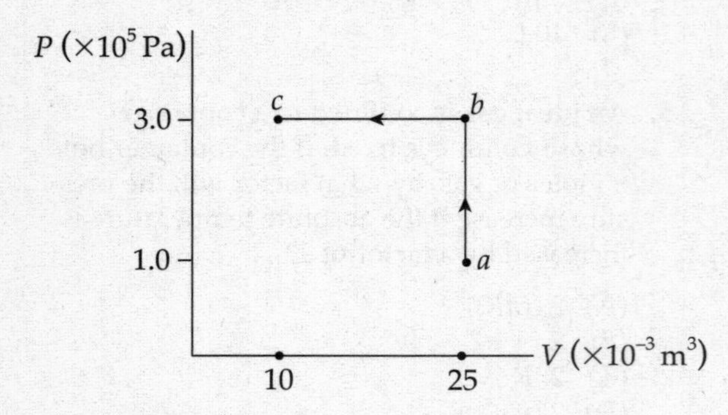

    (A) 1,500 J
    (B) 3,000 J
    (C) 4,500 J
    (D) 5,000 J
    (E) 9,500 J

9. In one of the steps of the Carnot cycle, the gas undergoes an isothermal expansion. Which of the following statements is true concerning this step?

    (A) No heat is exchanged between the gas and its surroundings, because the process is isothermal.
    (B) The temperature decreases because the gas expands.
    (C) This step violates the Second Law of Thermodynamics because all the heat absorbed is transformed into work.
    (D) The internal energy of the gas remains constant.
    (E) The internal energy of the gas decreases due to the expansion.

10. What's the maximum possible efficiency for a heat engine operating between heat reservoirs whose temperatures are 800°C and 200°C?

    (A) 25%
    (B) 33%
    (C) 50%
    (D) 56%
    (E) 75%

SECTION II: FREE RESPONSE

1. When a system is taken from state a to state b along the path acb shown in the figure below, 70 J of heat flows into the system, and the system does 30 J of work.

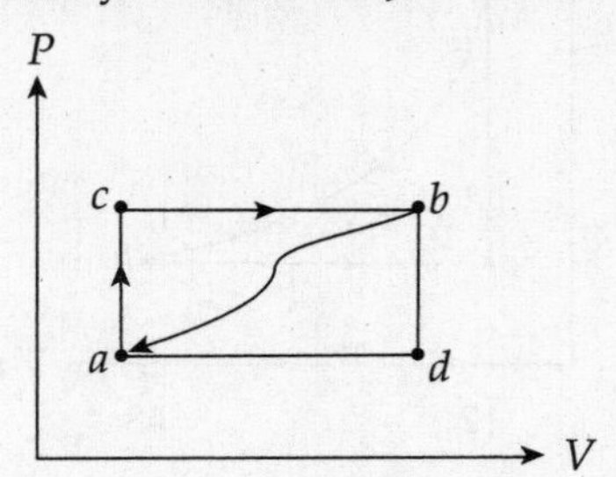

(a) When the system is returned from state b to state a along the curved path shown, 60 J of heat flows out of the system. Does the system perform work on its surroundings or do the surroundings perform work on the system? How much work is done?

(b) If the system does 10 J of work in transforming from state a to state b along path adb, does the system absorb or does it emit heat? How much heat is transferred?

(c) If $U_a = 0$ J and $U_d = 30$ J, determine the heat absorbed in the processes db and ad.

(d) For the process $adbca$, identify each of the following quantities as positive, negative, or zero:

W = ________________ Q = ________________ ΔU = ________________

2. A 0.4 mol sample of an ideal diatomic gas undergoes slow changes from state *a* to state *b* to state *c* and back to *a* along the cycle shown in the *P–V* diagram below:

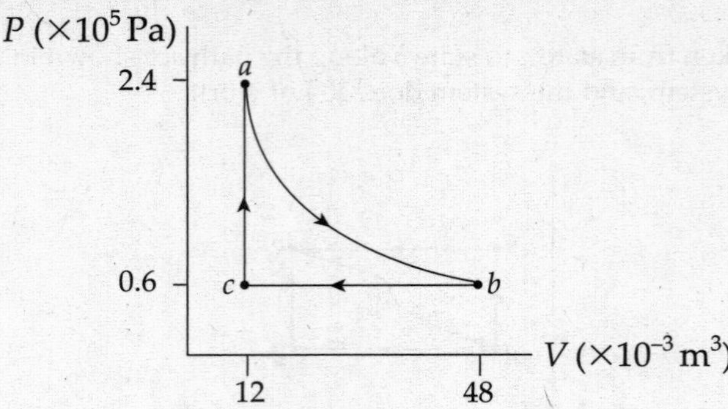

Path *ab* is an isotherm, and it can be shown that the work done by the gas as it changes isothermally from state *a* to state *b* is given by the equation

$$W_{ab} = -nRT \cdot \ln \frac{V_b}{V_a}$$

The molar heat capacities for the gas are $C_V = 20.8$ J/mol·K and $C_P = 29.1$ J/mol·K.

(a) What's the temperature of

(i) state *a*?

(ii) state *b*?

(iii) state *c*?

(b) Determine the change in the internal energy of the gas for

(i) step *ab*

(ii) step *bc*

(iii) step *ca*

(c) How much work, W_{ab}, is done by the gas during step *ab*?

(d) What is the total work done over cycle *abca*?

(e) (i) Is heat absorbed or discarded during step *ab*?

(ii) If so, how much?

(f) What is the maximum possible efficiency (without violating the Second Law of Thermodynamics) for a cyclical heat engine that operates between the temperatures of states *a* and *c*?

SUMMARY

- The amount of heat added to a substance and its corresponding change in temperature is related by the formula $\Delta Q = mc\Delta T$. In an isolated system $Q_{gained} + Q_{lost} = 0$
- For phase changes there is no temperature change and the equation that relates the amount of heat needed to change a certain mass to its new state is given by $Q = mL$
- For most objects as heat is added they will expand (the notable exception is water between 0 and 4°C). For solids the equations associated with this include

  $\Delta \ell = \alpha \ell_0 \Delta T$ for length changes

  $\Delta V = \beta V_0 \Delta T$ for volume changes

- The rate at which heat is transferred is given by $H = \frac{Q}{t}$ or $H = \frac{kA\Delta T}{L}$ where k is the thermal conductivity (a property of the material), A is the cross-sectional area, ΔT is the temperature difference between the two sides and L is the thickness or distance between the two ends of the material.
- For gases there are a few common equations to understand:

  Pressure is defined as the fore per unit area ($P = \frac{F}{A}$)

  The ideal gas law is expressed as either $PV = nRT$ or $PV = Nk_BT$

  The average kinetic energy of the gas molecules is given by $K_{ave} = \frac{3}{2}k_bT$

  The average speed of a molecule of the gas is given by $v_{rms} = \sqrt{\frac{3RT}{M}}$ or $v_{rms} = \sqrt{\frac{k_BT}{\mu}}$

- The work done on a gas is given by $W = -P\Delta V$ or can be found by the area under a pressure-vs.-volume graph.
- The First Law of Thermodynamics is $\Delta U = Q+W$ where ΔU depends only on the temperature change.
- The thermal efficiency is given by $e = \left|\frac{W}{Q_H}\right|$ or $e = \frac{T_H - |T_C|}{T_H}$ for a Carnot engine
- W positive (or negative) means energy is being added to (or subtracted from) the system by means of work done on the system by the surroundings (or on the surroundings by the system).

- Q positive (or negative) means energy is being added to (or subtracted from) the system by means of a flow of heat from the higher temperature surroundings (or system) to the lower temperature system (or surroundings).

  Note: Some textbooks define work in thermodynamics in a different way: Work is considered to be positive when work is done on the surroundings. This is consistent with the idea that the overall objective of a heat engine is to produce external (positive) work. Under this definition, the First Law must be written as $U = Q - W$ (or $U + W = Q$), and W must then be interpreted differently from Q. That is, while Q is still positive when heat is being added to the system, W is now positive when work is being done by the system on the surroundings (thus decreasing the internal energy of the system).

10

Electric Forces and Fields

ELECTRIC CHARGE

The basic components of atoms are protons, neutrons, and electrons. Protons and neutrons form the nucleus (and are referred to collectively as *nucleons*), while the electrons keep their distance, swarming around the nucleus. Most of an atom consists of empty space. In fact, if a nucleus were the size of the period at the end of this sentence, then the electrons would be 5 meters away. So what holds such an apparently tenuous structure together? One of the most powerful forces in nature: the *electromagnetic force*. Protons and electrons have a quality called **electric charge** that gives them an attractive force. Electric charge comes in two varieties; positive and negative. A positive particle always attracts a negative particle, and particles of the same charge always repel each other. Protons are positively charged, and electrons are negatively charged.

Protons and electrons are intrinsically charged, but bulk matter is not. This is because the amount of charge on a proton exactly balances the charge on an electron, which is quite remarkable in light of the fact that protons and electrons are very different particles. Since most atoms contain an equal number of protons and electrons, their overall electric charge is 0, because the negative charges cancel out the positive charges. Therefore, in order for matter to be **charged**, an imbalance between the numbers of protons and electrons must exist. This can be accomplished by either the removal or addition of electrons (that is, by the **ionization** of some of the object's atoms). If you remove electrons, then the object becomes positively charged, while if you add electrons, then it becomes negatively charged. Furthermore, charge is **conserved**. For example, if you rub a glass rod with a piece of silk, then the silk will acquire a negative charge and the glass will be left with an *equal* positive charge. *Net charge cannot be created or destroyed.* (*Charge* can be created or destroyed—it happens all the time—but *net* charge cannot.)

The magnitude of charge on an electron (and therefore on a proton) is denoted e. This stands for **elementary charge**, because it's the basic unit of electric charge. The charge of an ionized atom must be a whole number times e, because charge can be added or subtracted only in lumps of size e. For this reason we say that charge is **quantized**. To remind us of the quantized nature of electric charge, the charge of a particle (or object) is denoted by the letter q. In the SI system of units, charge is expressed in **coulombs** (abbreviated **C**). One coulomb is a tremendous amount of charge; the value of e is about 1.6×10^{-19} C.

COULOMB'S LAW

The electric force between two charged particles obeys a law that is very similar to that describing the gravitational force between two masses: they are both inverse-square laws. The **electric force** between two particles with charges of q_1 and q_2, separated by a distance r, is given by the equation

$$F_E = k\frac{q_1 q_2}{r^2}$$

This is **Coulomb's Law**. We interpret a negative F_E as an attraction between the charges and a positive F_E as a repulsion. The value of the proportionality constant, k, depends on the material between the charged particles. In empty space (vacuum)—or air, for all practical purposes—it is called **Coulomb's constant** and has the approximate value $k_0 = 9 \times 10^9\ \text{N} \cdot \text{m}^2/\text{C}^2$. For reasons that will become clear later in this chapter, k_0 is usually written in terms of a fundamental constant known as the **permittivity of free space**, denoted ε_0, whose numerical value is approximately $8.85 \times 10^{-12}\ \text{C}^2/\text{N}\cdot\text{m}^2$. The equation that gives k_0 in terms of ε_0 is:

$$k_0 = \frac{1}{4\pi\varepsilon_0}$$

Coulomb's Law for the force between two point charges is then written as

$$F_E = \frac{1}{4\pi\varepsilon_0}\frac{q_1 q_2}{r^2}$$

Recall that the value of the universal gravitational constant, G, is $6.67 \times 10^{-11}\ \text{N} \cdot \text{m}^2/\text{kg}^2$. The relative sizes of these fundamental constants show the relative strengths of the electric and gravitational forces. The value of k_0 is twenty orders of magnitude larger than G.

> **Example 10.1** Consider two small spheres, one carrying a charge of +1.5 nC and the other a charge of –2.0 nC, separated by a distance of 1.5 cm. Find the electric force between them. ("n" is the abbreviation for "nano," which means 10^{-9}.)

Solution. The electric force between the spheres is given by Coulomb's Law:

$$F_E = \frac{1}{4\pi\varepsilon_0}\frac{q_1 q_2}{r^2} = (9\times10^9\ \text{N}\cdot\text{m}^2/\text{C}^2)\frac{(1.5\times10^{-9}\ \text{C})(-2.0\times10^{-9}\ \text{C})}{(1.5\times10^{-2}\ \text{m})^2} = -1.2\times10^{-4}\ \text{N}$$

The fact that F_E is negative means that the force is one of *attraction*, which we naturally expect, since one charge is positive and the other is negative. The force between the spheres is along the line that joins the charges, as we've illustrated below. The two forces shown form an action/reaction pair.

$q_1 \oplus \longrightarrow \mathbf{F}_E \quad \mathbf{F}_E \longleftarrow \ominus q_2$

SUPERPOSITION

Consider three point charges: q_1, q_2, and q_3. The total electric force acting on, say, q_2 is simply the sum of $\mathbf{F}_{\text{1-on-2}}$, the electric force on q_2 due to q_1, and $\mathbf{F}_{\text{3-on-2}}$, the electric force on q_2 due to q_3:

$$\mathbf{F}_{\text{on 2}} = \mathbf{F}_{\text{1-on-2}} + \mathbf{F}_{\text{3-on-2}}$$

The fact that electric forces can be added in this way is known as **superposition**.

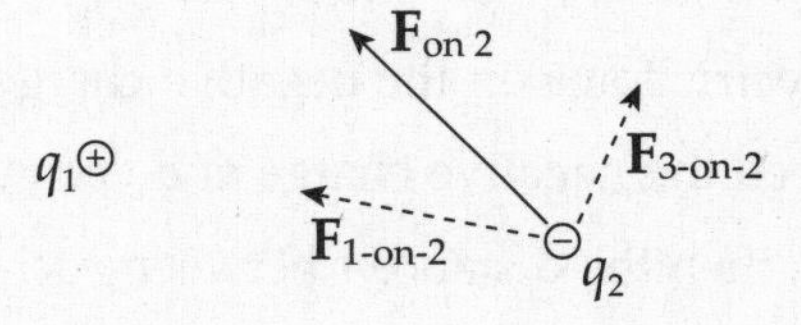

> **Example 10.2** Consider four equal, positive point charges that are situated at the vertices of a square. Find the net electric force on a negative point charge placed at the square's center.

Solution. Refer to the diagram on the next page. The attractive forces due to the two charges on each diagonal cancel out: $\mathbf{F}_1 + \mathbf{F}_3 = \mathbf{0}$, and $\mathbf{F}_2 + \mathbf{F}_4 = \mathbf{0}$, because the distances between the negative charge and the positive charges are all the same and the positive charges are all equivalent. Therefore, by symmetry, the net force on the center charge is zero.

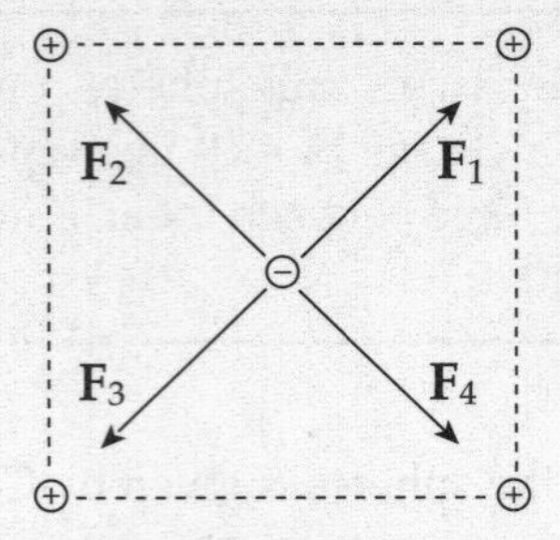

> **Example 10.3** If the two positive charges on the bottom side of the square in the previous example were removed, what would be the net electric force on the negative charge? Assume that each side of the square is 4.0 cm, each positive charge is 1.5 μC, and the negative charge is –6.2 nC. (μ is the symbol for "micro-," which equals 10^{-6}.)

Solution. If we break down $\mathbf{F}_1$ and $\mathbf{F}_2$ into horizontal and vertical components, then by symmetry the two horizontal components will cancel each other out, and the two vertical components will add:

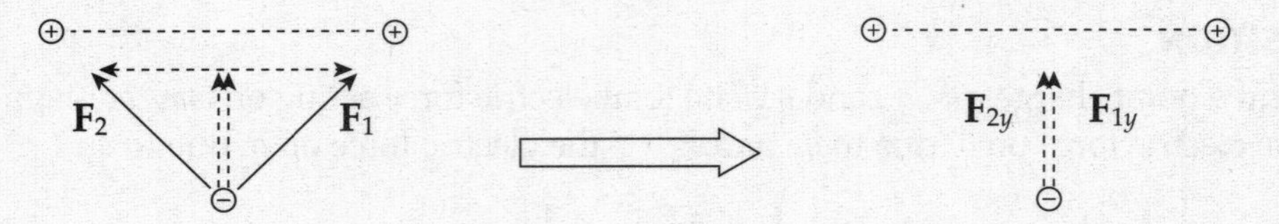

Since the diagram on the left shows the components of $\mathbf{F}_1$ and $\mathbf{F}_2$ making right triangles with legs each of length 2 cm, it must be that $F_{1y} = F_1 \sin 45°$ and $F_{2y} = F_2 \sin 45°$. Also, the magnitude of $\mathbf{F}_1$ equals that of $\mathbf{F}_2$. So the net electric force on the negative charge is $F_{1y} + F_{2y} = 2F \sin 45°$, where F is the strength of the force between the negative charge and each of the positive charges. If s is the length of each side of the square, then the distance r between each positive charge and the negative charge is $r = \frac{1}{2}s\sqrt{2}$ and

$$
\begin{aligned}
F_E &= 2F\sin 45° = 2\frac{1}{4\pi\varepsilon_0}\frac{q_1 q_2}{r^2}\sin 45° \\
&= 2(9\times10^9\ \text{N}\cdot\text{m}^2/\text{C}^2)\frac{(1.5\times10^{-6}\ \text{C})(6.2\times10^{-9}\ \text{C})}{(\frac{1}{2}\cdot 4.0\times10^{-2}\cdot\sqrt{2}\ \text{m})^2}\sin 45° \\
&= 0.15\ \text{N}
\end{aligned}
$$

The direction of the net force is straight upward, toward the center of the line that joins the two positive charges.

Example 10.4 Two pith balls of mass m are each given a charge of $+q$. They are hung side-by-side from two threads each of length L, and move apart as a result of their electrical repulsion. Find the equilibrium separation distance x in terms of m, q, and L. (Use the fact that if θ is small, then $\tan\theta \approx \sin\theta$.)

Solution. Three forces act on each ball: weight, tension, and electrical repulsion:

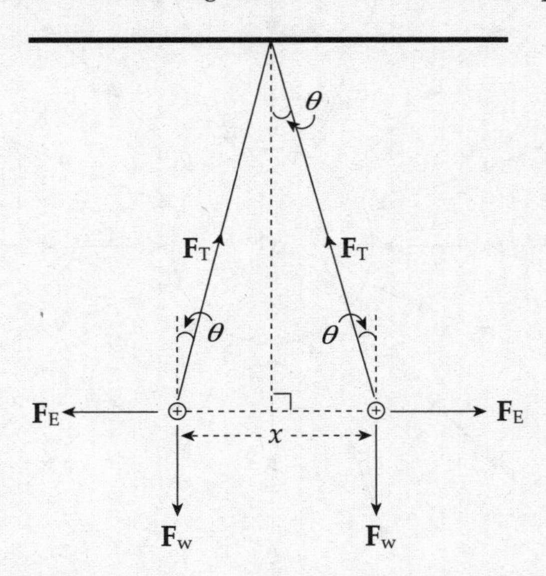

When the balls are in equilibrium, the net force each feels is zero. Therefore, the vertical component of $\mathbf{F}_T$ must cancel out $\mathbf{F}_w$ and the horizontal component of $\mathbf{F}_T$ must cancel out $\mathbf{F}_E$:

$$F_T \cos\theta = F_w \qquad \text{and} \qquad F_T \sin\theta = F_E$$

Dividing the second equation by the first, we get $\tan\theta = F_E/F_w$. Therefore,

$$\tan\theta = \frac{k\dfrac{q^2}{x^2}}{mg} = \frac{kq^2}{mgx^2}$$

Now, to approximate: If θ is small, the $\tan\theta \approx \sin\theta$ and, from the diagram $\sin\theta = \frac{1}{2}x/L$. Therefore, the equation above becomes

$$\frac{\frac{1}{2}x}{L} = \frac{kq^2}{mgx^2} \quad\Rightarrow\quad \frac{1}{2}mgx^3 = kq^2L \quad\Rightarrow\quad x = \sqrt[3]{\frac{2kq^2L}{mg}}$$

THE ELECTRIC FIELD

The presence of a massive body such as the earth causes objects to experience a gravitational force directed toward the earth's center. For objects located outside the earth, this force varies inversely with the square of the distance and directly with the mass of the gravitational source. A vector diagram of the gravitational field surrounding the earth looks like this:

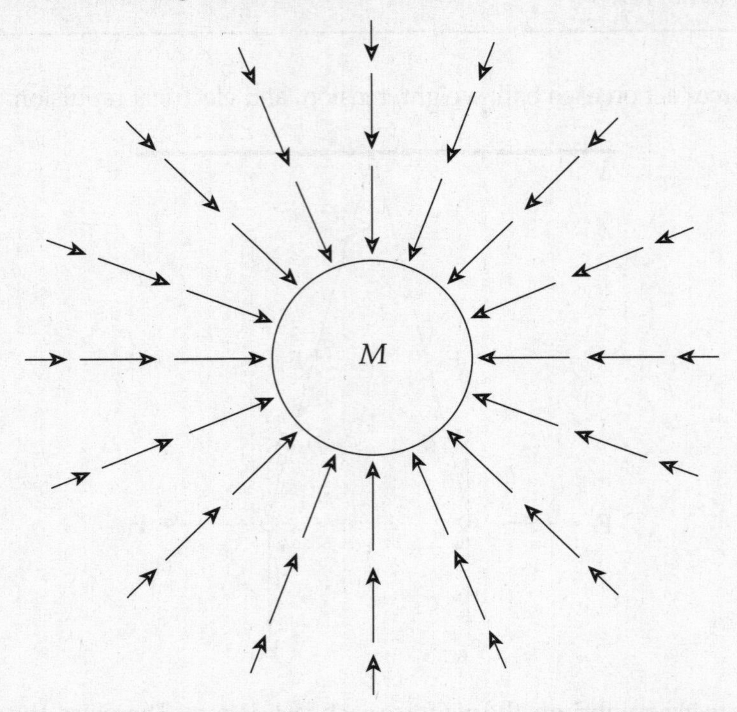

We can think of the space surrounding the earth as permeated by a **gravitational field** that's created by the earth. Any mass that's placed in this field then experiences a gravitational force due to this field.

The same process is used to describe the electric force. Rather than having two charges reach out across empty space to each other to produce a force, we can instead interpret the interaction in the following way: The presence of a charge creates an **electric field** in the space that surrounds it. Another charge placed in the field created by the first will experience a force due to the field.

Consider a point charge Q in a fixed position and assume that it's positive. Now imagine moving a tiny positive test charge q around to various locations near Q. At each location, measure the force that the test charge experiences, and call it $\mathbf{F}_{\text{on } q}$. Divide this force by the test charge q; the resulting vector is the **electric field vector, E**, at that location:

$$\mathbf{E} = \frac{\mathbf{F}_{\text{on } q}}{q}$$

The reason for dividing by the test charge is simple. If we were to use a different test charge with, say, twice the charge of the first one, then each of the forces **F** we'd measure would be twice as much as before. But when we divided this new, stronger force by the new, greater test charge, the factors of 2 would cancel, leaving the same ratio as before. So this ratio tells us the intrinsic strength of the field due to the source charge, independent of whatever test charge we may use to measure it.

What would the electric field of a positive charge Q look like? Since the test charge used to measure the field is positive, every electric field vector would point radially away from the source charge. *If the source charge is positive, the electric field vectors point away from it; if the source charge is negative, then the field vectors point toward it.* And, since the force decreases as we get farther away from the charge (as $1/r^2$), so does the electric field. This is why the electric field vectors farther from the source charge are shorter than those that are closer.

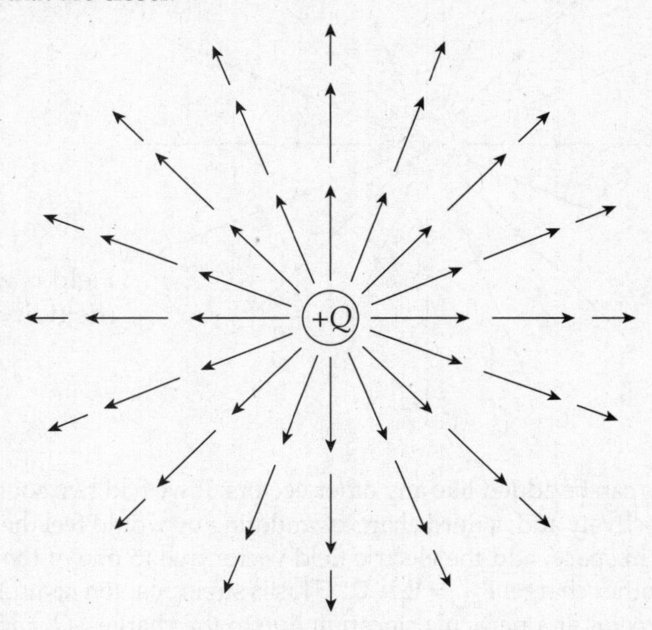

Since the force on the test charge q has a strength of $qQ/4\pi\varepsilon_0 r^2$, when we divide this by q, we get the expression for the strength of the electric field created by a point-charge source of magnitude Q:

$$E = \frac{1}{4\pi\varepsilon_0}\frac{Q}{r^2}$$

To make it easier to sketch an electric field, lines are drawn through the vectors such that the electric field vector is tangent to the line everywhere it's drawn.

Now, your first thought might be that obliterating the individual field vectors deprives us of information, since the length of the field vectors told us how strong the field was. Well, although the individual field vectors are gone, the strength of the field can be figured out by looking at the density of the field lines. Where the field lines are denser, the field is stronger.

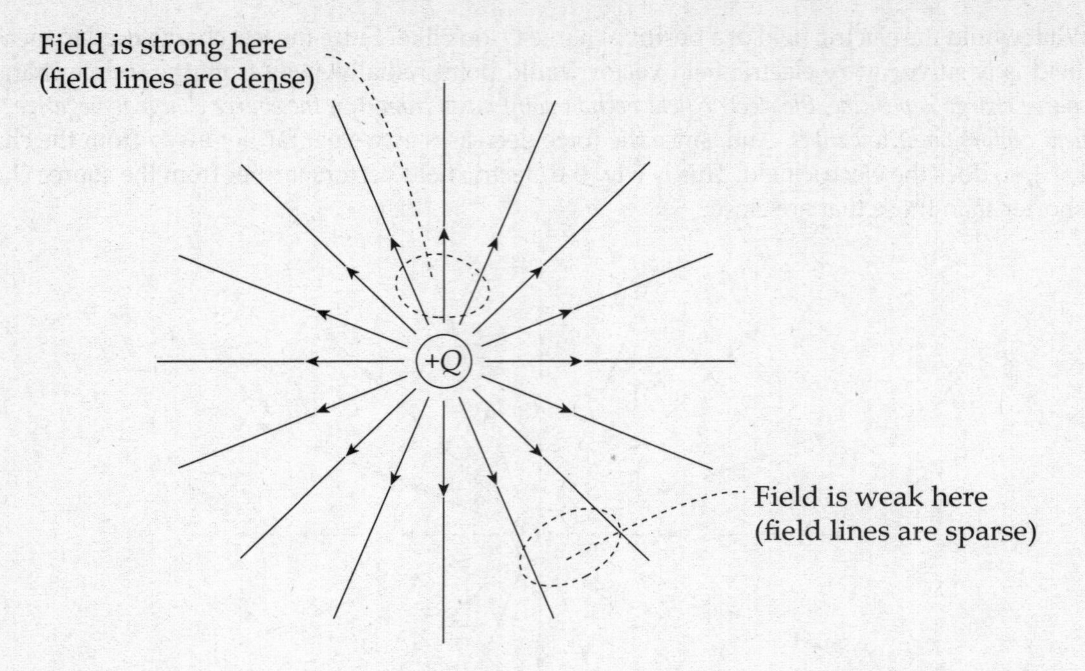

Electric field vectors can be added like any other vectors. If we had two source charges, their fields would overlap and effectively add; a third charge wandering by would feel the effect of the combined field. At each position in space, add the electric field vector due to one of the charges to the electric field vector due to the other charge: $\mathbf{E}_{total} = \mathbf{E}_1 + \mathbf{E}_2$. (This is superposition again.) In the diagram below, $\mathbf{E}_1$ is the electric field vector at a particular location due to the charge $+Q$, and $\mathbf{E}_2$ is the electric field vector at that same location due to the other charge, $-Q$. Adding these vectors gives the overall field vector $\mathbf{E}_{total}$ at that location.

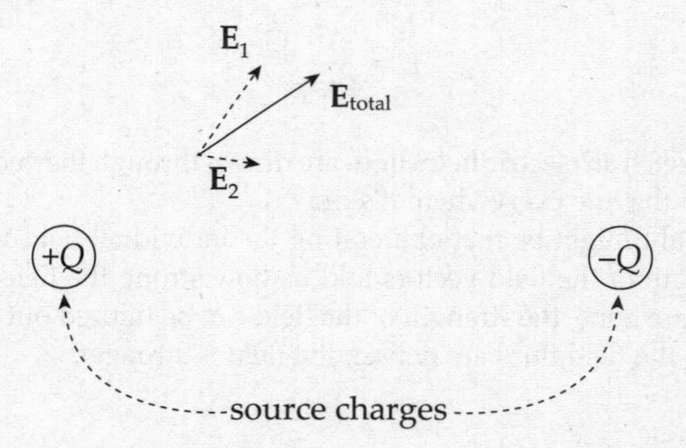

If this is done at enough locations, the electric field lines can be sketched.

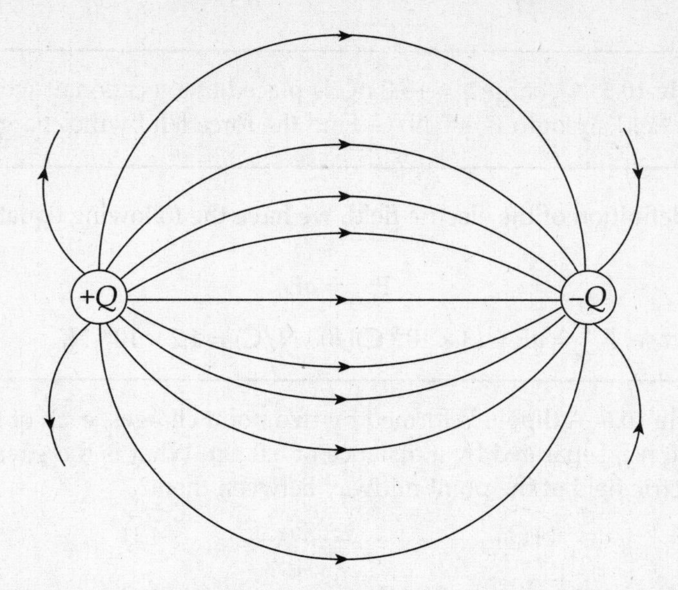

Note that, like electric field vectors, electric field lines always point away from positive source charges and toward negative ones. Two equal but opposite charges, like the ones shown in the diagram above, form a pair called an **electric dipole**.

If a positive charge $+q$ were placed in the electric field above, it would experience a force that is tangent to, and in the same direction as, the field line passing through $+q$'s location. After all, electric fields are sketched from the point of view of what a positive test charge would do. On the other hand, if a negative charge $-q$ were placed in the electric field, it would experience a force that is tangent to, but in the direction opposite from, the field line passing through $-q$'s location.

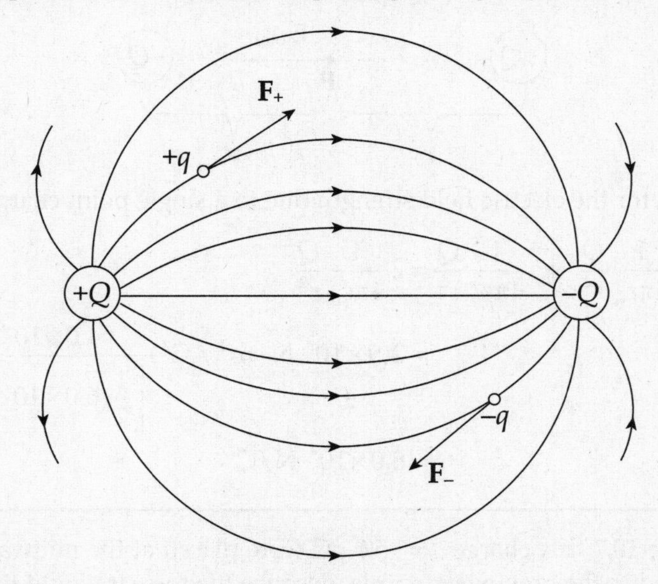

Finally, notice that electric field lines never cross.

Example 10.5 A charge $q = +3.0$ nC is placed at a location at which the electric field strength is 400 N/C. Find the force felt by the charge q.

Solution. From the definition of the electric field, we have the following equation:

$$\mathbf{F}_{\text{on } q} = q\mathbf{E}$$

Therefore, in this case, $F_{\text{on } q} = qE = (3 \times 10^{-9}\text{ C})(400\text{ N/C}) = 1.2 \times 10^{-6}\text{ N}$.

Example 10.6 A dipole is formed by two point charges, each of magnitude 4.0 nC, separated by a distance of 6.0 cm. What is the strength of the electric field at the point midway between them?

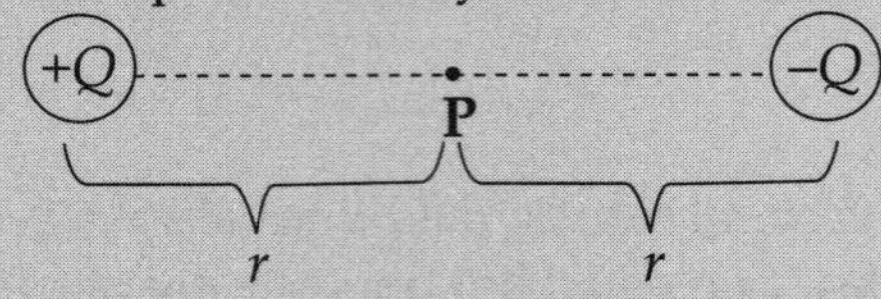

Solution. Let the two source charges be denoted $+Q$ and $-Q$. At Point P, the electric field vector due to $+Q$ would point directly away from $+Q$, and the electric field vector due to $-Q$ would point directly toward $-Q$. Therefore, these two vectors point in the same direction (from $+Q$ to $-Q$), so their magnitudes would add.

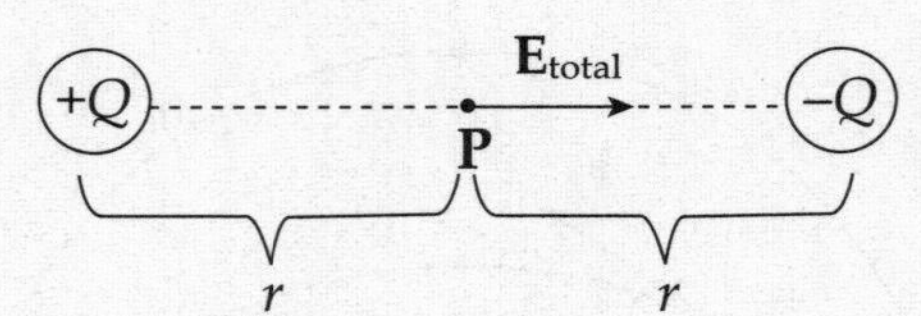

Using the equation for the electric field strength due to a single point charge, we find that

$$E_{\text{total}} = \frac{1}{4\pi\varepsilon_0}\frac{Q}{r^2} + \frac{1}{4\pi\varepsilon_0}\frac{Q}{r^2} = 2\frac{1}{4\pi\varepsilon_0}\frac{Q}{r^2}$$

$$= 2(9\times10^9\ \text{N}\cdot\text{m}^2/\text{C}^2)\frac{4.0\times10^{-9}\ \text{C}}{\left[\frac{1}{2}(6.0\times10^{-2}\ \text{m})\right]^2}$$

$$= 8.0\times10^4\ \text{N/C}$$

Example 10.7 If a charge $q = -5.0$ pC were placed at the midway point described in the previous example, describe the force it would feel. ("p" is the abbreviation for "pico-," which means 10^{-12}.)

Solution. Since the field **E** at this location is known, the force felt by q is easy to calculate:

$$\mathbf{F}_{\text{on } q} = q\mathbf{E} = (-5.0 \times 10^{-12}\ \text{C})(8.0 \times 10^{4}\ \text{N/C to the right}) = 4.0 \times 10^{-7}\ \text{N to the } \textit{left}$$

> **Example 10.8** What can you say about the electric force that a charge would feel if it were placed at a location at which the electric field was zero?

Solution. Remember that $\mathbf{F}_{\text{on } q} = q\mathbf{E}$. So if $\mathbf{E} = \mathbf{0}$, then $\mathbf{F}_{\text{on } q} = \mathbf{0}$. (Zero field means zero force.)

An important subset of problems with electric fields deals with a uniform electric field. One method of creating this uniform field is to have two large conducting sheets, each storing charge, some distance apart. Near the edges of each sheet, the field may not be uniform, but near the middle, for all practical purposes the field is uniform. Having a uniform field means you can use kinematics equations and Newton's Laws just as if you had a uniform gravitational field (as in previous chapters).

> **Example 10.9** Positive charge is distributed uniformly over a large, horizontal plate, which then acts as the source of a vertical electric field. An object of mass 5 g is placed at a distance of 2 cm above the plate. If the strength of the electric field at this location is 10^6 N/C, how much charge would the object need to have in order for the electrical repulsion to balance the gravitational pull?

Solution. Clearly, since the plate is positively charged, the object would also have to carry a positive charge so that the electric force would be repulsive.

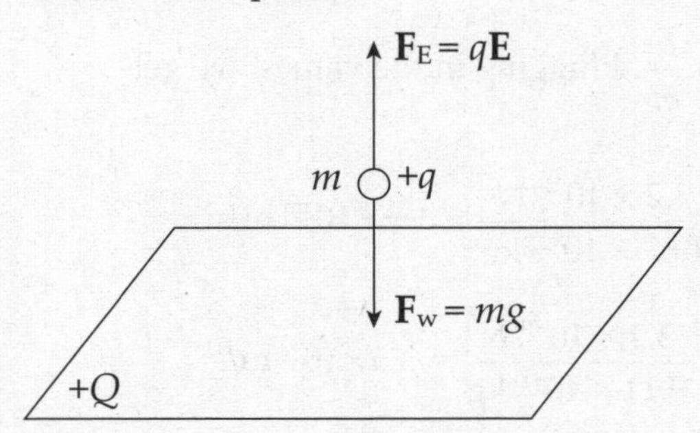

Let q be the charge on the object. Then, in order for F_E to balance mg, we must have

$$qE = mg \quad \Rightarrow \quad q = \frac{mg}{E} = \frac{(5 \times 10^{-3}\ \text{kg})(10\ \text{N/kg})}{10^{6}\ \text{N/C}} = 5 \times 10^{-8}\ \text{C} = 50\ \text{nC}$$

Example 10.10 A proton, neutron, and electron are in a uniform electric field of 20 N/C that is caused by two large charged plates that are 30 cm apart. The particles are far enough apart so that they don't interact with each other. They are released from rest equidistant from each plate.

(a) What is the magnitude of the net force acting on each particle?

(b) What is the magnitude of the acceleration of each particle?

(c) How much work will be done on the particle when it collides with one of the charged plates?

(d) What is the speed of each particle when it strikes the plate?

(e) How long does it take to reach the plate?

Solution.

(a) Since $E = \frac{F}{q}$, $F = qE$. Plugging in the values, we get

$$\text{proton: } F = (1.6 \times 10^{-19}\text{ C})\ (20\text{ N/C}) = 3.2 \times 10^{-18}\text{ N}$$

$$\text{electron: } F = (1.6 \times 10^{-19}\text{ C})\ (20\text{ N/C}) = 3.2 \times 10^{-18}\text{ N}$$

$$\text{neutron: } F = (0\text{ C})\ (20\text{ N/C}) = 0\text{ N}$$

Note: Because the proton and electron have the same magnitude, they will experience the same force. If you're asked for the direction, the proton travels in the same direction as the electric field and the electron travels in the opposite direction as the electric field.

(b) Since $F = ma$, $a = \frac{F}{m}$. Plugging in the values, we get

$$\text{proton: } a = \frac{3.2 \times 10^{-18}\text{ N}}{1.67 \times 10^{-27}\text{ kg}} = 1.9 \times 10^{-18}\text{ m/s}^2$$

$$\text{electron: } a = \frac{3.2 \times 10^{-18}\text{ N}}{9.11 \times 10^{-31}\text{ kg}} = 3.5 \times 10^{12}\text{ m/s}^2$$

$$\text{neutron: } a = \frac{0\text{ N}}{1.67 \times 10^{-27}\text{ kg}} = 0\text{ m/s}^2$$

Notice that although the charges have the same magnitude of force, the electron experiences an acceleration almost 2000 times greater due to its mass being almost 2000 times smaller than the proton's mass.

(c) Since $W = Fd$, we get $W = qEd$. Plugging in the values, we get

proton: $W = (1.6 \times 10^{-19}\ \text{C})\ (20\ \text{N/C})\ (0.15\ \text{m}) = 4.8 \times 10^{-19}\ \text{N}$

electron: $W = (1.6 \times 10^{-19}\ \text{C})\ (20\ \text{N/C})\ (0.15\ \text{m}) = 4.8 \times 10^{-19}\ \text{N}$

neutron: $W = (0\ \text{C})\ (20\ \text{N/C})\ (0.15\ \text{m}) = 0\ \text{N}$

(d) Recall one of the big five kinematics equations:

$$v_f^{\,2} = v_i^{\,2} + 2a(x - x_o) \rightarrow v_f = \sqrt{2a(x - x_o)}$$

If the particles are midway between the 30 cm plates, they will travel 0.15 m.

proton: $v_f = \sqrt{2(1.9 \times 10^{9}\ \text{m/s}^2)(0.15\ \text{m})} \rightarrow v_f = 24{,}000\ \text{m/s}$

electron: $v_f = \sqrt{2(3.5 \times 10^{12}\ \text{m/s}^2)(0.15\ \text{m})} \rightarrow v_f = 1.0 \times 10^{6}$

neutron: $v_f = \sqrt{2(0\ \text{m/s}^2)(0.15\ \text{m})} \rightarrow v_f = 0\ \text{m/s}$

Notice that, even though the force is the same and the same work is done on both charges, there is a significant difference in final velocities due to the large mass difference. An alternative solution to this would be using $W = \Delta KE \rightarrow W = \frac{1}{2}mv^2$. You would have obtained the same answers.

(e) Recall one of the big five kinematics equations:

$$v_f = v_i + at \rightarrow t = \frac{v_f - v_i}{a} \rightarrow t = \frac{v_f}{a}$$

proton: $t = \dfrac{24{,}000\ \text{m/s}}{1.9 \times 10^{9}\ \text{m/s}^2} \rightarrow 1.3 \times 10^{-5}\ \text{s}$

electron: $t = \dfrac{1 \times 10^{6}\ \text{m/s}}{-3.5 \times 10^{12}\ \text{m/s}^2} \rightarrow 2.9 \times 10^{-7}\ \text{s}$

neutron: The neutron never accelerates, so it will never hit the plate

CONDUCTORS AND INSULATORS

Materials can be classified into broad categories based on their ability to permit the flow of charge. If electrons were placed on a metal sphere, they would quickly spread out and cover the outside of the sphere uniformly. These electrons would be free to flow through the metal and redistribute themselves, moving to get as far away from each other as they could. Materials that permit the flow of excess charge are called **conductors**; they conduct electricity. Metals are the best examples of conductors, but other conductors are aqueous solutions that contain dissolved electrolytes (such as salt water). Metals conduct electricity because the structure of a typical metal consists of a lattice of nuclei and inner-shell electrons, with about one electron per atom not bound to its nucleus. Electrons are free to move about the lattice, creating a sort of sea of mobile (or conduction) electrons. This freedom allows excess charge to flow freely.

Insulators, on the other hand, closely guard their electrons—and even extra ones that might be added. Electrons are not free to roam throughout the atomic lattice. Examples of insulators are glass, wood, rubber, and plastic. If excess charge is placed on an insulator, it stays put.

Midway between conductors and insulators is a class of materials known as **semiconductors**. As the name indicates, they're more or less conductors. That is, they are less conducting than most metals, but more conducting than most insulators. Examples of semiconducting materials are silicon and germanium.

An extreme example of a conductor is the **superconductor**. This is a material that offers absolutely no resistance to the flow of charge; it is a *perfect* conductor of electric charge. Many metals and ceramics become superconducting when they are brought to extremely low temperatures.

Example 10.11 A solid sphere of copper is given a negative charge. Discuss the electric field inside and outside the sphere.

Solution. The electric field inside a conductor is zero. Therefore, the excess electrons that are deposited on the sphere move quickly to the outer surface (copper is a great conductor). *Any excess charge on a conductor resides entirely on the outer surface.*

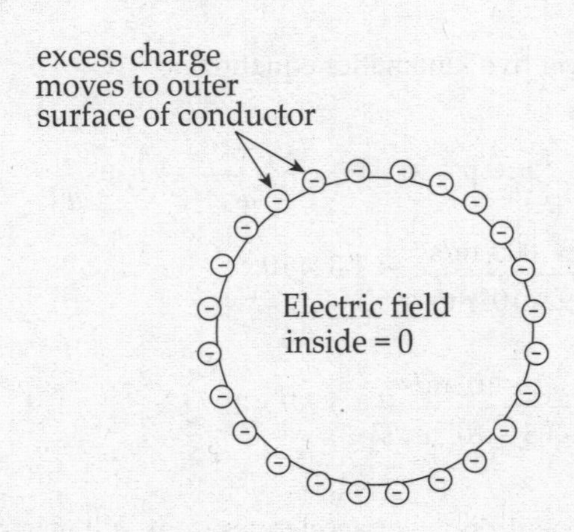

Once these excess electrons establish a uniform distribution on the outer surface of the sphere, there will be no net electric field within the sphere. Why not? Since there is no additional excess charge inside the conductor, there are no excess charges to serve as a source or sink of an electric field line cutting down into the sphere, because field lines begin or end on excess charges. *There can be no electrostatic field within the body of a conductor.*

In fact, you can shield yourself from electric fields simply by surrounding yourself with metal. Charges may move around on the outer surface of your cage, but within the cage, the electric field will be zero. For points outside the sphere, it can be shown that the sphere behaves as if all its excess charge were concentrated at its center. (Remember that this is just like the gravitational field due to a uniform spherical mass.) Also, *the electric field is always perpendicular to the surface, no matter what shape the surface may be.* See the diagram below.

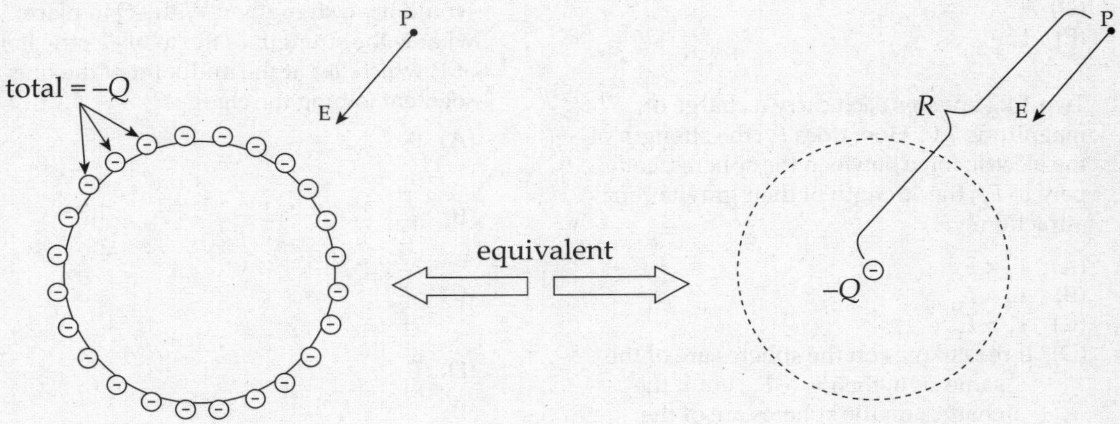

CHAPTER 10 REVIEW QUESTIONS

Solutions can be found in Chapter 18.

SECTION I: MULTIPLE CHOICE

1. If the distance between two positive point charges is tripled, then the strength of the electrostatic repulsion between them will decrease by a factor of

(A) 3
(B) 6
(C) 8
(D) 9
(E) 12

2. Two 1 kg spheres each carry a charge of magnitude 1 C. How does F_E, the strength of the electric force between the spheres, compare to F_G, the strength of their gravitational attraction?

(A) $F_E < F_G$
(B) $F_E = F_G$
(C) $F_E > F_G$
(D) If the charges on the spheres are of the same sign, then $F_E > F_G$; but if the charges on the spheres are of the opposite sign, then $F_E < F_G$.
(E) Cannot be determined without knowing the distance between the spheres

3. The figure below shows three point charges, all positive. If the net electric force on the center charge is zero, what is the value of y/x?

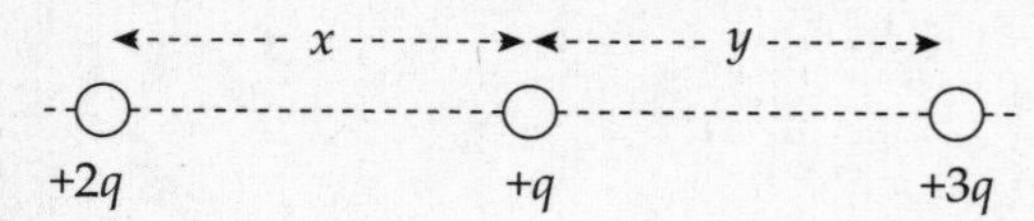

(A) $\frac{4}{9}$

(B) $\sqrt{\frac{2}{3}}$

(C) $\sqrt{\frac{3}{2}}$

(D) $\frac{3}{2}$

(E) $\frac{9}{4}$

4.

The figure above shows two point charges, +Q and –Q. If the negative charge were absent, the electric field at Point P due to +Q would have strength E. With –Q in place, what is the strength of the total electric field at P, which lies at the midpoint of the line segment joining the charges?

(A) 0

(B) $\frac{E}{4}$

(C) $\frac{E}{2}$

(D) E

(E) $2E$

5. A sphere of charge $+Q$ is fixed in position. A smaller sphere of charge $+q$ is placed near the larger sphere and released from rest. The small sphere will move away from the large sphere with

(A) decreasing velocity and decreasing acceleration
(B) decreasing velocity and increasing acceleration
(C) decreasing velocity and constant acceleration
(D) increasing velocity and decreasing acceleration
(E) increasing velocity and increasing acceleration

6. An object of charge $+q$ feels an electric force $\mathbf{F}_E$ when placed at a particular location in an electric field, **E**. Therefore, if an object of charge $-2q$ were placed at the same location where the first charge was, it would feel an electric force of

(A) $\frac{-\mathbf{F}_E}{2}$

(B) $-2\mathbf{F}_E$

(C) $-2q\mathbf{F}_E$

(D) $\frac{-2\mathbf{F}_E}{q}$

(E) $\frac{-\mathbf{F}_E}{2q}$

7. A charge of $-3Q$ is transferred to a solid metal sphere of radius r. Where will this excess charge reside?

(A) $-Q$ at the center, and $-2Q$ on the outer surface

(B) $-2Q$ at the center, and $-Q$ on the outer surface

(C) $-3Q$ at the center

(D) $-3Q$ on the outer surface

(E) $-Q$ at the center, $-Q$ in a ring of radius $\frac{1}{2}r$, and $-Q$ on the outer surface

SECTION II: FREE RESPONSE

1. In the figure shown, all four charges ($+Q$, $+Q$, $-q$, and $-q$) are situated at the corners of a square. The net electric force on each charge $+Q$ is zero.

    (a) Express the magnitude of q in terms of Q.

    (b) Is the net electric force on each charge $-q$ also equal to zero? Justify your answer.

    (c) Determine the electric field at the center of the square.

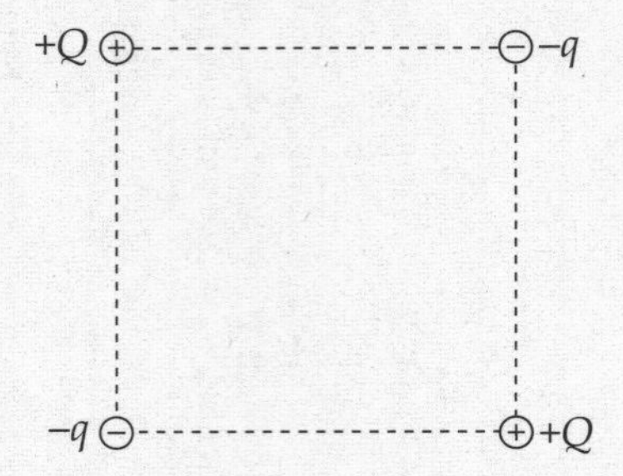

2. Two charges, $+Q$ and $+2Q$, are fixed in place along the y axis of an x-y coordinate system as shown in the figure below. Charge 1 is at the point $(0, a)$, and Charge 2 is at the point $(0, -2a)$.

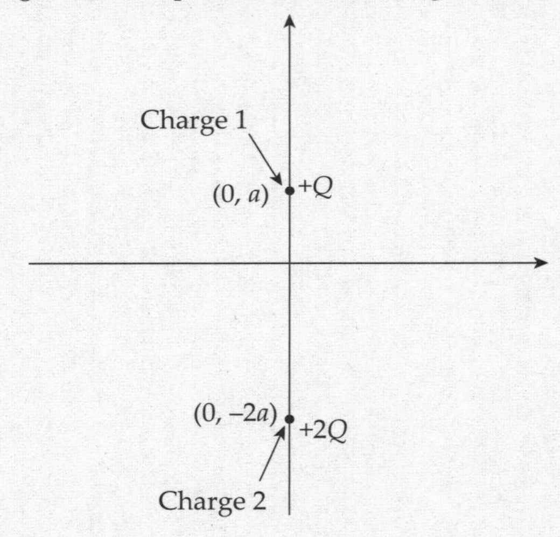

    (a) Find the electric force (magnitude and direction) felt by Charge 1 due to Charge 2.

    (b) Find the electric field (magnitude and direction) at the origin created by both Charges 1 and 2.

    (c) Is there a point on the x axis where the total electric field is zero? If so, where? If not, explain briefly.

    (d) Is there a point on the y axis where the total electric field is zero? If so, where? If not, explain briefly.

    (e) If a small negative charge, $-q$, of mass m were placed at the origin, determine its initial acceleration (magnitude and direction).

SUMMARY

- Coulomb's Law describes the force acting on two point charges and is given by

$$F_e = \frac{1}{4\pi\varepsilon_0}\frac{q_1 q_2}{r^2}$$

where $\frac{1}{4\pi\varepsilon_0}$ is a constant equal to 9.0×10^9 Nm²/C². You can use all the strategies you used in the Newton's Laws chapter to solve these types of problems.

- The electric field is given by $E = \frac{F}{q}$ or $E = \frac{1}{4\pi\varepsilon_0}\frac{q}{r^2}$
- Both the electric force and field are vector quantities and therefore all the rules for vector addition apply.

11

Electric Potential and Capacitance

INTRODUCTION

When an object moves in a gravitational field, it usually experiences a change in kinetic energy and in gravitational potential energy due to the work done on the object by gravity. Similarly, when a charge moves in an electric field, it generally experiences a change in kinetic energy and in electrical potential energy due to the work done on it by the electric field. By exploring the idea of electric potential, we can simplify our calculations of work and energy changes within electric fields.

ELECTRICAL POTENTIAL ENERGY

When a charge moves in an electric field, then unless its displacement is always perpendicular to the field, the electric force does work on the charge. If W_E is the work done by the electric force, then the change in the charge's **electrical potential energy** is defined by

$$\Delta U_E = -W_E$$

Notice that this is the same equation that defined the change in the gravitational potential energy of an object of mass m undergoing a displacement in a gravitational field ($\Delta U_G = -W_G$).

Example 11.1 A positive charge $+q$ moves from position A to position B in a uniform electric field **E**:

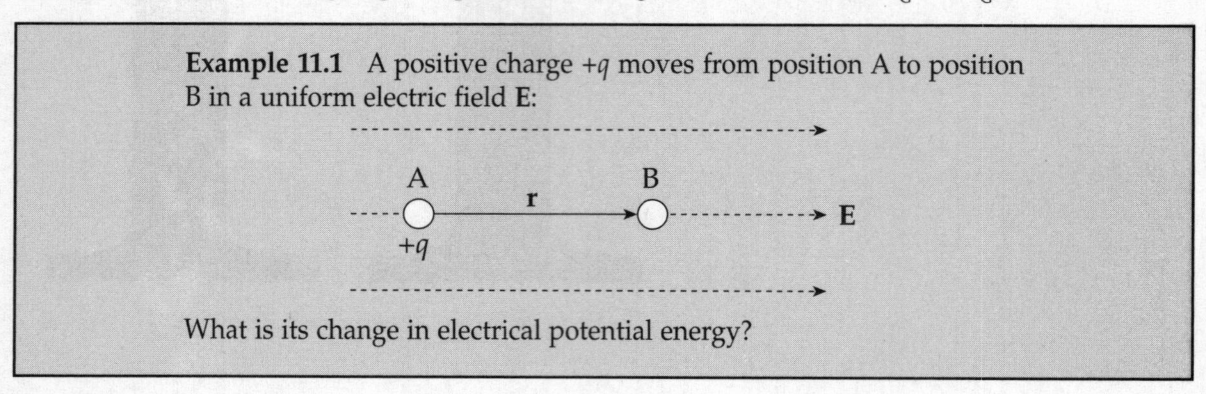

What is its change in electrical potential energy?

Solution. Since the field is uniform, the electric force that the charge feels, $\mathbf{F}_E = q\mathbf{E}$, is constant. Since q is positive, $\mathbf{F}_E$ points in the same direction as **E**, and, as the figure shows, they point in the same direction as the displacement, **r**. This makes the work done by the electric field equal to $W_E = F_E r = qEr$, so the change in the electrical potential energy is

$$\Delta U_E = -qEr$$

Note that the change in potential energy is negative, which means that potential energy has decreased; this always happens when the field does positive work. It's just like dropping a rock to the ground: Gravity does positive work, and the rock loses gravitational potential energy.

Example 11.2 Do the previous problem, but consider the case of a negative charge, $-q$.

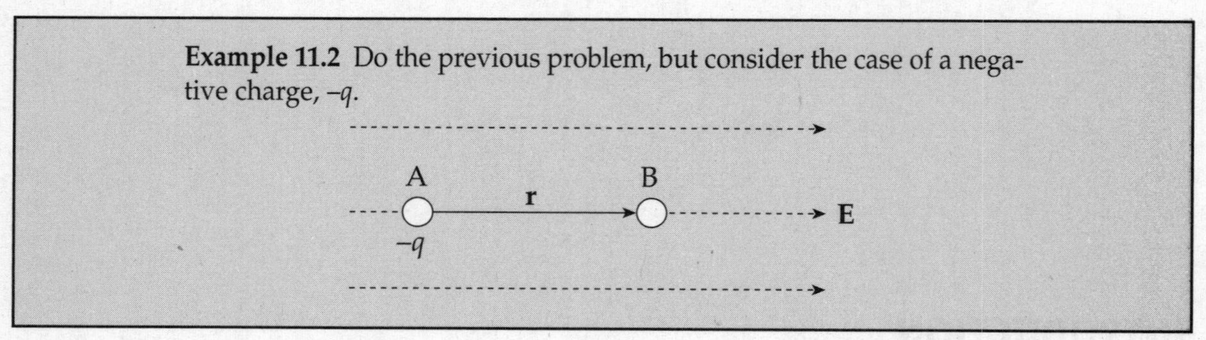

Solution. In this case, an outside agent must be pushing the charge to make it move, because the electric force *naturally* pushes negative charges against field lines. Therefore, we expect that the work done by the electric field is negative. The electric force, $\mathbf{F}_E = (-q)\mathbf{E}$, points in the direction opposite to the displacement, so the work it does is $W_E = -F_E r = -qEr = -qEr$. Thus, the change in electrical potential energy is positive: $\Delta U_E = -W_E = -(-qEr) = qEr$. Because the change in potential energy is positive, the potential energy increases; this always happens when the field does negative work. It's like lifting a rock off the ground: Gravity does negative work, and the rock gains gravitational potential energy.

Example 11.3 A positive charge $+q$ moves from position A to position B in a uniform electric field **E**:

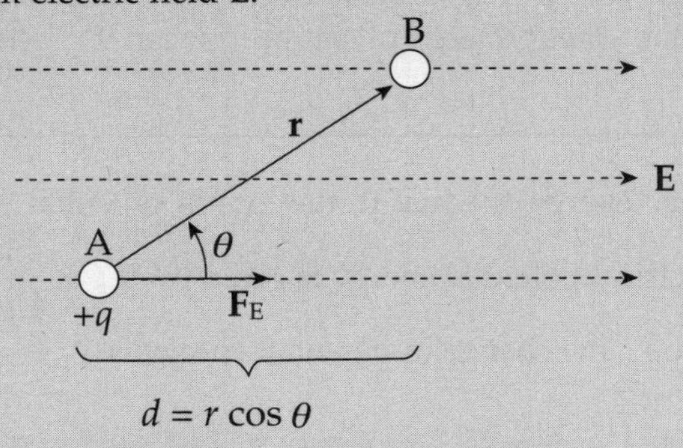

$d = r \cos \theta$

What is its change in electrical potential energy?

Solution. The electric force felt by the charge q is $\mathbf{F}_E = q\mathbf{E}$ and this force is parallel to **E**, because q is positive. In this case, because $\mathbf{F}_E$ is not parallel to **r** (as it was in Example 11.1), we will use the more general definition of work:

$$W_E = \mathbf{F}_E \cdot \mathbf{r} = F_E\, r \cos \theta = qEr \cos \theta$$

But $r \cos \theta = d$, so

$$W_E = qEd \text{ and } \Delta U_E = -W_E = -qEd$$

Because the electric force is a conservative force, which means that the work done does not depend on the path that connects the positions A and B, the work calculated above could have been figured out by considering the path from A to B composed of the segments $\mathbf{r}_1$ and $\mathbf{r}_2$:

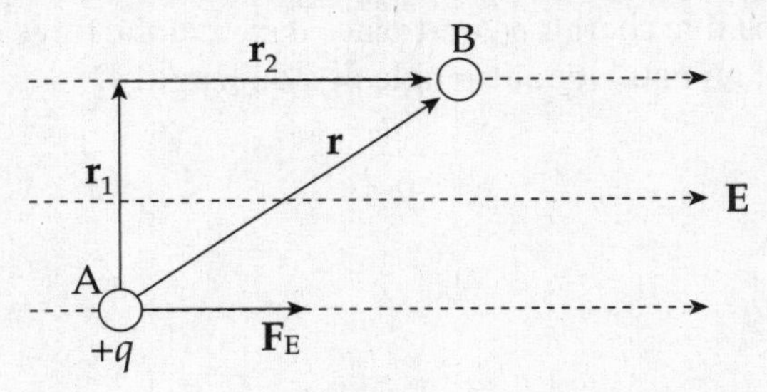

Along $\mathbf{r}_1$, the electric force does no work since this displacement is perpendicular to the force. Thus, the work done by the electric field as q moves from A to B is simply equal to the work it does along $\mathbf{r}_2$. And since the length of $\mathbf{r}_2$ is $d = r \cos \theta$, we have $W_E = F_E d = qEd$, just as before.

Example 11.4 A positive charge $q_1 = +2 \times 10^{-6}$ C is held stationary, while a negative charge, $q_2 = -1 \times 10^{-8}$ C, is released from rest at a distance of 10 cm from q_1. Find the kinetic energy of charge q_2 when it's 1 cm from q_1.

Solution. The gain in kinetic energy is equal to the loss in potential energy; you know this from Conservation of Energy. Electrical potential energy is given by $U_E = \frac{1}{4\pi\varepsilon_0}\left(\frac{q_1 q_2}{r}\right)$. Therefore, if q_1 is fixed and q_2 moves from r_A to r_B, the change in potential energy is

$$\begin{aligned}
\Delta U_E &= U_B - U_A \\
&= \frac{q_1}{4\pi\varepsilon_0}\left(\frac{q_2}{r_B} - \frac{q_2}{r_A}\right) \\
&= \frac{q_1 q_2}{4\pi\varepsilon_0}\left(\frac{1}{r_B} - \frac{1}{r_A}\right) \\
&= (9 \times 10^9 \text{ N}\cdot\text{m}^2/\text{C}^2)(+2 \times 10^{-6}\text{ C})(-1 \times 10^{-8}\text{ C})\left(\frac{1}{0.01\text{ m}} - \frac{1}{0.10\text{ m}}\right) \\
&= -0.016 \text{ J}
\end{aligned}$$

So the gain in kinetic energy is +0.016 J. Since q_2 started from rest (with no kinetic energy), this is the kinetic energy of q_2 when it's 1 cm from q_1.

Example 11.5 Two positive charges, q_1 and q_2, are held in the positions shown below. How much work would be required to bring (from infinity) a third positive charge, q_3, and place it so that the three charges form the corners of an equilateral triangle of side length s?

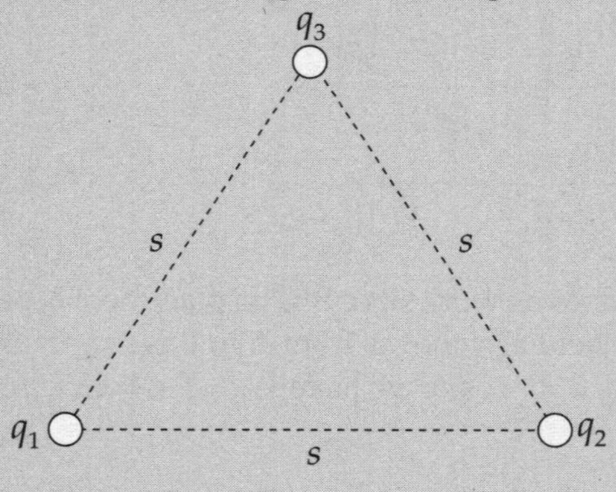

Solution. An external agent would need to do positive work, equal in magnitude to the negative work done by the electric force on q_3 as it is brought into place, so let's first compute this quantity. Let's first compute the work done *by the electric force* as q_3 is brought in. Since q_3 is fighting against both q_1's and q_2's electric field, the total work done on q_3 by the electric force, W_E, is equal to the work done on q_3 by q_1 ($W_{1\text{-}3}$) plus the work done on q_3 by q_2 ($W_{2\text{-}3}$). Using the equation $W_E = -\Delta U_E$ and the one we gave above for ΔU_E, we have

$$W_{1\text{-}3} + W_{2\text{-}3} = -\Delta U_{1\text{-}3} + -\Delta U_{2\text{-}3}$$

$$= -\frac{q_1 q_3}{4\pi\varepsilon_0}\left(\frac{1}{s} - 0\right) + -\frac{q_2 q_3}{4\pi\varepsilon_0}\left(\frac{1}{s} - 0\right)$$

$$= -\left(\frac{1}{4\pi\varepsilon_0}\frac{q_1 q_3}{s} + \frac{1}{4\pi\varepsilon_0}\frac{q_2 q_3}{s}\right)$$

Therefore, the work that an external agent must do to bring q_3 into position is

$$-W_E = \frac{1}{4\pi\varepsilon_0}\frac{q_1 q_3}{s} + \frac{1}{4\pi\varepsilon_0}\frac{q_2 q_3}{s}$$

In general, the work required by an external agent to assemble a collection of point charges $q_1, q_2, \ldots, q_n$ (bringing each one from infinity), such that the final fixed distance between q_i and q_j is r_{ij}, is equal to the total electrical potential energy of the arrangement:

$$W_{\text{by external agent}} = U_{\text{total}} = \frac{1}{4\pi\varepsilon_0}\sum_{i<j}\frac{q_i q_j}{r_{ij}}$$

ELECTRIC POTENTIAL

Let W_E be the work done by the electric field on a charge q as it undergoes a displacement. If another charge, say $2q$, were to undergo the same displacement, the electric force would be twice as great on this second charge, and the work done by the electric field would be twice as much, $2W_E$. Since the work would be twice as much in the second case, the change in electrical potential energy would be twice as great as well, but the ratio of the change in potential energy to the charge would be the same: $U_E/q = U_E/2q$. This ratio says something about the *field* and the *displacement* but not the charge that made the move. The change in **electric potential** ΔV, is defined as this ratio:

$$\Delta V = \frac{\Delta U_E}{q}$$

Electric potential is electrical potential energy *per unit charge*; the units of electric potential are joules per coulomb. One joule per coulomb is called one **volt** (abbreviated V); so 1 J/C = 1 V.

Consider the electric field that's created by a point source charge Q. If a charge q moves from a distance r_A to a distance r_B from Q, then the change in the potential energy is

$$U_B - U_A = \frac{Qq}{4\pi\varepsilon_0}\left(\frac{1}{r_B} - \frac{1}{r_A}\right)$$

The difference in electric potential between positions A and B in the field created by Q is

$$V_B - V_A = \frac{U_B - U_A}{q} = \frac{Q}{4\pi\varepsilon_0}\left(\frac{1}{r_B} - \frac{1}{r_A}\right)$$

If we designate $V_A \to 0$ as $r_A \to \infty$ (an assumption that's stated on the AP Physics Exam), then the electric potential at a distance r from Q is

$$V = \frac{1}{4\pi\varepsilon_0}\frac{Q}{r}$$

Note that the potential depends on the source charge making the field and the distance from it.

Example 11.6 Let $Q = 2 \times 10^{-9}$ C. What is the potential at a Point P that is 2 cm from Q?

Solution. Relative to $V = 0$ at infinity, we have

$$V = \frac{1}{4\pi\varepsilon_0}\frac{Q}{r} = \left(9\times10^9\ \text{N}\cdot\text{m}^2/\text{C}^2\right)\frac{2\times10^{-9}\ \text{C}}{0.02\ \text{m}} = 900\ \text{V}$$

This means that the work done by the electric field on a charge of q coulombs brought to a point 2 cm from Q would be $-900q$ joules.

Note that, like potential energy, electric potential is a *scalar*. In the preceding example, we didn't have to specify the direction of the vector from the position of Q to the Point P, because it didn't matter. At *any* point on a sphere that's 2 cm from Q, the potential will be 900 V. These spheres around Q are called **equipotential surfaces**, and they're surfaces of constant potential. Their cross sections in any plane are circles and are (therefore) perpendicular to the electric field lines. The equipotentials are always perpendicular to the electric field line.

Example 11.7 How much work is done by the electric field as a charge moves along an equipotential surface?

Solution. If the charge always remains on a single equipotential, then, by definition, the potential, V, never changes. Therefore, $\Delta V = 0$, so $\Delta U_E = 0$. Since $W_E = -\Delta U_E$, the work done by the electric field is zero.

Example 11.8 If charges $q_1 = 4 \times 10^{-9}$ C and $q_2 = -6 \times 10^{-9}$ C are stationary, calculate the potential at Point A in the figure below:

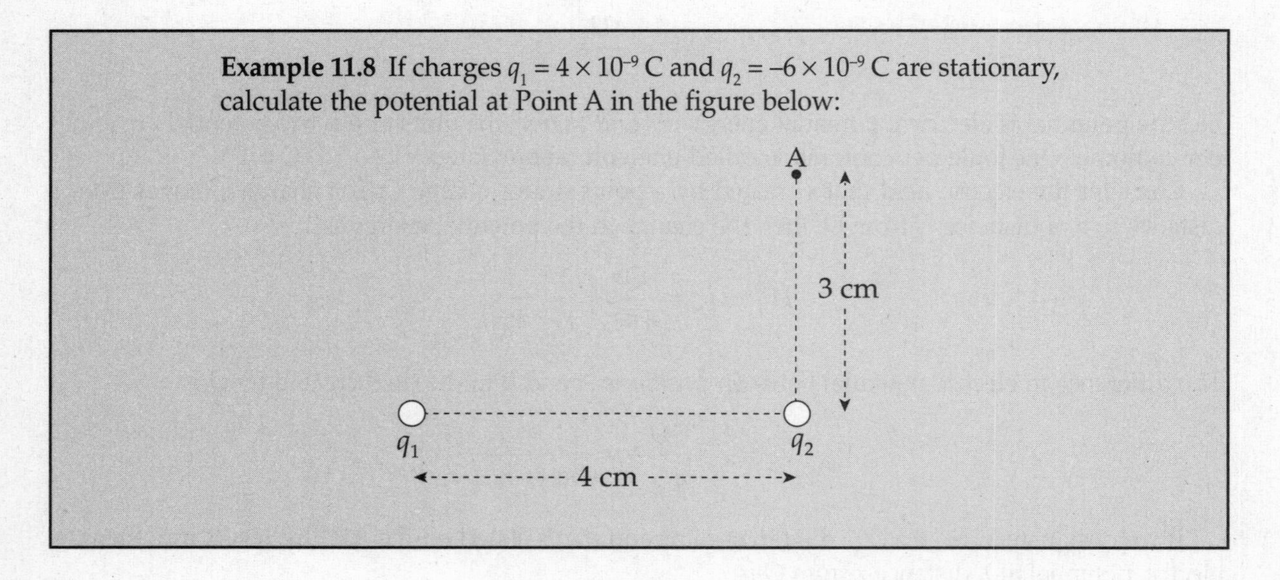

Solution. Potentials add like ordinary numbers. Therefore, the potential at A is just the sum of the potentials at A due to q_1 and q_2. Note that the distance from q_1 to A is 5 cm.

$$\begin{aligned} V &= \frac{1}{4\pi\varepsilon_0}\frac{q_1}{r_{1A}} + \frac{1}{4\pi\varepsilon_0}\frac{q_2}{r_{2A}} \\ &= \frac{1}{4\pi\varepsilon_0}\left(\frac{q_1}{r_{1A}} + \frac{q_2}{r_{2A}}\right) \\ &= (9\times10^9\ \text{N}\cdot\text{m}^2/\text{C}^2)\left(\frac{4\times10^{-9}\ \text{C}}{0.05\ \text{m}} + \frac{-6\times10^{-9}\ \text{C}}{0.03\ \text{m}}\right) \\ &= -1080\ \text{V} \end{aligned}$$

Example 11.9 How much work would it take to move a charge $q = +1 \times 10^{-2}$ C from Point A to Point B (the point midway between q_1 and q_2?

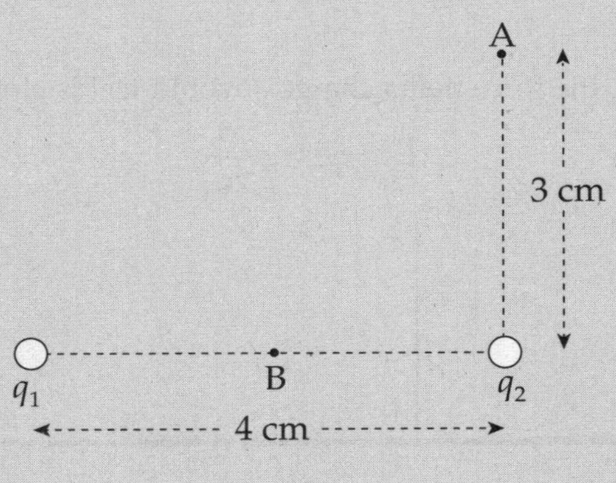

Solution. $\Delta U_E = q\Delta V$, so if we calculate the potential difference between Points A and B and multiply by q, we will have found the change in the electrical potential energy: $\Delta U_{A\to B} = q\Delta V_{A\to B}$. Then, since the work by the electric field is $-\Delta U$, the work required by an external agent is ΔU. In this case, the potential at Point B is

$$\begin{aligned} V_B &= \frac{1}{4\pi\varepsilon_0}\left(\frac{q_1}{r_{1B}} + \frac{q_2}{r_{2B}}\right) \\ &= (9\times10^9\ \text{N}\cdot\text{m}^2/\text{C}^2)\left(\frac{4\times10^{-9}\ \text{C}}{0.02\ \text{m}} + \frac{-6\times10^{-9}\ \text{C}}{0.02\ \text{m}}\right) \\ &= -900\ \text{V} \end{aligned}$$

In the preceding example, we calculated the potential at Point A: $V_A = -1080$ V, so $\Delta V_{A\to B} = V_B - V_A = (-900\ \text{V}) - (-1080\ \text{V}) = +180$ V. This means that the change in electrical potential energy as q moves from A to B s

$$\Delta U_{A\to B} = qV_{A\to B} = (+1 \times 10^{-2}\ \text{C})(+180\ \text{V}) = 1.8\ \text{J}$$

This is the work required by an external agent to move q from A to B.

THE POTENTIAL IN A UNIFORM FIELD

Example 11.10 Consider a very large, flat plate that contains a uniform surface charge density σ. At points that are not too far from the plate, the electric field is uniform and given by the equation

$$E = \frac{\sigma}{2\varepsilon_0}$$

What is the potential at a point which is a distance d from the sheet, relative to the potential of the sheet itself?

Solution. Let A be a point on the plate and let B be a point a distance d from the sheet. Then

$$V_B - V_A = \frac{-W_{E,\,A\to B} \text{ on } q}{q}$$

Since the field is constant, the force that a charge q would feel is also constant, and is equal to

$$F_E = qE = q\frac{\sigma}{2\varepsilon_0}$$

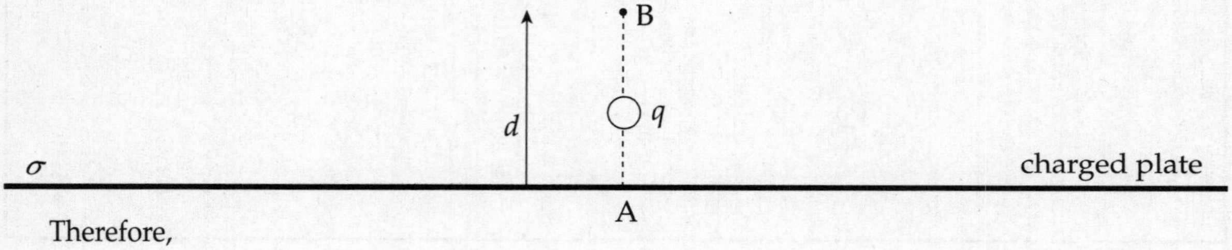

Therefore,

$$W_{E,\,A\to B} = F_E d = \frac{q\sigma}{2\varepsilon_0}d$$

so applying the definition gives us

$$V_B - V_A = \frac{-W_{E,\,A\to B}}{q} = -\frac{\sigma}{2\varepsilon_0}d$$

This says that the potential decreases linearly as we move away from the plate.

Example 11.11 Two large flat plates—one carrying a charge of $+Q$, the other $-Q$—are separated by a distance d. The electric field between the plates, **E**, is uniform. Determine the potential difference between the plates.

Solution. Imagine a positive charge q moving from the positive plate to the negative plate:

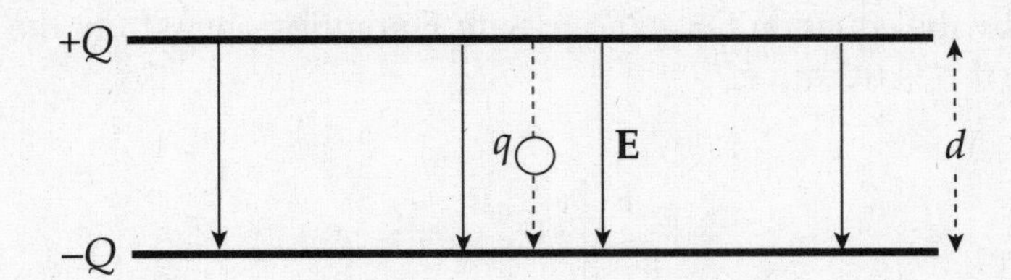

Since the work done by the electric field is

$$W_{E,+\to-} = F_E d = qEd$$

the potential difference between the plates is

$$V_- - V_+ = \frac{-W_{E,+\to-}}{q} = \frac{-qEd}{q} = -Ed$$

This tells us that the potential of the positive plate is greater than the potential of the negative plate, by the amount Ed. This equation can also be written as

$$E = -\frac{V_- - V_+}{d}$$

Therefore, if the potential difference and the distance between the plates are known, then the magnitude of the electric field can be determined quickly.

CAPACITANCE

Consider two conductors, separated by some distance, that carry equal but opposite charges, $+Q$ and $-Q$. Such a pair of conductors comprise a system called a **capacitor**. Work must be done to create this separation of charge, and, as a result, potential energy is stored. Capacitors are basically storage devices for electrical potential energy.

The conductors may have any shape, but the most common conductors are parallel metal plates or sheets. These types of capacitors are called **parallel-plate capacitors**. We'll assume that the distance d between the plates is small compared to the dimensions of the plates since, in this case, the electric field between the plates is uniform. The electric field due to *one* such plate, if its surface charge density is $\sigma = Q/A$, is given by the equation $E = \sigma/(2\varepsilon_0)$, with **E** pointing away from the sheet if σ is positive and toward the plate if σ is negative.

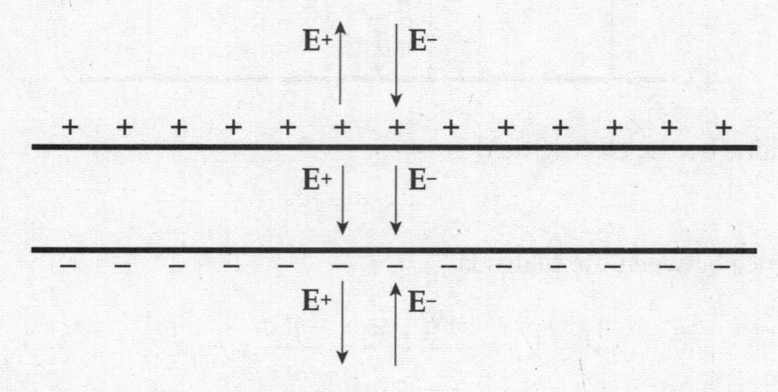

Therefore, with two plates, one with surface charge density $+\sigma$ and the other $-\sigma$, the electric fields combine to give a field that's zero outside the plates and that has the magnitude

$$E_{\text{total}} = \frac{\sigma}{2\varepsilon_0} + \frac{\sigma}{2\varepsilon_0} = \frac{\sigma}{\varepsilon_0}$$

in between. In Example 11.11, we learned that the magnitude of the potential difference, ΔV, between the plates satisfies the relationship $\Delta V = Ed$, so combining this with the previous equation, we get

$$E = \frac{\sigma}{\varepsilon_0} \quad \Rightarrow \quad \frac{\Delta V}{d} = \frac{\sigma}{\varepsilon_0} \quad \Rightarrow \quad \frac{\Delta V}{d} = \frac{Q/A}{\varepsilon_0} \quad \Rightarrow \quad \frac{Q}{\Delta V} = \frac{\varepsilon_0 A}{d}$$

The ratio of Q to ΔV, for *any* capacitor, is called its **capacitance** (C),

$$C = Q/\Delta V$$

so for a parallel-plate capacitor, we get

$$C = \frac{\varepsilon_0 A}{d}$$

The capacitance measures the capacity for holding charge. The greater the capacitance, the more charge can be stored on the plates at a given potential difference. The capacitance of any capacitor depends only on the size, shape, and separation of the conductors. From the definition, $C = Q/\Delta V$, the units of C are coulombs per volt. One coulomb per volt is renamed one **farad** (abbreviated F): 1 C/V = 1 F.

Example 11.12 A 10-nanofarad parallel-plate capacitor holds a charge of magnitude 50 μC on each plate.

(a) What is the potential difference between the plates?

(b) If the plates are separated by a distance of 0.2 mm, what is the area of each plate?

Solution.

(a) From the definition, $C = Q/\Delta V$, we find that

$$\Delta V = \frac{Q}{C} = \frac{50 \times 10^{-6}\ \text{C}}{10 \times 10^{-9}\ \text{F}} = 5000\ \text{V}$$

(b) From the equation $C = \varepsilon_0 A/d$, we can calculate the area, A, of each plate:

$$A = \frac{Cd}{\varepsilon_0} = \frac{(10 \times 10^{-9}\ \text{F})(0.2 \times 10^{-3}\ \text{m})}{8.85 \times 10^{-12}\ \text{C}^2/\text{N}\cdot\text{m}^2} = 0.23\ \text{m}^2$$

COMBINATIONS OF CAPACITORS

Capacitors are often arranged in combination in electric circuits. Here we'll look at two types of arrangements, the parallel combination and the series combination.

A collection of capacitors are said to be in **parallel** if they all share the same potential difference. The following diagram shows two capacitors wired in parallel:

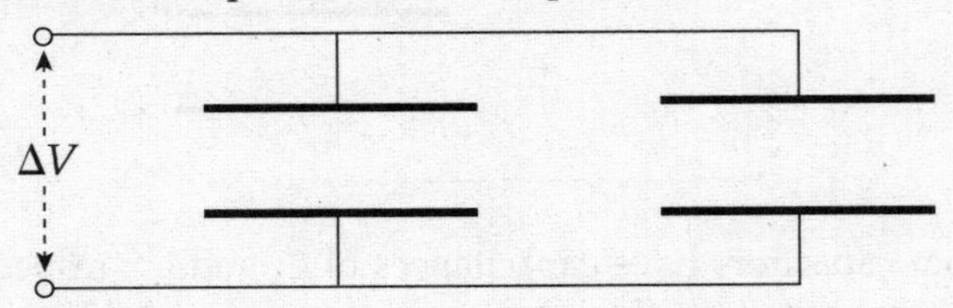

The top plates are connected by a wire and form a single equipotential; the same is true for the bottom plates. Therefore, the potential difference across one capacitor is the same as the potential difference across the other capacitor.

We want to find the capacitance of a *single* capacitor that would perform the same function as this combination. If the capacitances are C_1 and C_2, then the charge on the first capacitor is $Q_1 = C_1\Delta V$ and the charge on the second capacitor is $Q_2 = C_2\Delta V$. The total charge on the combination is $Q_1 + Q_2$, so the equivalent capacitance, C_P, must be

$$C_P = \frac{Q}{\Delta V} = \frac{Q_1 + Q_2}{\Delta V} = \frac{Q_1}{\Delta V} + \frac{Q_2}{\Delta V} = C_1 + C_2$$

So the **equivalent** capacitance of a collection of capacitors in parallel is found by adding the individual capacitances.

A collection of capacitors are said to be in **series** if they all share the same charge magnitude. The following diagram shows two capacitors wired in series:

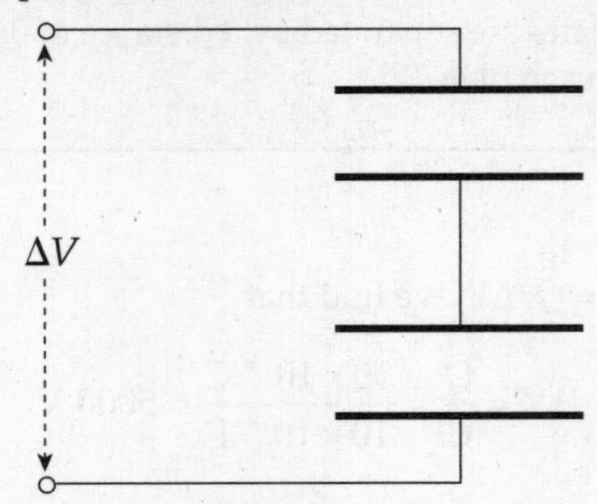

When a potential difference is applied, as shown, negative charge will be deposited on the bottom plate of the bottom capacitor; this will push an equal amount of negative charge away from the top plate of the bottom capacitor toward the bottom plate of the top capacitor. When the system has reached equilibrium, the charges on all the plates will have the same magnitude:

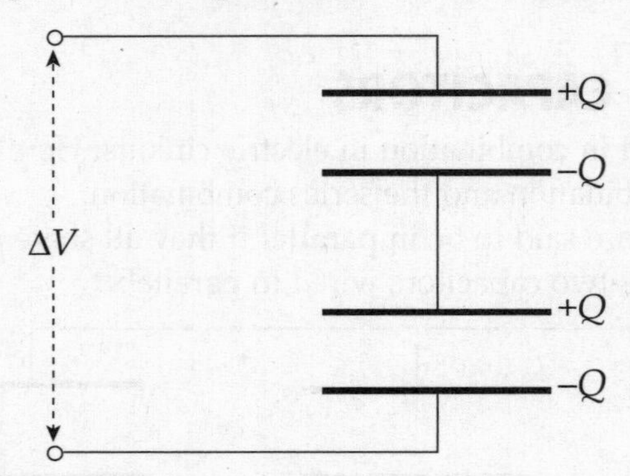

If the top and bottom capacitors have capacitances of C_1 and C_2, respectively, then the potential difference across the top capacitor is $\Delta V_1 = Q/C_1$, and the potential difference across the bottom capacitor is $\Delta V_2 = Q/C_2$. The total potential difference across the combination is $\Delta V_1 + \Delta V_1$, which must equal ΔV. Therefore, the equivalent capacitance, C_S, must be

$$C_S = \frac{Q}{\Delta V} = \frac{Q}{\Delta V_1 + \Delta V_2} = \frac{Q}{\frac{Q}{C_1} + \frac{Q}{C_2}} = \frac{1}{\frac{1}{C_1} + \frac{1}{C_2}}$$

We can write this in another form:

$$\frac{1}{C_S} = \frac{1}{C_1} + \frac{1}{C_2}$$

In words, the *reciprocal* of the capacitance of a collection of capacitors in series is found by adding the reciprocals of the individual capacitances.

Example 11.13 Given that $C_1 = 2\ \mu F$, $C_2 = 4\ \mu F$, an $C_3 = 6\ \mu F$, calculate the equivalent capacitance for the following combination:

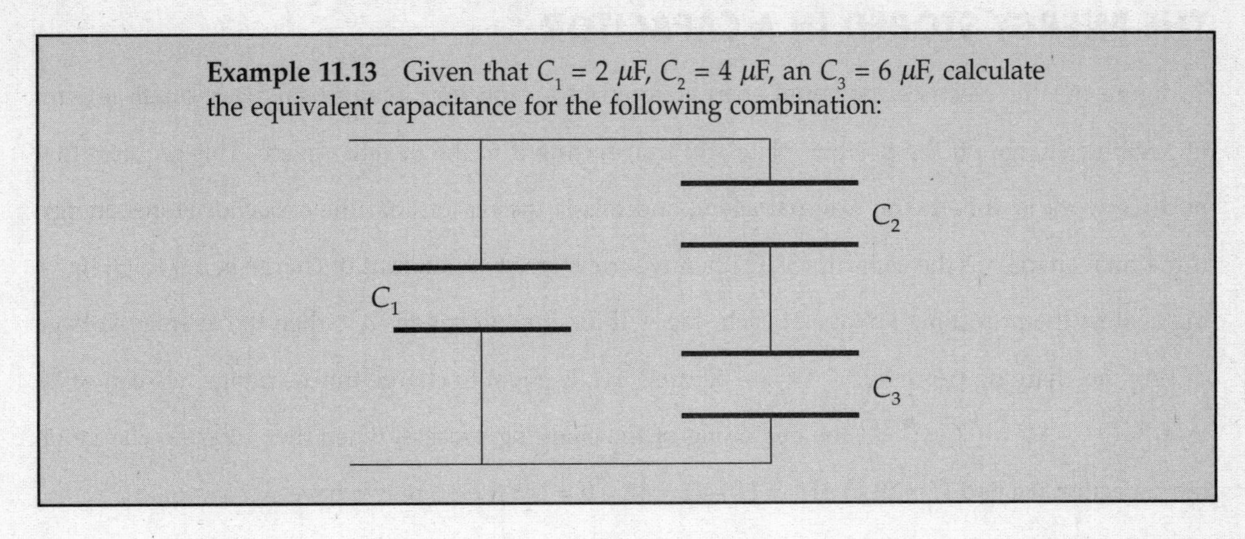

Solution. Notice that C_2 and C_3 are in series, and they are in parallel with C_1. That is, the capacitor equivalent to the series combination of C_2 and C_3 (which we'll call $C_{2\text{-}3}$) is in parallel with C_1. We can represent this as follows:

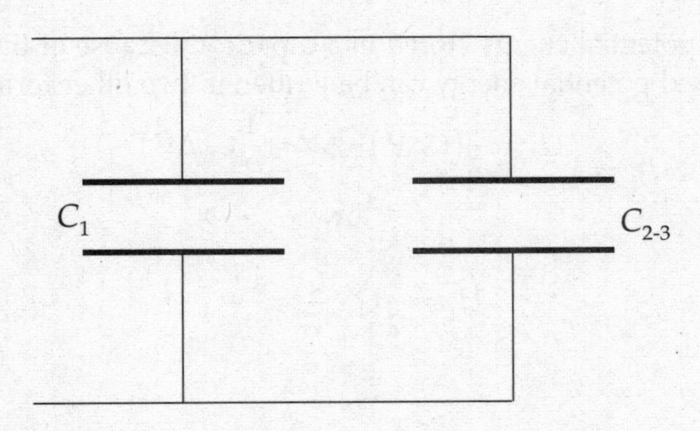

So, the first step is to find $C_{2\text{-}3}$:

$$\frac{1}{C_{2\text{-}3}} = \frac{1}{C_2} + \frac{1}{C_3} \quad \Rightarrow \quad C_{2\text{-}3} = \frac{C_2 C_3}{C_2 + C_3}$$

Now this is in parallel with C_1, so the overall equivalent capacitance ($C_{1\text{-}2\text{-}3}$) is

$$C_{1\text{-}2\text{-}3} = C_1 + C_{2\text{-}3} = C_1 + \frac{C_2 C_3}{C_2 + C_3}$$

Substituting in the given numerical values, we get

$$C_{1\text{-}2\text{-}3} = (2\ \mu\text{F}) + \frac{(4\ \mu\text{F})(6\ \mu\text{F})}{(4\ \mu\text{F}) + (6\ \mu\text{F})} = 4.4\ \mu\text{F}$$

THE ENERGY STORED IN A CAPACITOR

To figure out the electrical potential energy stored in a capacitor, imagine taking a small amount of negative charge off the positive plate and transferring it to the negative plate. This requires that positive work is done by an external agent, and this is the reason that the capacitor stores energy. If the final charge on the capacitor is Q, then we transferred an amount of charge equal to Q, fighting against the prevailing voltage at each stage. If the final voltage is ΔV, then the average voltage during the charging process is $\frac{1}{2}\Delta V$; so, because ΔU_E is equal to charge times voltage, we can write $\Delta U_E = Q \cdot \frac{1}{2}\Delta V = \frac{1}{2}Q\Delta V$. At the beginning of the charging process, when there was no charge on the capacitor, we had $U_i = 0$, so $\Delta U_E = U_f - U_i = U_f - 0 = U_f$; therefore, we have

$$U_E = \frac{1}{2}Q\Delta V$$

This is the electrical potential energy stored in a capacitor. Because of the definition $C = Q / \Delta V$, the equation for the stored potential energy can be written in two other forms:

$$U_E = \frac{1}{2}(C\Delta V) \cdot \Delta V = \frac{1}{2}C(\Delta V)^2$$

or

$$U_E = \frac{1}{2}Q \cdot \frac{Q}{C} = \frac{Q^2}{2C}$$

CHAPTER 11 REVIEW QUESTIONS

Solutions can be found in Chapter 18.

SECTION I: MULTIPLE CHOICE

1. Which of the following statements is/are true?

    I. If the electric field at a certain point is zero, then the electric potential at the same point is also zero.
    II. If the electric potential at a certain point is zero, then the electric field at the same point is also zero.
    III. The electric potential is inversely proportional to the strength of the electric field.

    (A) I only
    (B) II only
    (C) I and II only
    (D) I and III only
    (E) None are true

2. If the electric field does negative work on a negative charge as the charge undergoes a displacement from Position A to Position B within an electric field, then the electrical potential energy

    (A) is negative
    (B) is positive
    (C) increases
    (D) decreases
    (E) Cannot be determined from the information given

3. 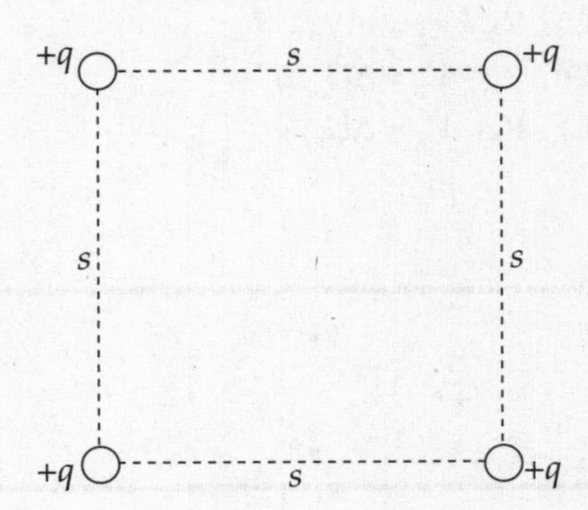

    The work required to assemble the system shown above, bringing each charge in from an infinite distance, is equal to

    (A) $\frac{1}{4\pi\varepsilon_0}\frac{4q^2}{s}$
    (B) $\frac{1}{4\pi\varepsilon_0}\frac{(4+\sqrt{2})q^2}{s}$
    (C) $\frac{1}{4\pi\varepsilon_0}\frac{6q^2}{s}$
    (D) $\frac{1}{4\pi\varepsilon_0}\frac{(4+2\sqrt{2})q^2}{s}$
    (E) $\frac{1}{4\pi\varepsilon_0}\frac{(8+2\sqrt{2})q^2}{s}$

4. Negative charges are accelerated by electric fields toward points

    (A) at lower electric potential
    (B) at higher electric potential
    (C) where the electric field is zero
    (D) where the electric field is weaker
    (E) where the electric field is stronger

5. A charge q experiences a displacement within an electric field from Position A to Position B. The change in the electrical potential energy is ΔU_E, and the work done by the electric field during this displacement is W_E. Then

(A) $V_B - V_A = W_E/q$
(B) $V_A - V_B = qW_E$
(C) $V_B - V_A = qW_E$
(D) $V_A - V_B = \Delta U_E/q$
(E) $V_B - V_A = \Delta U_E/q$

6.

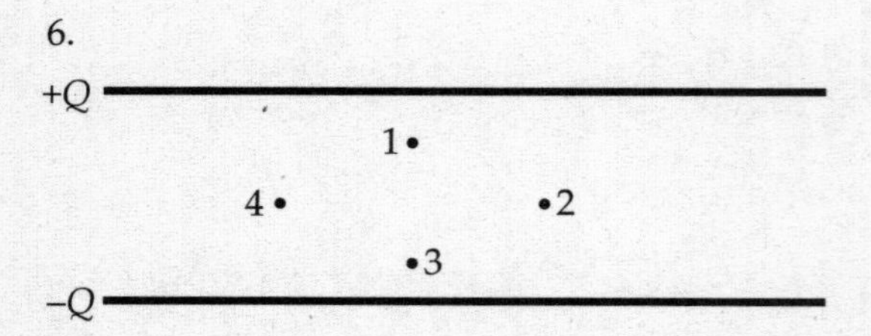

Which points in this uniform electric field (between the plates of the capacitor) shown above lie on the same equipotential?

(A) 1 and 2 only
(B) 1 and 3 only
(C) 2 and 4 only
(D) 3 and 4 only
(E) 1, 2, 3, and 4 all lie on the same equipotential since the electric field is uniform.

7.

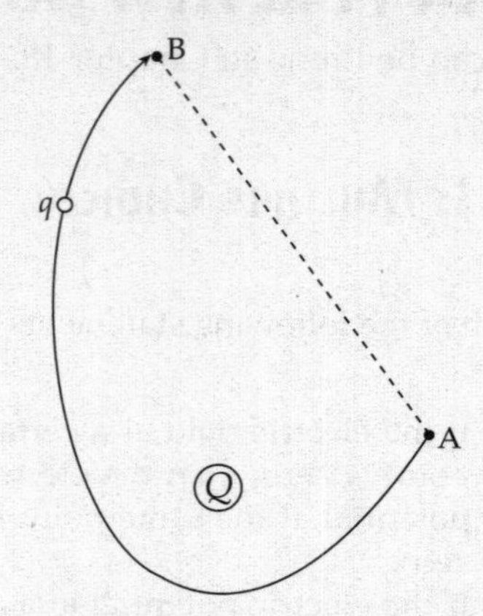

How much work would the electric field (created by the stationary charge Q) perform as a charge q is moved from Point A to B along the curved path shown?
$V_A = 200$ V, $V_B = 100$ V, $q = -0.05$ C, length of line segment AB = 10 cm, length of curved path = 20 cm.

(A) –10 J
(B) –5 J
(C) +5 J
(D) +10 J
(E) +2 J

8.

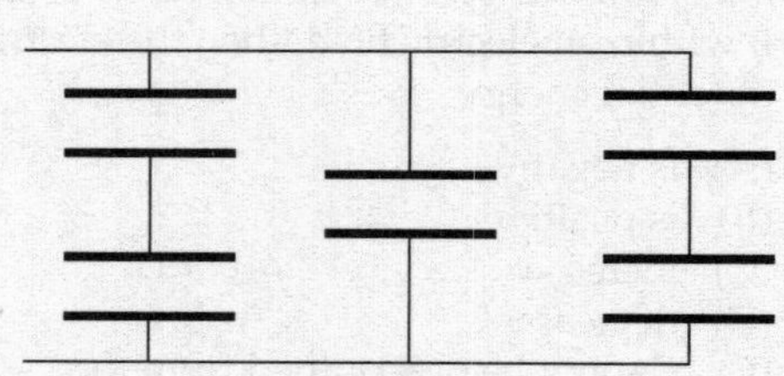

If each of the capacitors in the array shown above is C, what is the capacitance of the entire combination?

(A) C/2
(B) 2C/3
(C) 5C/6
(D) 2C
(E) 5C/3

SECTION II: FREE RESPONSE

1. In the figure shown, all four charges are situated at the corners of a square with sides *s*.

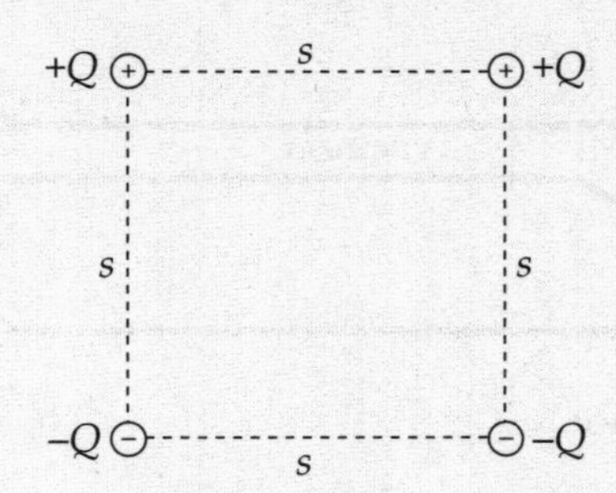

(a) What is the total electrical potential energy of this array of fixed charges?

(b) What is the electric field at the center of the square?

(c) What is the electric potential at the center of the square?

(d) Sketch (on the diagram) the portion of the equipotential surface that lies in the plane of the figure and passes through the center of the square.

(e) How much work would the electric field perform on a charge q as it moved from the midpoint of the right side of the square to the midpoint of the top of the square?

2. The figure below shows a parallel-plate capacitor. Each rectangular plate has length L and width w, and the plates are separated by a distance d.

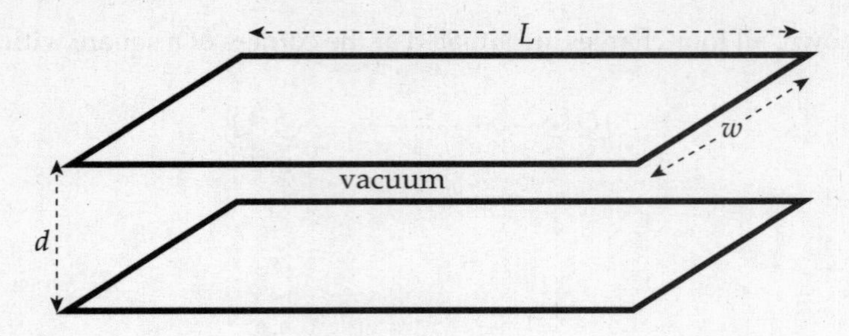

(a) Determine the capacitance.

An electron (mass m, charge $-e$) is shot horizontally into the empty space between the plates, midway between them, with an initial velocity of magnitude v_0. The electron just barely misses hitting the end of the top plate as it exits. (Ignore gravity.)

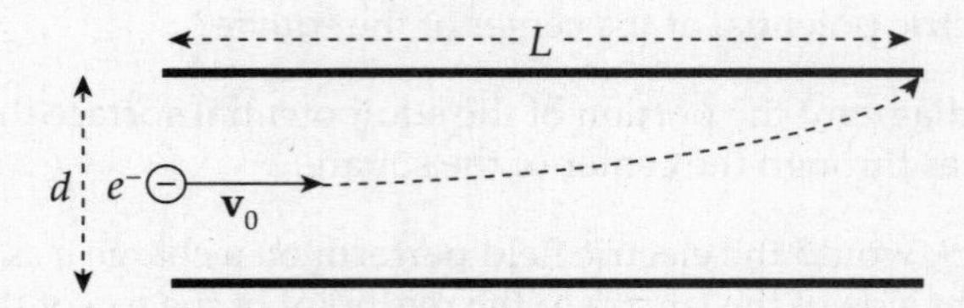

(b) In the diagram, sketch the electric field vector at the position of the electron when it has traveled a horizontal distance of $L/2$.

(c) In the diagram, sketch the electric force vector on the electron at the same position as in part (b).

(d) Determine the strength of the electric field between the plates. Write your answer in terms of L, d, m, e, and v_0.

(e) Determine the charge on the top plate.

(f) How much potential energy is stored in the capacitor?

3. A solid conducting sphere of radius a carries an excess charge of Q.

   (a) Determine the electric field magnitude, $E(r)$, as a function of r, the distance from the sphere's center.

   (b) Determine the potential, $V(r)$, as a function of r. Take the zero of potential at $r = \infty$.

   (c) On the diagrams below, sketch $E(r)$ and $V(r)$. (Cover at least the range $0 < r < 2a$.)

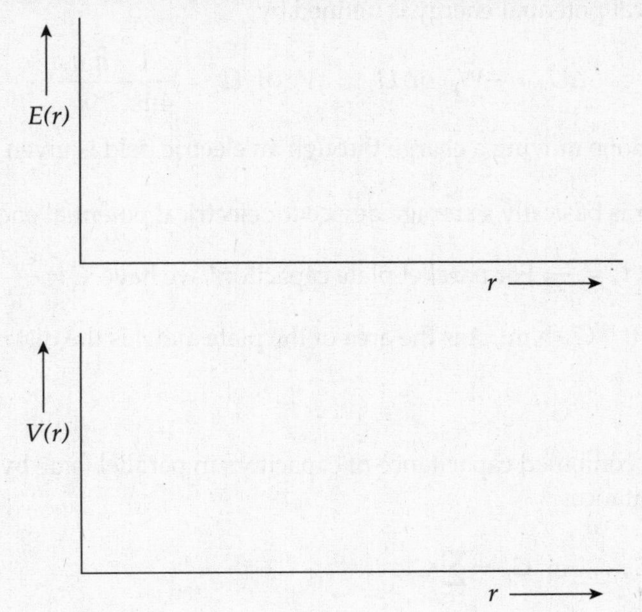

SUMMARY

- The electric potential (commonly referred to as the voltage) is defined as $V = \frac{\Delta U_E}{q}$.
- The electrical potential energy is defined by

$$\Delta U_E = -W_E \text{ or } U_E = qV \text{ or } U_E = \frac{1}{4\pi\varepsilon_0}\frac{q_1 q_2}{r}$$

- The work done moving a charge through an electric field is given by $W = qEd$.
- A capacitor is basically a storage device for electrical potential energy. Capacitance is given by $C = \frac{Q}{V}$. For parallel plate capacitors, we have $C = \frac{\varepsilon_0 A}{d}$ where $\varepsilon_0 = 8.85 \times 10^{-12}$ C²/Nm², A is the area of the plate and d is the distance between the plates.
- To find the combined capacitance of capacitors in parallel (side by side), simply add their capacitances

  $C_p = C_1 + C_2 + \ldots$ or $C_P = \sum_i C_i$.
- To find the capacitance of capacitors in series (one after another), add their inverses:

  $\frac{1}{C_S} = \frac{1}{C_1} + \frac{1}{C_2} + \ldots$ or $\frac{1}{C_S} = \sum_i \frac{1}{C_i}$.
- The electrical energy stored in a capacitor is given by either

  $U_C = \frac{1}{2}QV$ or $U_C = \frac{1}{2}CV^2$.

12

Direct Current Circuits

INTRODUCTION

In Chapter 10, when we studied electrostatic fields, we learned that, within a conductor, an electrostatic field cannot be sustained; the source charges move to the surface and the conductor forms a single equipotential. We will now look at conductors within which an electric field can be sustained because a power source maintains a potential difference across the conductor, allowing charges to continually move through it. This ordered motion of charge through a conductor is called **electric current**.

ELECTRIC CURRENT

Picture a piece of metal wire. Within the metal, electrons are zooming around at speeds of about a million m/s in random directions, colliding with other electrons and positive ions in the lattice. This constitutes charge in motion, but it doesn't constitute *net* movement of charge, because the electrons move randomly. If there's no net motion of charge, there's no current. However, if we were to create a potential difference between the ends of the wire, meaning if we set up an electric field, the electrons would experience an electric force, and they would start to drift through the wire. This is current. Although the electric field would travel through the wire at nearly the speed of light, the electrons themselves would still have to make their way through a crowd of atoms and other free electrons, so their **drift speed**, v_d, would be relatively slow, on the order of a millimeter per second.

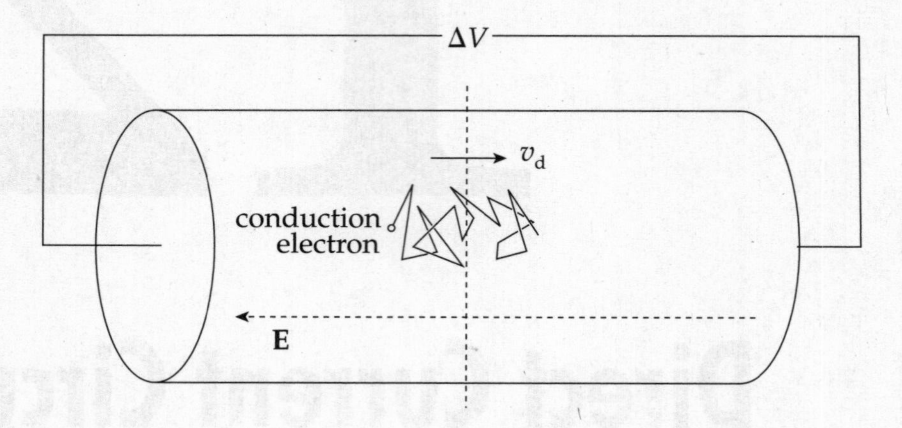

To measure the current, we have to measure how much charge crosses a plane per unit time. If an amount of charge of magnitude ΔQ crosses an imaginary plane in a time interval Δt, then the **average current** is

$$I_{avg} = \frac{\Delta Q}{\Delta t}$$

Because current is charge per unit time, it's expressed in coulombs per second. One coulomb per second is an **ampere** (abbreviated **A**), or amp. So 1 C/s = 1 A.

Although the charge carriers that constitute the current within a metal are electrons, the direction of the current is taken to be the direction that *positive* charge carriers would move. (This is explicitly stated on the AP Physics Exam.) So, if the conduction electrons drift to the right, we'd say the current points toward the left.

RESISTANCE

Let's say we had a copper wire and a glass fiber that had the same length and cross-sectional area, and that we hooked up the ends of the metal wire to a source of potential difference and measured the resulting current. If we were to do the same thing with the glass fiber, the current would probably be too small to measure, but why? Well, the glass provided more resistance to the flow of charge. If the potential difference is V and the current is I, then the **resistance** is

$$R = \frac{V}{I} \text{ or } V = IR$$

This is known as Ohm's Law. Not all devices are ohmic, but many are. Notice for the same voltage if the current is large, the resistance is low, and if the current is small, then resistance is high. The Δ in the equation above is often omitted, but you should always assume that, in this context $V = \Delta V =$ potential difference, also called voltage.

Because resistance is voltage divided by current, it is expressed in volts per amp. One volt per amp is one **ohm** (Ω, *omega*). So, $1 \text{ V/A} = 1 \ \Omega$.

RESISTIVITY

The resistance of an object depends on two things: the material it's made of and its shape. For example, again think of the copper wire and glass fiber of the same length and area. They have the same shape, but their resistances are different because they're made of different materials. Glass has a much greater intrinsic resistance than copper does; it has a greater **resistivity**. Each material has its own characteristic resistivity, and resistance depends on how the material is shaped. For a wire of length L and cross-sectional area A made of a material with resistivity ρ, resistance is given by:

$$R = \frac{\rho \ell}{A}$$

The resistivity of copper is around 10^{-8} Ω·m, while the resistivity of glass is *much* greater, around 10^{12} Ω·m.

> **Example 12.1** A wire of radius 1 mm and length 2 m is made of platinum (resistivity = 1×10^{-7} Ω·m). If a voltage of 9 V is applied between the ends of the wire, what will be the resulting current?

Solution. First, the resistance of the wire is given by the equation

$$R = \frac{\rho \ell}{A} = \frac{\rho \ell}{\pi r^2} = \frac{(1 \times 10^{-7} \ \Omega \cdot \text{m})(2 \text{ m})}{\pi (0.001 \text{ m})^2} = 0.064 \ \Omega$$

Then, from $I = V/R$, we get

$$I = \frac{V}{R} = \frac{9 \text{ V}}{0.064 \ \Omega} = 140 \text{ A}$$

ELECTRIC CIRCUITS

An electric current is maintained when the terminals of a voltage source (a battery, for example) are connected by a conducting pathway, in what's called a **circuit**. If the current always travels in the same direction through the pathway, it's called a **direct current**.

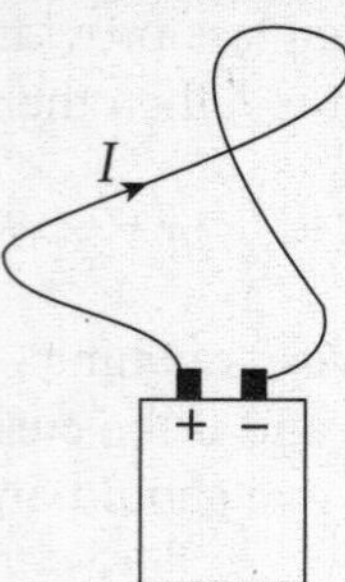

The job of the voltage source is to provide a potential difference called an **emf**, or **electromotive force**, which drives the flow of charge. The emf isn't really a force, it's the work done per unit charge, and it's measured in volts.

To try to imagine what's happening in a circuit in which a steady-state current is maintained, let's follow one of the charge carriers that's drifting through the pathway. (Remember we're pretending that the charge carriers are positive.) The charge is introduced by the positive terminal of the battery and enters the wire, where it's pushed by the electric field. It encounters resistance, bumping into the relatively stationary atoms that make up the metal's lattice and setting them into greater motion. So the electrical potential energy that the charge had when it left the battery is turning into heat. By the time the charge reaches the negative terminal, all of its original electrical potential energy is lost. In order to keep the current going, the voltage source must do positive work on the charge, forcing it to move from the negative terminal toward the positive terminal. The charge is now ready to make another journey around the circuit.

ENERGY AND POWER

When a carrier of positive charge q drops by an amount V in potential, it loses potential energy in the amount qV. If this happens in time t, then the rate at which this energy is transformed is equal to $(qV)/t = (q/t)V$. But q/t is equal to the current, I, so the rate at which electrical energy is transferred is given by the equation

$$P = IV$$

This equation works for the power delivered by a battery to the circuit as well as for resistors. The power dissipated in a resistor, as electrical potential energy is turned into heat, is given by $P = IV$, but, because of the relationship $V = IR$, we can express this in two other ways:

$$P = IV = I(IR) = I^2R$$

or

$$P = IV = \frac{V}{R} \cdot V = \frac{V^2}{R}$$

Resistors become hot when current passes through them.

CIRCUIT ANALYSIS

We will now develop a way of specifying the current, voltage, and power associated with each element in a circuit. Our circuits will contain three basic elements: batteries, resistors, and connecting wires. As we've seen, the resistance of an ordinary metal wire is negligible; resistance is provided by devices that control the current: **resistors**. All the resistance of the system is concentrated in the resistors, which are symbolized in a circuit diagram by this symbol:

Batteries are denoted by the symbol:

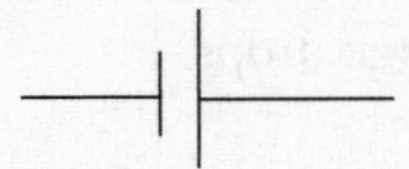

where the longer line represents the **positive** (higher potential) terminal, and the shorter line is the **negative** (lower potential) terminal. Sometimes a battery is denoted by more than one pair of such lines:

Here's a simple circuit diagram:

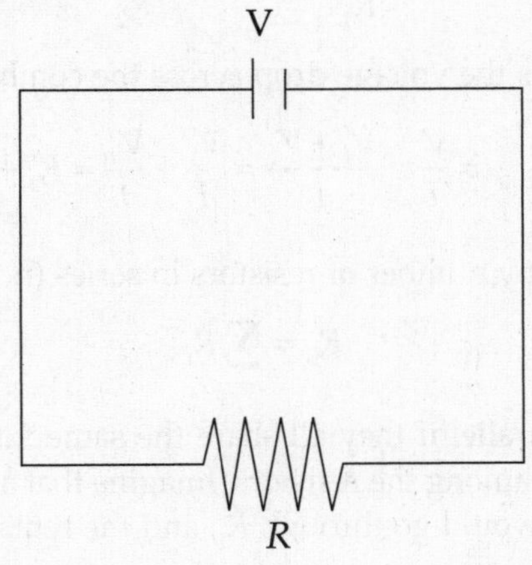

The electric potential (V) of the battery is indicated, as is the resistance (*R*) of the resistor. Determining the current in this case is straightforward, because there's only one resistor. The equation $V = IR$ gives us

$$I = \frac{V}{R}$$

Combinations of Resistors

Two common ways of combining resistors within a circuit is to place them either in **series** (one after the other),

R_1 R_2

or in **parallel** (that is, side-by-side):

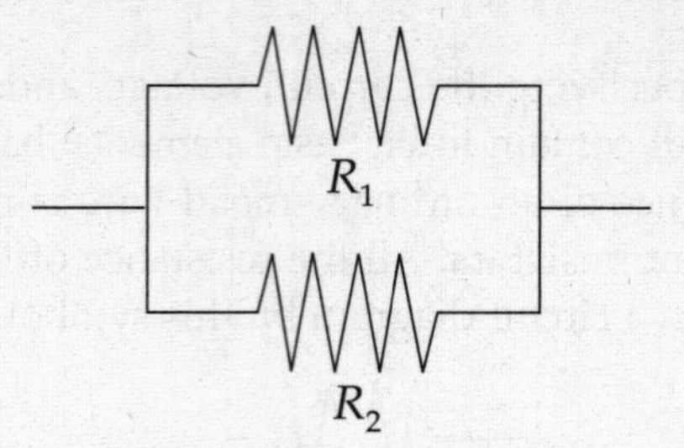

In order to simplify the circuit, our goal is to find the equivalent resistance of combinations. Resistors are said to be in series if they all share the same current and if the total voltage drop across them is equal to the sum of the individual voltage drops.

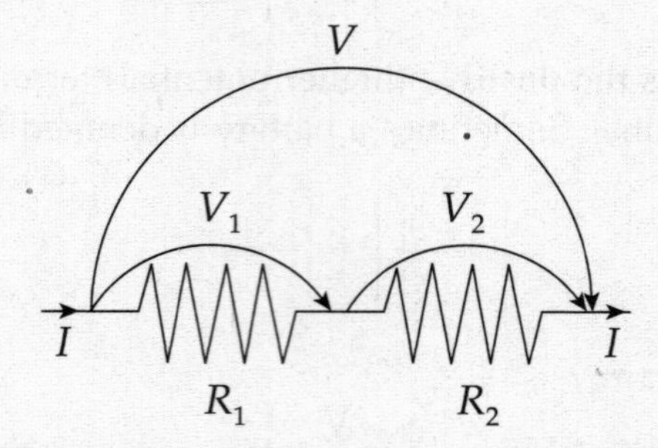

In this case, then, if V denotes the voltage drop across the combination, we have

$$R_{\text{equiv}} = \frac{V}{I} = \frac{V_1 + V_2}{I} = \frac{V_1}{I} + \frac{V_2}{I} = R_1 + R_2$$

This idea can be applied to any number of resistors in series (not just two):

$$R_S = \sum_i R_i$$

Resistors are said to be in parallel if they all share the same voltage drop, and the total current entering the combination is split among the resistors. Imagine that a current I enters the combination. It splits; some of the current, I_1, would go through R_1, and the remainder, I_2, would go through R_2.

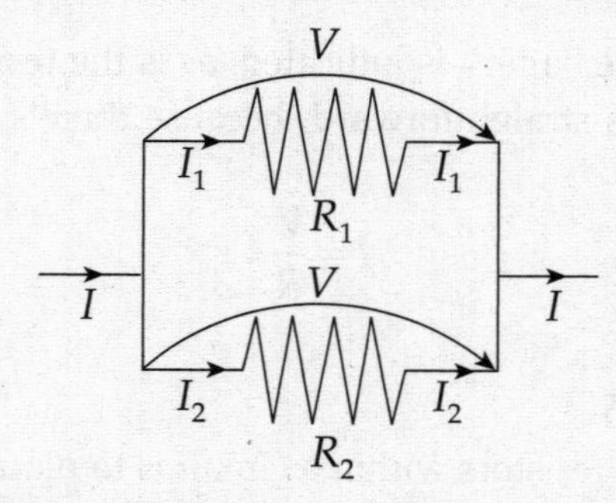

So if V is the voltage drop across the combination, we have

$$I = I_1 + I_2 \quad \Rightarrow \quad \frac{V}{R_{\text{equiv}}} = \frac{V}{R_1} + \frac{V}{R_2} \quad \Rightarrow \quad \frac{1}{R_{\text{equiv}}} = \frac{1}{R_1} + \frac{1}{R_2}$$

This idea can be applied to any number of resistors in parallel (not just two): The reciprocal of the equivalent resistance for resistors in parallel is equal to the sum of the reciprocals of the individual resistances:

$$\frac{1}{R_P} = \sum_i \frac{1}{R_i}$$

Example 12.2 Calculate the equivalent resistance for the following circuit:

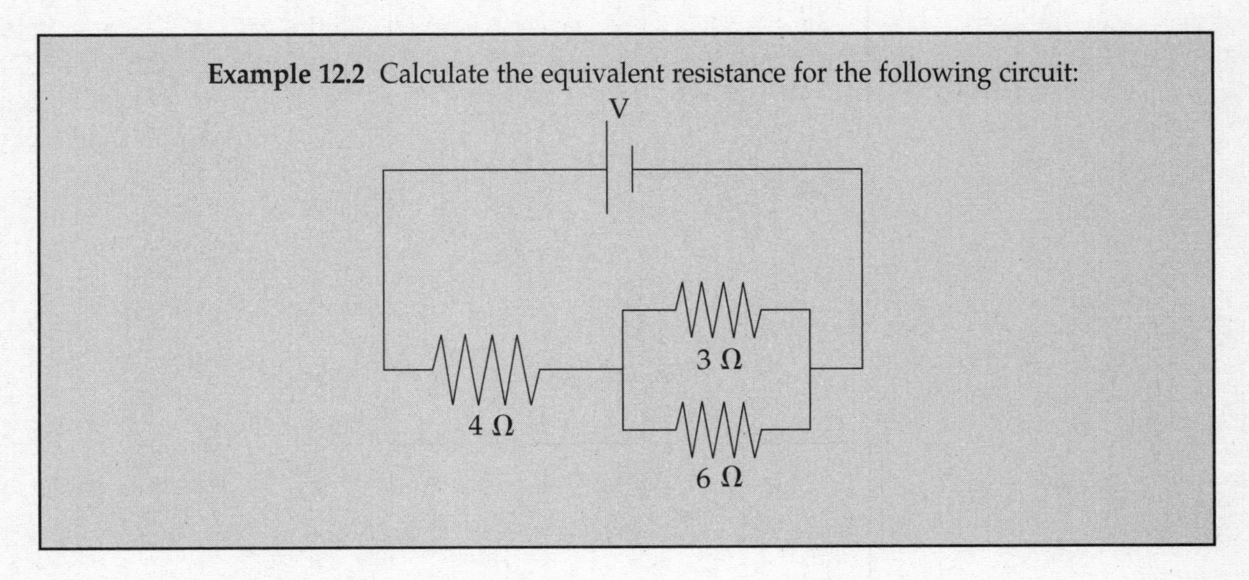

Solution. First find the equivalent resistance of the two parallel resistors:

$$\frac{1}{R_P} = \frac{1}{3\ \Omega} + \frac{1}{6\ \Omega} \quad \Rightarrow \quad \frac{1}{R_P} = \frac{1}{2\ \Omega} \quad \Rightarrow \quad R_P = 2\ \Omega$$

This resistance is in series with the 4 Ω resistor, so the overall equivalent resistance in the circuit is $R = 4\ \Omega + 2\ \Omega = 6\ \Omega$.

Example 12.3 Determine the current through each resistor, the voltage drop across each resistor, and the power given off (dissipated) as heat in each resistor of this circuit:

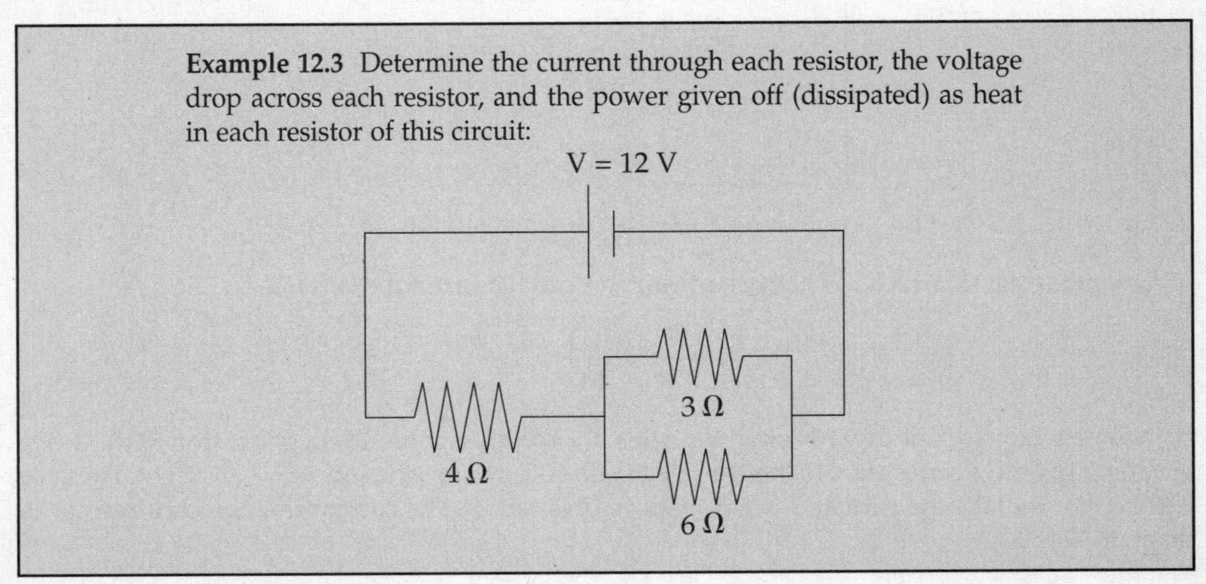

Solution. You might want to redraw the circuit each time we replace a combination of resistors by its equivalent resistance. From our work in the preceding example, we have

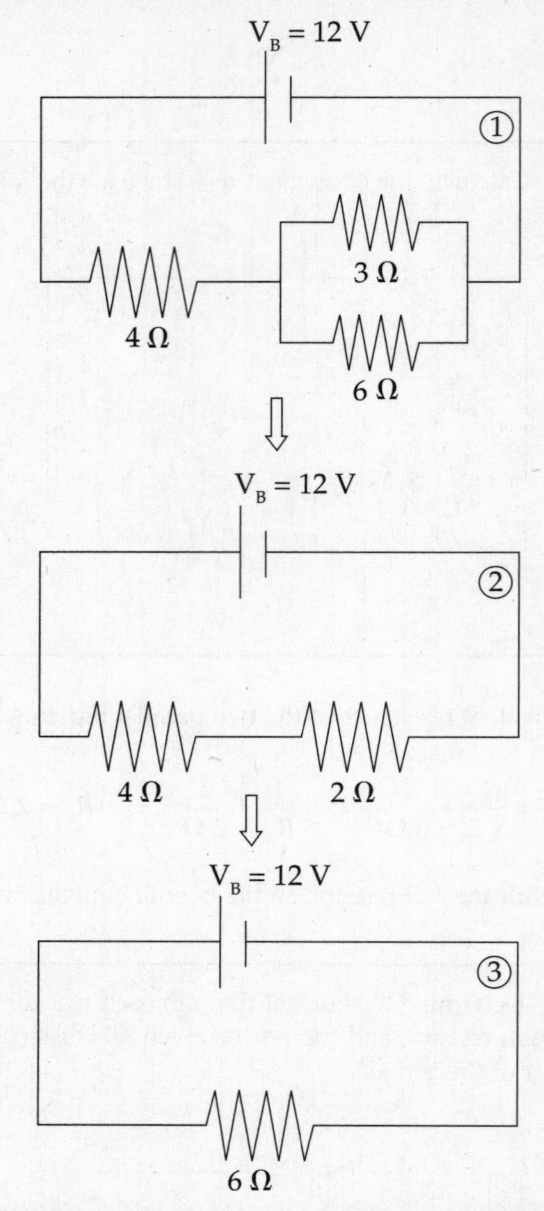

From diagram ③, which has just one resistor, we can figure out the current:

$$I = \frac{V_B}{R} = \frac{12 \text{ V}}{6\ \Omega} = 2 \text{ A}$$

Now we can work our way back to the original circuit (diagram ①). In going from ③ to ②, we are going back to a series combination, and what do resistors in series share? That's right, the same current. So, we take the current, I = 2 A, back to diagram ②. The current through each resistor in diagram ② is 2 A.

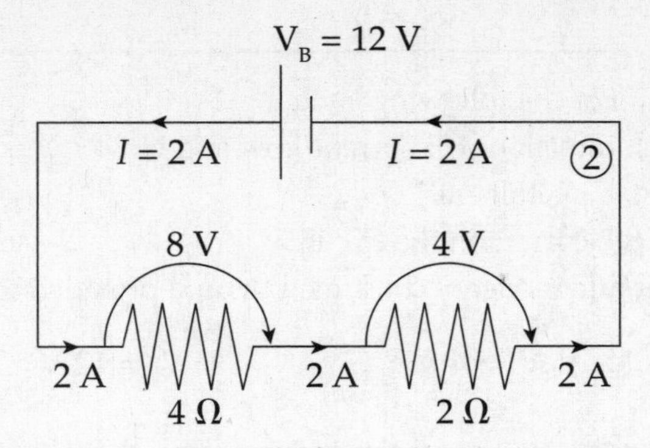

Since we know the current through each resistor, we can figure out the voltage drop across each resistor using the equation $V = IR$. The voltage drop across the 4 Ω resistor is (2 A)(4 Ω) = 8 V, and the voltage drop across the 2 Ω resistor is (2 A)(2 Ω) = 4 V. Notice that the total voltage drop across the two resistors is 8 V + 4 V = 12 V, which matches the emf of the battery.

Now for the last step; going from diagram ② back to diagram ①. Nothing needs to be done with the 4 Ω resistor; nothing about it changes in going from diagram ② to ①, but the 2 Ω resistor in diagram ② goes back to the parallel combination. And what do resistors in parallel share? The same voltage drop. So we take the voltage drop, $V = 4$ V, back to diagram ①. The voltage drop across each of the two parallel resistors in diagram ① is 4 V.

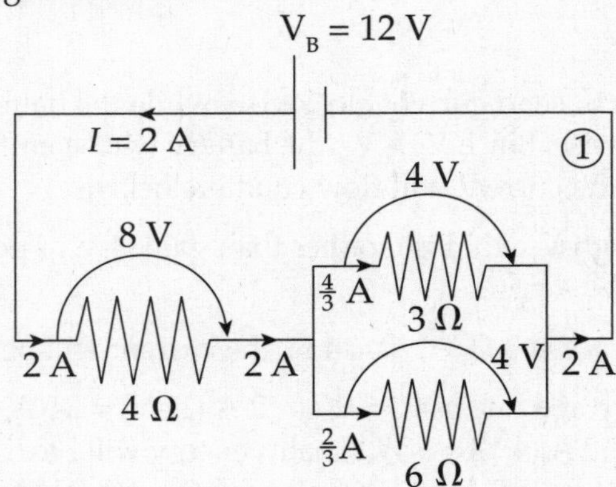

Since we know the voltage drop across each resistor, we can figure out the current through each resistor by using the equation $I = V/R$. The current through the 3 Ω resistor is (4 V)/(3 Ω) = $\frac{4}{3}$ A, and the current through the 6 Ω resistor is (4 V)/(6 Ω) = $\frac{2}{3}$ A. Note that the current entering the parallel combination (2 A) equals the total current passing through the individual resistors $\left(\frac{4}{3}\text{ A} + \frac{2}{3}\text{ A}\right)$. Again this was expected.

Finally, we will calculate the power dissipated as heat by each resistor. We can use any of the equivalent formulas: $P = IV$, $P = I^2R$, or $P = V^2/R$.

$$\text{For the 4 }\Omega\text{ resistor: } P = IV = (2\text{ A})(8\text{ V}) = 16\text{ W}$$

$$\text{For the 3 }\Omega\text{ resistor: } P = IV = (\tfrac{4}{3}\text{ A})(4\text{ V}) = \tfrac{16}{3}\text{ W}$$

$$\text{For the 6 }\Omega\text{ resistor: } P = IV = (\tfrac{2}{3}\text{ A})(4\text{ V}) = \tfrac{8}{3}\text{ W}$$

So, the resistors are dissipating a total of

$$16\text{ W} + \tfrac{16}{3}\text{ W} + \tfrac{8}{3}\text{ W} = 24\text{ W}$$

If the resistors are dissipating a total of 24 J every second, then they must be provided with that much power. This is easy to check: $P = IV = (2\text{ A})(12\text{ V}) = 24$ W.

Example 12.4 For the following circuit,

(a) In which direction will current flow and why?

(b) What's the overall emf?

(c) What's the current in the circuit?

(d) At what rate is energy consumed by, and provided to, this circuit?

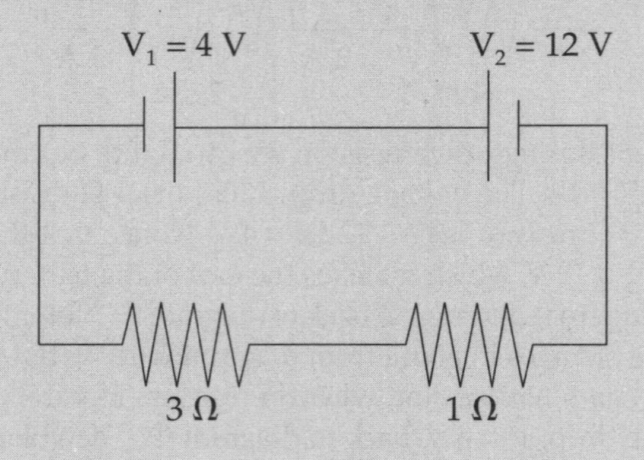

Solution.

(a) The battery V_1 wants to send current clockwise, while the battery V_2 wants to send current counterclockwise. Since $V_2 > V_1$, the battery whose emf is V_2 is the more powerful battery, so the current will flow counterclockwise.

(b) Charges forced through V_1 will lose, rather than gain, 4 V of potential, so the overall emf of this circuit is $V_2 - V_1 = 8$ V.

(c) Since the total resistance is 3 Ω + 1 Ω = 4 Ω, the current will be $I = (8 \text{ V})/(4\ \Omega) = 2$ A.

(d) V_2 will provide energy at a rate of $P_2 = IV_2 = (2 \text{ A})(12 \text{ V}) = 24$ W, while V_1 will absorb at a rate of $P_1 = IV_1 = (2 \text{ A})(4 \text{ V}) = 8$ W. Finally, energy will be dissipated in these resistors at a rate of $I^2R_1 + I^2R_2 = (2 \text{ A})^2(3\ \Omega) + (2 \text{ A})^2(1\ \Omega) = 16$ W. Once again, energy is conserved; the power delivered (24 W) equals the power taken (8 W + 16 W = 24 W).

Example 12.5 All real batteries contain **internal resistance**, *r*. Determine the current in the following circuit when the switch *S* is closed:

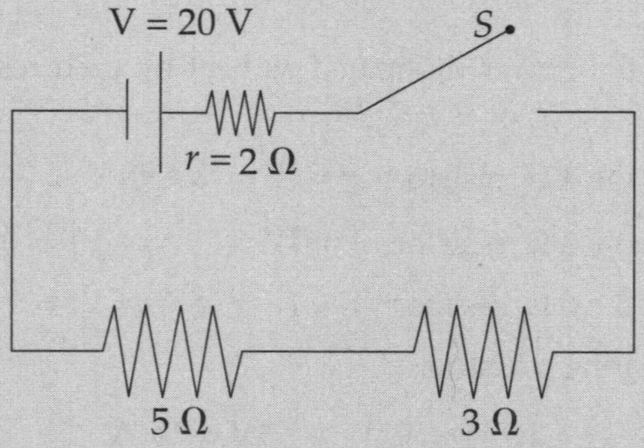

V = 20 V

S

r = 2 Ω

5 Ω

3 Ω

Solution. Before the switch is closed, there is no complete conducting pathway from the positive terminal of the battery to the negative terminal, so no current flows through the resistors. However, once the switch is closed, the resistance of the circuit is 2 Ω + 3 Ω + 5 Ω = 10 Ω, so the current in the circuit is I = (20 V)/(10 Ω) = 2 A. Often the battery and its internal resistance are enclosed in a dashed box:

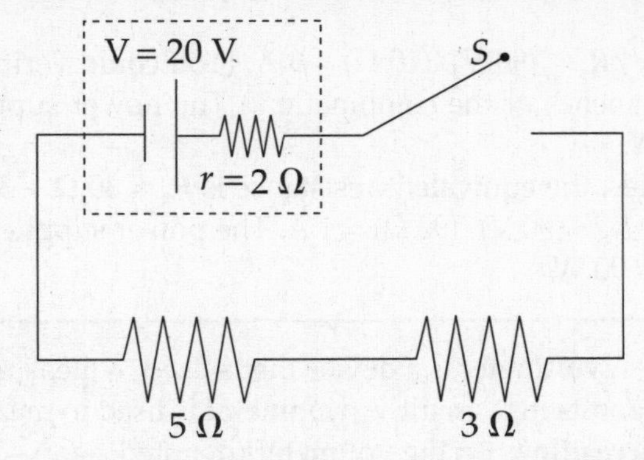

In this case, a distinction can be made between the emf of the battery and the actual voltage it provides once the current has begun. Since I = 2 A, the voltage drop across the internal resistance is Ir = (2 A)(2 Ω) = 4 V, so the effective voltage provided by the battery to the rest of the circuit—called the **terminal voltage**—is lower than the ideal voltage. It is $V = V_B - Ir$ = 20 V – 4 V = 16 V.

> **Example 12.6** A student has three 30 Ω resistors and an ideal 90 V battery. (A battery is *ideal* if it has a negligible internal resistance.) Compare the current drawn from—and the power supplied by—the battery when the resistors are arranged in parallel versus in series.

Solution. Resistors in series always provide an equivalent resistance that's greater than any of the individual resistances, and resistors in parallel always provide an equivalent resistance that's smaller than their individual resistances. So, hooking up the resistors in parallel will create the smallest resistance and draw the greatest total current:

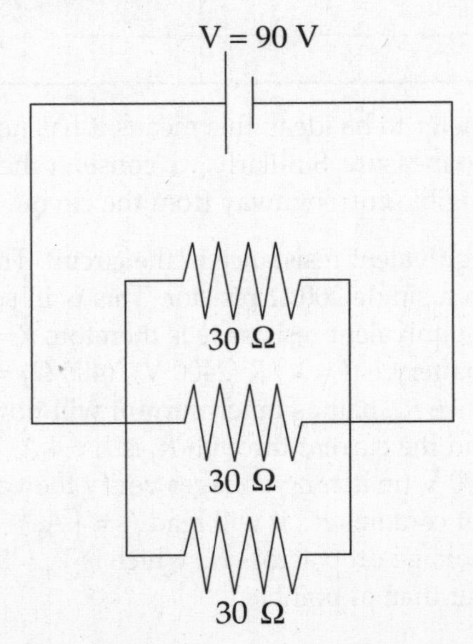

In this case, the equivalent resistance is

$$\frac{1}{R_P} = \frac{1}{30\ \Omega} + \frac{1}{30\ \Omega} + \frac{1}{30\ \Omega} \Rightarrow \frac{1}{R_P} = \frac{1}{10\ \Omega} \Rightarrow R_P = 10\ \Omega$$

and the total current is $I = V/R_P = (90\text{ V})/(10\ \Omega) = 9$ A. (You could verify that 3 A of current would flow in each of the three branches of the combination.) The power supplied by the battery will be $P = IV = (9\text{ A})(90\text{ V}) = 810$ W.

If the resistors are in series, the equivalent resistance is $R_S = 30\ \Omega + 30\ \Omega + 30\ \Omega = 90\ \Omega$, and the current drawn is only $I = V/R_S = (90\text{ V})/(90\ \Omega) = 1$ A. The power supplied by the battery in this case is just $P = IV = (1\text{ A})(90\text{ V}) = 90$ W.

Example 12.7 A **voltmeter** is a device that's used to measure the voltage between two points in a circuit. An **ammeter** is used to measure current. Determine the readings on the voltmeter (denoted —Ⓥ—) and the ammeter (denoted —Ⓐ—) in the circuit below.

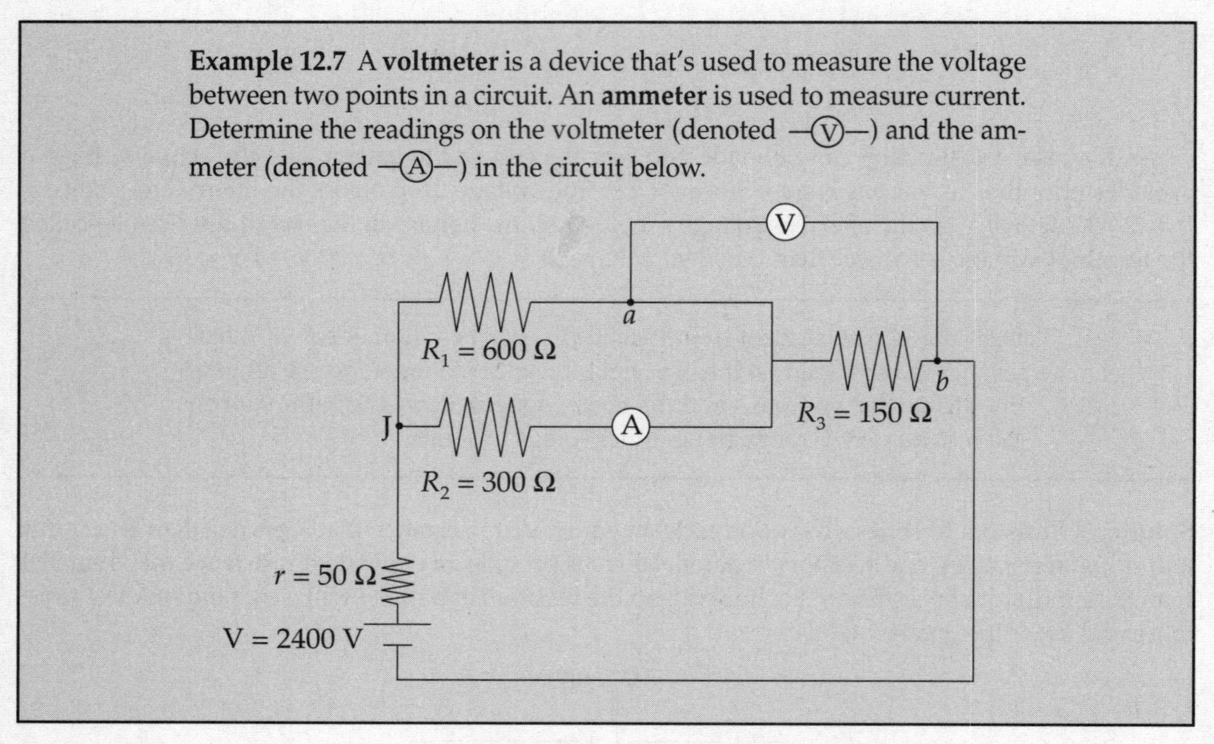

Solution. We consider the ammeter to be ideal; this means it has negligible resistance, so it doesn't alter the current that it's trying to measure. Similarly, we consider the voltmeter to have an extremely high resistance, so it draws negligible current away from the circuit.

Our first goal is to find the equivalent resistance in the circuit. The 600 Ω and 300 Ω resistors are in parallel; they're equivalent to a single 200 Ω resistor. This is in series with the battery's internal resistance, r, and R_3. The overall equivalent resistance is therefore $R = 50\ \Omega + 200\ \Omega + 150\ \Omega = 400\ \Omega$, so the current supplied by the battery is $I = V/R\ (2400\text{ V})/(400\ \Omega) = 6$ A. At the junction marked J, this current splits. Since R_1 is twice R_2, half as much current will flow through R_1 as through R_2; the current through R_1 is $I_1 = 2$ A, and the current through R_2 is $I_2 = 4$ A. The voltage drop across each of these resistors is $I_1R_1 = I_2R_2 = 1200$ V (matching voltages verify the values of currents I_1 and I_2). Since the ammeter is in the branch that contains R_2, it will read $I_2 = 4$ A.

The voltmeter will read the voltage drop across R_3, which is $V_3 = IR_3 = (6\text{ A})(150\ \Omega) = 900$ V. So the potential at point b is 900 V lower than at point a.

Example 12.8 The diagram below shows a point *a* at potential $V = 20$ V connected by a combination of resistors to a point (denoted G) that is **grounded**. *The* ***ground*** *is considered to be at potential zero.* If the potential at point *a* is maintained at 20 V, what is the current through R_3?

Solution. R_1 and R_2 are in parallel; their equivalent resistance is R_P, where

$$\frac{1}{R_P} = \frac{1}{4\ \Omega} + \frac{1}{6\ \Omega} \quad \Rightarrow \quad R_P = \frac{24}{10}\ \Omega = 2.4\ \Omega$$

R_P is in series with R_3, so the equivalent resistance is:

$$R = R_P + R_3 = (2.4\ \Omega) + (8\ \Omega) = 10.4\ \Omega$$

and the current that flows through R_3 is

$$I_3 = \frac{V}{R} = \frac{20\ \text{V}}{10.4\ \Omega} = 1.9\ \text{A}$$

KIRCHHOFF'S RULES

When the resistors in a circuit cannot be classified as either in series or in parallel, we need another method for analyzing the circuit. The rules of Gustav Kirchhoff (pronounced "Keer koff") can be applied to any circuit:

The Loop Rule. The sum of the potential differences (positive and negative) that traverse any closed loop in a circuit must be zero.

The Junction Rule. The total current that enters a junction must equal the total current that leaves the junction. (This is also known as the **Node Rule.**)

The Loop Rule just says that, starting at any point, by the time we get back to that same point by following any closed loop, we have to be back to the same potential. Therefore, the total drop in potential must equal the total rise in potential. Put another way, the Loop Rule says that all the decreases in electrical potential energy (for example, caused by resistors in the direction of the current) must be balanced by all the increases in electrical potential energy (for example, caused by a source of emf from the negative to positive terminal). So the Loop Rule is basically a re-statement of the Law of Conservation of Energy.

Similarly, the Junction Rule simply says that the charge (per unit time) that goes into a junction must equal the charge (per unit time) that comes out. This is basically a statement of the Law of Conservation of Charge.

In practice, the Junction Rule is straightforward to apply. The most important things to remember about the Loop Rule can be summarized as follows:

- When going across a resistor in the *same* direction as the current, the potential *drops* by *IR*.
- When going across a resistor in the *opposite* direction from the current, the potential *increases* by *IR*.
- When going from the negative to the positive terminal of a source of emf, the potential *increases* by V.
- When going from the positive to the negative terminal of a source of emf, the potential *decreases* by V.

Example 12.9 Use Kirchhoff's Rules to determine the current through R_2 in the following circuit:

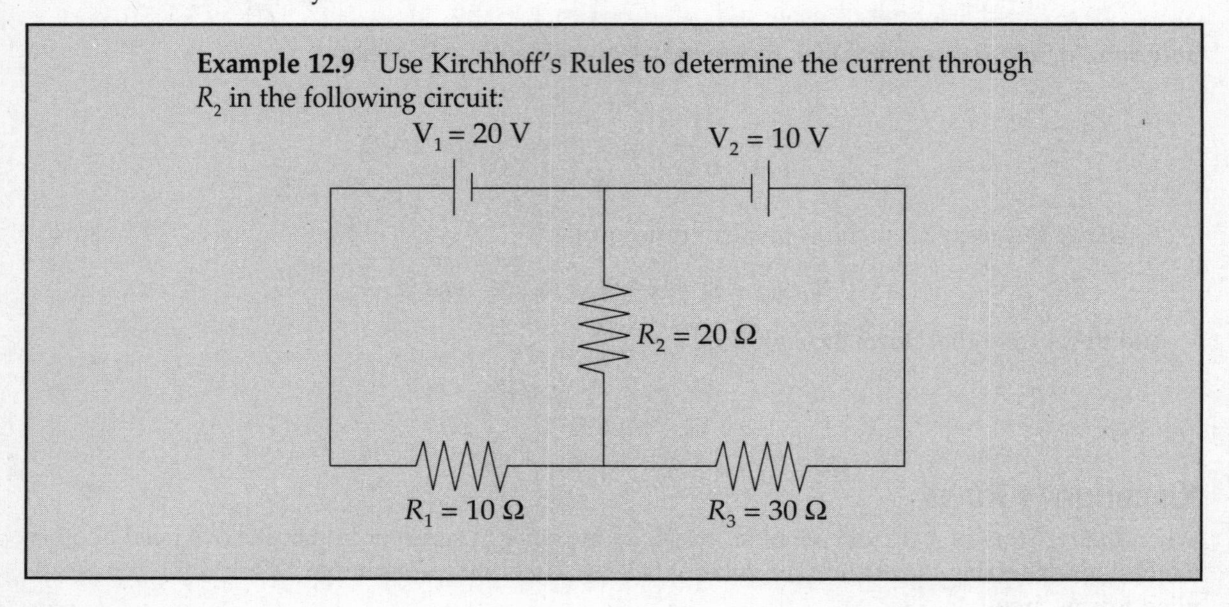

Solution. First let's label some points in the circuit.

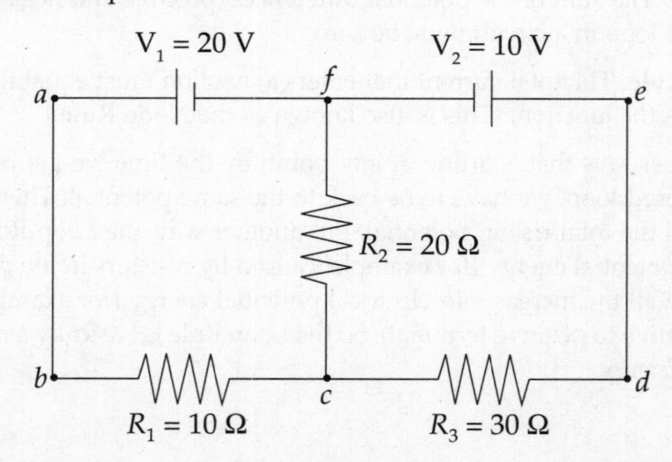

The points c and f are junctions (nodes). We have two nodes and three branches: one branch is *fabc*, another branch is *cdef*, and the third branch is *cf*. Each branch has one current throughout. If we label the current in *fabc* I_1 and the current in branch *cdef* I_2 (with the directions as shown in the diagram below), then the current in branch *cf* must be $I_1 - I_2$, by the Junction Rule: I_1 comes into c, and a total of $I_2 + (I_1 - I_2) = I_1$ comes out.

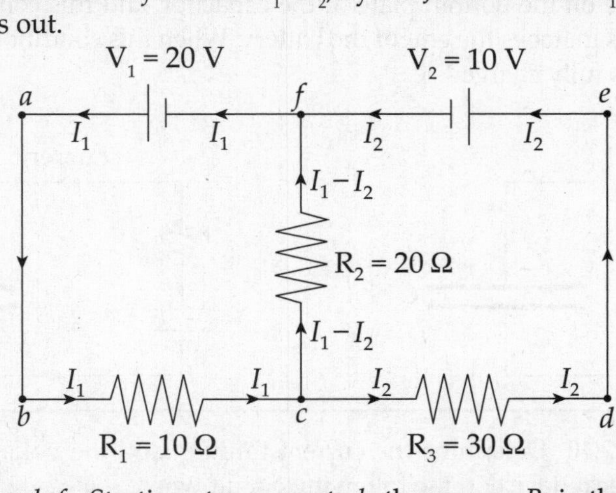

Now pick a loop; say, *abcfa*. Starting at a, we go to b, then across R_1 in the direction of the current, so the potential drops by I_1R_1. Then we move to c, then up through R_2 in the direction of the current, so the potential drops by $(I_1 - I_2)R_2$. Then we reach f, turn left and travel through V_1 from the negative to the positive terminal, so the potential increases by V_1. We now find ourselves back at a. By the Loop Rule, the total change in potential around this closed loop must be zero, and

$$-I_1R_1 - (I_1 - I_2)R_2 + V_1 = 0 \quad (1)$$

Since we have two unknowns (I_1 and I_2), we need two equations, so now pick another loop; let's choose *cdefc*. From c to d, we travel across the resistor in the direction of the current, so the potential drops by I_2R_3. From e to f, we travel through V_2 from the positive to the negative terminal, so the potential *drops* by V_2. Heading down from f to c, we travel across R_2 but in the direction opposite to the current, so the potential *increases* by $(I_1 - I_2)R_2$. At c, our loop is completed, so

$$-I_2R_3 - V_2 + (I_1 - I_2)R_2 = 0 \quad (2)$$

Substituting in the given numerical values for R_1, R_2, R_3, V_1, and V_2, and simplifying, these two equations become

$$3I_1 - 2I_2 = 2 \quad (1')$$

$$2I_1 - 5I_2 = 1 \quad (2')$$

Solving this pair of simultaneous equations, we get

$$I_1 = \frac{8}{11}\text{ A} = 0.73\text{ A} \quad \text{and} \quad I_2 = \frac{1}{11}\text{ A} = 0.09\text{ A}$$

So the current through R_2 is $I_1 - I_2 = \frac{7}{11}$ A = 0.64 A.

The choice of directions of the currents at the beginning of the solution was arbitrary. Don't worry about trying to guess the actual direction of the current in a particular branch. Just pick a direction, stick with it, and obey the Junction Rule. At the end, when you solve for the values of the branch current, a negative value will alert you that the direction of the current is actually opposite to the direction you originally chose for it in your diagram.

RESISTANCE–CAPACITANCE (RC) CIRCUITS

Capacitors are typically charged by batteries. Once the switch in the diagram on the left is closed, electrons are attracted to the positive terminal of the battery and leave the top plate of the capacitor. Electrons also accumulate on the bottom plate of the capacitor, and this continues until the voltage across the capacitor plates matches the emf of the battery. When this condition is reached, the current stops and the capacitor is fully charged.

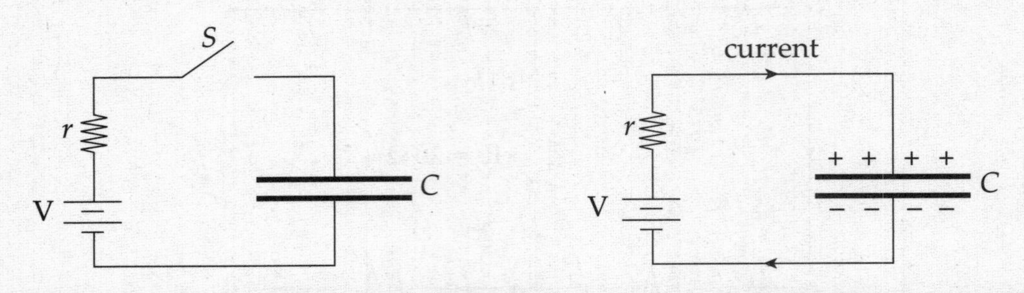

Example 12.10 Determine the current through and the voltage across each electrical device in the following circuit when

(a) the switch has just been closed

(b) the switch has been thrown for a long period of time

12 V

3 Ω

6 Ω

Solution.

(a) When the switch has just been closed, the capacitor is treated like a wire with no resistance. The circuit behaves like a simple parallel circuit. There are 12 V across the 6 Ω resistor so 2 amps of current flow through it. There are 12 V across the 3 Ω resistor so 4 amps flow through it. By Kirchhoff's Law, the currents add to 6 amps through the battery and the equivalent resistance of the circuit is 2 Ω.

(b) After the switch has been closed for a long period of time, the capacitor behaves like an infinite resistor. There are still 12 V across the 6 Ω resistor so 2 amps of current still flow through it. If the capacitor behaves like an infinite resistor, no current can flow through that branch. There will be no current through the 3 Ω resistor. The capacitor will have 12 V across it and there will be no voltage across the 3 Ω resistor (otherwise current would be flowing through it). The current will be 2 amps through the battery and the equivalent resistance of the circuit is 6 Ω.

Example 12.11 Determine the current through and the voltage across each electrical device in the following circuit when

(a) the switch has just been closed

(b) the switch has been thrown for a long period of time

Solution.

(a) Immediately after one throws the switch, the capacitor exhibits behavior similar to that of a wire with no resistance through it. The circuit can be redrawn and thought of as

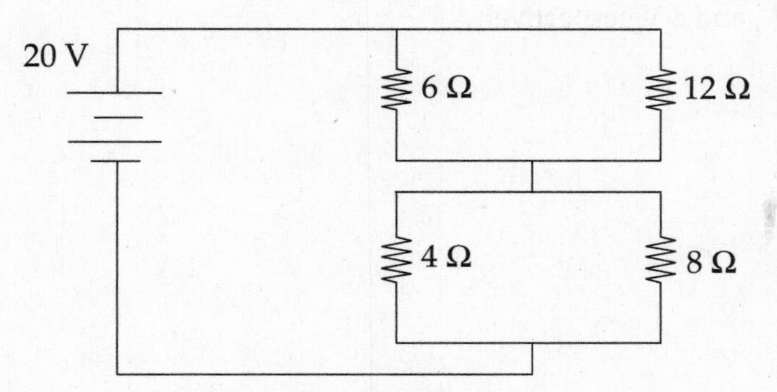

The 6 Ω and 12 Ω resistors are in parallel and can be thought of as 4 Ω (from $1/R_{eq} = 1/R_1 + 1/R_2$). Similarly, the 4 Ω and 8 Ω resistors are in parallel and can be thought of as 2.67 Ω (again, from $1/R_{eq} = 1/R_1 + 1/R_2$). The equivalent resistance of these two branches, which are in series with each other, is 6.67 Ω (from $R_{eq} = R_1 + R_2$).

The current through the battery is therefore 3 amps (from Ohm's Law, $V = IR$).

The current through the 6 Ω and 12 Ω (top two resistors) is 3 amps traveling through an effective resistance of 4 Ω. The voltage drop across the top branch is 12 V (from Ohm's Law). The currents through the 6 Ω and 12 Ω resistors are 2 amps and 1 amp, respectively (again from Ohm's Law).

The current through the 4 Ω and 8 Ω (bottom two resistors) is also 3 amps through an effective resistance of 2.67 Ω. The voltage drop across the branch is 8 V (from Ohm's Law or from the fact that the voltage across each series section must sum to the total voltage across the battery). The currents through the 4 Ω and 8 Ω resistors are 2 amps and 1 amp, respectively (again from Ohm's Law).

(b) When the switch has been closed for a long period of time, the capacitor behaves just like an infinite resistor. The circuit can be redrawn and thought of as

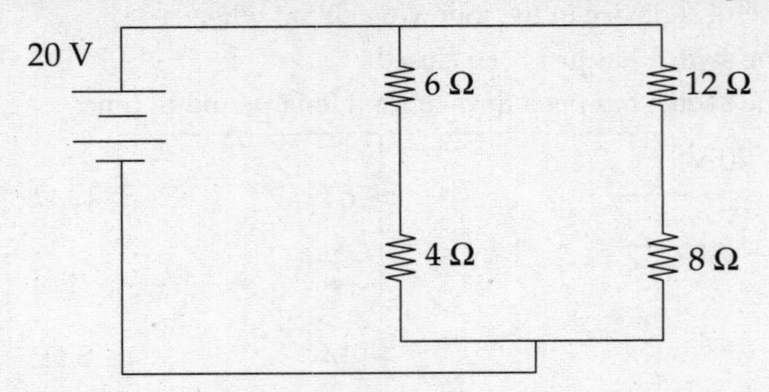

The 6 Ω and 4 Ω resistors are in series and can be thought of as a 10 Ω resistor (from $R_{eq} = R_1 + R_2$). Similarly, the 12 Ω and 8 Ω resistors can be thought of as 20 Ω (again, from $R_{eq} = R_1 + R_2$). Since the branches are in parallel, the voltage is 20 volts across each path and the current can be calculated by Ohm's Law ($V = IR$); 2 amps go through the left branch (the 6 Ω and 4 Ω resistor) and 1 amp goes through the right branch (the 12 Ω and 8 Ω resistor).

The current through the battery is therefore 3 amps (Kirchhoff's Node Rule).

The voltage across each of the 6 Ω, 4 Ω, 12 Ω, and 8 Ω resistors can be obtained from Ohm's Law. They are 12 V, 8 V, 12 V, and 8 V, respectively.

CHAPTER 12 REVIEW QUESTIONS

Solutions can be found in Chapter 18.

Section I: Multiple Choice

1. A wire made of brass and a wire made of silver have the same length, but the diameter of the brass wire is 4 times the diameter of the silver wire. The resistivity of brass is 5 times greater than the resistivity of silver. If R_B denotes the resistance of the brass wire and R_S denotes the resistance of the silver wire, which of the following is true?

   (A) $R_B = \frac{5}{16} R_S$
   (B) $R_B = \frac{4}{5} R_S$
   (C) $R_B = \frac{5}{4} R_S$
   (D) $R_B = \frac{5}{2} R_S$
   (E) $R_B = \frac{16}{5} R_S$

2. For an ohmic conductor, doubling the voltage without changing the resistance will cause the current to

   (A) decrease by a factor of 4
   (B) decrease by a factor of 2
   (C) remain unchanged
   (D) increase by a factor of 2
   (E) increase by a factor of 4

3. If a 60-watt light bulb operates at a voltage of 120 V, what is the resistance of the bulb?

   (A) 2 Ω
   (B) 30 Ω
   (C) 240 Ω
   (D) 720 Ω
   (E) 7200 Ω

4. A battery whose emf is 40 V has an internal resistance of 5 Ω. If this battery is connected to a 15 Ω resistor R, what will the voltage drop across R be?

   (A) 10 V
   (B) 30 V
   (C) 40 V
   (D) 50 V
   (E) 70 V

5. 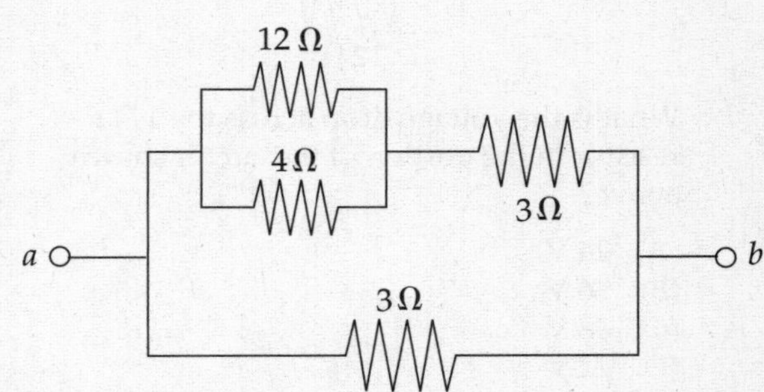

   Determine the equivalent resistance between points a and b.

   (A) 0.167 Ω
   (B) 0.25 Ω
   (C) 0.333 Ω
   (D) 1.5 Ω
   (E) 2 Ω

6. 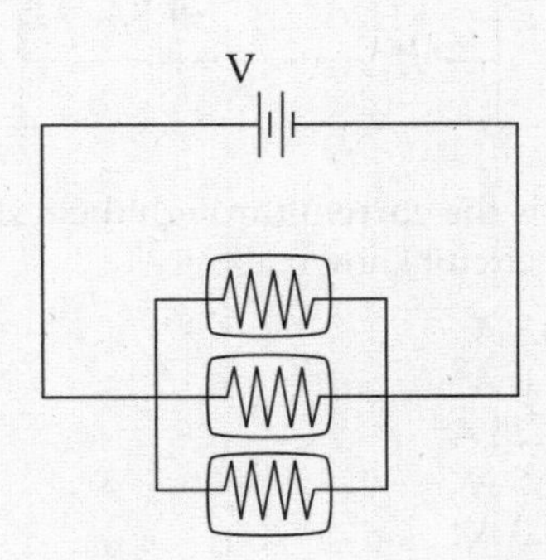

   Three identical light bulbs are connected to a source of emf, as shown in the diagram above. What will happen if the middle bulb burns out?

   (A) All the bulbs will go out.
   (B) The light intensity of the other two bulbs will decrease (but they won't go out).
   (C) The light intensity of the other two bulbs will increase.
   (D) The light intensity of the other two bulbs will remain the same.
   (E) More current will be drawn from the source of emf.

7.

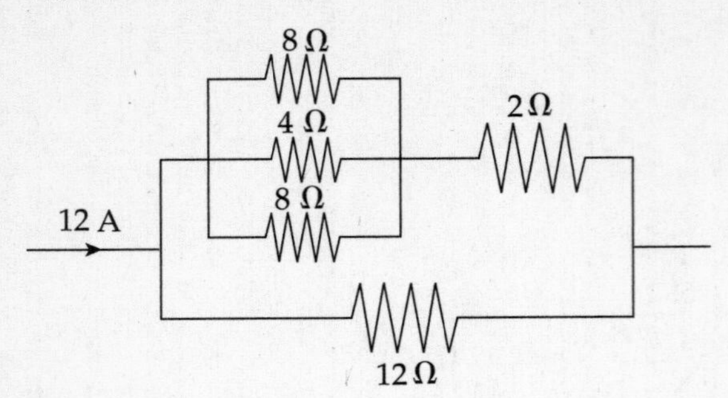

What is the voltage drop across the 12 Ω resistor in the portion of the circuit shown above?

(A) 24 V
(B) 36 V
(C) 48 V
(D 72 V
(E) 144 V

8.

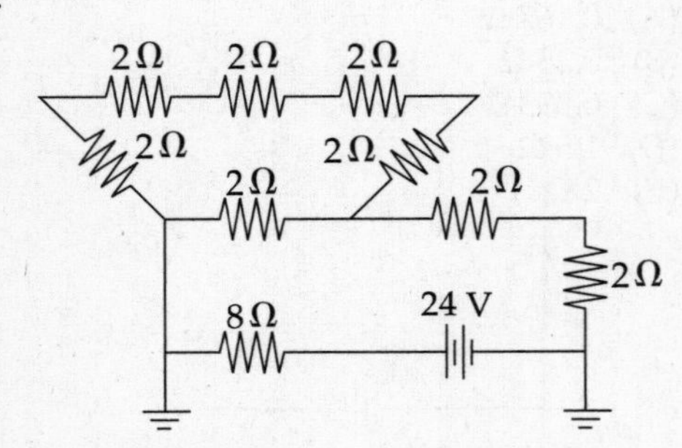

What is the current through the 8 Ω resistor in the circuit shown above?

(A) 0.5 A
(B) 1.0 A
(C) 1.25 A
(D) 1.5 A
(E) 3.0 A

9. How much energy is dissipated as heat in 20 s by a 100 Ω resistor that carries a current of 0.5 A?

(A) 50 J
(B) 100 J
(C) 250 J
(D) 500 J
(E) 1000 J

SECTION II: FREE RESPONSE

1. Consider the following circuit:

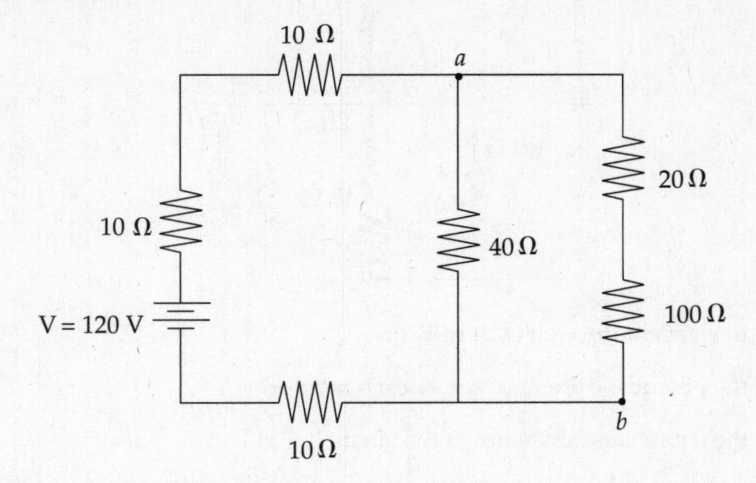

(a) At what rate does the battery deliver energy to the circuit?

(b) Find the current through the 40 Ω resistor.

(c) (i) Determine the potential difference between points *a* and *b*.

(ii) At which of these two points is the potential higher?

(d) Find the energy dissipated by the 100 Ω resistor in 10 s.

(e) Given that the 100 Ω resistor is a solid cylinder that's 4 cm long, composed of a material whose resistivity is 0.45 Ω·m, determine its radius.

2. Consider the following circuit:

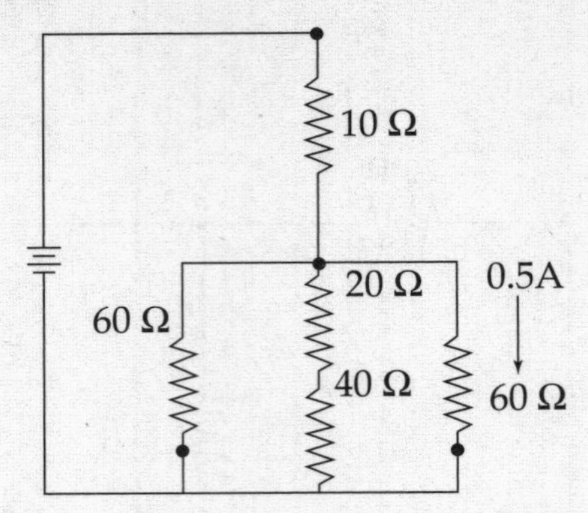

(a) What is the current through each resistor?

(b) What is the potential difference across each resistor?

(c) What is the equivalent resistance of the circuit?

SUMMARY

- The resistance of an object can be determined by $R = \frac{\rho \ell}{A}$ where ρ is the resistivity (a property of the material), ℓ is the length, and A is the cross-sectional area.
- The current is the rate at which charge is transferred and given by $I_{AVG} = \frac{\Delta Q}{\Delta t}$.
- Many objects obey Ohm's Law, which is given by $V = IR$.
- The electrical power in a circuit is given by

$$P = IV \text{ or } P = I^2R \text{ or } P = \frac{V^2}{R}$$

  This is the same power we've encountered in our discussion of energy $P = \frac{W}{t}$.

- In a series circuit

  $V_B = V_1 + V_2 + \ldots$

  $I_B = I_1 = I_2 = \ldots$

  $R_{eq} = R_1 + R_2 + \ldots$

  $R_S = \sum_i R_i$

  In a parallel circuit

  $V_B = V_1 = V_2 = \ldots$

  $I_B = I_1 + I_2 + \ldots$

  $1/R_{eq} = 1/R_1 + 1/R_2 + \ldots$

  $\frac{1}{R_P} = \sum_i \frac{1}{R_i}$

- Kirchhoff's Loop Rule tells us that the sum of the potential differences in any closed loop in a circuit must be zero.
- Kirchhoff's Junction Rule (Node Rule) tells us the total current that enters a junction must equal the total current that leaves the junction.

13

Magnetic Forces and Fields

INTRODUCTION

In Chapter 10, we learned that electric charges are the sources of electric fields and that other charges experience an electric force in those fields. The charges generating the field were assumed to be at rest, because if they weren't, then another force field would have been generated in addition to the electric field. Electric charges *that move* are the sources of **magnetic fields**, and other charges that move can experience a magnetic force in these fields.

MAGNETIC FIELDS

Similar to our discussion about electric fields, the space surrounding a magnet is permeated by a magnetic field. The direction of the magnetic field is defined as pointing out of the north end of a magnet and into the south end of a magnet as illustrated below.

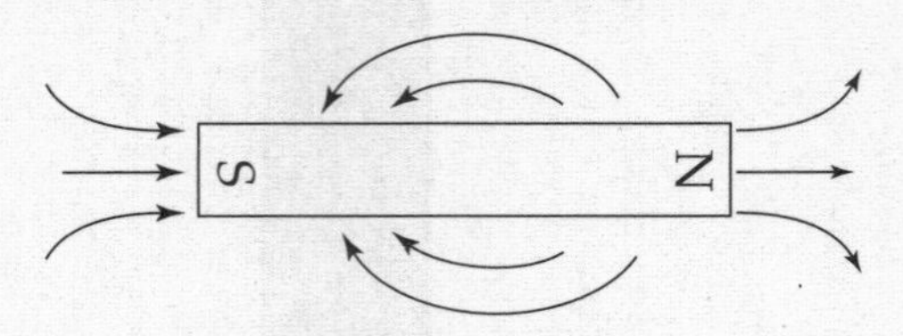

When two magnets get near each other the magnetic fields interfere with each other and can be drawn as follows. Note that, for simplicity's sake, only the field lines closest to the poles are shown.

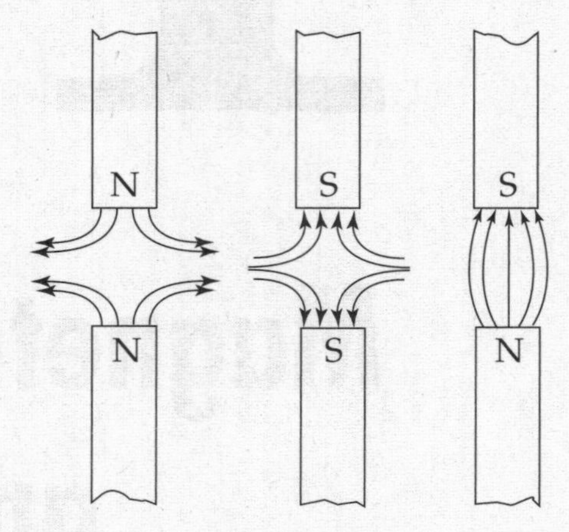

Notice there is a curve to the above fields. We call the field uniform if the field lines are parallel and of equal strength. It is easy to recognize a uniform magnetic field to the right, left, top of the page, or bottom of the page. But you will also see a field going into or out of the page. A field into the page looks as if there were a north pole of a magnet above the page pointing down at the south pole of a magnet that is below the page. It is represented by an area with X's going into the page. A field coming out of the page looks as if there were a north pole of a magnet below the page pointing up at the south pole of a magnet that is above the page. It is represented by an area with dots (•) coming out of the page.

INTO	OUT OF
X X X X	• • • •
X X X X	• • • •
X X X X	• • • •
X X X X	• • • •

THE MAGNETIC FORCE ON A MOVING CHARGE

If a particle with charge q moves with velocity **v** through a magnetic field **B**, it will experience a magnetic force, $\mathbf{F}_B$, with magnitude:

$$F_B = |q| vB \sin\theta \quad (1)$$

where θ is the angle between **v** and **B**. From this equation, we can see that if the charge is at rest, then $v = 0$ immediately gives us $F_B = 0$. This tells us that magnetic forces only act on moving charges. Also, if **v** is parallel (or antiparallel) to **B**, then $F_B = 0$ since, in either of these cases, $\sin\theta = 0$. So, only charges that cut across the magnetic field lines will experience a magnetic force. Furthermore, the magnetic force is maximized when **v** is perpendicular to **B**, since if $\theta = 90°$, then $\sin\theta$ is equal to 1, its maximum value.

The direction of $\mathbf{F}_B$ is always perpendicular to both **v** and **B** and depends on the sign of the charge q and the direction of $\mathbf{v} \times \mathbf{B}$ (which can be found by using the right-hand rule).

If q is *positive*, use your *right* hand and the *right*-hand rule.

If q is *negative*, use your *left* hand and the *left*-hand rule.

Whether you use the right-hand rule or the left-hand rule, you will always follow these steps:

1. Orient your hand so that your thumb points in the direction of the velocity **v**.
2. Point your fingers in the direction of **B**.
3. The direction of $\mathbf{F}_B$ will then be perpendicular to your palm.

Think of your palm pushing with the force $\mathbf{F}_B$; the direction it pushes is the direction of $\mathbf{F}_B$.

Right-Hand Rule:
For determining the direction of the magnetic force, $\mathbf{F}_B$, on a *positive* charge

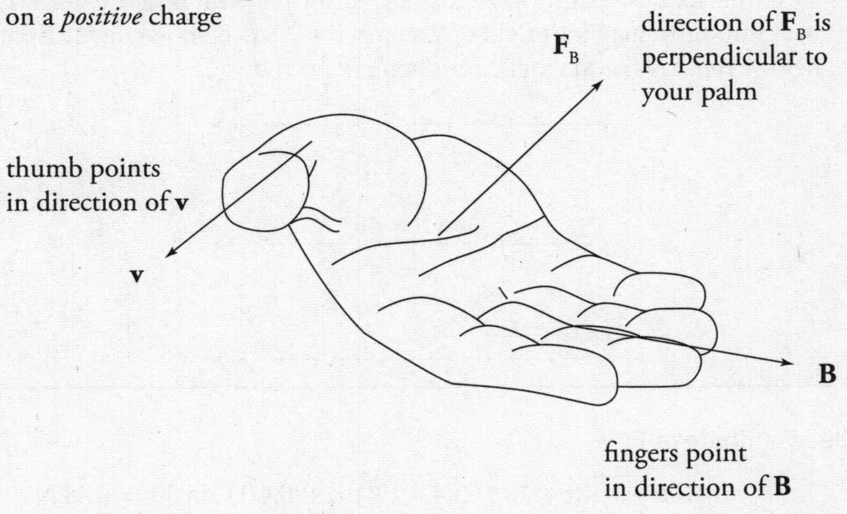

$\mathbf{F}_B$

direction of $\mathbf{F}_B$ is perpendicular to your palm

thumb points in direction of **v**

v

B

fingers point in direction of **B**

Note that there are fundamental differences between the electric force and magnetic force on a charge. First, a magnetic force acts on a charge only if the charge is moving; the electric force acts on a charge whether it moves or not. Second, the direction of the magnetic force is always perpendicular to the magnetic field, while the electric force is always parallel (or antiparallel) to the electric field.

Example 13.1 For each of the following charged particles moving through a magnetic field, determine the force acting on the charge.

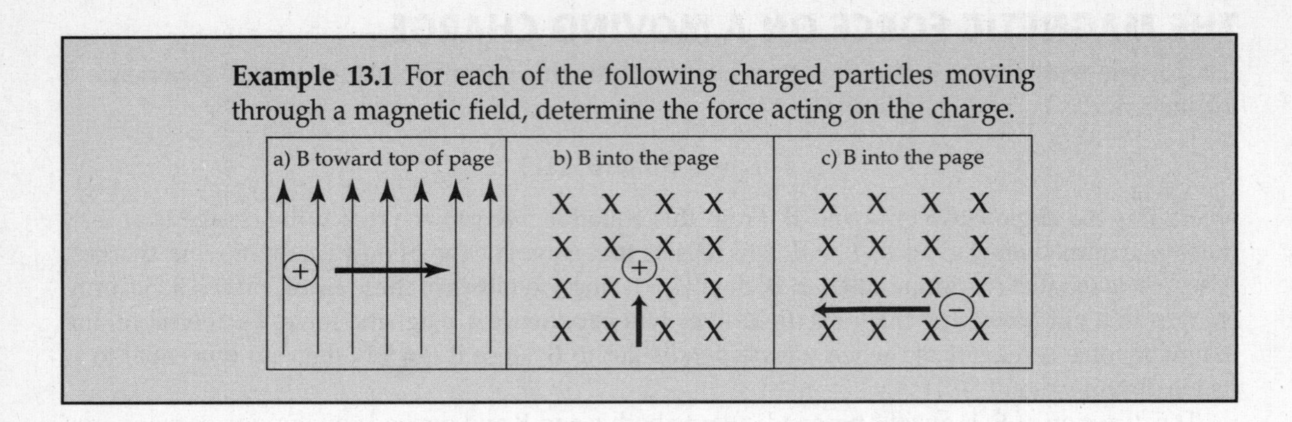

Solution.

a) If you point your fingers to the top of the page and thumb to the right of the page, your palm should point out of the page. The force is out of the page.

b) If you point your fingers into the page and thumb to the top of the page your palm should point to the left of the page. The force is to the left of the page.

c) If you point your fingers into the page and thumb to the left of the page, your palm should point to the bottom of the page. The force of a positive charge would be toward the bottom of the page, but because this is a negative charge, the force points up.

The SI unit for the magnetic field is the **tesla** (abbreviated **T**) which is one newton per ampere-meter. Another common unit for magnetic field strength is the **gauss** (abbreviated **G**); 1 G = 10^{-4} T.

Example 13.2 A charge $+q = +6 \times 10^{-6}$ C moves with speed $v = 4 \times 10^5$ m/s through a magnetic field of strength $B = 0.4$ T, as shown in the figure below. What is the magnetic force experienced by q?

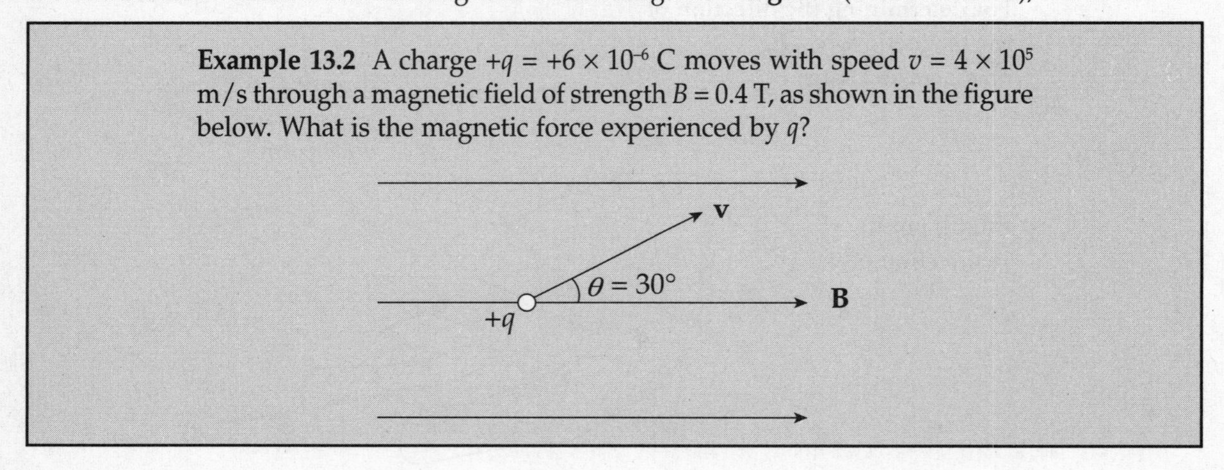

Solution. The magnitude of $\mathbf{F}_B$ is

$$F_B = qvB \sin\theta = (6 \times 10^{-6}\text{ C})(4 \times 10^5\text{ m/s})(0.4\text{ T}) \sin 30° = 0.48\text{ N}$$

By the right-hand rule, the direction is into the plane of the page, which is symbolized by ×.

Example 13.3 A particle of mass m and charge $+q$ is projected with velocity **v** (in the plane of the page) into a uniform magnetic field **B** that points into the page. How will the particle move?

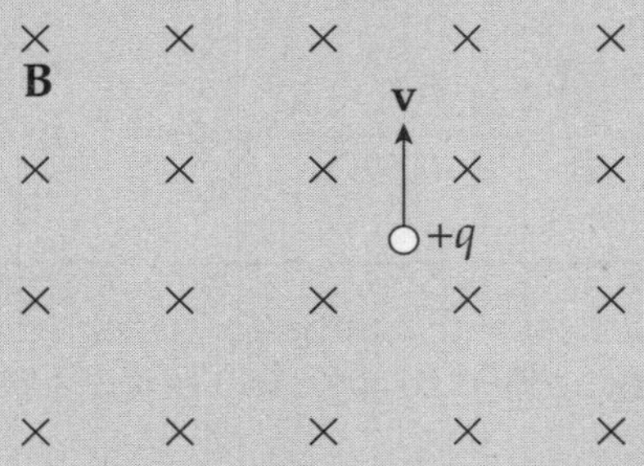

Solution. Since **v** is perpendicular to **B**, the particle will feel a magnetic force of strength qvB, which will be directed perpendicular to **v** (and to **B**) as shown:

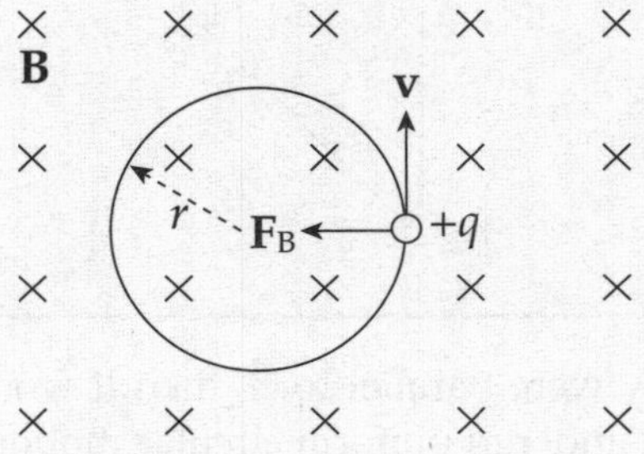

Since $\mathbf{F}_B$ is always perpendicular to **v**, the particle will undergo uniform circular motion; $\mathbf{F}_B$ will provide the centripetal force. Notice that, because $\mathbf{F}_B$ is always perpendicular to **v**, the magnitude of **v** will not change, just its direction. *Magnetic forces alone cannot change the speed of a charged particle, they can only change its direction of motion.* The radius of the particle's circular path is found from the equation $F_B = F_C$:

$$qvB = \frac{mv^2}{r} \quad \Rightarrow \quad r = \frac{mv}{qB}$$

Example 13.4 A particle of charge $-q$ is shot into a region that contains an electric field, **E**, crossed with a perpendicular magnetic field, **B**. If $E = 2 \times 10^4$ N/C and $B = 0.5$ T, what must be the speed of the particle if it is to cross this region without being deflected?

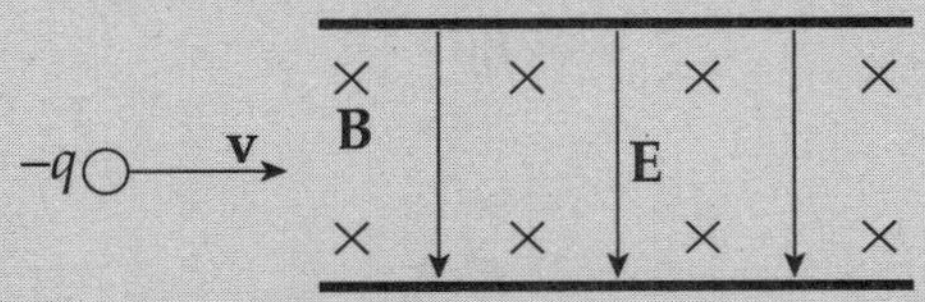

Solution. If the particle is to pass through undeflected, the electric force it feels has to be canceled by the magnetic force. In the diagram above, the electric force on the particle is directed upward (since the charge is negative and **E** is downward), and the magnetic force is directed downward by the right-hand rule. So $\mathbf{F}_E$ and $\mathbf{F}_B$ point in opposite directions, and in order for their magnitudes to balance, qE must equal qvB, so v must equal E/B, which in this case gives

$$v = \frac{E}{B} = \frac{2\times10^4 \text{ N/C}}{0.5 \text{ T}} = 4\times10^4 \text{ m/s}$$

Example 13.5 A particle with charge $+q$, traveling with velocity **v**, enters a uniform magnetic field **B**, as shown below. Describe the particle's subsequent motion.

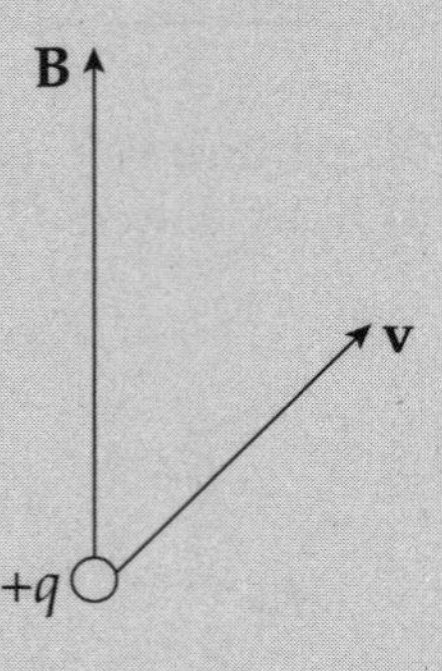

Solution. If the particle's velocity were parallel to **B**, then it would be unaffected by **B**. If **v** were perpendicular to **B**, then it would undergo uniform circular motion (as we saw in Example 13.2). In this case, **v** is neither purely parallel nor perpendicular to **B**. It has a component ($\mathbf{v}_1$) that's parallel to **B** and a component ($\mathbf{v}_2$) that's perpendicular to **B**.

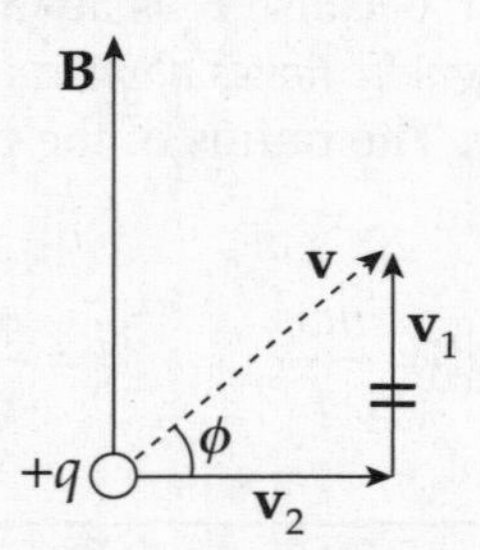

Component $\mathbf{v}_1$ will not be changed by **B**, so the particle will continue upward in the direction of **B**. However, the presence of $\mathbf{v}_2$ will create circular motion. The superposition of these two types of motion will cause the particle's trajectory to be a *helix*; it will spin in circular motion while traveling upward with the speed $v_1 = v \sin \phi$:

THE MAGNETIC FORCE ON A CURRENT-CARRYING WIRE

Since magnetic fields affect moving charges, they should also affect current-carrying wires. After all, a wire that contains a current contains charges that move.

Let a wire of length ℓ be immersed in magnetic field **B**. If the wire carries a current I, then the magnitude of the magnetic force it feels is

$$F_B = \mathbf{B}I\ell \sin \theta$$

where θ is the angle between ℓ and **B**. Here, the direction of ℓ is the direction of the current, I. The direction of $\mathbf{F}_B$ can be found using the right-hand rule and by letting your thumb point in the direction in which the current flows.

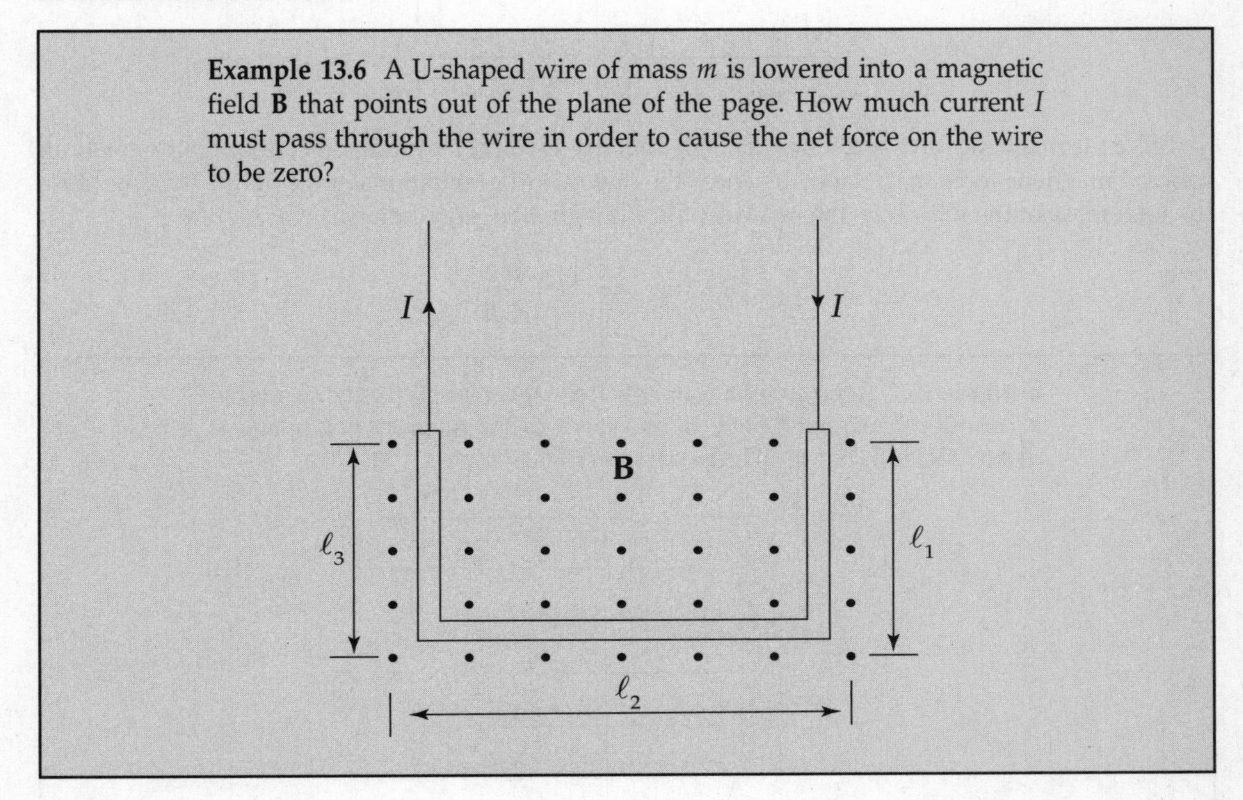

Example 13.6 A U-shaped wire of mass m is lowered into a magnetic field **B** that points out of the plane of the page. How much current I must pass through the wire in order to cause the net force on the wire to be zero?

Solution. The total magnetic force on the wire is equal to the sum of the magnetic forces on each of the three sections of wire. The force on the first section (the right, vertical one), $\mathbf{F}_{B1}$, is directed to the left (applying the right-hand rule, and the force on the third piece (the left, vertical one), $\mathbf{F}_{B3}$, is directed to the right. Since these pieces are the same length, these two oppositely directed forces have the same magnitude, $I\ell_1\mathbf{B} = I\ell_3\mathbf{B}$, and they cancel. So the net magnetic force on the wire is the magnetic force on the middle piece. Since I points to the left and **B** is out of the page, the right-hand rule tells us the force is upward.

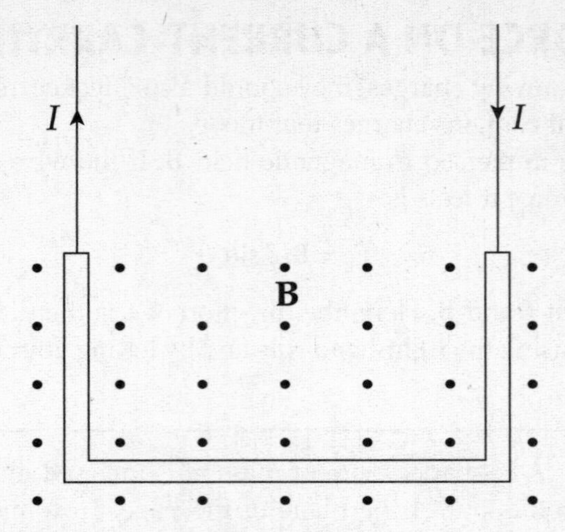

Since the magnetic force on the wire is $I\ell_2\mathbf{B}$, directed upward, the amount of current must create an upward magnetic force that exactly balances the downward gravitational force on the wire. Because the total mass of the wire is m, the resultant force (magnetic + gravitational) will be zero if

$$I\ell_2\mathbf{B} = mg \quad \Rightarrow \quad I = \frac{mg}{\ell_2\mathbf{B}}$$

Example 13.7 A rectangular loop of wire that carries a current I is placed in a uniform magnetic field, **B**, as shown in the diagram below and is free to rotate. What torque does it experience?

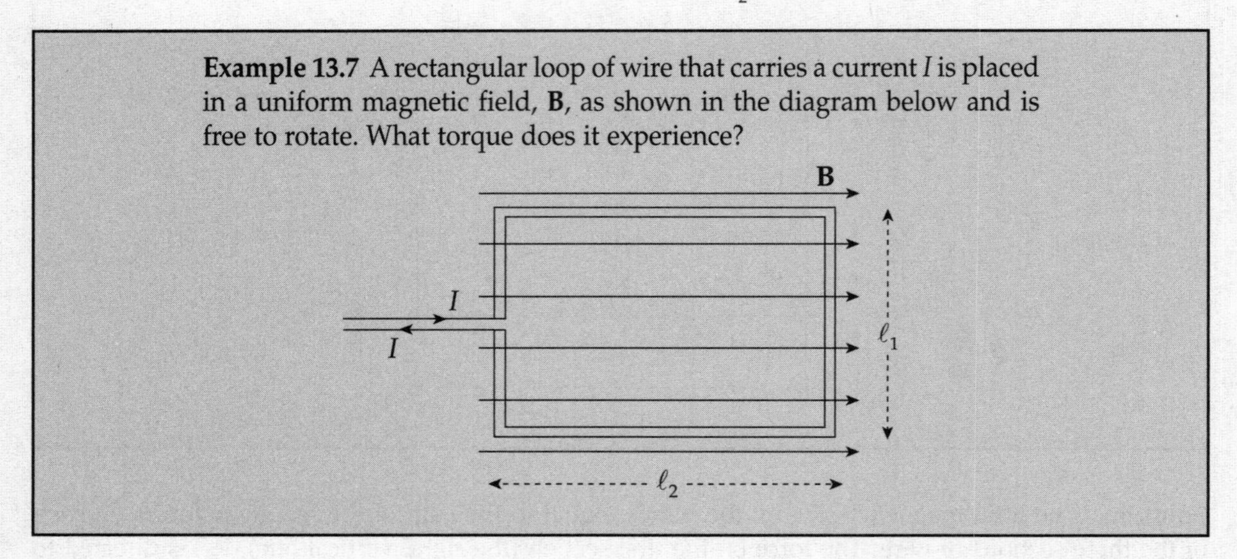

Solution. Ignoring the tiny gap in the vertical left-hand wire, we have two wires of length ℓ_1 and two of length ℓ_2. There is no magnetic force on either of the sides of the loop of length ℓ_2, because the current in the top side is parallel to **B** and the current in the bottom side is antiparallel to **B**. The magnetic force on the right-hand side points out of the plane of the page, while the magnetic force on the left-hand side points into the plane of the page.

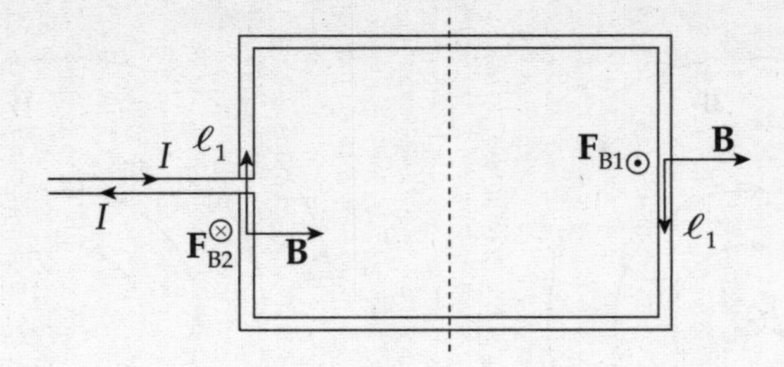

If the loop is free to rotate, then each of these two forces exerts a torque that tends to turn the loop in such a way that the right-hand side rises out of the plane of the page and the left-hand side rotates into the page. Relative to the axis shown above (which cuts the loop in half), the torque of $\mathbf{F}_{B1}$ is

$$\tau_1 = rF_{B1} \sin \theta = (\tfrac{1}{2}\ell_2)(I\ell_1 B)\sin 90° = \tfrac{1}{2} I\ell_1\ell_2 B$$

and the torque of $\mathbf{F}_{B2}$ is

$$\tau_2 = rF_{B2} \sin \theta = (\tfrac{1}{2}\ell_2)(I\ell_1 B)\sin 90° = \tfrac{1}{2} I\ell_1\ell_2 B$$

Since both these torques rotate the loop in the same direction, the net torque on the loop is

$$\tau_1 + \tau_2 = I\ell_1\ell_2 B$$

MAGNETIC FIELDS CREATED BY CURRENT-CARRYING WIRES

As we said at the beginning of this chapter, the sources of magnetic fields are electric charges that move; they may spin, circulate, move through space, or flow down a wire. For example, consider a long, straight wire that carries a current *I*. The current generates a magnetic field in the surrounding space, of magnitude

$$\mathbf{B} = \frac{\mu_0}{2\pi}\frac{I}{r}$$

where *r* is the distance from the wire. The symbol μ_0 denotes a fundamental constant called the permeability of free space. Its value is:

$$\mu_0 = 4\pi\times10^{-7}\,\mathrm{N/A^2} = 4\pi\times10^{-7}\,\mathrm{T\cdot m/A}$$

The magnetic field lines are actually circles whose centers are on the wire. The direction of these circles is determined by a variation of the right-hand rule. Imagine grabbing the wire in your right hand with your thumb pointing in the direction of the current. Then the direction in which your fingers curl around the wire gives the direction of the magnetic field line.

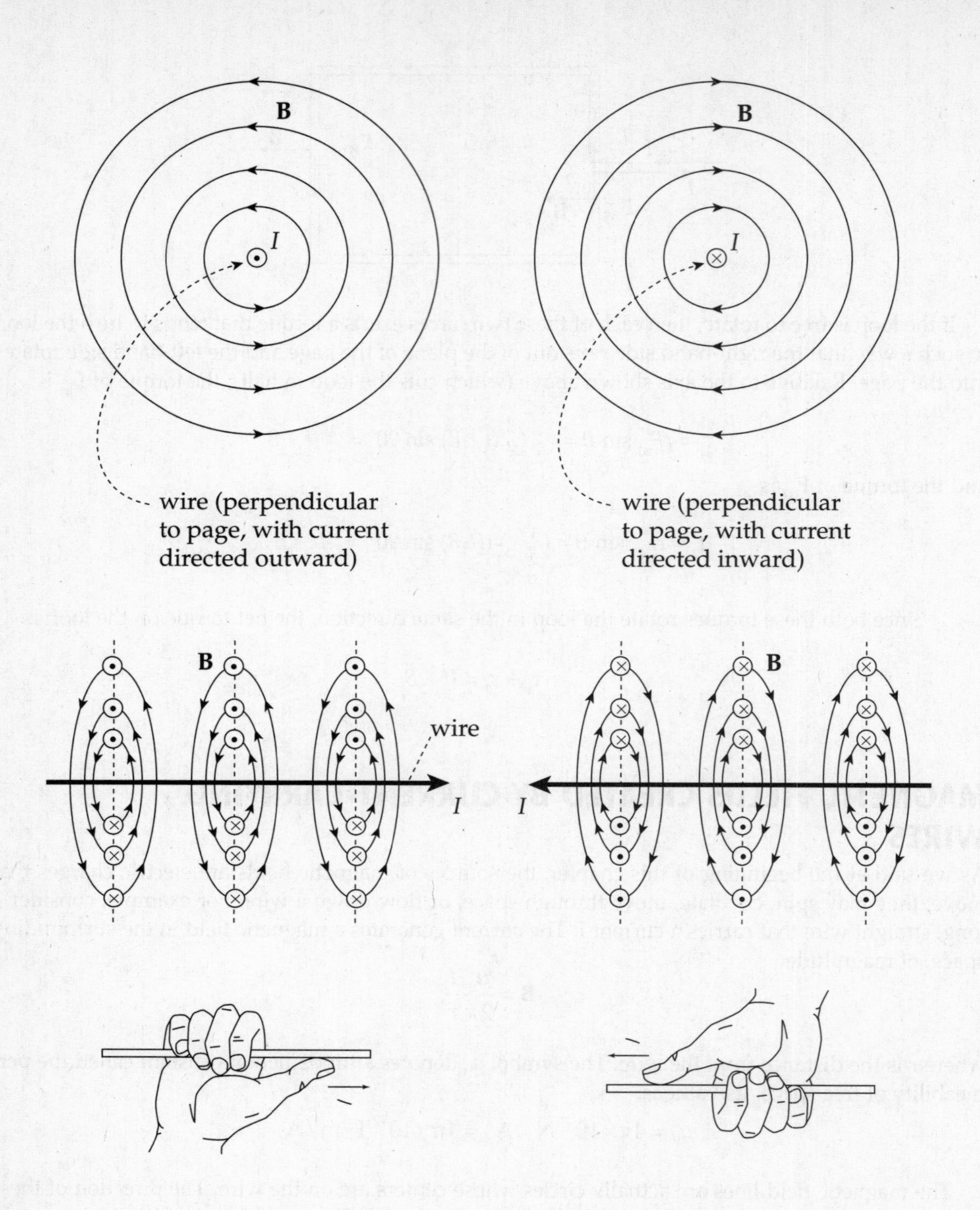
B
I
wire (perpendicular
to page, with current
directed outward)
B
I
wire (perpendicular
to page, with current
directed inward)
B
wire
I
B
I

Example 13.8 The diagram below shows a proton moving with a speed of 2×10^5 m/s, initially parallel to, and 4 cm from, a long, straight wire. If the current in the wire is 20 A, what's the magnetic force on the proton?

Solution. Above the wire (where the proton is), the magnetic field lines generated by the current-carrying wire point out of the plane of the page, so $\mathbf{v}_0 \times \mathbf{B}$ points downward. Since the proton's charge is positive, the magnetic force $\mathbf{F}_B = q(\mathbf{v}_0 \times \mathbf{B})$ is also directed down, toward the wire.

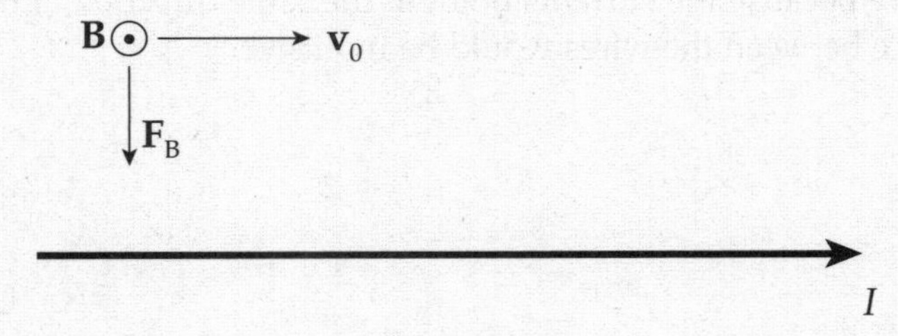

The strength of the magnetic force on the proton is

$$F_B = qv_0\mathbf{B} = ev_0 \frac{\mu_0}{2\pi}\frac{I}{r} = (1.6 \times 10^{-19}\ \text{C})(2 \times 10^5\ \text{m/s})\frac{4\pi \times 10^{-7}\ \text{N/A}^2}{2\pi}\frac{20\ \text{A}}{0.04\ \text{m}}$$

$$= 3.2 \times 10^{-18}\ \text{N}$$

Example 13.9 The diagram below shows a pair of long, straight, parallel wires, separated by a small distance, r. If currents I_1 and I_2 are established in the wires, what is the magnetic force per unit length they exert on each other?

Solution. To find the force on Wire 2, consider the current in Wire 1 as the source of the magnetic field. Below Wire 1, the magnetic field lines generated by Wire 1 point into the plane of the page. Therefore, the force on Wire 2, as given by the equation $\mathbf{F}_{B2} = I_2(\ell_2 \times \mathbf{B}_1)$, points upward.

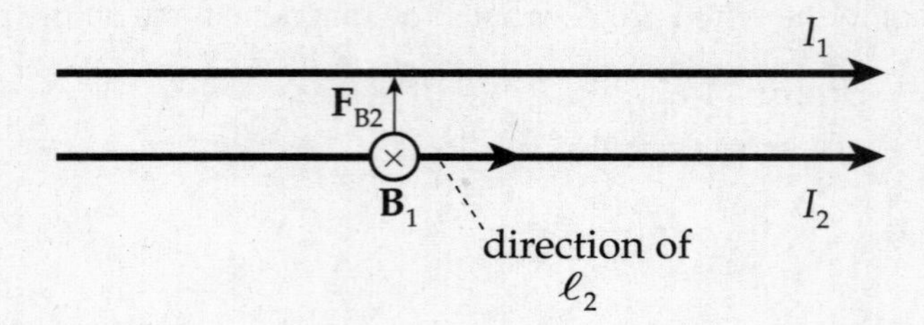

The magnitude of the magnetic force per unit length felt by Wire 2, due to the magnetic field generated by Wire 1, is found this way:

$$F_{B2} = I_2 \ell_2 \mathbf{B}_1 = I_2 \ell_2 \frac{\mu_0}{2\pi} \frac{I_1}{r} \quad \Rightarrow \quad \frac{F_{B2}}{\ell_2} = \frac{\mu_0}{2\pi} \frac{I_1 I_2}{r}$$

By Newton's Third Law, this is the same force that Wire 1 feels due to the magnetic field generated by Wire 2. The force is attractive because the currents point in the same direction; if one of the currents were reversed, then the force between the wires would be repulsive.

CHAPTER 13 REVIEW QUESTIONS

Solutions can be found in Chapter 18.

SECTION I: MULTIPLE CHOICE

1. Which of the following is/are true concerning magnetic forces and fields?

    I. The magnetic field lines due to a current-carrying wire radiate away from the wire.
    II. The kinetic energy of a charged particle can be increased by a magnetic force.
    III. A charged particle can move through a magnetic field without feeling a magnetic force.

    (A) I only
    (B) II and III only
    (C) I and II only
    (D) III only
    (E) I and III only

2. The velocity of a particle of charge $+4.0 \times 10^{-9}$ C and mass 2×10^{-4} kg is perpendicular to a 0.1-tesla magnetic field. If the particle's speed is 3×10^4 m/s, what is the acceleration of this particle due to the magnetic force?

    (A) 0.0006 m/s^2
    (B) 0.006 m/s^2
    (C) 0.06 m/s^2
    (D) 0.6 m/s^2
    (E) None of the above

3. In the figure below, what is the direction of the magnetic force $\mathbf{F}_B$?

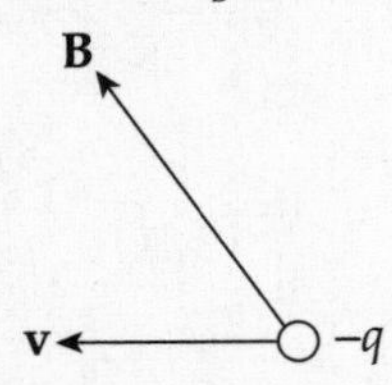

    (A) To the right
    (B) Downward, in the plane of the page
    (C) Upward, in the plane of the page
    (D) Out of the plane of the page
    (E) Into the plane of the page

4. In the figure below, what must be the direction of the particle's velocity, **v**?

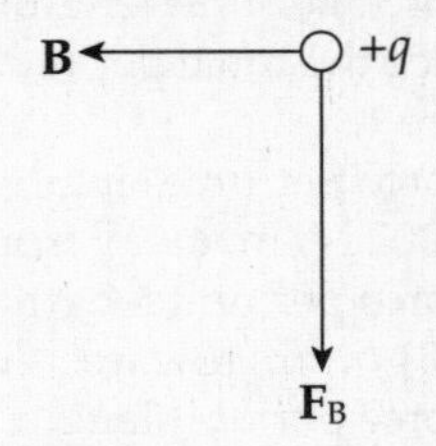

    (A) To the right
    (B) Downward, in the plane of the page
    (C) Upward, in the plane of the page
    (D) Out of the plane of the page
    (E) Into the plane of the page

5. Due to the magnetic force, a positively charged particle executes uniform circular motion within a uniform magnetic field, **B**. If the charge is q and the radius of its path is r, which of the following expressions gives the magnitude of the particle's linear momentum?

    (A) $q\mathbf{B}r$
    (B) $q\mathbf{B}/r$
    (C) $q/(\mathbf{B}r)$
    (D) $\mathbf{B}/(qr)$
    (E) $r/(q\mathbf{B})$

6. A straight wire of length 2 m carries a 10-amp current. How strong is the magnetic field at a distance of 2 cm from the wire?

    (A) 1×10^{-6} T
    (B) 1×10^{-5} T
    (C) 2×10^{-5} T
    (D) 1×10^{-4} T
    (E) 2×10^{-4} T

7. Two long, straight wires are hanging parallel to each other and are 1 cm apart. The current in Wire 1 is 5 A, and the current in Wire 2 is 10 A, in the same direction. Which of the following best describes the magnetic force per unit length felt by the wires?

   (A) The force per unit length on Wire 1 is twice the force per unit length on Wire 2.
   (B) The force per unit length on Wire 2 is twice the force per unit length on Wire 1.
   (C) The force per unit length on Wire 1 is 0.0003 N/m, away from Wire 2.
   (D) The force per unit length on Wire 1 is 0.001 N/m, toward Wire 2.
   (E) The force per unit length on Wire 1 is 0.001 N/m, away from Wire 2.

8. In the figure below, what is the magnetic field at the Point P, which is midway between the two wires?

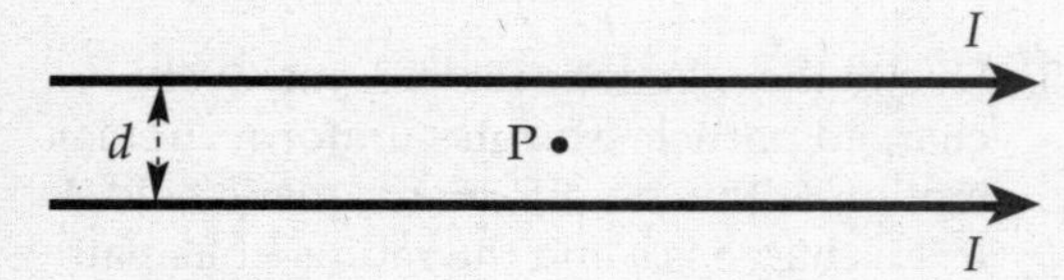

   (A) $2\mu_0 I/(\pi d)$, out of the plane of the page
   (B) $2\mu_0 I/(\pi d)$, into the plane of the page
   (C) $\mu_0 I/(2\pi d)$, out of the plane of the page
   (D) $\mu_0 I/(2\pi d)$, into the plane of the page
   (E) Zero

9. Here is a section of a wire with a current moving to the right. Where is the magnetic field strongest and pointing INTO the page?

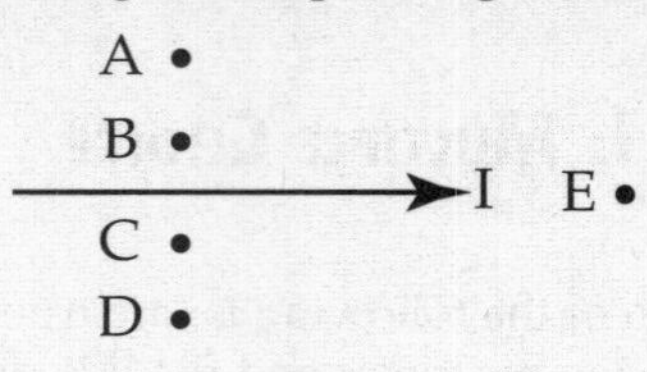

   (A) A
   (B) B
   (C) C
   (D) D
   (E) E

10. What is the direction of force acting on the current-carrying wire as shown below?

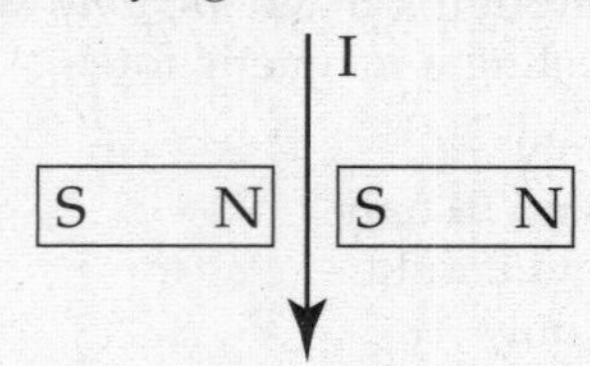

   (A) to the top of the page
   (B) to the bottom of the page
   (C) into the page
   (D) out of the page
   (E) to the right of the page

SECTION II: FREE RESPONSE

1. The diagram below shows a simple mass spectrograph. It consists of a source of ions (charged atoms) that are accelerated (essentially from rest) by the voltage V and enter a region containing a uniform magnetic field, **B**. The polarity of V may be reversed so that both positively charged ions (cations) and negatively charged ions (anions) can be accelerated. Once the ions enter the magnetic field, they follow a semicircular path and strike the front wall of the spectrograph, on which photographic plates are constructed to record the impact. Assume that the ions have mass m.

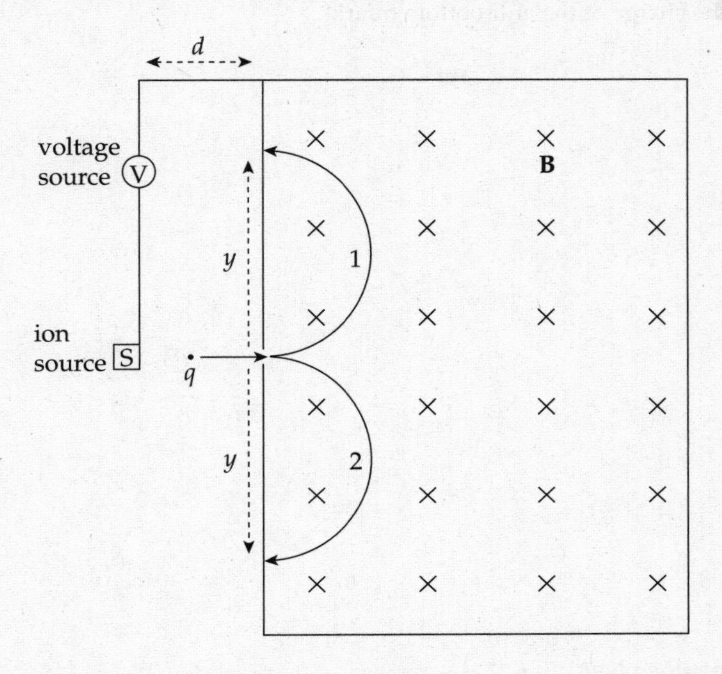

(a) What is the acceleration of an ion of charge q just before it enters the magnetic field?

(b) Find the speed with which an ion of charge q enters the magnetic field.

(c) (i) Which semicircular path, 1 or 2, would a cation follow?

(ii) Which semicircular path, 1 or 2, would an anion follow?

(d) Determine the mass of a cation entering the apparatus in terms of y, q, **B**, and V.

(e) Once a cation of charge q enters the magnetic field, how long does it take to strike the photographic plate?

(f) What is the work done by the magnetic force in the spectrograph on a cation of charge q?

2. A particle accelerator has a collision that results in a photon, an anti-bottom quark, and a charm quark. The magnetic field is 6.00×10^{-8} T and can be described as into the page. A photon has no charge and has an upper theoretical mass of 3.6×10^{-52} kg. The charm quark has a mass of 2.23×10^{-27} kg, a charge of 1.07×10^{-19} C and velocity of 40.1 m/s. The anti-bottom quark has a mass of the 7.49×10^{-27} kg and orbits with a radius of 92.7 m at a velocity of 41.5 m/s in a clockwise manner.

(a) What is the orbital radius of the photon?

(b) What is the orbital radius of the charm quark?

(c) What is the charge of the anti-bottom quark?

SUMMARY

- Charges moving though a magnetic field experience a force whose magnitude is given by $F_B = qv\mathbf{B}\sin\theta$ and whose direction is given by the right-hand rule.
- Because the force is always perpendicular to the direction of velocity, the charge may experience uniform circular motion. It would then follow all the appropriate circular motion relationships and orbit in a radius given by $r = \frac{mv}{q\mathbf{B}}$.
- Because wires have charges moving though them, a wire will experience a force if placed in a magnetic field. This is expressed by $F_B = \mathbf{B}I\ell\sin\theta$. Also, a current-carrying wire will produce a magnetic field whose strength is given by $\mathbf{B} = \frac{\mu_o}{2\pi}\frac{I}{r}$, where $\mu_0 = 4\pi \times 10^{-7}\,\mathrm{T \cdot m/A}$, I is the current through the wire, and r is the radial distance from the wire.

14

Electromagnetic Induction

INTRODUCTION

In Chapter 13, we learned that electric currents generate magnetic fields, and we will now see how magnetism can generate electric currents.

MOTIONAL EMF

The figure below shows a conducting wire of length ℓ, moving with constant velocity **v** in the plane of the page through a uniform magnetic field **B** that's perpendicular to the page. The magnetic field exerts a force on the moving conduction electrons in the wire. With **B** pointing into the page, the direction of **v** is to the right, so the magnetic force, $\mathbf{F}_B$, for positive charges, would be upward by the right-hand rule, so $\mathbf{F}_B$ on these electrons (which are negatively charged) is downward.

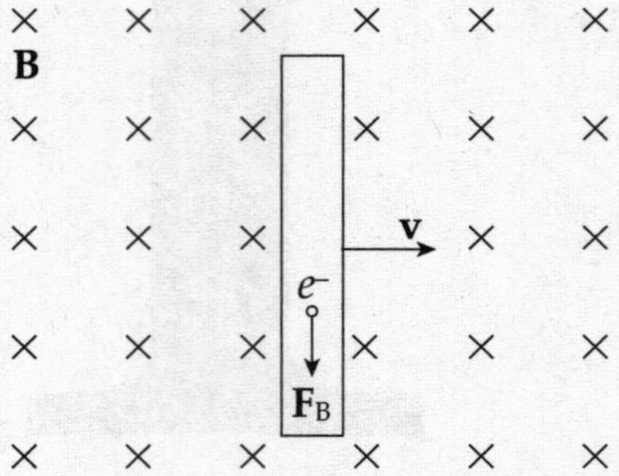

As a result, electrons will be pushed to the lower end of the wire, which will leave an excess of positive charge at its upper end. This separation of charge creates a uniform electric field, **E**, within the wire (pointing downward).

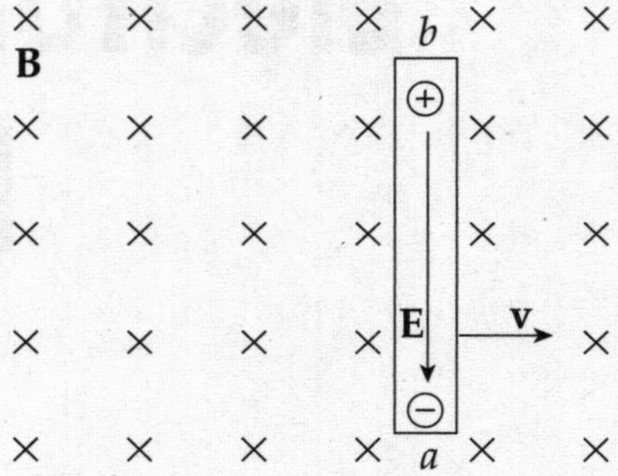

A charge q in the wire feels two forces: an electric force, $\mathbf{F}_E = q\mathbf{E}$, and a magnetic force, $F_B = qvB\sin\theta = qvB$, because $\theta = 90°$.

If q is negative, $\mathbf{F}_E$ is upward and $\mathbf{F}_B$ is downward; if q is positive, $\mathbf{F}_E$ is downward and $\mathbf{F}_B$ is upward. So, in both cases, the forces act in opposite directions. Once the magnitude of $\mathbf{F}_E$ equals the magnitude of $\mathbf{F}_B$, the charges in the wire are in electromagnetic equilibrium. This occurs when $qE = qvB$; that is, when $E = vB$.

The presence of the electric field creates a potential difference between the ends of the rod. Since negative charge accumulates at the lower end (which we'll call point a) and positive charge accumulates at the upper end (point b), point b is at a higher electric potential.

The potential difference V_{ba} is equal to $E\ell$ and, since $E = vB$, the potential difference can be written as $vB\ell$.

Now, imagine that the rod is sliding along a pair of conducting rails connected at the left by a stationary bar. The sliding rod now completes a rectangular circuit, and the potential difference V_{ba} causes current to flow.

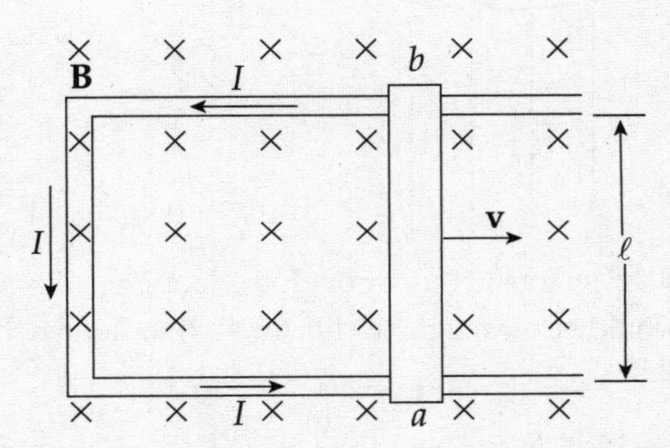

The motion of the sliding rod through the magnetic field creates an electromotive force, called **motional emf**:

$$\mathcal{E} = B\ell v$$

The existence of a current in the sliding rod causes the magnetic field to exert a force on it. Using the formula $F_B = BI\boldsymbol{\ell}$, the fact that $\boldsymbol{\ell}$ points upward (in the direction of the current) and **B** is into the page, tells us that the direction of $\mathbf{F}_B$ on the rod is to the left. An external agent must provide this same amount of force to the right to maintain the rod's constant velocity and keep the current flowing. The power that the external agent must supply is $P = Fv = I\ell Bv$, and the electrical power delivered to the circuit is $P = IV_{ba} = I\mathcal{E} = IvB\ell$. Notice that these two expressions are identical. The energy provided by the external agent is transformed first into electrical energy and then thermal energy as the conductors making up the circuit dissipate heat.

FARADAY'S LAW OF ELECTROMAGNETIC INDUCTION

Electromotive force can be created by the motion of a conductor through a magnetic field, but there's another way to create an emf from a magnetic field. (We have used **A** to represent the area vector, which points in a direction perpendicular to the plane of the given surface.)

The **magnetic flux**, Φ_B, through an area A is equal to the product of A and the magnetic field perpendicular to it: $\Phi_B = B_\perp A = \mathbf{B} \cdot \mathbf{A} = BA \cos \theta$.

Example 14.1 The figure below shows two views of a circular loop of radius 3 cm placed within a uniform magnetic field, **B** (magnitude 0.2 T).

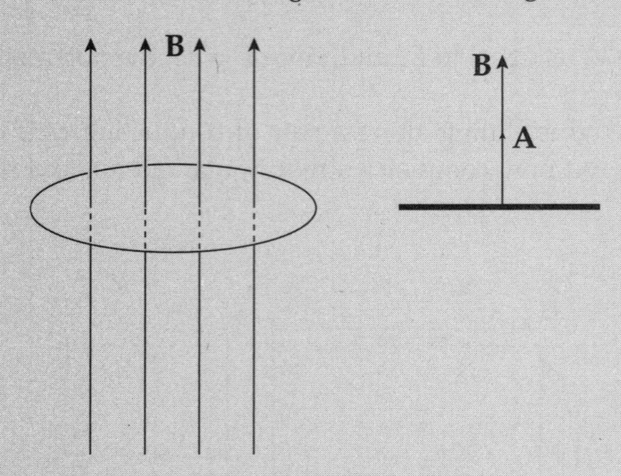

(a) What's the magnetic flux through the loop?

(b) What would be the magnetic flux through the loop if the loop were rotated 45°?

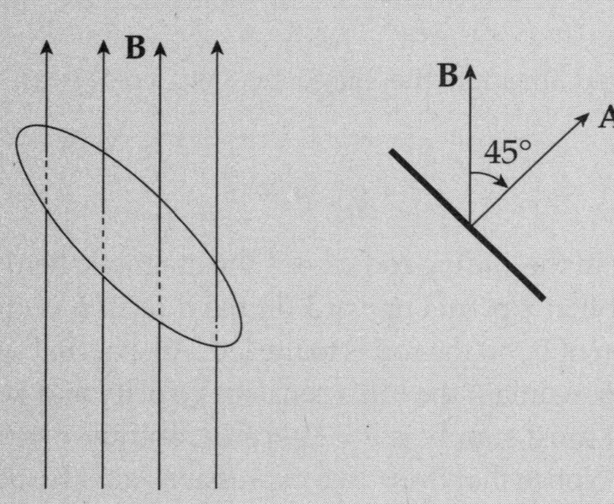

(c) What would be the magnetic flux through the loop if the loop were rotated 90°?

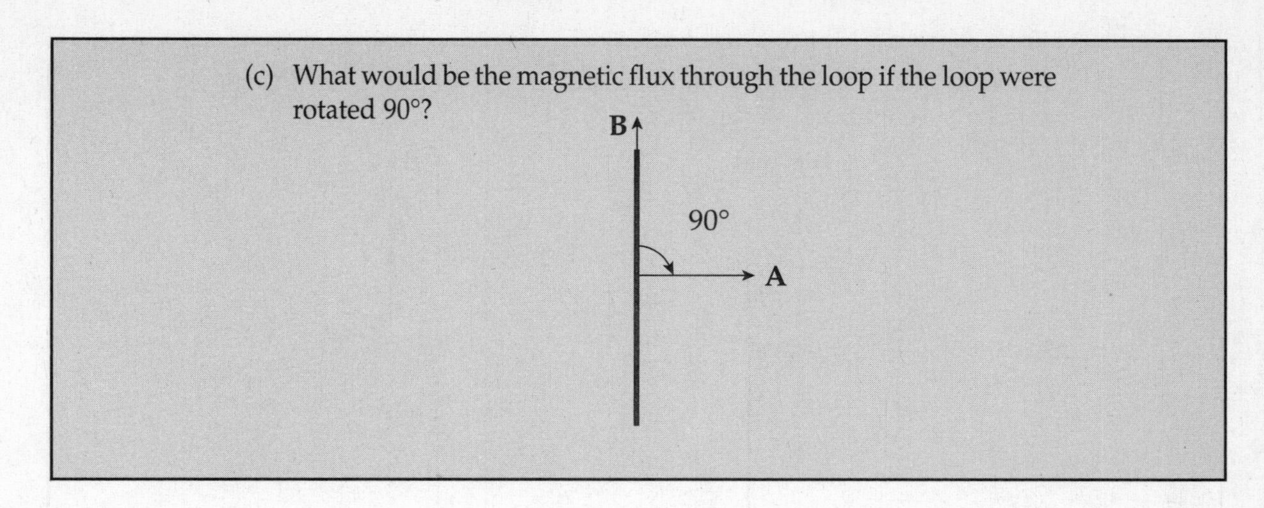

Solution.

(a) Since **B** is parallel to **A**, the magnetic flux is equal to BA:

$$\Phi_B = BA = B \cdot \pi r^2 = (0.2\text{ T}) \cdot \pi(0.03\text{ m})^2 = 5.7 \times 10^{-4}\text{ T·m}^2$$

The SI unit for magnetic flux, the tesla·meter², is called a **weber** (abbreviated **Wb**). So $\Phi_B = 5.7 \times 10^{-4}$ Wb.

(b) Since the angle between **B** and **A** is 45°, the magnetic flux through the loop is

$$\Phi_B = BA \cos 45° = B \cdot \pi r^2 \cos 45° = (0.2\text{ T}) \cdot \pi(0.03\text{ m})^2 \cos 45° = 4.0 \times 10^{-4}\text{ Wb}$$

(c) If the angle between **B** and **A** is 90°, the magnetic flux through the loop is zero, since $\cos 90° = 0$.

The concept of magnetic flux is crucial, because changes in magnetic flux induce emf. According to **Faraday's Law of Electromagnetic Induction**, the magnitude of the emf induced in a circuit is equal to the rate of change of the magnetic flux through the circuit. This can be written mathematically in the form

$$|\mathcal{E}_{avg}| = \left|\frac{\Delta\Phi_B}{\Delta t}\right|$$

This induced emf can produce a current, which will then create its own magnetic field. The direction of the induced current is determined by the polarity of the induced emf and is given by **Lenz's Law**: The induced current will always flow in the direction that opposes the change in magnetic flux that produced it. If this were not so, then the magnetic flux created by the induced current would magnify the change that produced it, and energy would not be conserved. Lenz's Law can be included mathematically with Faraday's Law by the introduction of a minus sign; this leads to a single equation that expresses both results:

$$\mathcal{E}_{avg} = -\frac{\Delta\Phi_B}{\Delta t}$$

Example 14.2 The circular loop of Example 14.1 rotates at a constant angular speed through 45° in 0.5 s.

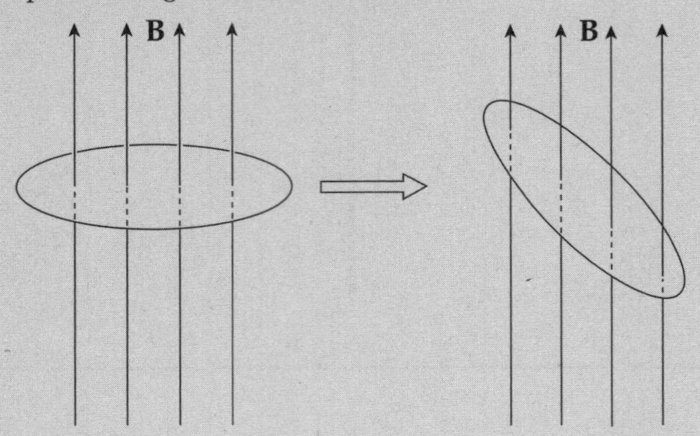

(a) What's the induced emf in the loop?

(b) In which direction will current be induced to flow?

Solution.

(a) As we found in Example 14.1, the magnetic flux through the loop changes when the loop rotates. Using the values we determined earlier, Faraday's Law gives

$$\mathcal{E}_{\text{avg}} = -\frac{\Delta\Phi_B}{\Delta t} = -\frac{(4.0\times10^{-4}\ \text{Wb})-(5.7\times10^{-4}\ \text{Wb})}{0.5\ \text{s}} = 3.4\times10^{-4}\ \text{V}$$

(b) The original magnetic flux was 5.7×10^{-4} Wb upward, and was decreased to 4.0×10^{-4} Wb. So the change in magnetic flux is -1.7×10^{-4} Wb upward, or, equivalently, $\Delta\Phi_B = 1.7 \times 10^{-4}$ Wb, downward. To oppose this change we would need to create some magnetic flux upward. The current would be induced in the counterclockwise direction (looking down on the loop), because the right-hand rule tells us that then the current would produce a magnetic field that would point up.

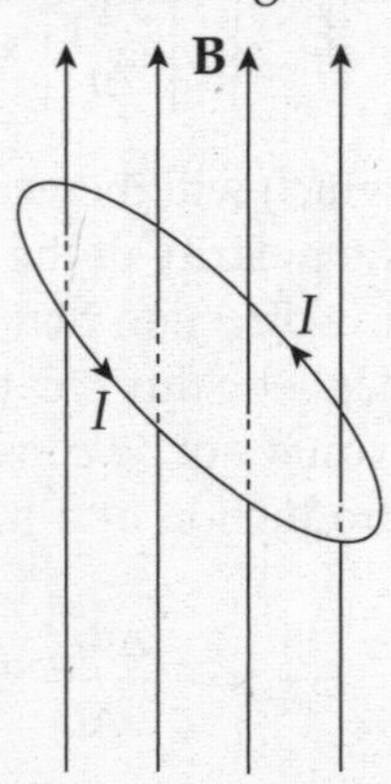

The current will flow only while the loop rotates, because emf is induced only when magnetic flux is changing. If the loop rotates 45° and then stops, the current will disappear.

Example 14.3 Again consider the conducting rod that's moving with constant velocity **v** along a pair of parallel conducting rails (separated by a distance ℓ), within a uniform magnetic field, **B**:

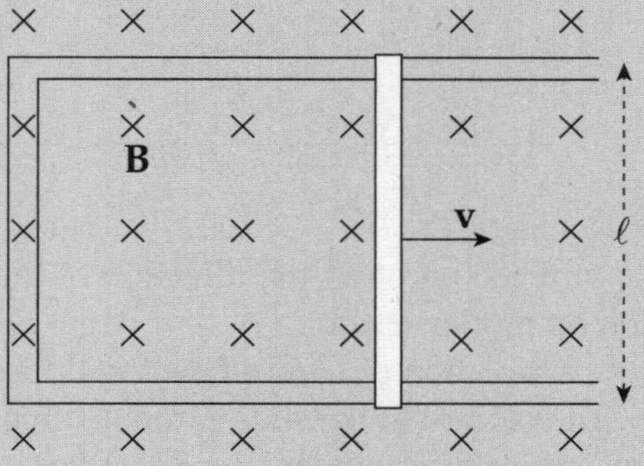

Find the induced emf and the direction of the induced current in the rectangular circuit.

Solution. The area of the rectangular loop is ℓx, where x is the distance from the left-hand bar to the moving rod:

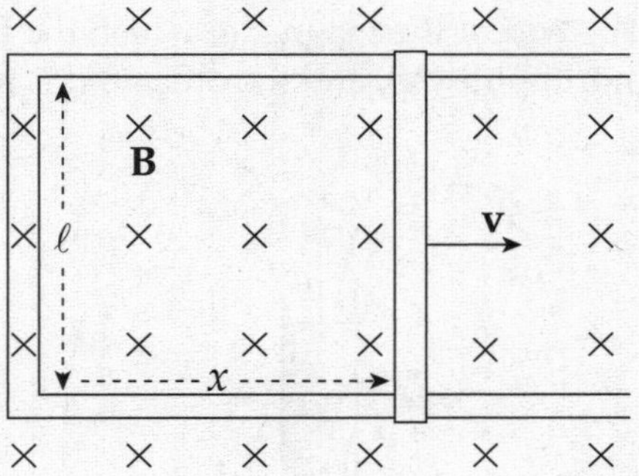

Because the area is changing, the magnetic flux through the loop is changing, which means that an emf will be induced in the loop. To calculate the induced emf, we first write $\Phi_B = BA = B\ell x$, then since $\Delta x/\Delta t = v$, we get

$$\mathcal{E}_{avg} = -\frac{\Delta\Phi_B}{\Delta t} = -\frac{\Delta(B\ell x)}{\Delta t} = -B\ell\frac{\Delta x}{\Delta t} = -B\ell v$$

We can figure out the direction of the induced current from Lenz's Law. As the rod slides to the right, the magnetic flux into the page increases. How do we oppose an increasing into-the-page flux? By producing out-of-the-page flux. In order for the induced current to generate a magnetic field that points out of the plane of the page, the current must be directed counterclockwise (according to the right-hand rule).

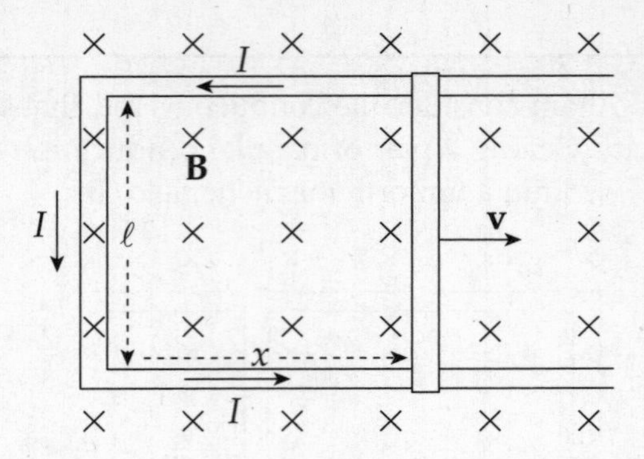

Note that the magnitude of the induced emf and the direction of the current agree with the results we derived earlier, in the section on motional emf.

This example also shows how a violation of Lenz's Law would lead directly to a violation of the Law of Conservation of Energy. The current in the sliding rod is directed upward, as given by Lenz's Law, so the conduction electrons are drifting downward. The force on these drifting electrons—and thus, the rod itself—is directed to the left, opposing the force that's pulling the rod to the right. If the current were directed downward, in violation of Lenz's Law, then the magnetic force on the rod would be to the right, causing the rod to accelerate to the right with ever-increasing speed and kinetic energy, without the input of an equal amount of energy from an external agent.

Example 14.4 A permanent magnet creates a magnetic field in the surrounding space. The end of the magnet at which the field lines emerge is designated the **north pole** (N), and the other end is the **south pole** (S):

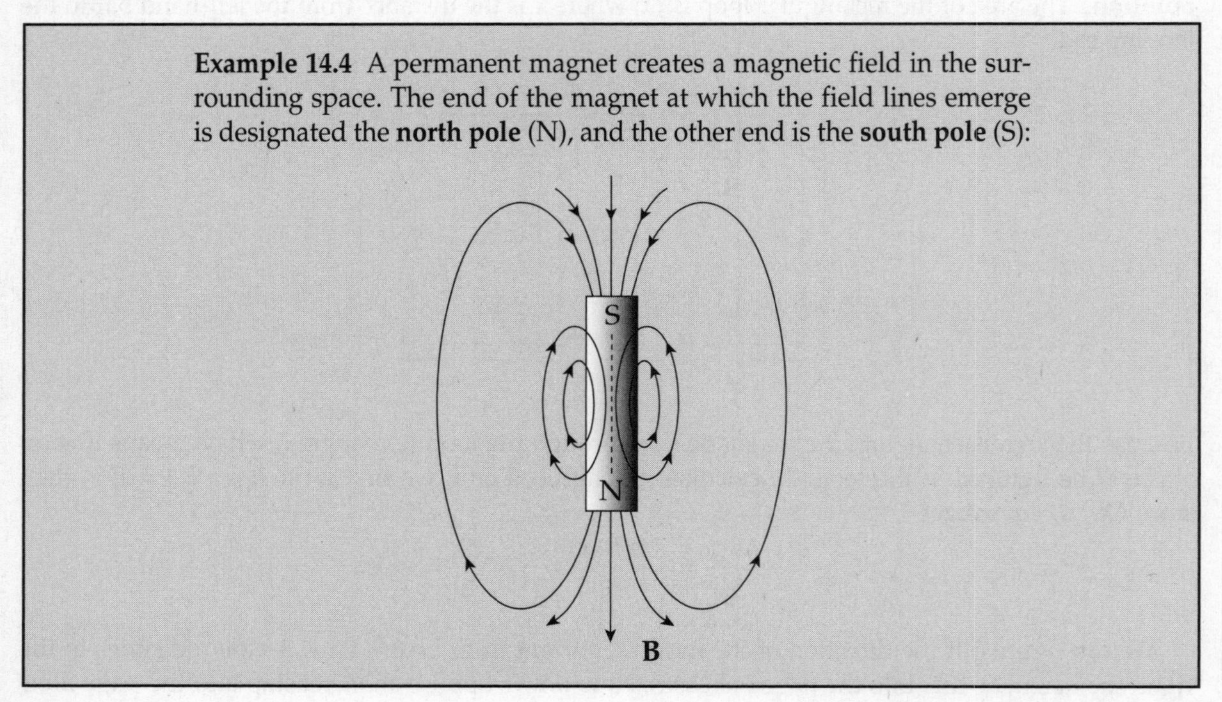

(a) The figure below shows a bar magnet moving down, through a circular loop of wire. What will be the direction of the induced current in the wire?

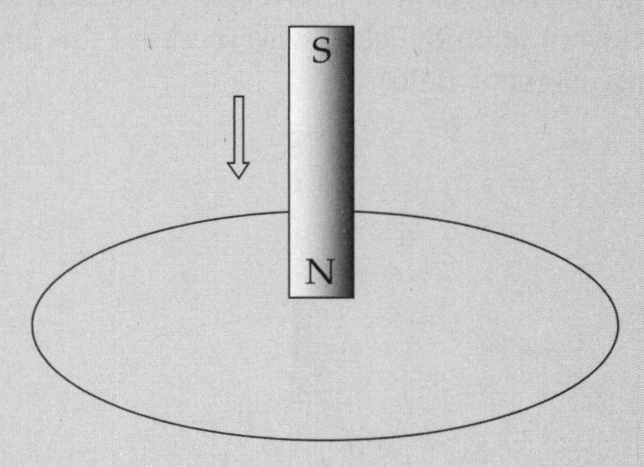

(b) What will be the direction of the induced current in the wire if the magnet is moved as shown in the following diagram?

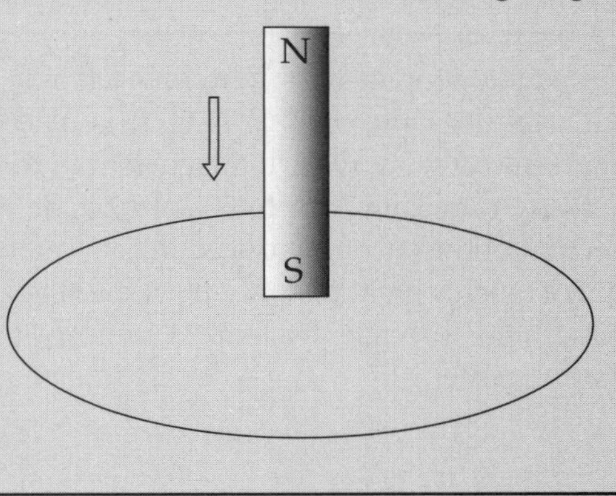

Solution.

(a) The magnetic flux down, through the loop, increases as the magnet is moved. By Lenz's Law, the induced emf will generate a current that opposes this change. How do we oppose a change of *more flux downward*? By creating flux *upward*. So, according to the right-hand rule, the induced current must flow counterclockwise (because this current will generate an upward-pointing magnetic field):

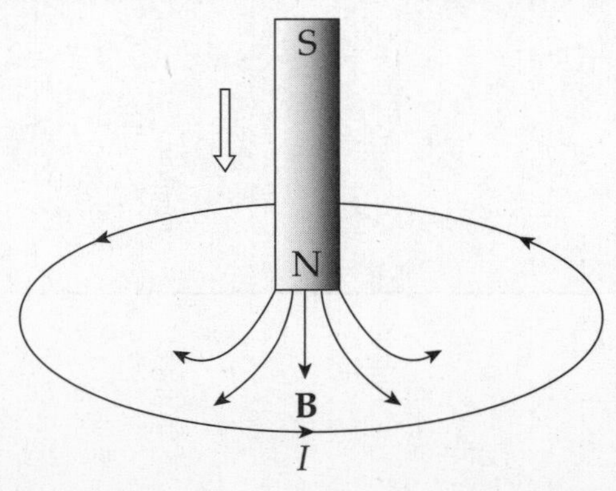

(b) In this case, the magnetic flux through the loop is upward and, as the south pole moves closer to the loop, the magnetic field strength increases so the magnetic flux through the loop increases upward. How do we oppose a change of *more flux upward*? By creating flux *downward*. Therefore, in accordance with the right-hand rule, the induced current will flow clockwise (because this current will generate a downward-pointing magnetic field):

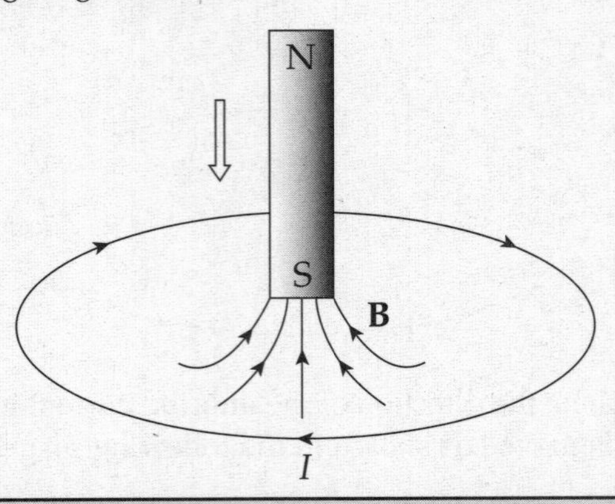

Example 14.5 A square loop of wire 2 cm on each side contains 5 tight turns and has a total resistance of 0.0002 Ω. It is placed 20 cm from a long, straight, current-carrying wire. If the current in the straight wire is increased at a steady rate from 20 A to 50 A in 2 s, determine the magnitude and direction of the current induced in the square loop. (Because the square loop is at such a great distance from the straight wire, assume that the magnetic field through the loop is uniform and equal to the magnetic field at its center.)

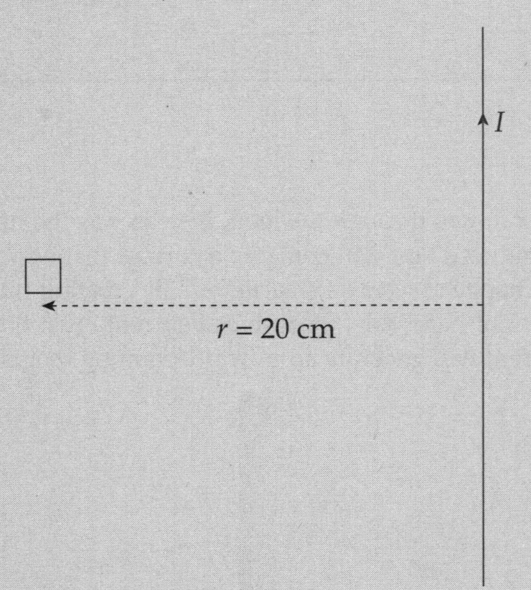

Solution. At the position of the square loop, the magnetic field due to the straight wire is directed out of the plane of the page and its strength is given by the equation $B = (\mu_0/2\pi)(I/r)$. As the current in the straight wire increases, the magnetic flux through the turns of the square loop changes, inducing an emf and current. There are $N = 5$ turns; Faraday's Law becomes $\mathcal{E}_{avg} = -N(\Delta\Phi_B/\Delta t)$, and

$$\mathcal{E}_{avg} = -N\frac{\Delta\Phi_B}{\Delta t} = -N\frac{\Delta(BA)}{\Delta t} = -NA\frac{\Delta B}{\Delta t} = -NA\frac{\mu_0}{2\pi r}\frac{\Delta I}{\Delta t}$$

Substituting the given numerical values, we get

$$\begin{aligned}\mathcal{E}_{avg} &= -NA\frac{\mu_0}{2\pi r}\frac{\Delta I}{\Delta t}\\ &= -(5)(0.02\text{ m})^2\frac{4\pi\times10^{-7}\text{ T}\cdot\text{m/A}}{2\pi(0.20\text{ m})}\frac{(50\text{ A}-20\text{ A})}{2\text{ s}}\\ &= -3\times10^{-8}\text{ V}\end{aligned}$$

The magnetic flux through the loop is out of the page and increases as the current in the straight wire increases. To oppose an increasing out-of-the-page flux, the direction of the induced current should be clockwise, thereby generating an into-the-page magnetic field (and flux).

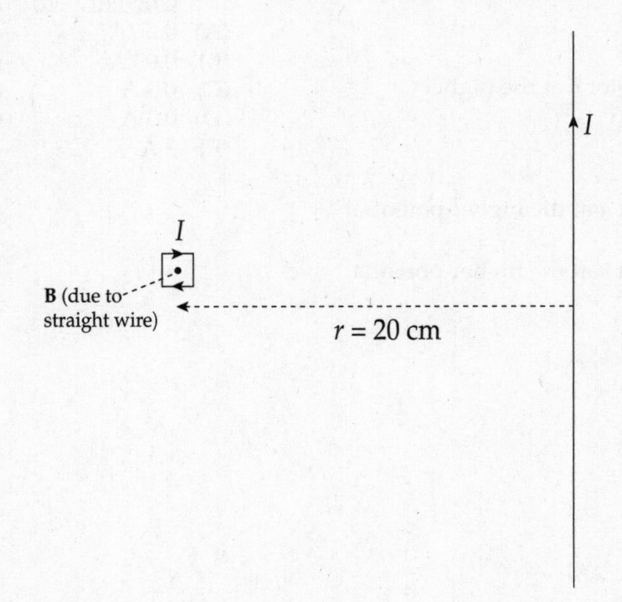

The value of the current in the loop will be

$$I = \frac{\mathcal{E}}{R} = \frac{3\times10^{-8}\text{ V}}{0.0002\ \Omega} = 1.5\times10^{-4}\text{ A}$$

CHAPTER 14 REVIEW QUESTIONS

Solutions can be found in Chapter 18.

SECTION I: MULTIPLE CHOICE

1. A metal rod of length L is pulled upward with constant velocity **v** through a uniform magnetic field **B** that points out of the plane of the page.

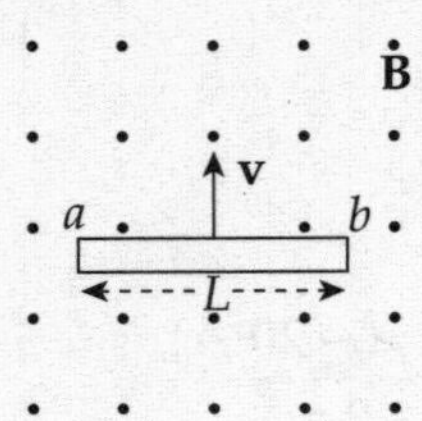

What is the potential difference between points a and b?

(A) 0

(B) $\frac{1}{2}vBL$, with point a at the higher potential

(C) $\frac{1}{2}vBL$, with point b at the higher potential

(D) vBL, with point a at the higher potential

(E) vBL, with point b at the higher potential

2. A conducting rod of length 0.2 m and resistance 10 ohms between its endpoints slides without friction along a U-shaped conductor in a uniform magnetic field B of magnitude 0.5 T perpendicular to the plane of the conductor, as shown in the diagram below.

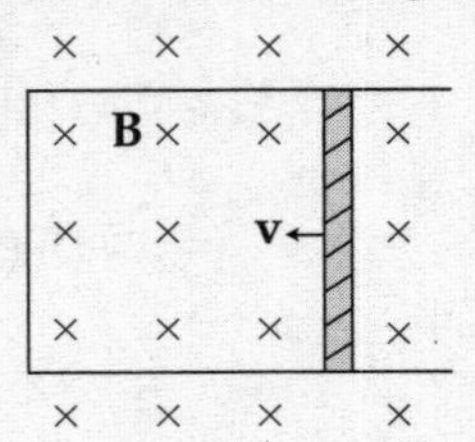

If the rod is moving with velocity **v** = 3 m/s to the left, what is the magnitude and direction of the current induced in the rod?

	Current	Direction
(A)	0.03 A	down
(B)	0.03 A	up
(C)	0.3 A	down
(D)	0.3 A	up
(E)	3 A	down

3. In the figure below, a small, circular loop of wire (radius r) is placed on an insulating stand inside a hollow solenoid of radius R. The solenoid has n turns per unit length and carries a counterclockwise current I. If the current in the solenoid is decreased at a steady rate of a amps/s, determine the induced emf, $\mathcal{E}$, and the direction of the induced current in the loop.

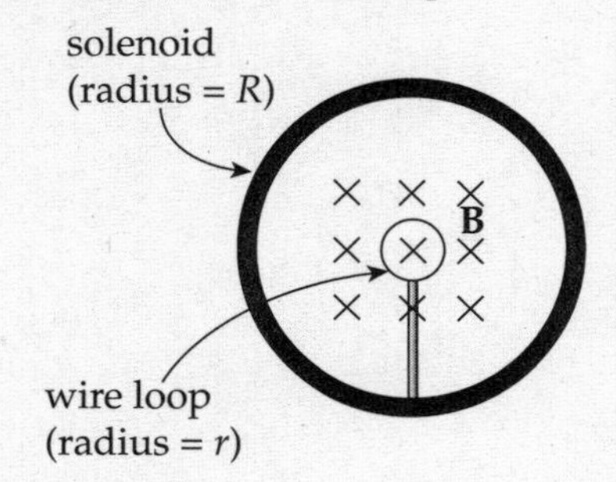

(A) $\mathcal{E} = \mu_0 \pi n r^2 a$; induced current is clockwise

(B) $\mathcal{E} = \mu_0 \pi n r^2 a$; induced current is counterclockwise

(C) $\mathcal{E} = \mu_0 \pi n R^2 a$; induced current is clockwise

(D) $\mathcal{E} = \mu_0 \pi n R^2 a$; induced current is counterclockwise

(E) $\mathcal{E} = \mu_0 \pi I n R^2 a$; induced current is counterclockwise

4. In the figures below, a permanent bar magnet is below a loop of wire. It is pulled upward with a constant velocity through the loop of wire as shown in figure b.

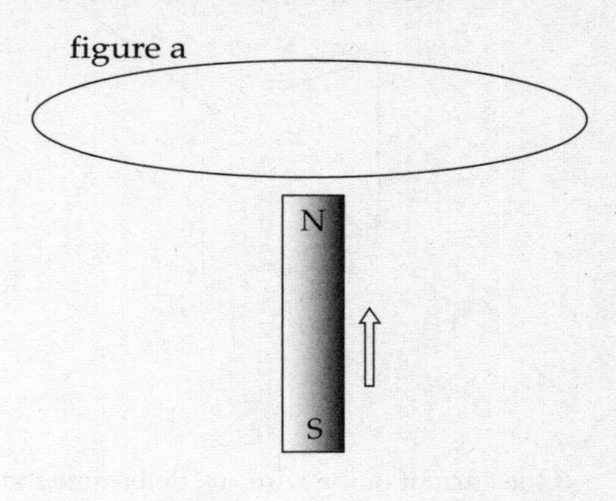

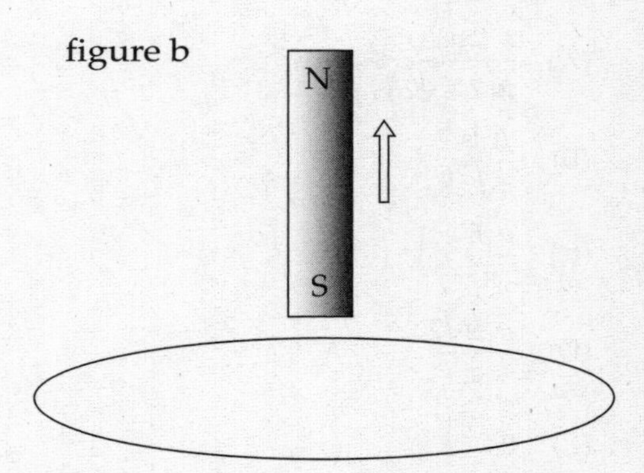

Which of the following best describes the direction(s) of the current induced in the loop (looking down on the loop from above)?

(A) Always clockwise
(B) Always counterclockwise
(C) First clockwise, then counterclockwise
(D) First counterclockwise, then clockwise
(E) No current will be induced in the loop

5. A square loop of wire (side length = s) surrounds a long, straight wire such that the wire passes through the center of the square.

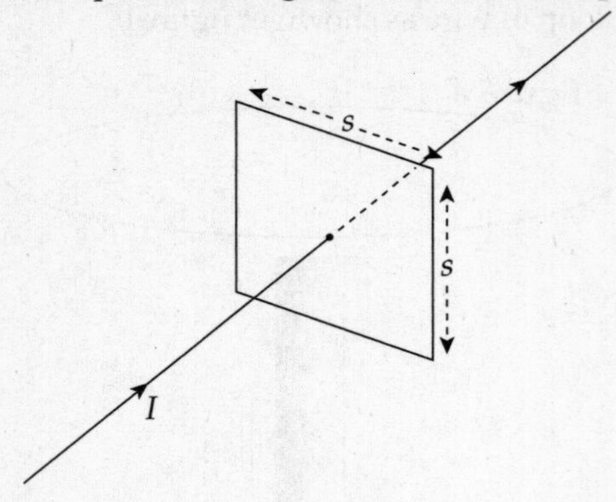

If the current in the wire is I, determine the current induced in the square loop.

(A) $\dfrac{2\mu_0 Is}{\pi(1+\sqrt{2})}$

(B) $\dfrac{\mu_0 Is}{\pi\sqrt{2}}$

(C) $\dfrac{\mu_0 Is}{\pi}$

(D) $\dfrac{\mu_0 Is\sqrt{2}}{\pi}$

(E) 0

Section II: Free Response

1. A rectangular wire is pulled through a uniform magnetic field of 2T going into the page as shown. The resistor has a resistance of 20 Ω.

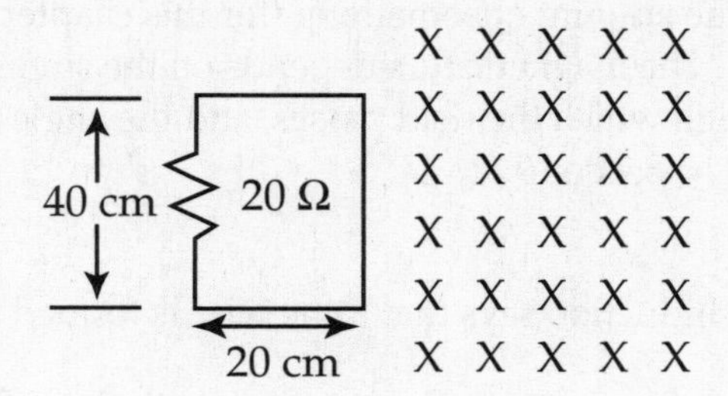

   (a) What is the voltage across the resistor as the wire is pulled horizontally at a velocity of 1 m/s and it just enters the field?

   (b) What is the current through the circuit in the above case and in what direction does it flow?

   (c) What would the voltage be if the wire were rotated 90 degrees and pulled horizontally at the same velocity?

   (d) What would the velocity have to be in order to maintain the same voltage as in part (a) but with the orientation of part (c)?

SUMMARY

- An electromotive force is produced as a conducting wire's position changes or as the magnetic field changes. This idea is summarized by $\varepsilon = B\ell v$.
- The flux (ϕ) tells the amount of something (for this chapter the magnetic field) that goes through a surface. The magnetic flux depends on the strength of the magnetic field, the surface area through which the field passes, and the angle between the two. This idea is summed up by $\phi_m = BA\cos\theta$.
- Faraday's Law of Induction says that if the wire is formed in a loop, an electromagnetic force is produced if the magnetic flux changes with time. This idea can be summarized by $\varepsilon_{ave} = -\frac{\Delta\phi_m}{\Delta t}$.

15
Waves

INTRODUCTION

Imagine holding the end of a long rope in your hand, with the other end attached to a wall. Move your hand up and down, and you'll create a wave that travels along the rope, from your hand to the wall. This simple example displays the basic idea of a **mechanical wave**: a disturbance transmitted by a medium from one point to another, without the medium itself being transported. In the case of water waves, wind or an earthquake can cause a disturbance in the ocean, and the resulting waves can travel thousands of miles. Now, of course, no water actually makes that journey; the water is only the medium that conducts the disturbance.

TRANSVERSE TRAVELING WAVES

Let's return to our long rope. Someone standing near the system would see peaks and valleys actually moving along the rope, in what's called a **traveling wave**.

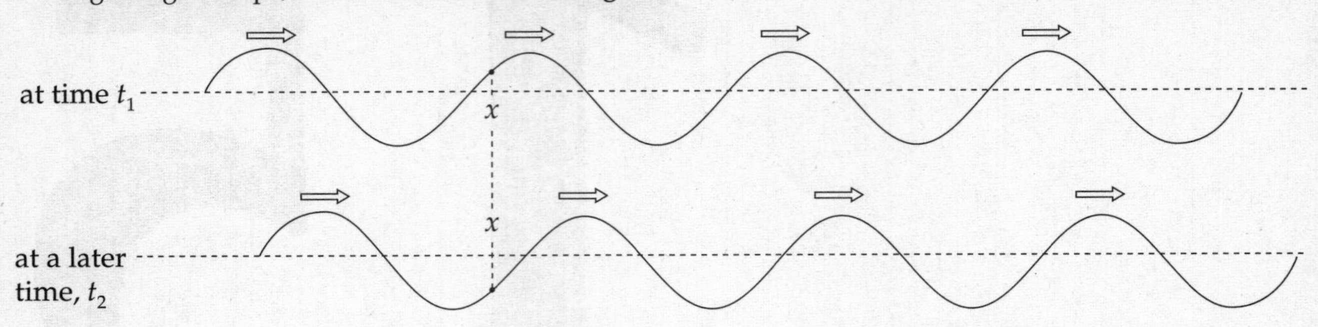

At any point (x) along the rope, the rope has a certain, and varying, vertical displacement. It's this variation in the vertical displacement that defines the shape of the wave. In the figure above, one particular location along the rope—marked x—is shown. Notice that, at time t_1, the vertical position y of the rope at x is slightly positive (that is, it's above the horizontal). But at a later time, t_2, the vertical position of the rope at this same x is slightly negative (that is, it's below the horizontal). It's clear that for a traveling wave, the displacement y of each point depends not only on x but also on t. An equation that gives y must therefore be a function of both position (x) and time (t). Because y depends on *two* independent variables, wave analysis can be difficult. But there's a way of looking at waves that simplifies things.

Instead of looking at a wave in which both variables (x and t) are changing, we'll look at a wave and allow *only one* of the variables to change. We'll use two points of view:

Point of View #1: x varies, t does not

Point of View #2: t varies, x does not

Point of View #1: *x* Varies, *t* Does Not

In order to keep t from varying, we must freeze time. How do we do this? By imagining a photograph of the wave. In fact, the figure above shows two snapshots; one taken at time t_1 and the second taken at a slightly later time, t_2. All the x's along the rope are visible (that is, x varies but t does not). What features of the wave can we see in this point of view? Well, we can see the points at which the rope has its maximum vertical displacement above the horizontal; these points are called **crests**. The points at which the rope has its maximum vertical displacement below the horizontal are called **troughs**. These crests and troughs repeat themselves at regular intervals along the rope, and the distance between two adjacent crests (or two adjacent troughs) is the length of one wave, and is called the **wavelength** (λ, *lambda*). Also, the maximum displacement from the horizontal equilibrium position of the rope is also measurable; this is known as the **amplitude** (A) of the wave.

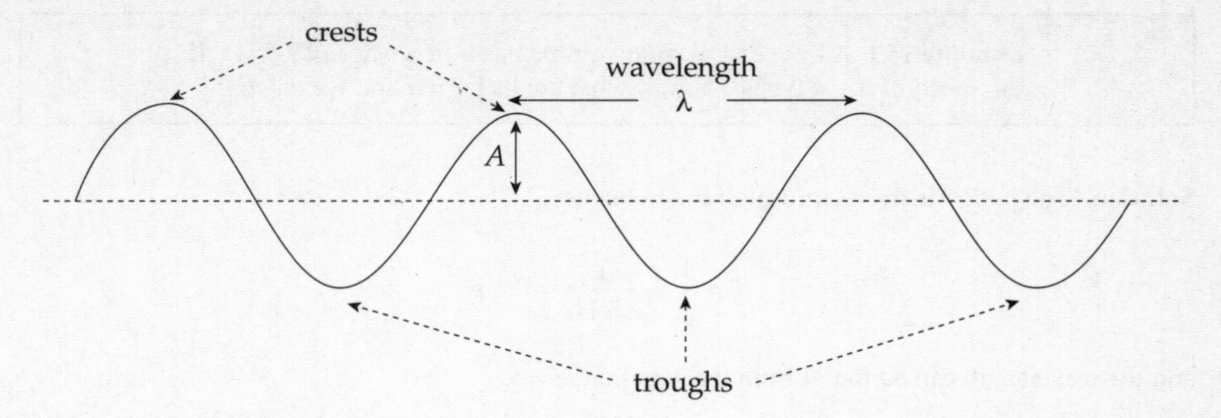

POINT OF VIEW #2: *t* VARIES, *x* DOES NOT

Now we will designate one position x along the rope to watch as time varies. One way to do this is to visualize two screens in front of the rope, with only a narrow gap between them. We can then observe how a single point on the rope varies as a wave travels behind the screen (the wave is traveling because we're not freezing time here). This point on the rope moves up and down. Since the direction in which the rope oscillates (vertically) is perpendicular to the direction in which the wave **propagates** (or travels, horizontally), this wave is **transverse**. The time it takes for one complete vertical oscillation of a point on the rope is called the **period**, T, of the wave, and the number of cycles it completes in one second is called its **frequency**, f. The period and frequency are established by the source of the wave and, of course, $T = 1/f$.

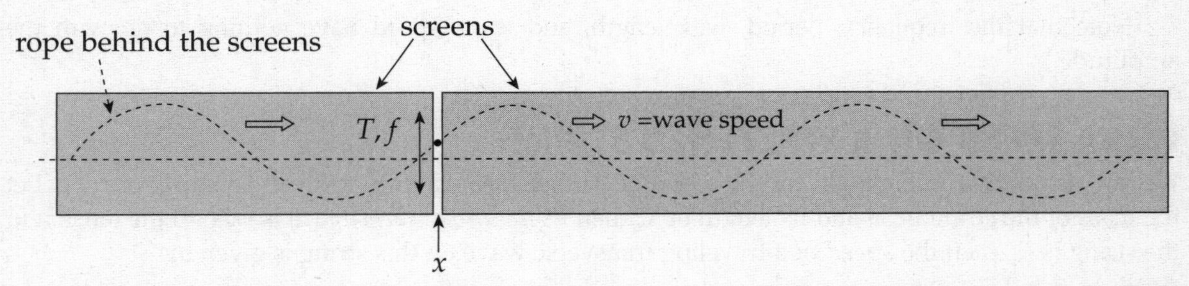

Four of the most important characteristics of any wave are its wavelength, amplitude, period, and frequency. The first two of these were identified in Point of View #1, and the second two were identified in Point of View #2.

A fifth important characteristic of a traveling wave is its speed, v. Look at the figure above and imagine the visible point on the rope moving from its crest position, down to its trough position, and then back up to the crest position. How long did this take? The period, T. Behind the screen, the wave moved a distance of one wavelength. T is the time required for one wave to travel by a point, and λ is the distance traveled by one wave. Therefore, the equation *distance* = *rate* × *time* becomes

$$\lambda = vT$$

$$\lambda \cdot \frac{1}{T} = v$$

$$\lambda f = v$$

The simple equation $v = \lambda f$ shows how the wave speed, wavelength, and frequency are interconnected. *It's the most basic equation in wave theory.*

Example 15.1 A traveling wave on a rope has a frequency of 2.5 Hz. If the speed of the wave is 1.5 m/s, what are its period and wavelength?

Solution. The period is the reciprocal of the frequency:

$$T = \frac{1}{f} = \frac{1}{2.5 \text{ Hz}} = 0.4 \text{ s}$$

and the wavelength can be found from the equation $\lambda f = v$:

$$\lambda = \frac{v}{f} = \frac{1.5 \text{ m/s}}{2.5 \text{ Hz}} = 0.6 \text{ m}$$

Example 15.2 The period of a traveling wave is 0.5 s, its amplitude is 10 cm, and its wavelength is 0.4 m. What are its frequency and wave speed?

Solution. The frequency is the reciprocal of the period: $f = 1/T = 1/(0.5 \text{ s}) = 2$ Hz. The wave speed can be found from the equation $v = \lambda f$:

$$v = \lambda f = (0.4 \text{ m})(2 \text{ Hz}) = 0.8 \text{ m/s}$$

(Note that the frequency, period, wavelength, and wave speed have nothing to do with the amplitude.)

WAVE SPEED ON A STRETCHED STRING

We can also derive an equation for the speed of a transverse wave on a stretched string or rope. Let the mass of the string be m and its length be L; then its *linear mass density* (μ) is m/L. If the tension in the string is F_T, then the speed of a traveling transverse wave on this string is given by

$$v = \sqrt{\frac{F_T}{\mu}}$$

Note that v depends only on the physical characteristics of the string: its tension and linear density. So, because $v = \lambda f$ for a given stretched string, varying f will create different waves that have different wavelengths, but v will not vary.

Example 15.3 A horizontal rope with linear mass density $\mu = 0.5$ kg/m sustains a tension of 60 N. The non-attached end is oscillated vertically with a frequency of 4 Hz.

(a) What are the speed and wavelength of the resulting wave?

(b) How would you answer these questions if f were increased to 5 Hz?

Solution.

(a) Wave speed is established by the physical characteristics of the rope:

$$v = \sqrt{\frac{F_T}{\mu}} = \sqrt{\frac{60 \text{ N}}{0.5 \text{ kg/m}}} = 11 \text{ m/s}$$

With v, we can find the wavelength: $\lambda = v/f = (11 \text{ m/s})/(4 \text{ Hz}) = 2.8$ m.

(b) If f were increased to 5 Hz, then v would not change, but λ would; the new wavelength would be

$$\lambda' = v/f' = (11 \text{ m/s})/(5 \text{ Hz}) = 2.2 \text{ m}$$

Example 15.4 Two ropes of unequal linear densities are connected, and a wave is created in the rope on the left, which propagates to the right, toward the interface with the heavier rope.

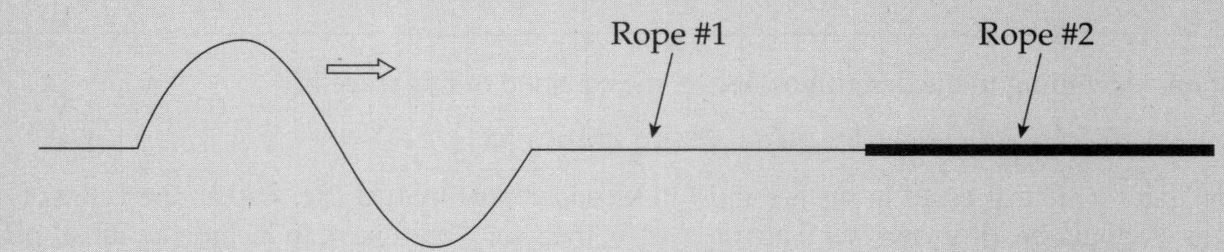

When a wave strikes the boundary to a new medium (in this case, the heavier rope), some of the wave's energy is reflected and some is transmitted. The frequency of the transmitted wave is the same, but the speed and wavelength are not. How do the speed and wavelength of the incident wave compare to the speed and wavelength of the wave as it travels through Rope #2?

Solution. Since the wave enters a new medium, it will have a new wave speed. Because Rope #2 has a greater linear mass density than Rope #1, and because v is inversely proportional to the square root of the linear mass density, the speed of the wave in Rope #2 will be less than the speed of the wave in Rope #1. Since $v = \lambda f$ must always be satisfied and f does not change, the fact that v changes means that λ must change, too. In particular, since v decreases upon entering Rope #2, so will λ.

The Mathematical Description of a Traveling Wave

Remember that the location of a point in a medium that conducts a wave depends on two variables, position and time (x and t). It has the basic form

$$y = A \sin(\omega t \pm \kappa x)$$

To show that y depends on both x and t, the dependent variable, y, is often written as $y(x, t)$, and read "y of x and t." If the minus sign is used in the equation above, the wave travels in the $+x$ direction; if the plus sign is used, the wave travels in the $-x$ direction.

What do the coefficients A, ω, and κ stand for? Well, A is the amplitude, ω (*omega*) is the **angular frequency** and is defined as 2π times the frequency: $\omega = 2\pi f$. Note that the units for angular frequency are s^{-1}. The coefficient κ (*kappa*) is called the **angular wave number**;

it's equal to 2π times the **wave number** (or **propagation constant**), k, which is the number of wavelengths per unit distance. By definition, then $k = 1/\lambda$, and $\kappa = 2\pi k$. So, the equation given above for y contains quantities that characterize all of the important features of a wave: the amplitude, the frequency f (through ω), the period T (which is $1/f = 2\pi/\omega$), the wavelength ($\lambda = 2\pi/\kappa$), and the wave speed v (which is $\lambda f = \omega/k$). Since, at this point, it's easier to relate to f and λ than to ω and κ, it's helpful to rewrite the equation above in the following equivalent forms:

$$y = A\sin 2\pi(ft \pm \frac{1}{\lambda}x) \quad \text{or} \quad y = A\sin\frac{2\pi}{\lambda}(vt \pm x)$$

which now include the quantities f and λ (or, in the second form, v and λ).

Example 15.5 Write the equation for a transverse wave traveling in the $+x$ direction with a frequency of 6 Hz, wavelength 0.2 m, and amplitude 8 cm, given that the left-end of the rope (the point $x = 0$) has zero displacement at time $t = 0$.

Solution. According to the description above, the equation of this wave is

$$y = 0.08\sin\left[2\pi\left(6t - 5x\right)\right]$$

where x and y are expressed in meters and t in seconds. Note that, at $(x,t) = (0,0)$, the value of y is zero, as it should be. [If y were *not* 0 at $(x,t) = (0,0)$, then we would have to include an initial phase in the equation.]

Example 15.6 The vertical position y of any point on a rope that supports a transverse wave traveling horizontally is given by the equation

$$y = 0.1\sin(6\pi t + 8\pi x)$$

with x and y in meters and t in seconds. Find the amplitude, frequency, angular frequency, period, wavelength, wave speed, wave number, and angular wave number. Does this wave travel in the $+x$ or the $-x$ direction?

Solution. The coefficient of the sine term is the amplitude, so

amplitude: $A = 0.1$ m

Next, we recognize that the coefficient of t is the angular frequency (ω) and that the coefficient of x is the angular wave number (κ). Therefore,

angular frequency: $\omega = 6\pi\ \text{s}^{-1}$
frequency: $f = \omega/(2\pi) = (6\pi\ \text{s}^{-1})/(2\pi) = 3$ Hz
angular wave number: $\kappa = 8\pi\ \text{m}^{-1}$
wave number: $k = \kappa/(2\pi) = (8\pi\ \text{m}^{-1})/(2\pi) = 4\ \text{m}^{-1}$
wavelength: $\lambda = 1/k = 1/(4\ \text{m}^{-1}) = 0.25$ m
period: $T = 1/f = 1/(3\text{ Hz}) = 0.33$ s
wave speed: $v = \lambda f = (0.25\text{ m})(3\text{ Hz}) = 0.75$ m/s

Since the plus sign is used in the argument of the sine function, this wave is going in the $-x$ direction.

SUPERPOSITION OF WAVES

When two or more waves meet, the displacement at any point of the medium is equal to the algebraic sum of the displacements due to the individual waves. This is **superposition**. The figure below shows two wave pulses traveling toward each other along a stretched string. Note that when they meet and overlap (**interfere**), the displacement of the string is equal to the sum of the individual displacements, but after they pass, the wave pulses continue, unchanged by their meeting.

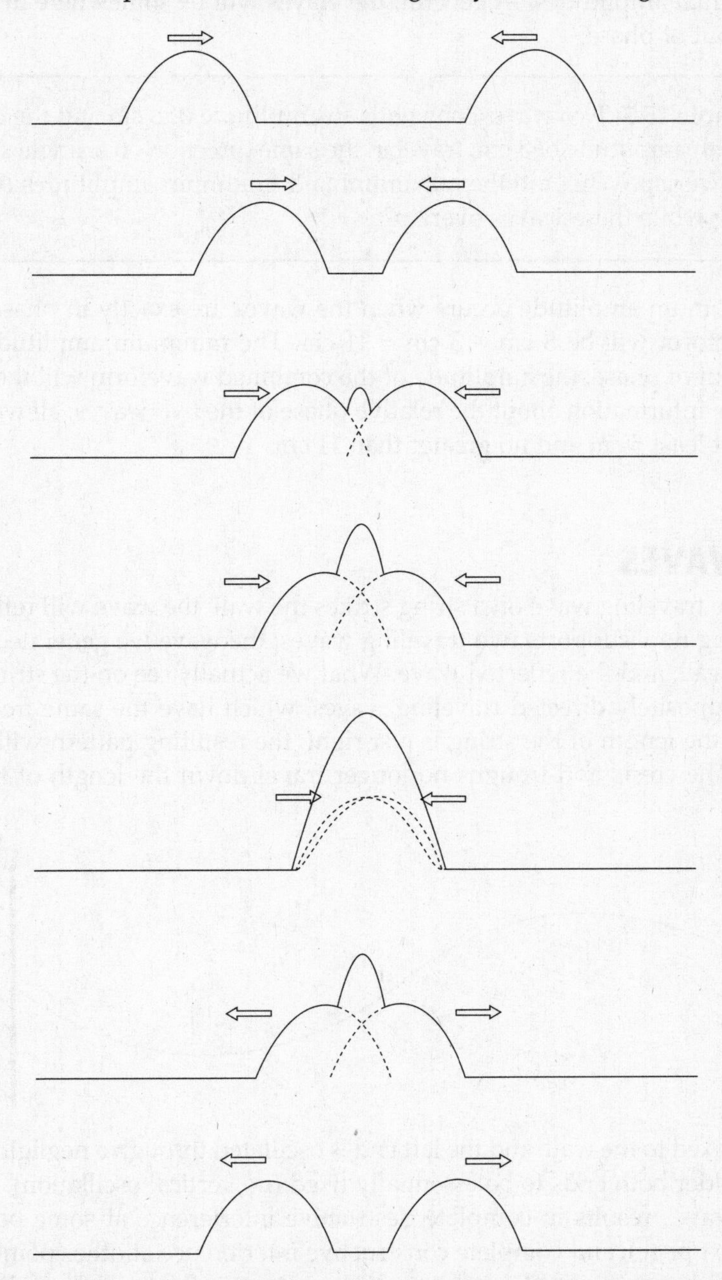

If the two waves have displacements of the same sign when they overlap, the combined wave will have a displacement of greater magnitude than either individual wave; this is called **constructive interference**. Similarly, if the waves have opposite displacements when they meet, the combined waveform will have a displacement of smaller magnitude than either individual wave; this is called

destructive interference. If the waves travel in the same direction, the amplitude of the combined wave depends on the relative phase of the two waves. If the waves are exactly **in phase**—that is, if crest meets crest and trough meets trough—then the waves will constructively interfere completely, and the amplitude of the combined wave will be the sum of the individual amplitudes. However, if the waves are exactly **out of phase**—that is, if crest meets trough and trough meets crest—then they will destructively interfere completely, and the amplitude of the combined wave will be the difference between the individual amplitudes. In general, the waves will be somewhere in between exactly in phase and exactly out of phase.

> **Example 15.7** Two waves, one with an amplitude of 8 cm and the other with an amplitude of 3 cm, travel in the same direction on a single string and overlap. What are the maximum and minimum amplitudes of the string while these waves overlap?

Solution. The maximum amplitude occurs when the waves are exactly in phase; the amplitude of the combined waveform will be 8 cm + 3 cm = 11 cm. The minimum amplitude occurs when the waves are exactly out of phase; the amplitude of the combined waveform will then be 8 cm – 3 cm = 5 cm. Without more information about the relative phase of the two waves, all we can say is that the amplitude will be at least 5 cm and no greater than 11 cm.

STANDING WAVES

When our prototype traveling wave on a string strikes the wall, the wave will reflect and travel back toward us. The string now supports two traveling waves; the wave we generated at our end, which travels toward the wall, and the reflected wave. What we actually see on the string is the superposition of these two oppositely directed traveling waves, which have the same frequency, amplitude, and wavelength. If the length of the string is just right, the resulting pattern will oscillate vertically and remain fixed. The crests and troughs no longer travel down the length of the string. This is a **standing wave**.

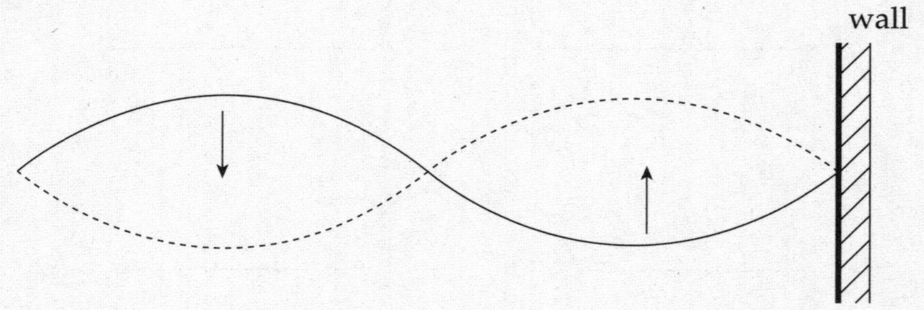

The right end is fixed to the wall, and the left end is oscillated through a negligibly small amplitude so that we can consider both ends to be essentially fixed (no vertical oscillation). The interference of the two traveling waves results in complete destructive interference at some points (marked N in the figure on the next page), and complete constructive interference at other points (marked A in the figure). Other points have amplitudes between these extremes. Note another difference between a traveling wave and a standing wave: While every point on the string had the same amplitude as the traveling wave went by, each point on a string supporting a standing wave has an individual amplitude. The points marked N are called **nodes**, and those marked A are called **antinodes**.

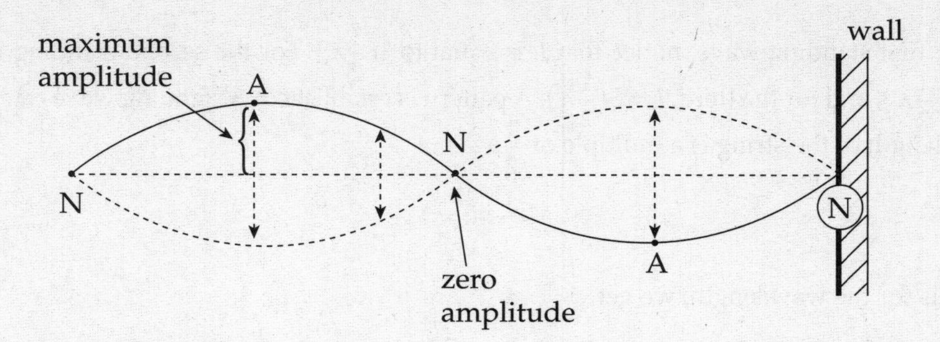

Nodes and antinodes always alternate, they're equally spaced, and the distance between two successive nodes (or antinodes) is equal to $\frac{1}{2}\lambda$. This information can be used to determine how standing waves can be generated. The following figures show the three simplest standing waves that our string can support. The first standing wave has one antinode, the second has two, and the third has three. The length of the string in all three diagrams is L.

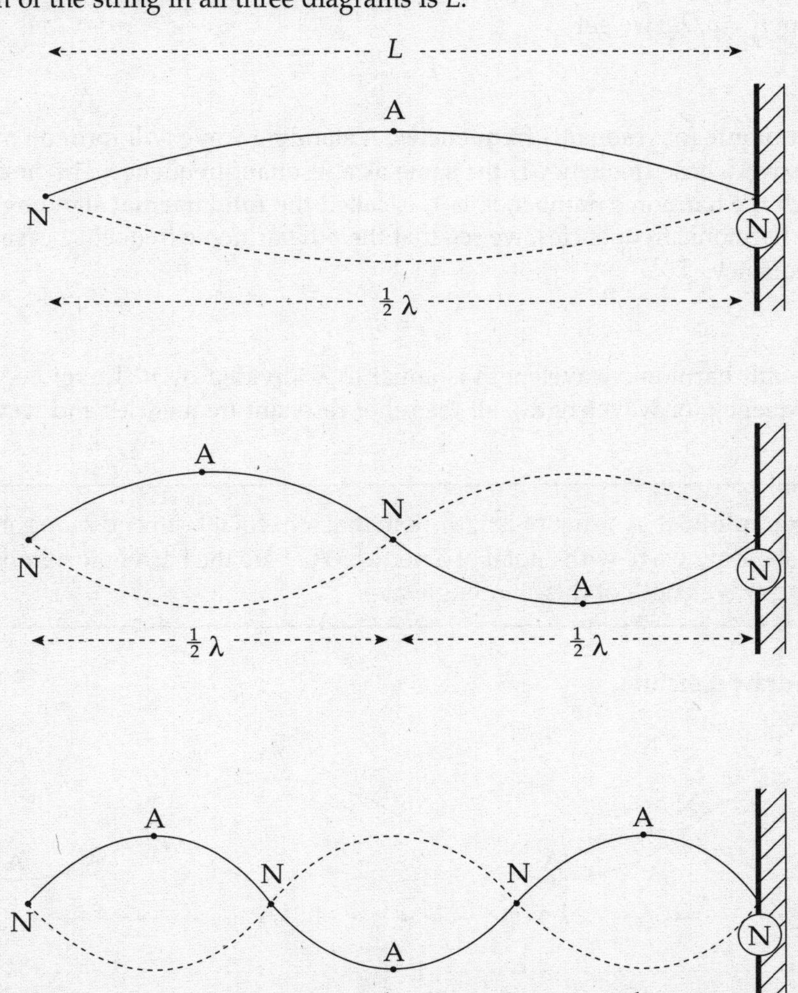

For the first standing wave, notice that L is equal to $1(\frac{1}{2}\lambda)$. For the second standing wave, L is equal to $2(\frac{1}{2}\lambda)$, and for the third, $L = 3(\frac{1}{2}\lambda)$. A pattern is established: A standing wave can only form when the length of the string is a multiple of $\frac{1}{2}\lambda$:

$$L = n(\frac{1}{2}\lambda)$$

Solving this for the wavelength, we get

$$\lambda_n = \frac{2L}{n}$$

These are called the **harmonic** (or **resonant**) **wavelengths**, and the integer n is known as the **harmonic number**.

Since we typically have control over the frequency of the waves we create, it's more helpful to figure out the *frequencies* that generate a standing wave. Because $\lambda f = v$, and because v is fixed by the physical characteristics of the string, the special λ's found above correspond to equally special frequencies. From $f_n = v/\lambda_n$, we get

$$f_n = \frac{nv}{2L}$$

These are the **harmonic** (or **resonant**) **frequencies**. A standing wave will form on a string if we create a traveling wave whose frequency is the same as a resonant frequency. The first standing wave, the one for which the harmonic number, n, is 1, is called the **fundamental** standing wave. From the equation for the harmonic frequencies, we see that the nth harmonic frequency is simply n times the fundamental frequency:

$$f_n = nf_1$$

Similarly, the nth harmonic wavelength is equal to λ_1 divided by n. Therefore, by knowing the fundamental frequency (or wavelength), all the other resonant frequencies and wavelengths can be determined.

> **Example 15.8** A string of length 12 m that's fixed at both ends supports a standing wave with a total of 5 nodes. What are the harmonic number and wavelength of this standing wave?

Solution. First, draw a picture.

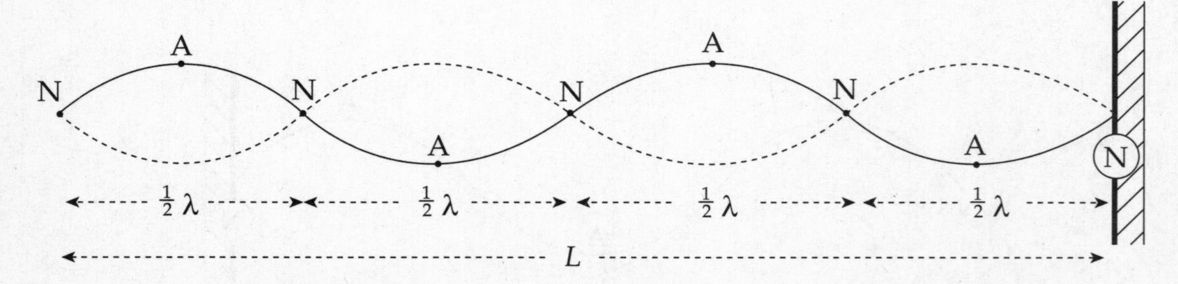

This shows that the length of the string is equal to $4(\frac{1}{2}\lambda)$, so

$$L = 4(\frac{1}{2}\lambda) \Rightarrow \lambda = \frac{2L}{4}$$

This is the fourth-harmonic standing wave, with wavelength λ_4 (because the expression above matches $\lambda_n = 2L/n$ for $n = 4$). Since $L = 12$ m, the wavelength is

$$\lambda_4 = \frac{2(12 \text{ m})}{4} = 6 \text{ m}$$

Example 15.9 A string of length 10 m and mass 300 g is fixed at both ends, and the tension in the string is 40 N. What is the frequency of the standing wave for which the distance between a node and the closest antinode is 1 m?

Solution. Because the distance between two successive nodes (or successive antinodes) is equal to $\frac{1}{2}\lambda$, the distance between a node and the closest antinode is half this, or $\frac{1}{4}\lambda$. Therefore, $\frac{1}{4}\lambda = 1$ m, so $\lambda = 4$ m. Since the harmonic wavelengths are given by the equation $\lambda_n = 2L/n$, we can find that

$$4 \text{ m} = \frac{2(10 \text{ m})}{n} \Rightarrow n = 5$$

The frequency of the fifth harmonic is

$$f_5 = \frac{5v}{2L} = \frac{5}{2L}\sqrt{\frac{F_T}{\mu}} = \frac{5}{2L}\sqrt{\frac{F_T}{m/L}} = \frac{5}{2(10 \text{ m})}\sqrt{\frac{40 \text{ N}}{0.3 \text{ kg}/10 \text{ m}}} = 9.1 \text{ Hz}$$

If you attached a rope or string to a ring that could slide up and down a pole without friction you would make a rope that is fixed at one end, but free at another end. This would create nodes at the closed end and antinodes at the open end. Below are some possible examples of this.

$$L = \frac{1}{4}\lambda \quad \text{or} \quad \lambda_1 = \frac{4L}{1}$$

$$L = \frac{3}{4}\lambda \quad \text{or} \quad \lambda_3 = \frac{4L}{3}$$

$$L = \frac{5}{4}\lambda \quad \text{or} \quad \lambda_5 = \frac{4L}{5}$$

L

If one end of the rope is free to move, standing waves can form for wavelengths of $\lambda_n = \frac{4L}{n}$ or by frequencies of $f_n = \frac{nv}{4L}$ where L is the length of the rope, v is the speed of the wave, and n must be an odd integer.

SOUND WAVES

Sound waves are produced by the vibration of an object, such as your vocal cords, a plucked string, or a jackhammer. The vibrations cause pressure variations in the conducting medium (which can be gas, liquid, or solid), and if the frequency is between 20 Hz and 20,000 Hz, the vibrations may be detected by human ears and perceived as sound. The variations in the conducting medium can be positions at which the molecules of the medium are bunched together (where the pressure is above normal), which are called **compressions**, and positions where the pressure is below normal, called **rarefactions**. In the figure below, a vibrating diaphragm sets up a sound wave in an air-filled tube. Each dot represents many, many air molecules:

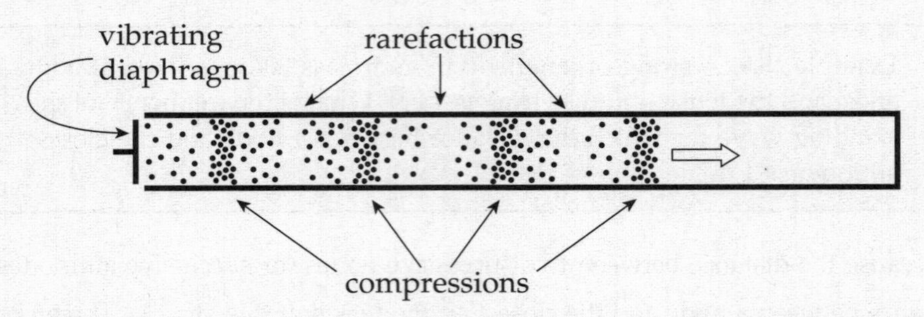

An important difference between sound waves and the waves we've been studying on stretched strings is that the molecules of the medium transmitting a sound wave move *parallel* to the direction of wave propagation, rather than perpendicular to it. For this reason, sound waves are said to be **longitudinal**. Despite this difference, all of the basic characteristics of a wave—amplitude, wavelength, period, frequency—apply to sound waves as they did for waves on a string. Furthermore, the all-important equation $\lambda f = v$ also holds true. However, because it's very difficult to draw a picture of a longitudinal wave, an alternate method is used: We can graph the pressure as a function of position:

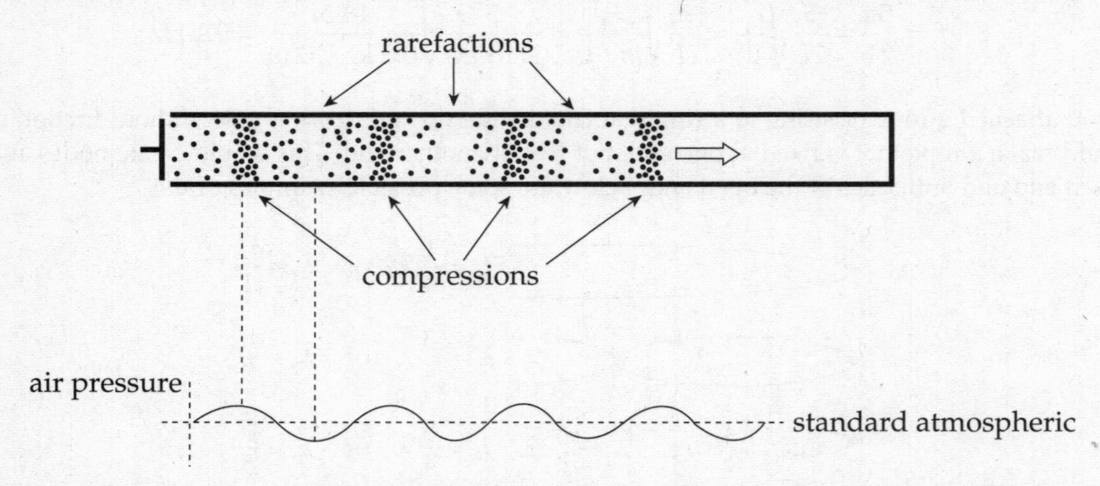

The speed of a sound wave depends on the medium through which it travels. In particular, it depends on the density (ρ) and on the **bulk modulus** (B), a measure of the medium's response to compression. A medium that is easily compressible, like a gas, has a low bulk modulus; liquids and solids, which are much less easily compressed, have significantly greater bulk modulus values. For this reason, sound generally travels faster through solids than through liquids and faster through liquids than through gases.

The equation that gives a sound wave's speed in terms of ρ and B is

$$v = \sqrt{\frac{B}{\rho}}$$

The speed of sound through air can also be written in terms of air's mean pressure, which depends on its temperature. At room temperature (approximately 20°C) and normal atmospheric pressure, sound travels at 343 m/s. This value increases as air warms or pressure increases.

Example 15.10 A sound wave with a frequency of 300 Hz travels through the air.

(a) What is its wavelength?

(b) If its frequency increased to 600 Hz, what is its wave speed and the wavelength?

Solution.

(a) Using v = 343 m/s for the speed of sound through air, we find that

$$\lambda = \frac{v}{f} = \frac{343 \text{ m/s}}{300 \text{ Hz}} = 1.14 \text{ m}$$

(b) Unless the ambient pressure or the temperature of the air changed, the speed of sound would not change. Wave speed depends on the characteristics of the medium, not on the frequency, so v would still be 343 m/s. However, a change in frequency would cause a change in wavelength. Since f increased by a factor of 2, λ would decrease by a factor of 2, to $\frac{1}{2}$(1.14 m) = 0.57 m.

Example 15.11 A sound wave traveling through water has a frequency of 500 Hz and a wavelength of 3 m. How fast does sound travel through water?

Solution. $v = \lambda f$ = (3 m)(500 Hz) = 1500 m/s.

SOUND LEVEL

Sound waves transmit energy; the rate at which they transmit energy, per unit area, is called the **intensity**. It is denoted by I and has the unit W/m^2. Intensity depends on the wave's amplitude: The greater the amplitude (A), the greater the intensity (and, therefore, the louder the sound); in fact, I is proportional to A^2. Also, since sound waves spread out to form a spherical wavefront whose area is proportional to r^2, the intensity is proportional to $1/r^2$, where r is the distance from the source. The loudness of a sound is usually expressed not in terms of W/m^2 but in **decibels** (abbreviated **dB**). The **loudness level** of an acoustic wave whose intensity is I is defined as:

$$\beta = 10 \log \frac{I}{I_0}$$

where $I_0 = 1 \times 10^{-12}$ W/m², a value known as the *threshold of hearing*. A sound wave at the threshold of hearing has $I = I_0$, so its sound level is $\beta = 0$ dB (because log 1 = 0). Since the decibel scale is logarithmic, when intensity increases by a factor of 10, the decibel level increases by *the addition of* 10. Similarly, if the intensity decreases by a factor of 10, the decibel level decreases by *the subtraction of* 10.

Example 15.12 The loudness level in a library is 30 dB. If the loudness level of normal conversation is 60 dB, how many times more intense are the sound waves produced by normal conversation than those that exist in a library?

Solution. The difference between the decibel levels is 60 – 30 = 30 dB. Each change by 10 in the decibel level corresponds to a factor of 10 in intensity. Since 30 dB represents 3 changes of 10, the intensity of normal conversation must be 3 *factors* of 10, or $10^3 = 1000$ times, greater than that in the library.

Beats

If two sound waves whose frequencies are close but not identical interfere, the resulting sound modulates in amplitude, becoming loud, then soft, then loud, then soft. This is due to the fact that as the individual waves travel, they are in phase, then out of phase, then in phase again, and so on. Therefore, by superposition, the waves interfere constructively, then destructively, then constructively, etc. When the waves interfere constructively, the amplitude increases, and the sound is loud; when the waves interfere destructively, the amplitude decreases, and the sound is soft. Each time the waves interfere constructively, producing an increase in sound level, we say that a **beat** has occurred. The number of beats per second, known as the **beat frequency**, is equal to the difference between the frequencies of the two combining sound waves:

$$f_{\text{beat}} = |f_1 - f_2|$$

If frequencies f_1 and f_2 match, then the combined waveform doesn't waver in amplitude, and no beats are heard.

Example 15.13 A piano tuner uses a tuning fork to adjust the key that plays the A note above middle C (whose frequency should be 440 Hz). The tuning fork emits a perfect 440 Hz tone. When the tuning fork and the piano key are struck, beats of frequency 3 Hz are heard.

(a) What is the frequency of the piano key?

(b) If it's known that the piano key's frequency is too high, should the piano tuner tighten or loosen the wire inside the piano in order to tune it?

Solution.

(a) Since $f_{\text{beat}} = 3$ Hz, the tuning fork and the piano string are off by 3 Hz. Since the fork emits a tone of 440 Hz, the piano string must emit a tone of either 437 Hz or 443 Hz. Without more information, we can't determine which.

(b) If we know that the frequency of the tone emitted by the out-of-tune string is too high (that is, it's 443 Hz), we need to find a way to lower the frequency. Remember that the resonant frequencies for a stretched string fixed at both ends are given by the equation $f = nv/2L$, and that $v = \sqrt{F_T / \mu}$. Since f is too high, v must be too high. To lower v, we must reduce F_T. The piano tuner should loosen the string and listen for beats again, adjusting the string until the beats disappear.

RESONANCE FOR SOUND WAVES

Just as standing waves can be set up on a vibrating string, standing *sound* waves can be established within an enclosure. In the figure below, a vibrating source at one end of an air-filled tube produces sound waves that travel the length of the tube.

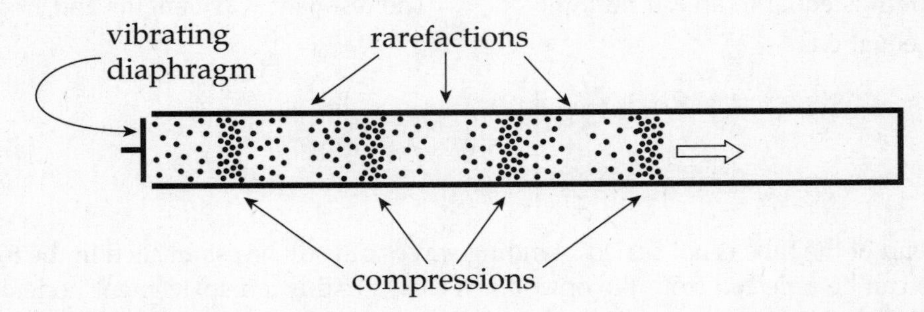

These waves reflect off the far end, and the superposition of the forward and reflected waves can produce a standing wave pattern if the length of the tube and the frequency of the waves are related in a certain way.

Notice that air molecules at the far end of the tube can't oscillate horizontally because they're up against a wall. So the far end of the tube is a displacement node. But the other end of the tube (where the vibrating source is located) is a displacement antinode. A standing wave with one antinode (A) and one node position (N) can be depicted as follows:

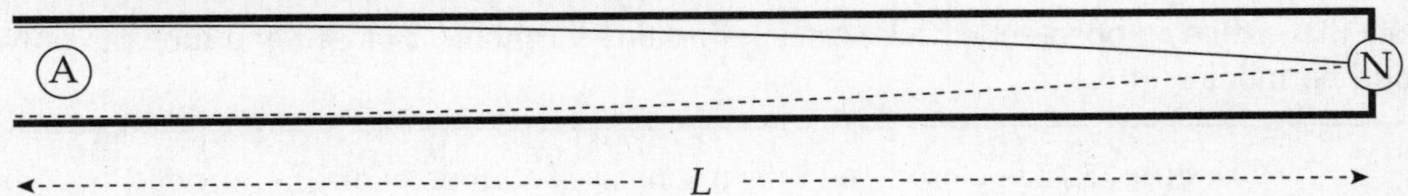

Although sound waves in air are longitudinal, here we'll show the wave as transverse so that it's easier to determine the wavelength. Since the distance between an antinode and an adjacent node is always $\frac{1}{4}$ of the wavelength, the length of the tube, L, in the figure above is $\frac{1}{4}$ the wavelength. This is the longest standing wavelength that can fit in the tube, so it corresponds to the lowest standing wave frequency, the fundamental:

$$L = \frac{\lambda_1}{4} \Rightarrow \lambda_1 = 4L \Rightarrow f_1 = \frac{v}{\lambda_1} = \frac{v}{4L}$$

Our condition for resonance was a node at the closed end and an antinode at the open end. Therefore, the next higher frequency standing wave that can be supported in this tube must have two antinodes and two nodes:

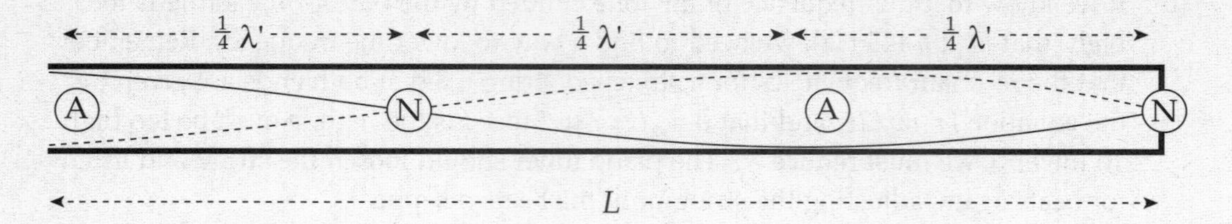

In this case, the length of the tube is equal to $3(\frac{1}{4}\lambda')$, so

$$L = \frac{3\lambda'}{4} \quad \Rightarrow \quad \lambda' = \frac{4L}{3} \quad \Rightarrow \quad f' = \frac{v}{\lambda'} = \frac{3v}{4L}$$

Here's the pattern: Standing sound waves can be established in a tube that's closed at one end if the tube's length is equal to an *odd* multiple of $\frac{1}{4}\lambda$. The resonant wavelengths and frequencies are given by the equations

$$\left.\begin{aligned} \lambda_n &= \frac{4L}{n} \\ f_n &= n\frac{v}{4L} \end{aligned}\right\} \text{ for any } odd \text{ integer } n$$

If the far end of the tube is not sealed, standing waves can still be established in the tube, because sound waves can be reflected from the open air. A closed end is a displacement node, but an open end is a displacement antinode. In this case, then, the standing waves will have two displacement antinodes (at the ends of the tube), and the resonant wavelengths and frequencies will be given by

$$\left.\begin{aligned} \lambda_n &= \frac{2L}{n} \\ f_n &= n\frac{v}{2L} \end{aligned}\right\} \text{ for any integer } n$$

Note that, while an open-ended tube can support any harmonic, a closed-end tube can only support odd harmonics.

Example 15.14 A closed-end tube resonates at a fundamental frequency of 440.0 Hz. The air in the tube is at a temperature of 20°C, and it conducts sound at a speed of 343 m/s.

(a) What is the length of the tube?

(b) What is the next higher harmonic frequency?

(c) Answer the questions posed in (a) and (b) assuming that the tube were open at its far end.

Solution.

(a) For a closed-end tube, the harmonic frequencies obey the equation $f_n = nv/(4L)$. The fundamental corresponds to $n = 1$, so

$$f_1 = \frac{v}{4L} \quad \Rightarrow \quad L = \frac{v}{4f_1} = \frac{343 \text{ m/s}}{4(440.0 \text{ Hz})} = 0.195 \text{ m} = 19.5 \text{ cm}$$

(b) Since a closed-end tube can support only *odd* harmonics, the next higher harmonic frequency (the first **overtone**) is the *third* harmonic, f_3, which is $3f_1 = 3(440.0 \text{ Hz}) = 1320 \text{ Hz}$.

(c) For an open-end tube, the harmonic frequencies obey the equation $f_n = nv/(2L)$. The fundamental corresponds to $n = 1$, so

$$f_1 = \frac{v}{2L'} \quad \Rightarrow \quad L' = \frac{v}{2f_1} = \frac{343 \text{ m/s}}{2(440.0 \text{ Hz})} = 0.390 \text{ m} = 39.0 \text{ cm}$$

And, since an open-end tube can support any harmonic, the first overtone would be the second harmonic, $f_2 = 2f_1 = 2(440.0 \text{ Hz}) = 880.0 \text{ Hz}$.

THE DOPPLER EFFECT

When a source of sound waves and a detector are not in relative motion, the frequency that the source emits matches the frequency that the detector receives.

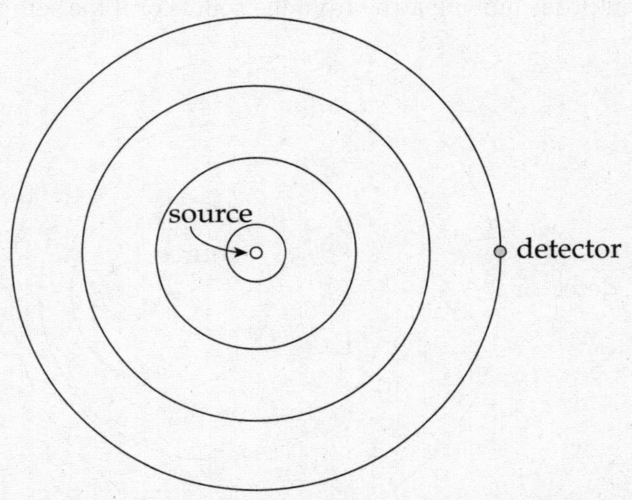

However, if there *is* relative motion between the source and the detector, then the waves that the detector receives are different in frequency (and wavelength). For example, if the detector moves toward the source, then the detector intercepts the waves at a rate higher than the one at which they were emitted; the detector hears a higher frequency than the source emitted. In the same way, if the source moves toward the detector, the wavefronts pile up, and this results in the detector receiving waves with shorter wavelengths and higher frequencies:

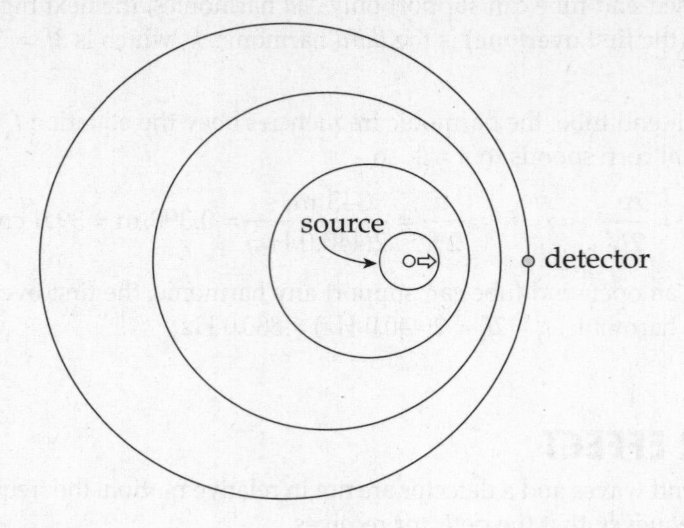

Conversely, if the detector is moving away from the source or if the source is moving away from the detector,

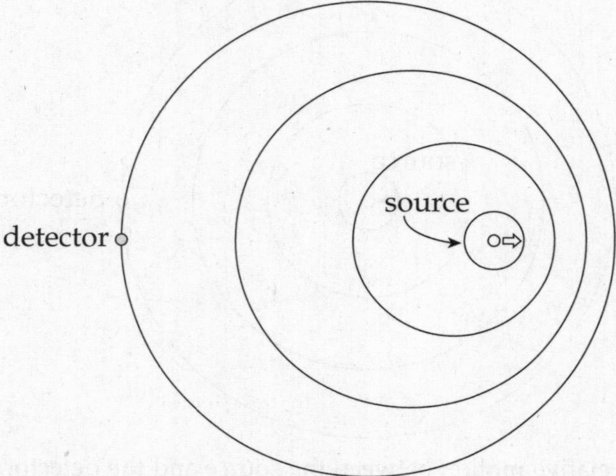

then the detected waves have a longer wavelength and a lower frequency than they had when they were emitted by the source. The shift in frequency and wavelength that occurs when the source and detector are in relative motion is known as the **Doppler effect**. In general, relative motion *toward* each other results in a frequency shift upward, and relative motion *away* from each other results in a frequency shift downward.

Let f_S be the frequency of waves that the source emits and f_D the frequency that the detector hears. To determine f_D from f_S, use the following equation:

$$f_D = \frac{v \pm v_D}{v \mp v_S} \cdot f_S$$

where v is the speed of sound, v_D is the speed of the detector, and v_S is the speed of the source. The signs in the numerator and denominator depend on the directions in which the source and detector are moving.

For example, consider the case of a detector moving toward a stationary source. Since the source doesn't move, $v_S = 0$, and there's no decision to be made about the sign in the denominator. But, because v_D is not zero (the detector is moving), we need to decide whether to use the + or the – in the numerator. Since the detector moves *toward* the source, we expect a frequency shift upward. Therefore, to give the higher frequency predicted by the Doppler effect, the + sign must be used:

$$f_D = \frac{v + v_D}{v} \cdot f_S$$

If the detector had been moving *away* from the source, we would have used the – sign in the numerator, to get the lower frequency

$$f_D = \frac{v - v_D}{v} \cdot f_S$$

If both the source and the detector are moving, simply make each decision (+ or – in the numerator for v_D and + or – in the denominator for v_S) separately. For example, consider a car following a police car whose siren emits a sound of constant frequency. In this case, both the source (the police car) and the detector (the driver of the pursuing car) are moving. Because the detector is moving toward the source, we should use the + sign on v_D in the numerator for an upward contribution to f_D. Since the source is moving away from the detector, we should use the + sign on v_S in the denominator for a downward contribution to f_D. Therefore, we get

$$f_D = \frac{v + v_D}{v + v_S} \cdot f_S$$

We can learn something else from this. What if the car were moving at the same speed as the police car? Then, although both are moving relative to the road, *they are not moving relative to each other*. If there is no relative motion between the source and the detector, then there should be no Doppler shift. This conclusion is supported by the equation above, since $v_D = v_S$ implies that $f_D = f_S$.

Example 15.15 A source of 4 kHz sound waves travels at $\frac{1}{9}$ the speed of sound toward a detector that's moving at $\frac{1}{9}$ the speed of sound, toward the source.

(a) What is the frequency of the waves as they're received by the detector?

(b) How does the wavelength of the detected waves compare to the wavelength of the emitted waves?

Solution.

(a) Because the detector moves toward the source, we use the + sign in the numerator of the Doppler effect formula, and since the source moves toward the detector, we use the – sign in the denominator. This gives us:

$$f_D = \frac{v + v_D}{v - v_S} \cdot f_S = \frac{v + \frac{1}{9}v}{v - \frac{1}{9}v} \cdot f_S = \frac{5}{4} f_S = \frac{5}{4}(4 \text{ kHz}) = 5 \text{ kHz}$$

(b) Since the frequency shifted up by a factor of $\frac{5}{4}$, the wavelength will shift down by the same factor. That is,

$$\lambda_D = \frac{\lambda_S}{\frac{5}{4}} = \frac{4}{5}\lambda_S$$

Example 15.16 A person yells, emitting a constant frequency of 200 Hz, as he runs at 5 m/s toward a stationary brick wall. When the reflected waves reach the person, how many beats per second will he hear? (Use 343 m/s for the speed of sound.)

Solution. We need to determine what the frequency of the reflected waves will be when they reach the runner. First, the person is the source of the sound, and the wall is the detector. So:

$$f_{\text{wall}} = \frac{v}{v - v_{\text{runner}}} f$$

Then, as the sound waves reflect off the wall (with no change in frequency upon reflection), the wall acts as the source and the runner acts as the detector:

$$f' = \frac{v + v_{\text{runner}}}{v} f_{\text{wall}}$$

Combining these two equations gives

$$f' = \frac{v + v_{\text{runner}}}{v - v_{\text{runner}}} f$$

With $v = 343$ m/s and $v_{\text{runner}} = 5$ m/s, we get

$$f' = \frac{(343 + 5)\text{ m/s}}{(343 - 5)\text{ m/s}}(200\text{ Hz}) = 206\text{ Hz}$$

Therefore, the beat frequency is

$$f_{\text{beat}} = f' - f = 206\text{ Hz} - 200\text{ Hz} = 6\text{ Hz}$$

THE DOPPLER EFFECT FOR LIGHT

Light, or electromagnetic waves, also experience the Doppler effect. The detected wavelength, λ_D, is related to the emitted wavelength, λ_S, by the equation $\lambda_D = (1 \pm u/c)\lambda_S$, where u is the relative speed of the source and detector, and $c = 3.00 \times 10^8$ m/s, the speed of light. This formula assumes that u is much less than c. Again, the choice of + and – depends on whether the source and detector are moving away from or toward each other. As with sound, motion *toward* corresponds to a frequency shift upward (and a wavelength shift downward), and motion *away* corresponds to a frequency shift downward (and to a wavelength shift upward).

Example 15.17 Light from a distant galaxy is received on Earth with a wavelength of 679 nm. It's known that the wavelength of this light upon emission was 657 nm. Is the galaxy moving toward or away from us? At what speed?

Solution. The fact that the light has a longer wavelength upon detection than it had at emission tells us that the source is receding. Its speed is determined as follows:

$$\lambda_D = \left(1 + \frac{u}{c}\right)\lambda_S \quad \Rightarrow \quad u = c\left(\frac{\lambda_D}{\lambda_S} - 1\right) = (3.00 \times 10^8 \text{ m/s})\left(\frac{679 \text{ nm}}{657 \text{ nm}} - 1\right) = 1.00 \times 10^7 \text{ m/s}$$

Astronomers would say that the light has been **red-shifted**; that is, the wavelength was shifted upward, toward the red end of the visible spectrum.

CHAPTER 15 REVIEW QUESTIONS

Solutions can be found in Chapter 18.

Section I: Multiple Choice

1. What is the wavelength of a 5 Hz wave that travels with a speed of 10 m/s?

   (A) 0.25 m
   (B) 0.5 m
   (C) 1 m
   (D) 2 m
   (E) 50 m

2. A rope of length 5 m is stretched to a tension of 80 N. If its mass is 1 kg, at what speed would a 10 Hz transverse wave travel down the string?

   (A) 2 m/s
   (B) 5 m/s
   (C) 20 m/s
   (D) 50 m/s
   (E) 200 m/s

3. A transverse wave on a long horizontal rope with a wavelength of 8 m travels at 2 m/s. At $t = 0$, a particular point on the rope has a vertical displacement of $+A$, where A is the amplitude of the wave. At what time will the vertical displacement of this same point on the rope be $-A$?

   (A) $t = \frac{1}{8}$ s
   (B) $t = \frac{1}{4}$ s
   (C) $t = \frac{1}{2}$ s
   (D) $t = 2$ s
   (E) $t = 4$ s

4. The vertical displacement, y, of a transverse traveling wave is given by the equation $y = 6 \sin(10\pi t - \frac{1}{2}\pi x)$, with x and y in centimeters and t in seconds. What is the wavelength?

   (A) 0.25 cm
   (B) 0.5 cm
   (C) 2 cm
   (D) 4 cm
   (E) 8 cm

5. A string, fixed at both ends, supports a standing wave with a total of 4 nodes. If the length of the string is 6 m, what is the wavelength of the wave?

   (A) 0.67 m
   (B) 1.2 m
   (C) 1.5 m
   (D) 3 m
   (E) 4 m

6. A string, fixed at both ends, has a length of 6 m and supports a standing wave with a total of 4 nodes. If a transverse wave can travel at 40 m/s down the rope, what is the frequency of this standing wave?

   (A) 6.7 Hz
   (B) 10 Hz
   (C) 13.3 Hz
   (D) 20 Hz
   (E) 26.7 Hz

7. A sound wave travels through a metal rod with wavelength λ and frequency f. Which of the following best describes the wave when it passes into the surrounding air?

	Wavelength	Frequency
(A)	Less than λ	Equal to f
(B)	Less than λ	Less than f
(C)	Greater than λ	Equal to f
(D)	Greater than λ	Less than f
(E)	Greater than λ	Greater than f

8. In the figure below, two speakers, S_1 and S_2, emit sound waves of wavelength 2 m, in phase with each other.

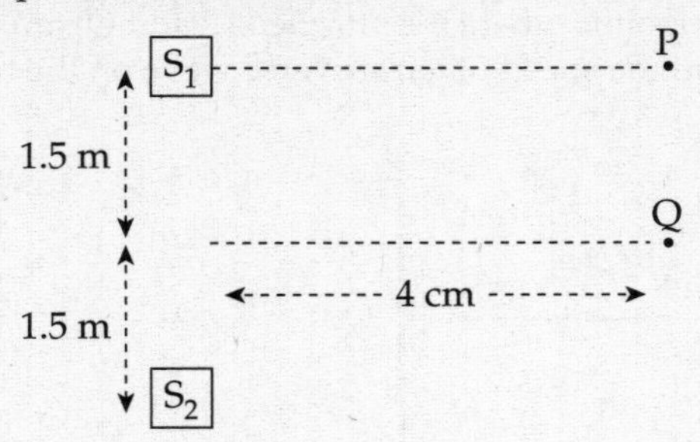

Let A_P be the amplitude of the resulting wave at Point P, and A_Q the amplitude of the resultant wave at Point Q. How does A_P compare to A_Q?

(A) $A_P < A_Q$
(B) $A_P = A_Q$
(C) $A_P > A_Q$
(D) $A_P < 0, A_Q > 0$
(E) A_P and A_Q vary with time, so no comparison can be made.

9. An observer is 2 m from a source of sound waves. By how much will the sound level decrease if the observer moves to a distance of 20 m?

(A) 1 dB
(B) 2 dB
(C) 10 dB
(D) 18 dB
(E) 20 dB

10. An organ pipe that's closed at one end has a length of 17 cm. If the speed of sound through the air inside is 340 m/s, what is the pipe's fundamental frequency?

(A) 250 Hz
(B) 500 Hz
(C) 1000 Hz
(D) 1500 Hz
(E) 2000 Hz

11. A bat emits a 40 kHz "chirp" with a wavelength of 8.75 mm toward a tree and receives an echo 0.4 s later. How far is the bat from the tree?

(A) 35 m
(B) 70 m
(C) 105 m
(D) 140 m
(E) 175 m

12. A car is traveling at 20 m/s away from a stationary observer. If the car's horn emits a frequency of 600 Hz, what frequency will the observer hear? (Use v = 340 m/s for the speed of sound.)

(A) (34/36)(600 Hz)
(B) (34/32)(600 Hz)
(C) (36/34)(600 Hz)
(D) (32/34)(600 Hz)
(E) (32/36)(600 Hz)

SECTION II: FREE RESPONSE

1. A rope is stretched between two vertical supports. The points where it's attached (P and Q) are fixed. The linear density of the rope, μ, is 0.4 kg/m, and the speed of a transverse wave on the rope is 12 m/s.

(a) What's the tension in the rope?

(b) With what frequency must the rope vibrate to create a traveling wave with a wavelength of 2 m?

The rope can support standing waves of lengths 4 m and 3.2 m, whose harmonic numbers are consecutive integers.

(c) Find

(i) the length of the rope;

(ii) the mass of the rope.

(d) What is the harmonic number of the 4 m standing wave?

(e) On the diagram above, draw a sketch of the 4 m standing wave, labeling the nodes and antinodes.

2. For a physics lab experiment on the Doppler effect, students are brought to a racetrack. A car is fitted with a horn that emits a tone with a constant frequency of 500 Hz, and the goal of the experiment is for the students to observe the pitch (frequency) of the car's horn.

   (a) If the car drives at 50 m/s directly away from the students (who remain at rest), what frequency will they hear? (Note: speed of sound = 345 m/s)

   (b) Would it make a difference if the car remained stationary and the students were driven in a separate vehicle at 50 m/s away from the car?

   The following diagram shows a long straight track along which the car is driven. The students stand 40 m from the track.

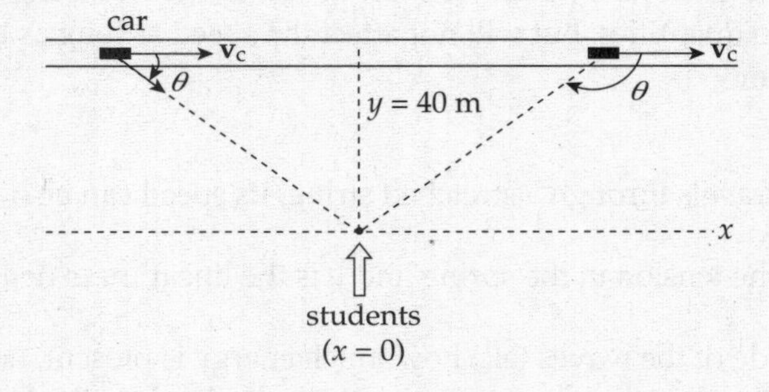

   As the car drives by, the frequency which the students hear, f', is related to the car horn's frequency, f, by the equation

$$f' = \frac{v}{v - v_C \cos\theta} \cdot f$$

   where v is the speed of sound, v_C is the speed of the car (50 m/s), and θ is the angle shown in the diagram.

   (c) Find f' when $\theta = 60°$.

   (d) Find f' when $\theta = 120°$.

   (e) On the axes below, sketch the graph of f' as a function of x, the horizontal position of the car relative to the students' location.

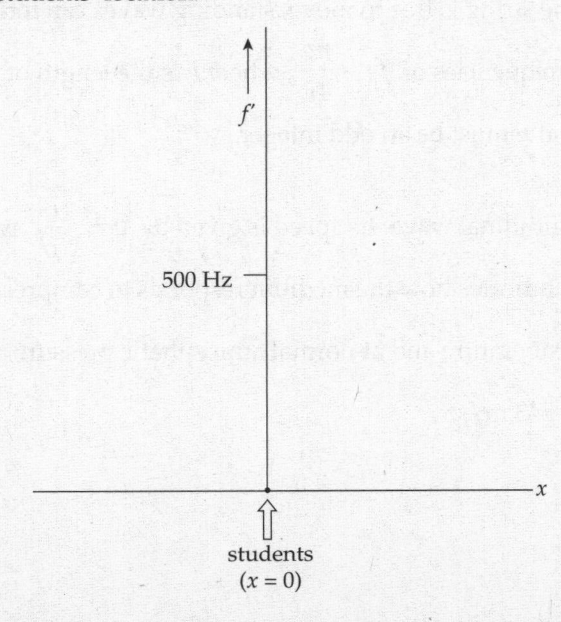

SUMMARY

- The speed of a wave depends on the medium through which it travels. Speed can be determined by $v = f\lambda$, where f is the frequency ($f = \frac{\#\,cycles}{time}$) and λ is the wavelength. Because $f = \frac{1}{T}$, the speed can also be written $v = \frac{\lambda}{T}$.
- Changing any one of the period, the frequency, or the wavelength of a wave will affect the other two quantities, but will not affect the speed as long as the wave stays in the same medium.
- If the wave travels through a stretched string, its speed can be determined by $v = \sqrt{\frac{F_T}{\mu}}$, where F_T is the tension in the spring and μ is the linear mass density (mass/length).
- The amplitude of the waves tells how much energy is present, not how fast the wave will travel. A traveling wave may be mathematically described by

$$y = A\sin(\omega t \pm \kappa x) \text{ or } y = A\sin 2\pi(ft \pm \frac{x}{\lambda}) \text{ or } y = A\sin\frac{2\pi}{\lambda}(vt \pm x).$$

- Superimposition is when parts of waves interact so that they constructively or destructively interfere (e.g., create larger or smaller amplitudes, respectively).
- Standing waves on a string that is fixed at both ends can form with wavelengths of $\lambda_n = \frac{2L}{n}$ and frequencies of $f_n = \frac{nv}{2L}$, where L is the length of the string, v is the speed of the wave, and n is a whole positive number.
- If one end of the string is free to move, standing waves can form with wavelengths of $\lambda_n = \frac{4L}{n}$ and frequencies of $f_n = \frac{nv}{4L}$, where L is the length of the string, v is the speed of the wave, and n must be an odd integer.
- Sound is a longitudinal wave. Its speed is given by $v = \sqrt{\frac{B}{\rho}}$, where B is the bulk modulus (this indicates how the medium responds to compression) and the density of air. At room temperature and at normal atmospheric pressure, the speed of sound is approximately 343 m/s.

- The loudness level of a sound of intensity I is given by $\beta = 10\log\frac{I}{I_0}$, where I_0 is 1×10^{-12} W/m^2 (the threshold of hearing). The intensity follows an inverse square law $I \propto \frac{1}{r^2}$, where r is the distance between the source and detector (observer).
- Two waves interfering at slightly different frequencies will form a beat frequency given by $f_{beat} = |f_1 - f_2|$.
- Resonance of sound in a tube depends on whether the tube is closed on one end or open on both ends. For a closed-ended tube resonance occurs for $f_n = n\frac{v}{4L}$ and $\lambda = \frac{4L}{n}$, where n must be an odd integer. For an open-ended tube, resonance occurs for $f_n = n\frac{v}{2L}$ and $\lambda = \frac{2L}{n}$, where n can be *any* integer.
- The Doppler effect tells you how the frequency of the sounds will vary if there is a relative motion between the source of the sound and the detector (observer). If the relative motion of the objects is toward each other, the frequency will be perceived as higher. If the relative motion is away from each other, the frequency will be perceived as lower. This idea is summed up in the equation $f_D = \frac{v \pm v_D}{v \mp v_S} \cdot f_S$.

16

Optics

INTRODUCTION

Light (or visible light) makes up only a small part of the entire spectrum of electromagnetic waves, which ranges from radio waves to gamma rays. The waves we studied in the preceding chapter required a material medium for their transmission, but electromagnetic waves can propagate through empty space. Electromagnetic waves consist of time-varying electric and magnetic fields that oscillate perpendicular to each other and to the direction of propagation of the wave. Through vacuum, all electromagnetic waves travel at a fixed speed:

$$c = 3.00 \times 10^8 \text{ m/s}$$

regardless of their frequency. Like all the waves in the previous chapter, we can say $v = f\lambda$.

THE ELECTROMAGNETIC SPECTRUM

Electromagnetic waves can be categorized by their frequency (or wavelength); the full range of waves is called the **electromagnetic** (or **EM**) **spectrum**. Types of waves include **radiowaves, microwaves, infrared**, **visible light**, **ultraviolet**, **X-rays**, and **Υ -rays** (**gamma rays**) and, although they've been delineated in the spectrum below, there's no universal agreement on all the boundaries, so many of these bands overlap. You should be familiar with the names of the major categories, and, in particular, memorize the order of the colors within the visible spectrum (which, as you can see, accounts for only a tiny sliver of the full EM spectrum). In order of increasing wave frequency, the colors are red, orange, yellow, green, blue, and violet, which is commonly remembered as ROYGBV ("roy-gee-biv"). The wavelengths of the colors in the visible spectrum are usually expressed in nanometers. For example, electromagnetic waves whose wavelengths are between 577 nm and 597 nm are seen as yellow light.

ELECTROMAGNETIC SPECTRUM

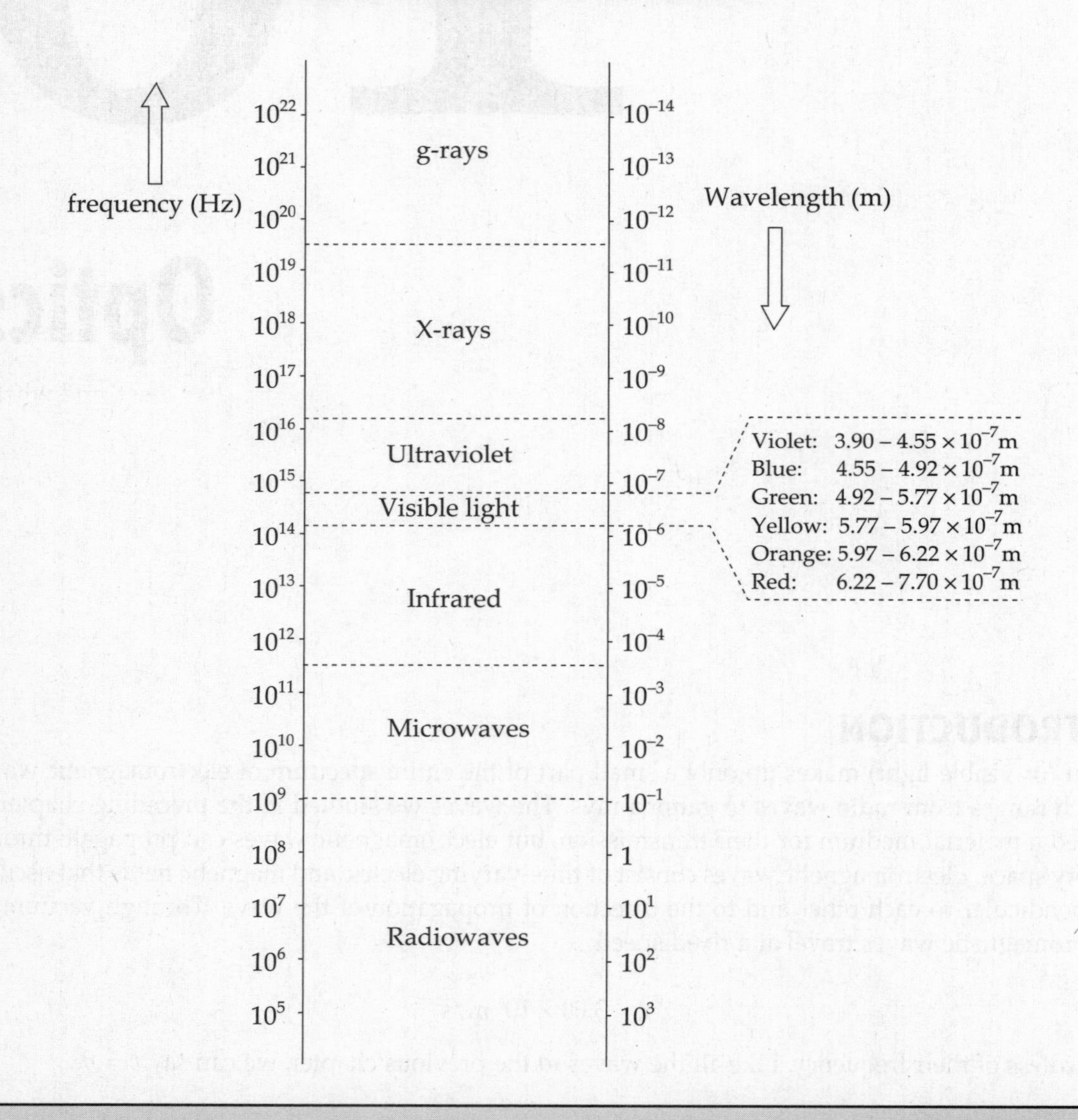

Example 16.1 What's the frequency range for green light?

Solution. According to the spectrum, light is green if its wavelength is between 4.92×10^{-7} m and 5.77×10^{-7} m. Using the equation $v = f\lambda$, we find that the upper end of this wavelength range corresponds to a frequency of

$$f_1 = \frac{v}{\lambda_1} = \frac{3.00 \times 10^8 \text{ m/s}}{5.77 \times 10^{-7} \text{ m}} = 5.20 \times 10^{14} \text{ Hz}$$

while the lower end corresponds to

$$f_2 = \frac{v}{\lambda_2} = \frac{3.00 \times 10^8 \text{ m/s}}{4.92 \times 10^{-7} \text{ m}} = 6.10 \times 10^{14} \text{ Hz}$$

So the frequency range for green light is

$$5.20 \times 10^{14} \text{ Hz} \le f_{\text{green}} \le 6.10 \times 10^{14} \text{ Hz}$$

Example 16.2 How would you classify electromagnetic radiation that has a wavelength of 1 cm?

Solution. According to the electromagnetic spectrum presented above, electromagnetic waves with $\lambda = 10^{-2}$ m are microwaves.

INTERFERENCE AND DIFFRACTION

As we learned in the preceding chapter, waves experience interference when they meet, and whether they interfere constructively or destructively depends on their relative phase. If they meet *in phase* (crest meets crest), they combine constructively, but if they meet *out of phase* (crest meets trough), they combine destructively. The key to the interference patterns we'll study in the next section rests on this observation. In particular, if waves that have the same wavelength meet, then the difference in the distances they've traveled determine whether the waves are in phase. Assuming that the waves are **coherent** (which means that their phase difference remains constant over time and does not vary), if the difference in their path lengths, $\Delta\ell$, is a whole number of wavelengths—0, $\pm\lambda$, $\pm 2\lambda$, etc.—they'll arrive in phase at the meeting point. On the other hand, if this difference is a whole number plus one-half a wavelength—$\pm\frac{1}{2}\lambda$, $\pm(1+\frac{1}{2})\lambda$, $\pm(2+\frac{1}{2})\lambda$, etc.—then they'll arrive exactly out of phase. That is,

$$\left.\begin{array}{ll}\text{constructive interference:} & \Delta\ell = m\lambda \\ \text{destructive interference:} & \Delta\ell = (m+\frac{1}{2})\lambda\end{array}\right\} \text{where } m = 0, 1, 2\ldots$$

YOUNG'S DOUBLE-SLIT INTERFERENCE EXPERIMENT

The following figure shows light incident on a barrier that contains two narrow slits (perpendicular to the plane of the page), separated by a distance *d*. On the right is a screen whose distance from the barrier, *L*, is much greater than *d*. The question is, What will we see on the screen? You might expect that we'll just see two bright narrow strips of light, directly opposite the slits in the barrier. As reasonable as this may sound, it doesn't take into account the wave nature of light.

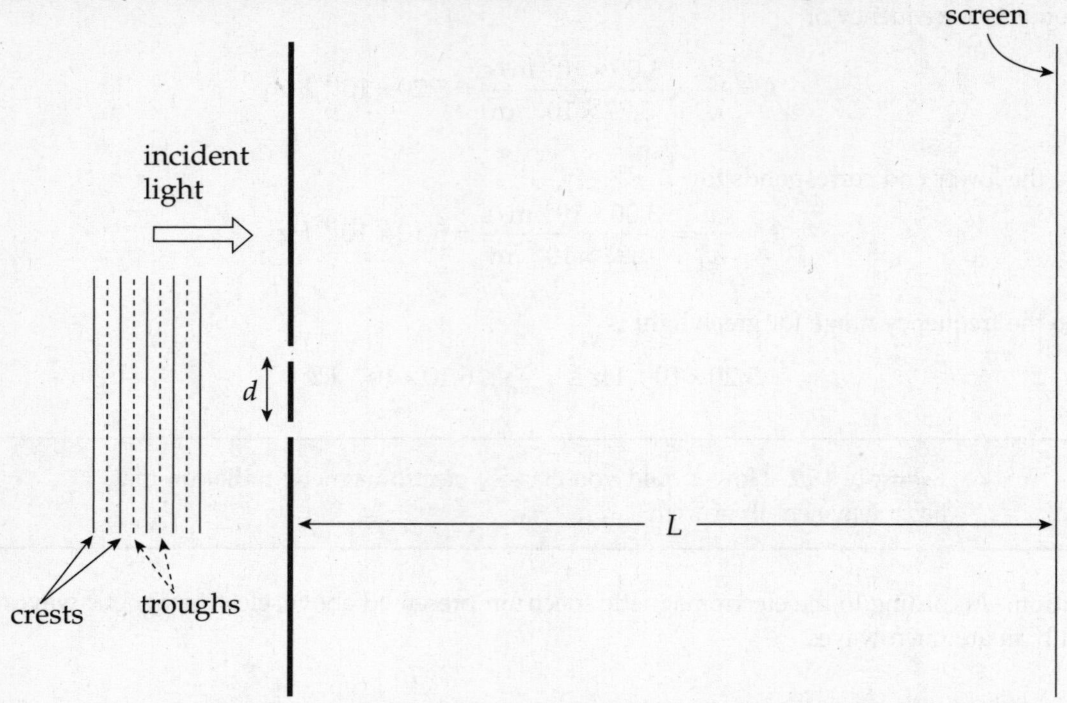

When a wave encounters an aperture whose width is comparable to its wavelength, the wave will fan out after it passes through. The alteration in the straight-line propagation of a wave when it encounters a barrier is called **diffraction**. In the set-up above, the waves will diffract through the slits, and spread out and interfere as they travel toward the screen.

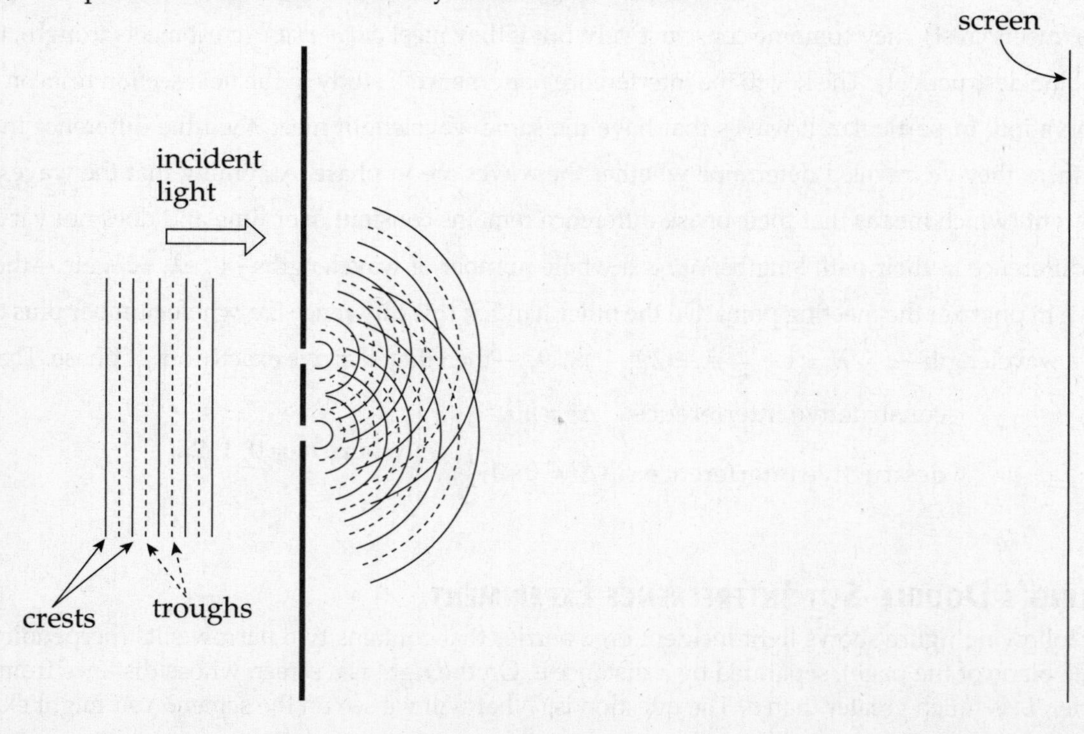

The screen will show the results of this interference: There will be bright bands (bright **fringes**) centered at those points at which the waves interfere constructively, alternating with dark fringes, where the waves interfere destructively. Let's determine the locations of these fringes.

In the figure below, we've labeled the slits S_1 and S_2. A Point P on the screen is selected, the path lengths from S_1 and S_2 to P are called ℓ_1 and ℓ_2, respectively, and the angle that the line from the midpoint of the slits to P makes with the horizontal is θ. Segment S_1Q is perpendicular to line S_2P. Because L is so much larger than d, the angle that line S_2P makes with the horizontal is also approximately θ, which tells us that $\angle S_2S_1Q$ is approximately θ and that the path difference, $\Delta \ell = \ell_2 - \ell_1$, is nearly equal to S_2Q.

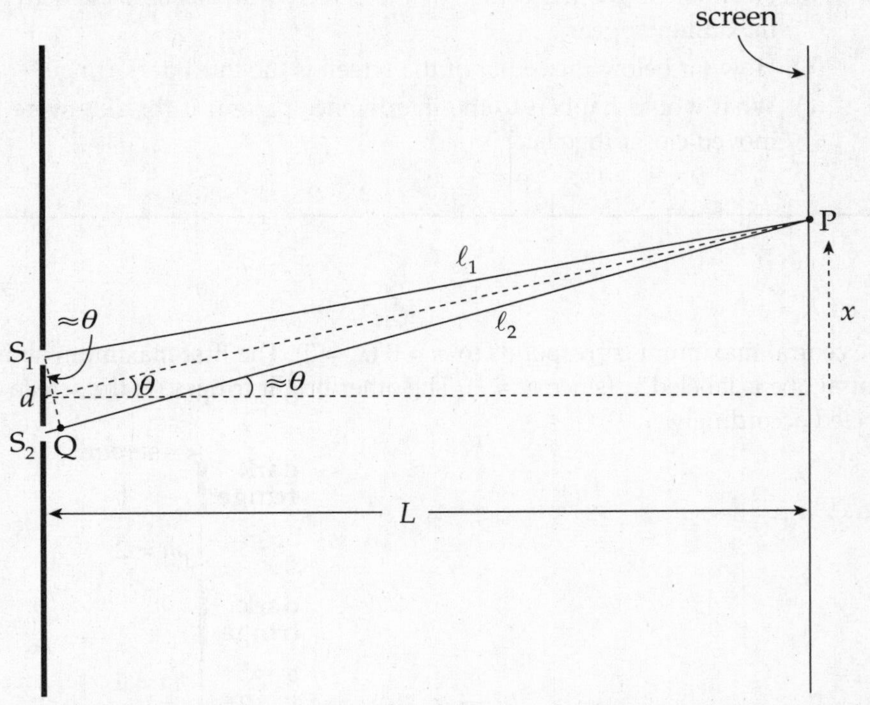

Because $S_2Q = d \sin \theta$, we get $\Delta\ell = d \sin \theta$. Now, using what we learned earlier about how constructive or destructive interference depends on $\Delta\ell$, we can write:

constructive interference (intensity maximum—bright fringe—on screen): $d \sin\theta = m\lambda$

destructive interference (intensity minimum—dark fringe—on screen): $d \sin\theta = (m + \frac{1}{2}\lambda)$

where $m = 0, 1, 2...$

To locate the positions of, say, the bright fringes on the screen, we use the fact that $x = L \tan \theta$. If θ is small, then $\tan \theta \approx \sin \theta$, so we can write $x = L \sin \theta$ (we can tell this from the figure). Since $\sin \theta = m\lambda/d$ for bright fringes, we get

$$x_m = \frac{m\lambda L}{d}$$

Also, the intensity of the bright fringes decreases as m increases in magnitude. The bright fringe directly opposite the midpoint of the slits—the **central maximum**—will have the greatest intensity when

$m = 0$. The bright fringes with $m = 1$ will have a lower intensity, those with $m = 2$ will be fainter still, and so on. If more than two slits are cut in the barrier, the interference pattern becomes sharper, and the distinction between dark and bright fringes becomes more pronounced. Barriers containing thousands of tiny slits per centimeter—called **diffraction gratings**—are used precisely for this purpose.

> **Example 16.3** For the experimental set-up we've been studying, assume that $d = 1.5$ mm, $L = 6.0$ m, and that the light used has a wavelength of 589 nm.
>
> (a) How far above the center of the screen will the first intensity maximum appear?
>
> (b) How far below the center of the screen is the third dark fringe?
>
> (c) What would happen to the interference pattern if the slits were moved closer together?

Solution.

(a) The central maximum corresponds to $m = 0$ ($x_0 = 0$). The first maximum above the central one is labeled x_1 (since $m = 1$). The other bright fringes on the screen are labeled accordingly.

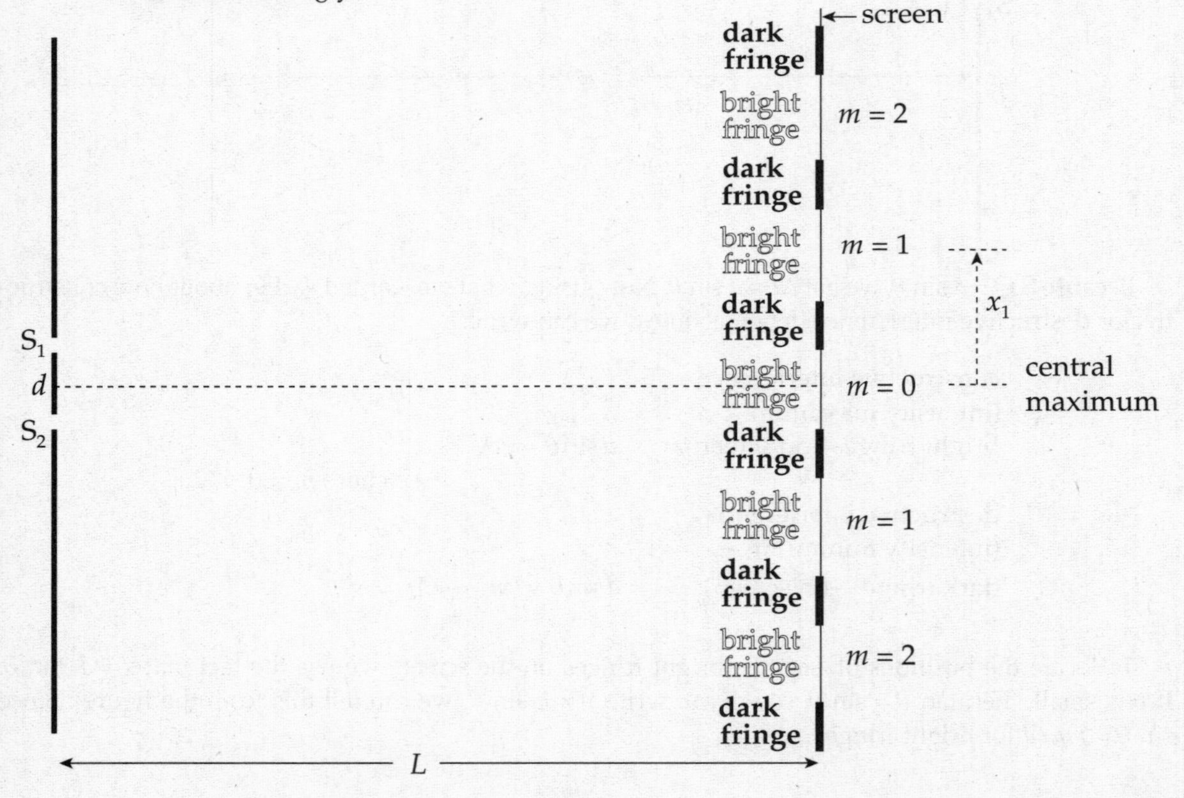

The value of x_1 is

$$x_1 = \frac{1 \cdot \lambda L}{d} = \frac{(589 \times 10^{-9}\ \text{m})(6.0\ \text{m})}{1.5 \times 10^{-3}\ \text{m}} = 2.4 \times 10^{-3}\ \text{m} = 2.4\ \text{mm}$$

(b) The first dark fringe occurs when the path difference is 0.5λ, the second dark fringe occurs when the path difference is 1.5λ, and the third dark fringe occurs when the path difference is 2.5λ, so

$$x_{\substack{\text{3rd minimum}\\ \text{below central max}}} = \frac{(2+\frac{1}{2})\lambda L}{d} = \frac{(2+\frac{1}{2})(589\times10^{-9}\text{ m})(6.0\text{ m})}{1.5\times10^{-3}\text{ m}} = -5.9\times10^{-3}\text{ m} = -5.9\text{ mm}$$

(c) Since $x_m = m\lambda L/d$, a decrease in d would cause an increase in x_m. That is, the fringes would become larger; the interference pattern would be more spread out.

Single-Aperture Diffraction

A diffraction pattern will also form on the screen if the barrier contains only one slit. The central maximum will be very pronounced, but lower-intensity maxima will also be seen because of interference from waves arriving from different locations within the slit itself. The width of the central maximum will become wider as the width of the slit is decreased.

For a circular pinhole aperture, the diffraction pattern will consist of a central, bright circular disk surrounded by rings of decreasing intensity.

REFLECTION AND REFRACTION

Imagine a beam of light directed toward a smooth transparent surface. When it hits this surface, some of its energy will be reflected off the surface and some will be transmitted into the new medium. We can figure out the directions of the reflected and transmitted beams by calculating the angles that the beams make with the normal to the interface. In the following figure, an incident beam strikes the boundary of another medium; it could be a beam of light in air striking a piece of glass. Notice all angles are measured from the normal.

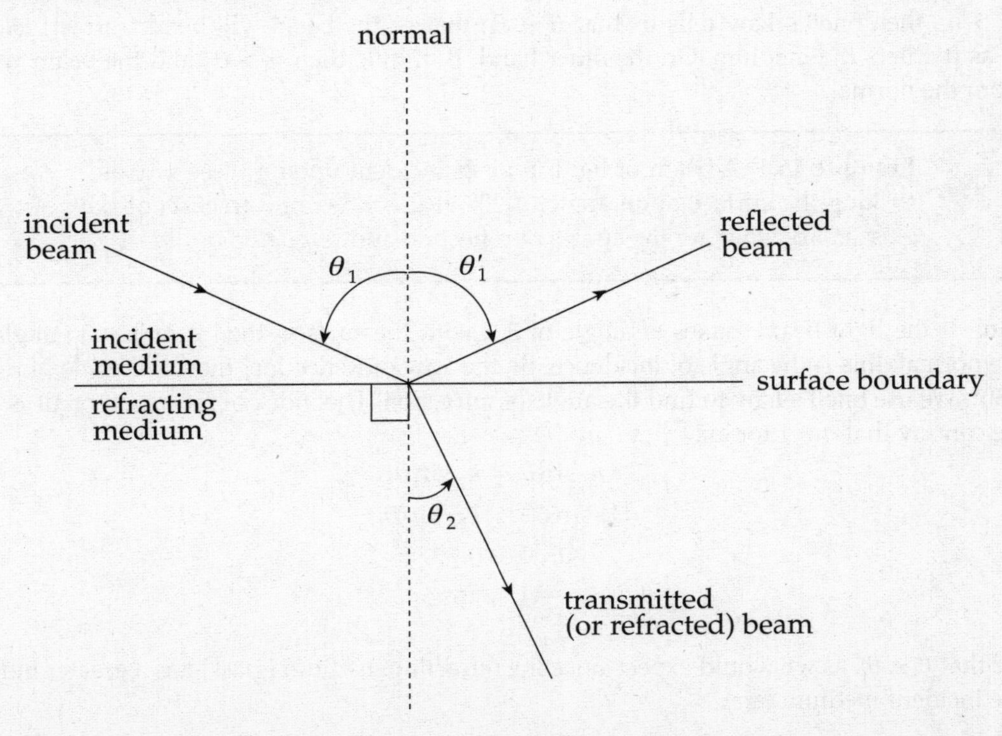

The angle that the **incident beam** makes with the normal is called the **angle of incidence**, or θ_1. The angle that the **reflected beam** makes with the normal is called the **angle of reflection**, θ_1', and the angle that the **transmitted beam** makes with the normal is called the **angle of refraction**, θ_2. The incident, reflected, and transmitted beams of light all lie in the same plane.

The relationship between θ_1 and θ_1' is pretty easy; it is called the **Law of Reflection**:

$$\theta_1 = \theta_1'$$

In order to describe how θ_1 and θ_2 are related, we first need to talk about a medium's index of refraction.

When light travels through empty space (vacuum), its speed is $c = 3.00 \times 10^8$ m/s; this is one of the fundamental constants of nature. But when light travels through a material medium (such as water or glass), it's constantly being absorbed and re-emitted by the atoms that compose the material and, as a result, its apparent speed, v, is some fraction of c. The reciprocal of this fraction,

$$n = \frac{c}{v}$$

is called the medium's **index of refraction**. For example, since the speed of light in water is $v = 2.25 \times 10^8$ m/s, the index of refraction of water is

$$n = \frac{3.00 \times 10^8 \text{ m/s}}{2.25 \times 10^8 \text{ m/s}} = 1.33$$

Note that n has no units; it's also never less than 1.

The equation that relates θ_1 and θ_2 involves the index of refraction of the incident medium (n_1) and the index of refraction of the refracting medium (n_2); it's called **Snell's Law**:

$$n_1 \sin\theta_1 = n_2 \sin\theta_2$$

If $n_2 > n_1$, then Snell's Law tells us that $\theta_2 < \theta_1$; that is, the beam will bend (**refract**) *toward* the normal as it enters the medium. On the other hand, if $n_2 < n_1$, then $\theta_2 > \theta_1$, and the beam will bend *away* from the normal.

Example 16.4 A beam of light in air is incident upon a piece of glass, striking the surface at an angle of 30°. If the index of refraction of the glass is 1.5, what are the angles of reflection and refraction?

Solution. If the light beam makes an angle of 30° with the surface, then it makes an angle of 60° with the normal; this is the angle of incidence. By the Law of Reflection, then, the angle of reflection is also 60°. We use Snell's Law to find the angle of refraction. The index of refraction of air is close to 1, so we can say that $n = 1$ for air.

$$n_1 \sin\theta_1 = n_2 \sin\theta_2$$
$$(1) \sin 60° = 1.5 \sin\theta_2$$
$$\sin\theta_2 = 0.5774$$
$$\theta_2 = 35°$$

Note that $\theta_2 < \theta_1$, as we would expect since the refracting medium (glass) has a greater index than does the incident medium (air).

Example 16.5 A fisherman drops a flashlight into a lake that's 10 m deep. The flashlight sinks to the bottom where its emerging light beam is directed almost vertically upward toward the surface of the lake, making a small angle (θ_1) with the normal. How deep will the flashlight appear to be to the fisherman? (Use the fact that tan θ is almost equal to sin θ if θ is small.)

Solution. Take a look at the figure below.

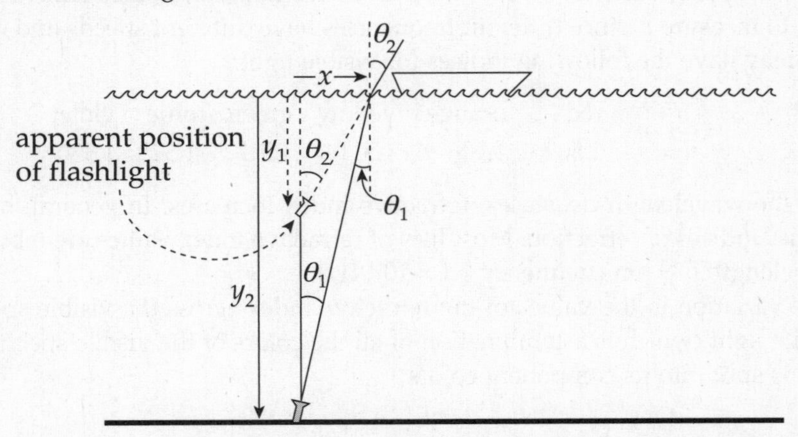

Since the refracting medium (air) has a lower index than the incident medium (water), the beam of light will bend away from the normal as it emerges from the water. As a result, the fisherman will think that the flashlight is at a depth of only y_1, rather than its actual depth of y_2 = 10 m. By simple trigonometry, we know that

$$\tan\theta_1 = \frac{x}{y_2} \quad \text{and} \quad \tan\theta_2 = \frac{x}{y_1}$$

So,

$$\frac{y_1}{y_2} = \frac{\tan\theta_1}{\tan\theta_2}$$

Snell's Law tells us that $n_1 \sin\theta_1 = \sin\theta_2$ (since $n_2 = n_{air} = 1$), so

$$\frac{\sin\theta_1}{\sin\theta_2} = \frac{1}{n_1}$$

Using the approximations $\sin\theta_1 \approx \tan\theta_1$ and $\sin\theta_2 \approx \tan\theta_2$, we can write

$$\frac{\tan\theta_1}{\tan\theta_2} \approx \frac{\sin\theta_1}{\sin\theta_2}$$

which means

$$\frac{y_1}{y_2} = \frac{1}{n_1}$$

Because y_2 = 10 m and $n_1 = n_{water}$ = 1.33,

$$y_1 = \frac{y_2}{n_1} = \frac{10 \text{ m}}{1.33} = 7.5 \text{ m}$$

DISPERSION OF LIGHT

One thing we learned when we studied waves is that wave speed is independent of frequency. For a given medium, different frequencies give rise to different wavelengths, because the equation $v = f\lambda$ must always be satisfied and v doesn't vary. But when light travels through a material medium, it displays **dispersion**, which is a variation in wave speed with frequency (or wavelength). So, the definition of the index of refraction, $n = c/v$, should be accompanied by a statement of the frequency of the light used to measure v, since different frequencies have different speeds and different indices. A piece of glass may have the following indices for visible light:

	red	orange	yellow	green	blue	violet
$n =$	1.502	1.506	1.511	1.517	1.523	1.530

Note that as the wavelength decreases, refractive index increases. In general, higher frequency waves have higher indices of refraction. Most lists of refractive index values are tabulated using yellow light of wavelength 589 nm (frequency 5.1×10^{14} Hz).

Although the variation in the values of the refractive index across the visible spectrum is pretty small, when white light (which is a combination of all the colors of the visible spectrum) hits a glass prism, the beam is split into its component colors.

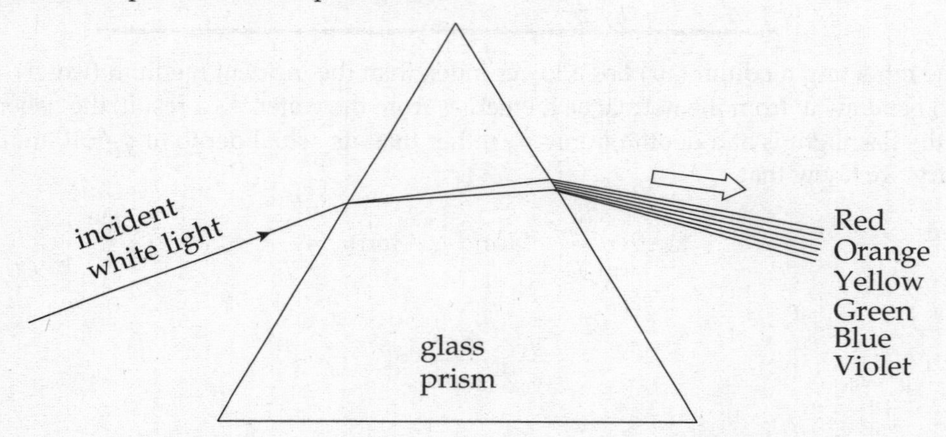

Why? Because each color has its own index, Snell's Law tells us that each color will have its own angle of refraction. Therefore, each color emerges from the prism at a slightly different angle, so the light disperses into its component colors.

Example 16.6 In the figure above, assume that the incident beam strikes the prism with an angle of incidence of 60.00°. If the top angle of the prism has a measure of 60.00°, what are the angles of refraction of red light and violet light when the beam emerges from the other side? (Use n = 1.502 for red light and 1.530 for violet light.)

Solution. The beam refracts twice, once when it enters the prism and again when it exits.

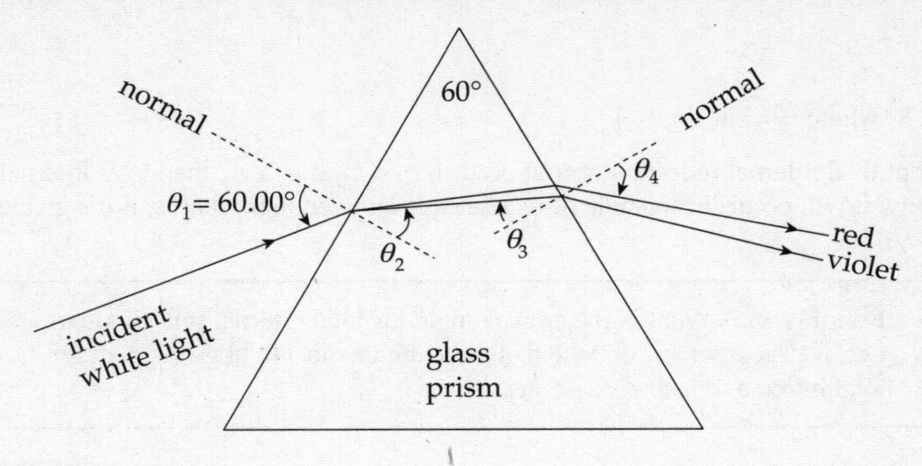

Snell's Law tells us that

$$(1)\sin 60.00° = n \sin\theta_2 \text{ and } n\sin\theta_3 = (1)\sin\theta_4 \lim_{x\to\infty}$$

where n is the index of the glass. We'll ignore the minuscule variation in n for light in air, and just use n_{air} = 1 for all colors. The relationship between θ_2 and θ_3 follows from the fact that the sum of the measures of the angles in any triangle is 180°, so

$$(90° - \theta_2) + 60° + (90° - \theta_3) = 180°$$

which implies that $\theta_3 = 60° - \theta_2$. We can now make the following table of values:

color	n	θ_2	θ_3	θ_4
red	1.502	35.210°	24.790°	39.03°
violet	1.530	34.474°	25.526°	41.25°

Total Internal Reflection

When a beam of light strikes the boundary to a medium that has a lower index of refraction, the beam bends away from the normal. As the angle of incidence increases, the angle of refraction becomes larger. At some point, when the angle of incidence reaches a **critical angle**, θ_c, the angle of refraction becomes 90°, which means the refracted beam is directed along the surface.

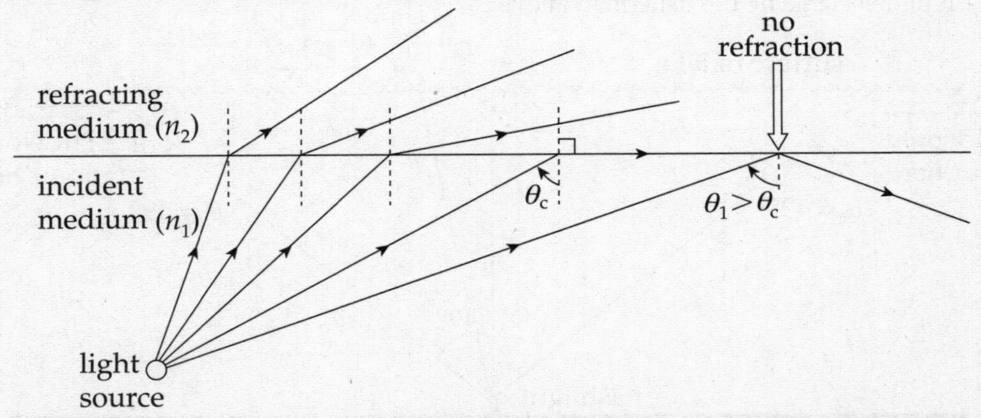

For angles of incidence that are greater than θ_c, there is *no* angle of refraction; the entire beam is reflected back into the original medium. This phenomenon is called **total internal reflection** (sometimes abbreviated **TIR**).

Total internal reflection occurs when:

1) $n_1 > n_2$

and

2) $\theta_1 > \theta_c$, where $\theta_c = \sin^{-1}(n_2/n_1)$

Notice that total internal reflection cannot occur if $n_1 < n_2$. If $n_1 > n_2$, then total internal reflection is a possibility; it will occur if the angle of incidence is large enough, that is, if it's greater than the critical angle, θ_c.

> **Example 16.7** What is the critical angle for total internal reflection between air and water? In which of these media must light be incident for total internal reflection to occur?

Solution. First, total internal reflection can only occur when the light is incident in the medium that has the greater refractive index and strikes the boundary to a medium that has a lower index. So, in this case, total internal reflection can occur only when the light source is in the water and the light is incident upon the water/air surface. The critical angle is found as follows:

$$\sin\theta_c = \frac{n_2}{n_1} \Rightarrow \sin\theta_c = \frac{n_{\text{air}}}{n_{\text{water}}} = \frac{1}{1.33} \Rightarrow \sin\theta_c = 0.75 \Rightarrow \theta_c = 49°$$

Total internal reflection will occur if the light from the water strikes the water/air boundary at an angle of incidence greater than 49°.

> **Example 16.8** How close must the fisherman be to the flashlight in Example 16.5 in order to see the light it emits?

Solution. In order for the fisherman to see the light, the light must be transmitted into the air from the water; that is, it cannot undergo total internal reflection. The figure below shows that, within a circle of radius *x*, light from the flashlight will emerge from the water. Outside this circle, the angle of incidence is greater than the critical angle, and the light would be reflected back into the water, rendering it undetectable by the fisherman above.

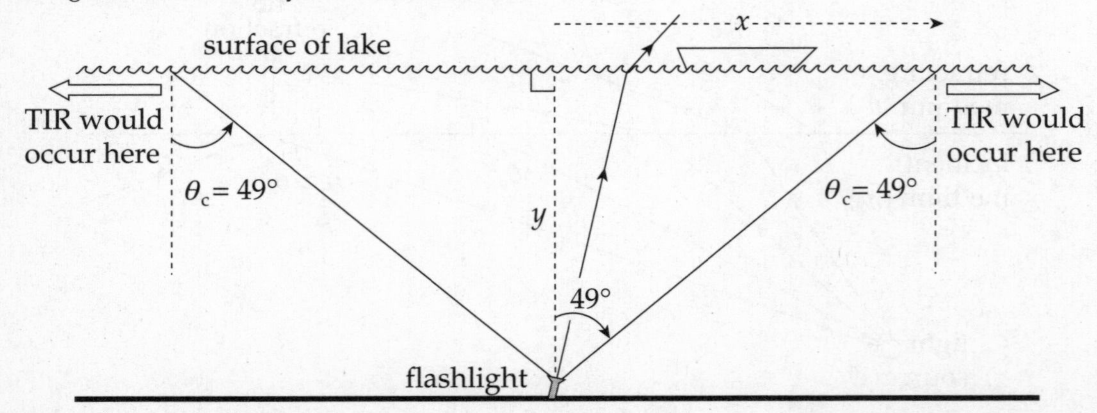

Because the critical angle for total internal reflection at a water/air interface is 49° (as we found in the preceding example), we can solve for x:

$$\tan 49° = \frac{x}{y} \Rightarrow x = y\tan 49° = (10\text{ m})\tan 49° = 11.5\text{ m}$$

Example 16.9 The refractive index for the glass prism shown below is 1.55. In order for a beam of light to experience total internal reflection at the right-hand face of the prism, the angle θ_1 must be smaller than what value?

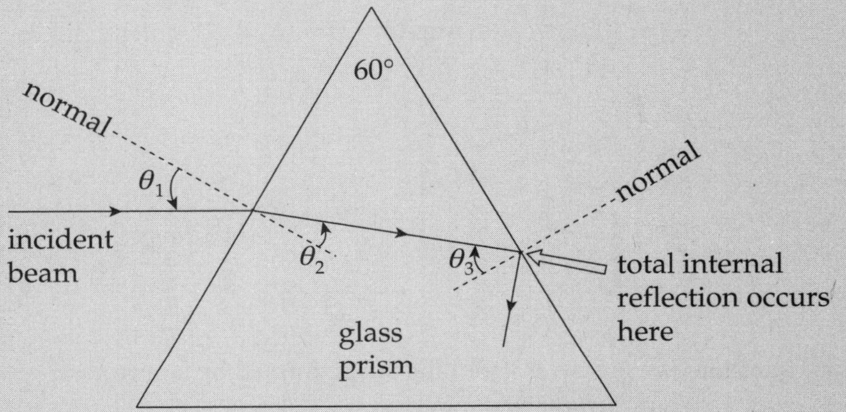

Solution. Total internal reflection will occur at the glass/air boundary if θ_3 is greater than the critical angle, θ_c, which we can calculate this way:

$$\sin\theta_c = \frac{n_{air}}{n_{glass}} = \frac{1}{1.55} \Rightarrow \theta_c = 40°$$

Because $\theta_3 = 60° - \theta_2$, total internal reflection will take place if θ_2 is smaller than 20°. Now, by Snell's Law, $\theta_2 = 20°$ if

$$n_{air}\sin\theta_1 = n_{glass}\sin\theta_2$$
$$\sin\theta_1 = 1.55\sin 20°$$
$$\theta_1 = 32°$$

Therefore, total internal reflection will occur at the right-hand face of the prism if θ_1 is smaller than 32°.

MIRRORS

A **mirror** is an optical device that forms an image by reflecting light. We've all looked into a mirror and seen images of nearby objects, and the purpose of this section is to analyze these images mathematically. We begin with a plane mirror, which is flat, and is the simplest type of mirror. Then we'll examine curved mirrors; we'll have to use geometrical methods or algebraic equations to analyze the patterns of reflection from these.

PLANE MIRRORS

The figure below shows an object (denoted by a bold arrow) in front of a flat mirror. Light that's reflected off of the object strikes the mirror and is reflected back to our eyes. The directions of the rays reflected off the mirror determine where we perceive the image to be.

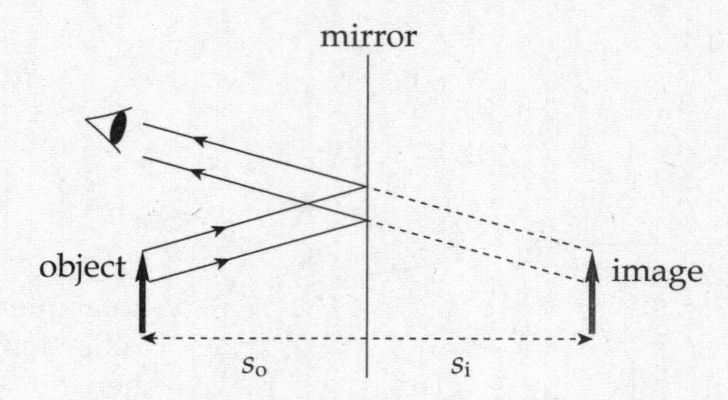

There are four questions we'll answer about the image formed by a mirror:

(1) Where is the image?

(2) Is the image real or is it virtual?

(3) Is the image upright or is it inverted?

(4) What is the height of the image (compared to that of the object)?

When we look at ourselves in a mirror, it seems like our image is *behind* the mirror, and, if we take a step back, our image also takes a step back. The Law of Reflection can be used to show that the image seems as far behind the mirror as the object is in front of the mirror. This answers question (1).

An image is said to be **real** if light rays actually focus at the image. A real image can be projected onto a screen. For a flat mirror, light rays bounce off the front of the mirror; so, of course, no light focuses behind it. Therefore, the images produced by a flat mirror are not real; they are **virtual**. This answers question (2).

When we look into a flat mirror, our image isn't upside down; flat mirrors produce upright images, and question (3) is answered.

Finally, the image formed by a flat mirror is neither magnified nor diminished (minified) relative to the size of the object. This answers question (4).

SPHERICAL MIRRORS

A **spherical mirror** is a mirror that's curved in such a way that its surface forms part of a sphere.

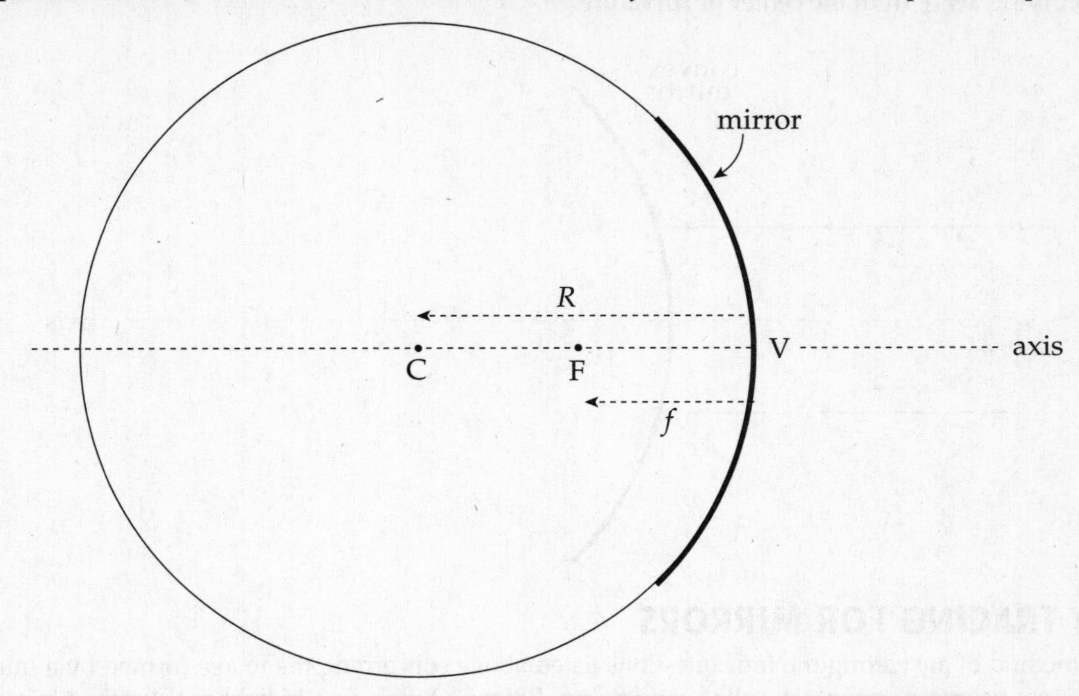

The center of this imaginary sphere is the mirror's **center of curvature**, and the radius of the sphere is called the mirror's **radius of curvature**, R. Halfway between the mirror and the center of curvature, C, is the **focus** (or **focal point**), F. The intersection of the mirror's optic **axis** (its axis of symmetry) with the mirror itself is called the **vertex**, V, and the distance from V to F is called the **focal length**, f, equal to one-half of the radius of curvature:

$$f = \frac{R}{2}$$

If the mirror had a parabolic cross-section, then any ray parallel to the axis would be reflected by the mirror through the focal point. Spherical mirrors do this for incident light rays near the axis (**paraxial rays**) because, in the region of the mirror that's close to the axis, the shapes of a parabolic mirror and a spherical mirror are nearly identical.

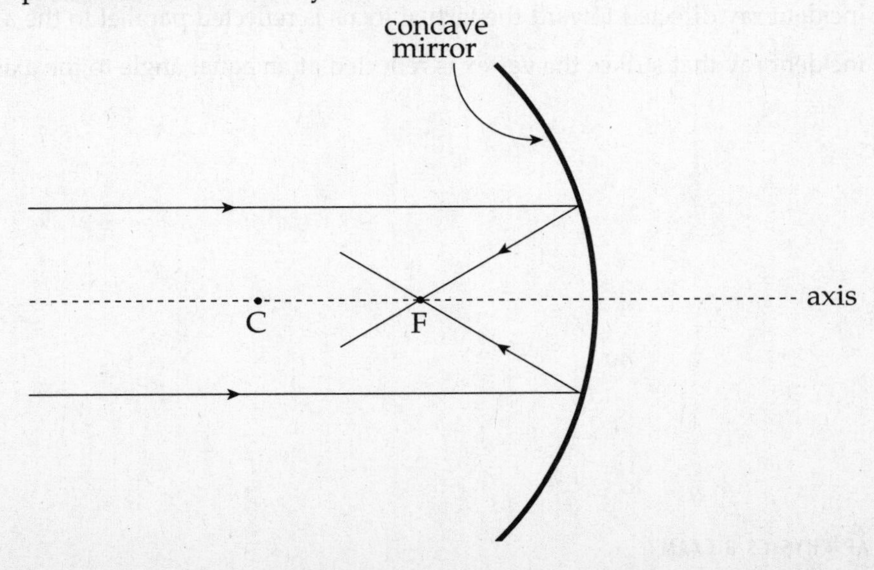

The previous two figures illustrate a **concave mirror**, a mirror whose reflective side is *caved in* toward the center of curvature. The following figure illustrates the **convex mirror**, which has a reflective side curving away from the center of curvature.

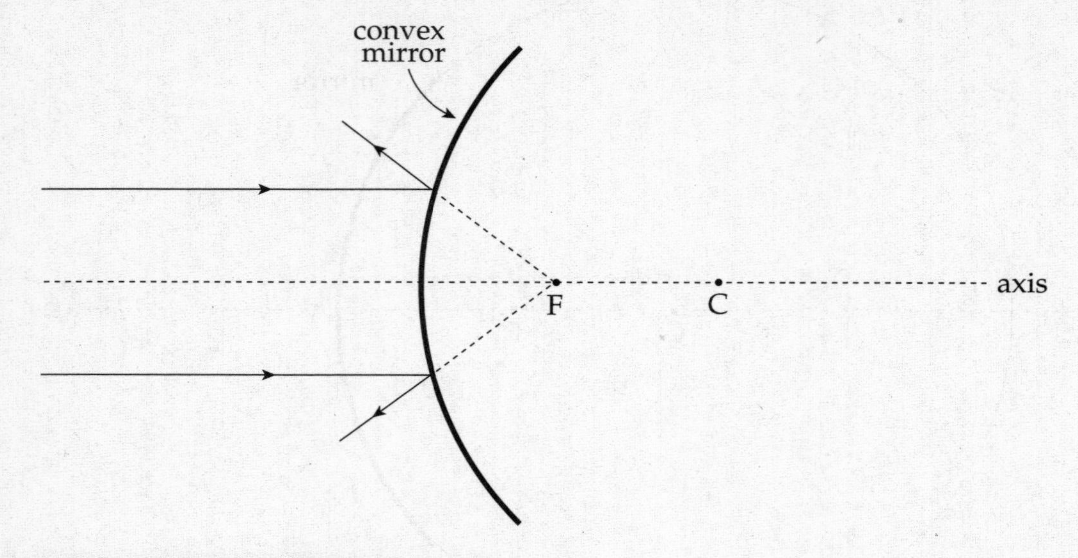

RAY TRACING FOR MIRRORS

One method of answering the four questions listed above concerning the image formed by a mirror involves a geometric approach called **ray tracing**. Representative rays of light are sketched in a diagram that depicts the object and the mirror; the point at which the reflected rays intersect (or appear to intersect) is the location of the image. Some rules governing rays are:

Concave mirrors:

- An incident ray parallel to the axis is reflected through the focus.
- An incident ray that passes through the focus is reflected parallel to the axis.
- An incident ray that strikes the vertex is reflected at an equal angle to the axis.

Convex mirrors:

- An incident ray parallel to the axis is reflected away from the virtual focus.
- An incident ray directed toward the virtual focus is reflected parallel to the axis.
- An incident ray that strikes the vertex is reflected at an equal angle to the axis.

Example 16.10 The figure below shows a concave mirror and an object (the bold arrow). Use a ray diagram to locate the image of the object.

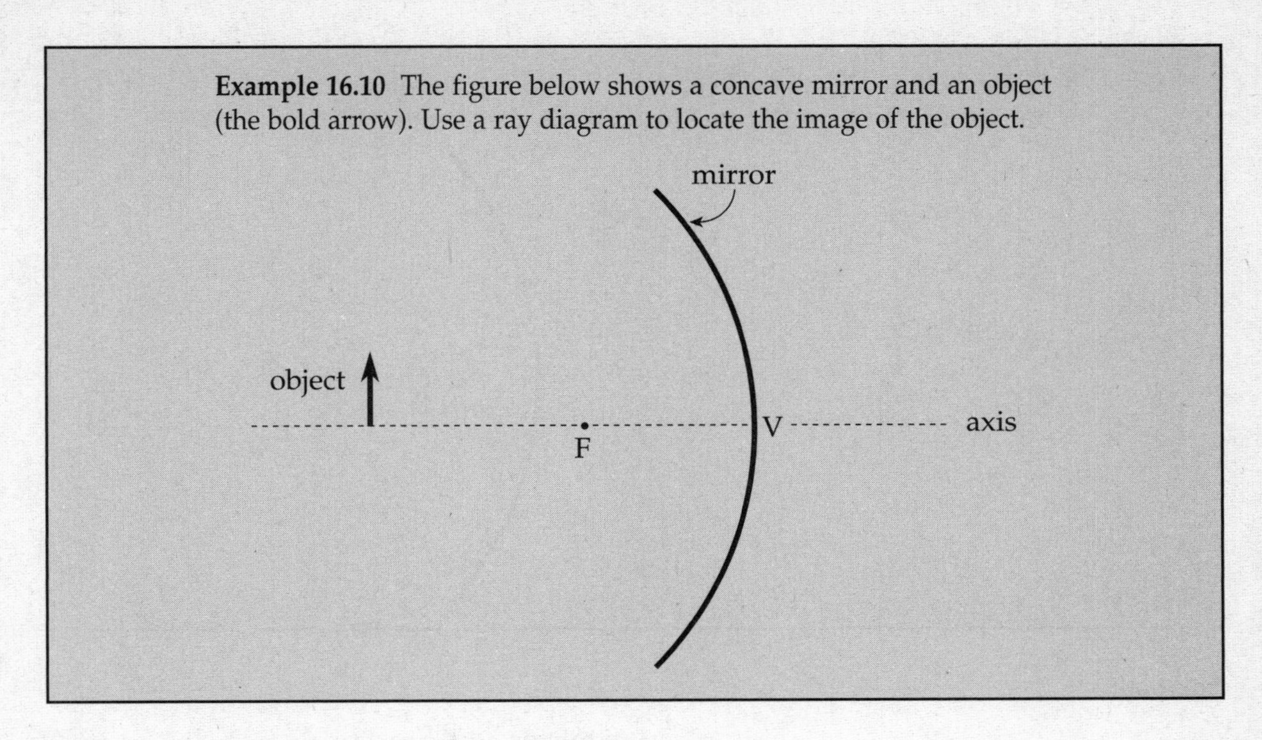

Solution. It only takes two distinct rays to locate the image:

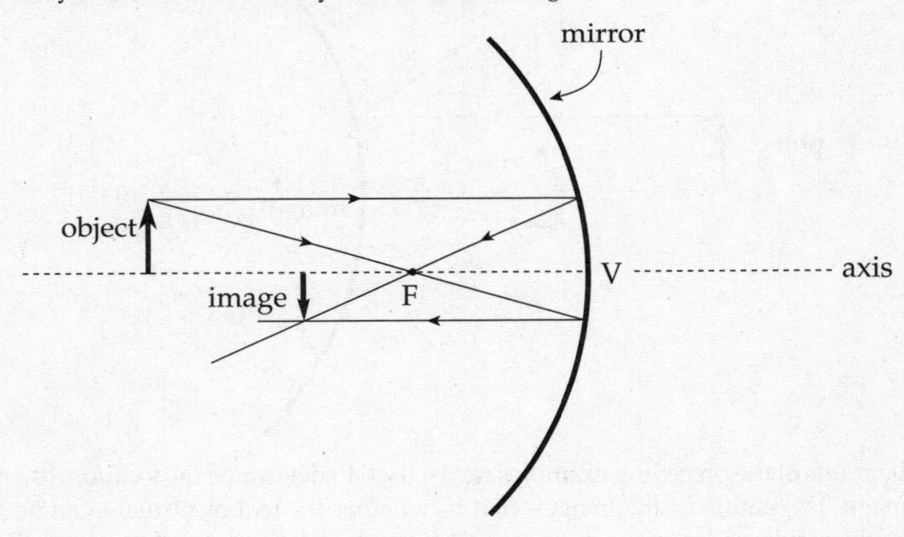

Example 16.11 The figure below shows a convex mirror and an object (the arrow). Use a ray diagram to locate the image of the object.

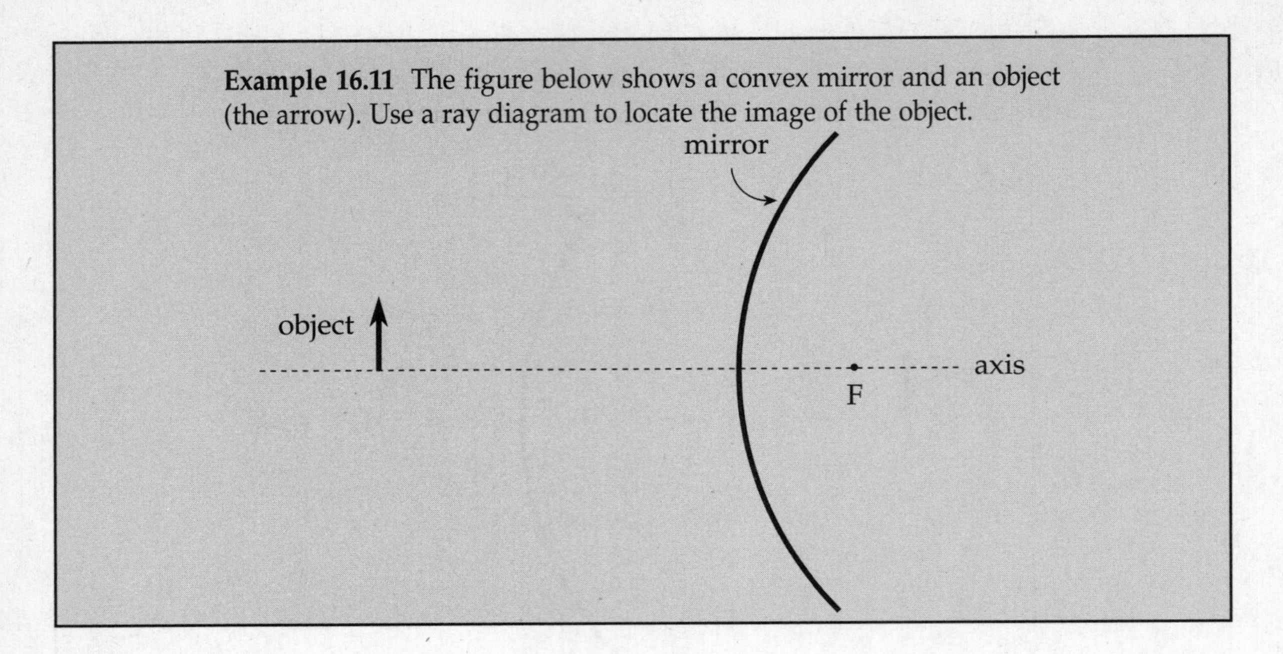

Solution.

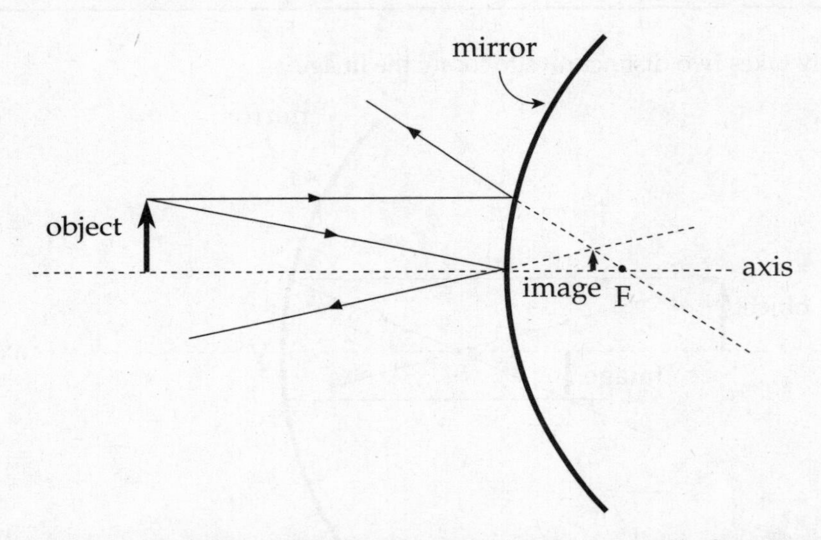

The ray diagrams of the preceding examples can be used to determine the location, orientation, and size of the image. The nature of the image—that is, whether it's real or virtual—can be determined by seeing on which side of the mirror the image is formed. If the image is formed on the same side of the mirror as the object, then the image is real, but if the image is formed on the opposite side of the mirror, it's virtual. Another way of looking at this is: if you had to trace lines back to form an image, that image is virtual. Therefore, the image in Example 16.10 is real, and the image in Example 16.11 is virtual.

Using Equations to Answer Questions About the Image

The fastest and easiest way to get information about an image is to use two equations and some simple conventions. The first equation, called the **mirror equation**, is

$$\frac{1}{s_o}+\frac{1}{s_i}=\frac{1}{f}$$

where s_o is the object's distance from the mirror, s_i is the image's distance from the mirror, and f is the focal length of the mirror. The value of s_o is *always* positive for a real object, but s_i can be positive or negative. The sign of s_i tells us whether the image is real or virtual: If s_i is positive, the image is real; and if s_i is negative, the image is virtual.

The second equation is called the **magnification equation**:

$$M=\frac{h_i}{h_o}=\frac{-s_i}{s_o}$$

This gives the magnification; the height of the image, h_i, is $|M|$ times the height of the object, h_o. If M is positive, then the image is upright relative to the object; if M is negative, it's inverted. Because s_o is always positive, we can come to two important conclusions. If s_i is positive, then M is negative, so real images are always inverted, and, if s_i is negative, then M is positive, so virtual images are always upright.

Finally, to distinguish *mathematically* between concave and convex mirrors, we always write the focal length f as a positive value for concave mirrors and a negative value for convex mirrors. With these two equations and their accompanying conventions, all four questions about an image can be answered.

MIRRORS

converging—concave $\Leftrightarrow$ f positive
diverging—convex $\Leftrightarrow$ f negative

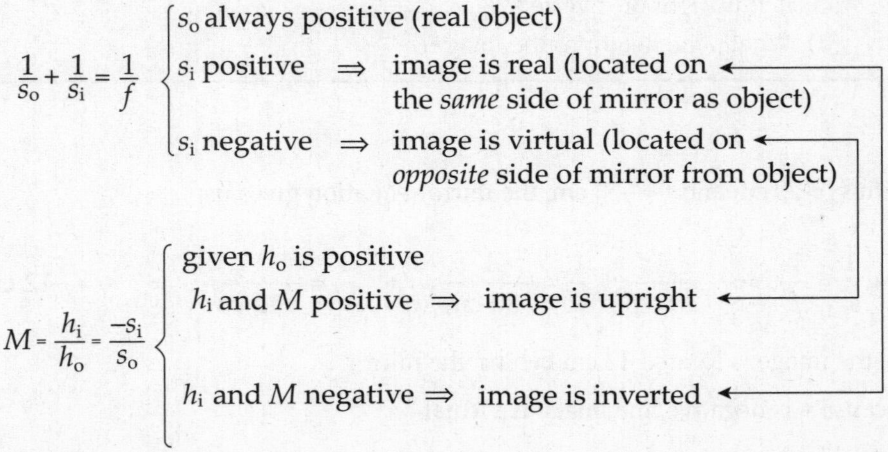

$\frac{1}{s_o}+\frac{1}{s_i}=\frac{1}{f}$
- s_o always positive (real object)
- s_i positive $\Rightarrow$ image is real (located on the *same* side of mirror as object)
- s_i negative $\Rightarrow$ image is virtual (located on *opposite* side of mirror from object)

$M=\frac{h_i}{h_o}=\frac{-s_i}{s_o}$
- given h_o is positive
- h_i and M positive $\Rightarrow$ image is upright
- h_i and M negative $\Rightarrow$ image is inverted

Example 16.12 An object of height 4 cm is placed 30 cm in front of a concave mirror whose focal length is 10 cm.

(a) Where's the image?
(b) Is it real or virtual?
(c) Is it upright or inverted?
(d) What's the height of the image?

Solution.

(a) With s_o = 30 cm and f = 10 cm, the mirror equation gives

$$\frac{1}{s_o}+\frac{1}{s_i}=\frac{1}{f} \Rightarrow \frac{1}{30\text{ cm}}+\frac{1}{s_i}=\frac{1}{10\text{ cm}} \Rightarrow \frac{1}{s_i}=\frac{1}{15\text{ cm}} \Rightarrow s_i = 15\text{ cm}$$

The image is located 15 cm in front of the mirror.

(b) Because s_i is positive, the image is real.

(c) Real images are inverted.

(d) $\frac{h_i}{h_o}=\frac{-s_i}{s_o} \rightarrow h_i=\frac{-s_i h_o}{s_o}$

$$\rightarrow h_i=\frac{(-15\text{ cm})(4\text{ cm})}{30\text{ cm}}=-2\text{ cm}$$

The –2 cm also confirms the image is inverted.

Example 16.13 An object of height 4 cm is placed 20 cm in front of a convex mirror whose focal length is –30 cm.

(a) Where's the image?
(b) Is it real or virtual?
(c) Is it upright or inverted?
(d) What's the height of the image?

Solution.

(a) With s_o = 20 cm and f = –30 cm, the mirror equation gives us:

$$\frac{1}{s_o}+\frac{1}{s_i}=\frac{1}{f} \Rightarrow \frac{1}{20\text{ cm}}+\frac{1}{s_i}=\frac{1}{-30\text{ cm}} \Rightarrow \frac{1}{s_i}=-\frac{1}{12\text{ cm}} \Rightarrow s_i=-12\text{ cm}$$

So, the image is located 12 cm behind the mirror.

(b) Because s_i is negative, the image is virtual.

(c) Virtual images are upright.

(d) $\frac{h_i}{h_o} = \frac{-s_i}{s_o} \rightarrow h_i = \frac{-s_i h_o}{s_o}$

$\rightarrow h_i = \frac{-(-12 \text{ cm})(4 \text{ cm})}{20 \text{ cm}} = 2.4 \text{ cm}$

[The fact that the magnification is positive tells us that the image is upright, as we said in part (c).]

Example 16.14 Show how the statements made earlier about plane mirrors can be derived from the mirror and magnification equations.

Solution. A plane mirror can be considered a spherical mirror with an infinite radius of curvature (and an infinite focal length). If $f = \infty$, then $1/f = 0$, so the mirror equation becomes

$$\frac{1}{s_o} + \frac{1}{s_i} = 0 \quad \Rightarrow \quad s_i = -s_o$$

So, the image is as far behind the mirror as the object is in front. Also, since s_o is always positive, s_i is negative, so the image is virtual. The magnification is

$$M = -\frac{s_i}{s_o} = -\frac{-s_o}{s_o} = 1$$

and the image is upright and has the same height as the object. The mirror and magnification equations confirm our description of images formed by plane mirrors.

Example 16.15 Show why convex mirrors can only form virtual images.

Solution. Because f is negative and s_o is positive, the mirror equation

$$\frac{1}{s_o} + \frac{1}{s_i} = \frac{1}{f}$$

immediately tells us that s_i cannot be positive (if it were, the left-hand side would be the sum of two positive numbers, while the right-hand side would be negative). Since s_i must be negative, the image must be virtual.

Example 16.16 An object placed 60 cm in front of a spherical mirror forms a real image at a distance of 30 cm from the mirror.

(a) Is the mirror concave or convex?

(b) What's the mirror's focal length?

(c) Is the image taller or shorter than the object?

Solution.

(a) The fact that the image is real tells us that the mirror cannot be convex, since convex mirrors form only virtual images. The mirror is concave.

(b) With $s_o = 60$ cm and $s_i = 30$ cm (s_i is positive since the image is real), the mirror equation tells us that

$$\frac{1}{s_o}+\frac{1}{s_i}=\frac{1}{f} \Rightarrow \frac{1}{60 \text{ cm}}+\frac{1}{30 \text{ cm}}=\frac{1}{f} \Rightarrow \frac{1}{f}=\frac{1}{20 \text{ cm}} \Rightarrow f=20 \text{ cm}$$

Note that f is positive, which confirms that the mirror is concave.

(c) The magnification is

$$M=-\frac{s_i}{s_o}=-\frac{30 \text{ cm}}{60 \text{ cm}}=-\frac{1}{2}$$

Since the absolute value of M is less than 1, the mirror makes the object look smaller (minifies the height of the object). The image is only half as tall as the object (and is upside down, because M is negative).

Example 16.17 A concave mirror with a focal length of 25 cm is used to create a real image that has twice the height of the object. How far is the image from the mirror?

Solution. Since h_i (the height of the image) is twice h_o (the height of the object) the value of the magnification is either +2 or −2. To figure out which, we just notice that the image is real; real images are inverted, so the magnification, M, must be negative. Therefore, $M = -2$, so

$$-\frac{s_i}{s_o}=-2 \Rightarrow s_o=\tfrac{1}{2}s_i$$

Substituting this into the mirror equation gives us

$$\frac{1}{s_o}+\frac{1}{s_i}=\frac{1}{f} \Rightarrow \frac{1}{\frac{1}{2}s_i}+\frac{1}{s_i}=\frac{1}{f} \Rightarrow \frac{3}{s_i}=\frac{1}{f} \Rightarrow s_i=3f=3(25 \text{ cm})=75 \text{ cm}$$

THIN LENSES

A lens is an optical device that forms an image by *refracting* light. We'll now talk about the equations and conventions that are used to analyze images formed by the two major categories of lenses: converging and diverging.

A **converging lens**—like the bi-convex one shown below—converges parallel paraxial rays of light to a focal point on the far side. (This lens is *bi-convex;* both of its faces are convex. Converging lenses all have at least one convex face.) Because parallel light rays actually focus at F, this point is called a **real focus.** Its distance from the lens is the focal length, f.

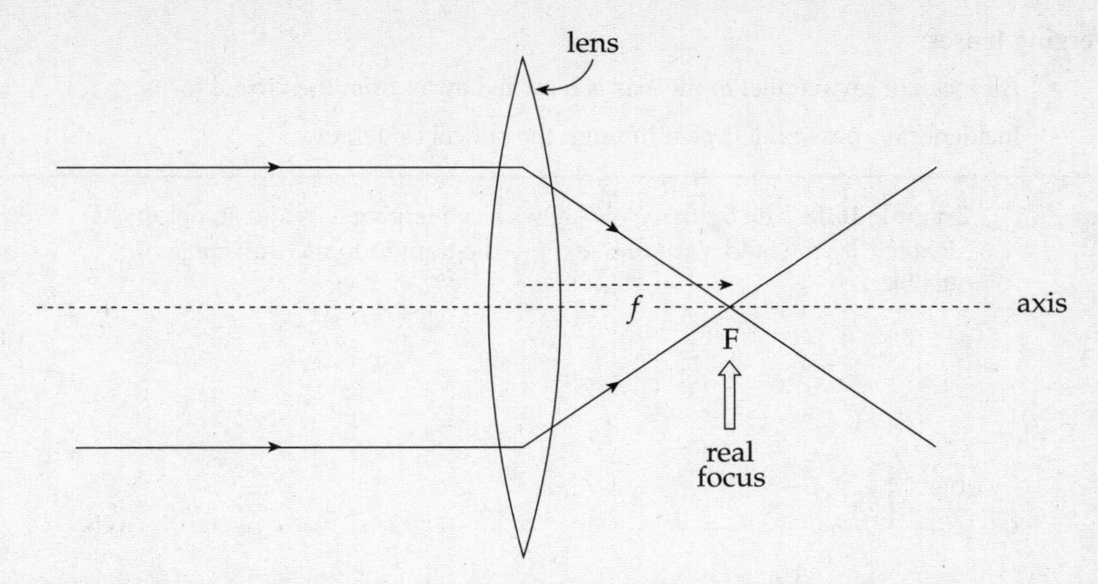

A **diverging lens**—like the *bi-concave* one shown below—causes parallel paraxial rays of light to diverge away from a **virtual focus**, F, on the same side as the incident rays. (Diverging lenses all have at least one concave face.)

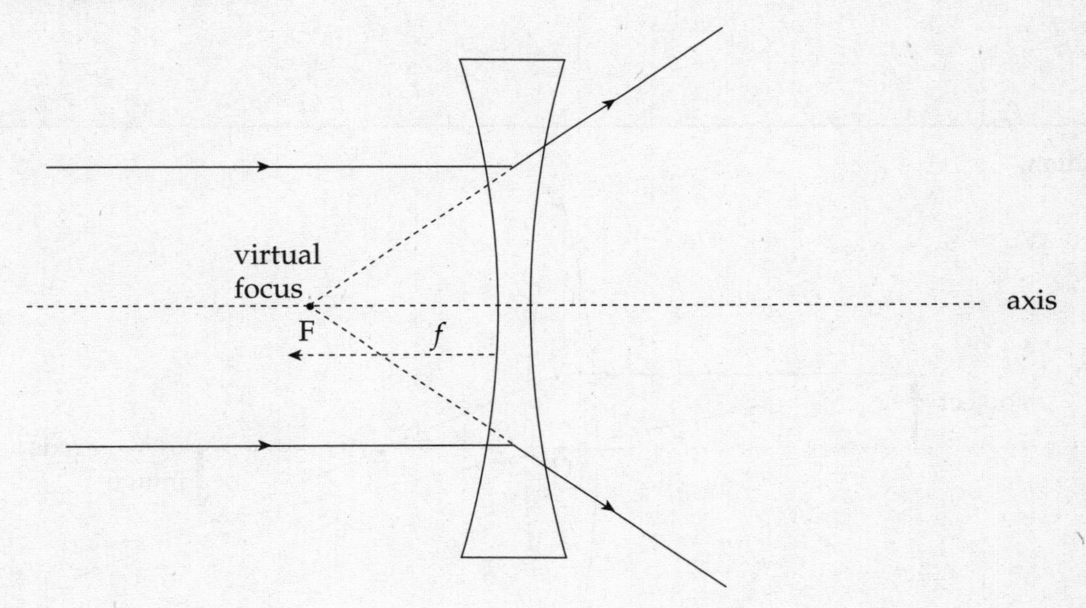

Ray Tracing for Lenses

Just as is the case with mirrors, representative rays of light can be sketched in a diagram along with the object and the lens; the point at which the reflected rays intersect (or appear to intersect) is the location of the image. The rules that govern these rays are as follows:

Converging lenses:

- An incident ray parallel to the axis is refracted through the real focus.
- Incident rays pass undeflected through the **optical center**, O (the central point within the lens where the axis intersects the lens).

Diverging lenses:

- An incident ray parallel to the axis is refracted away from the virtual focus.
- Incident rays pass undeflected through the optical center, O.

Example 16.18 The figure below shows a converging lens and an object (denoted by the bold arrow). Use a ray diagram to locate the image of the object.

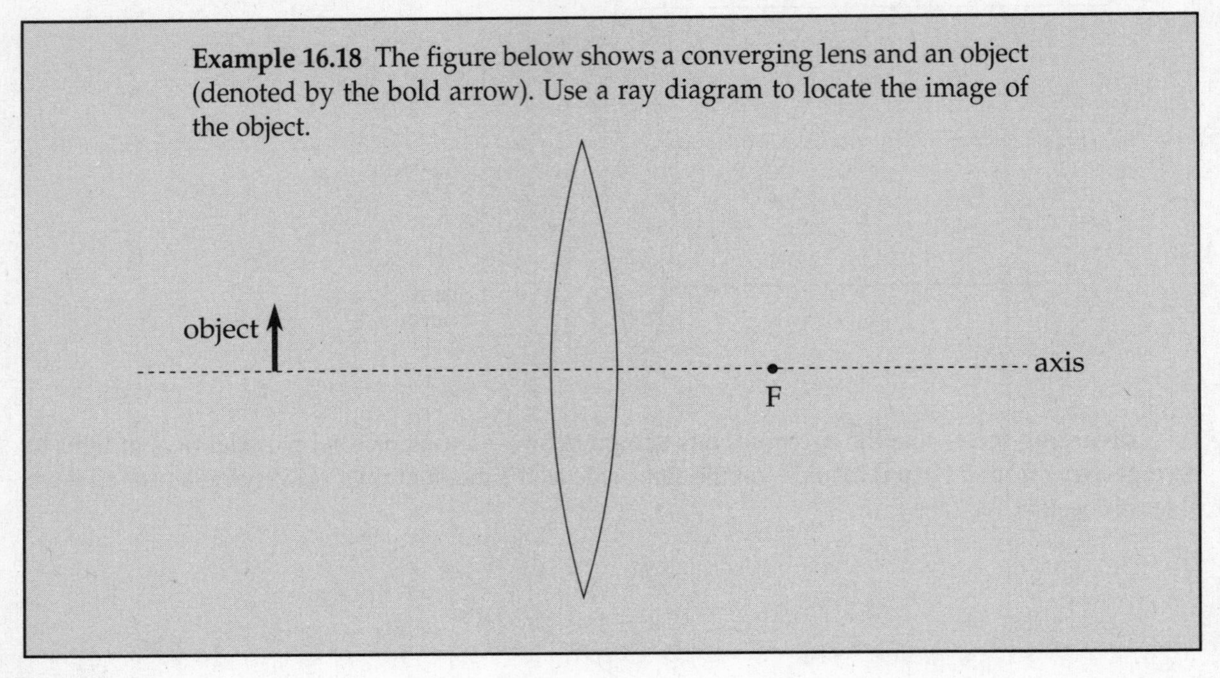

Solution.

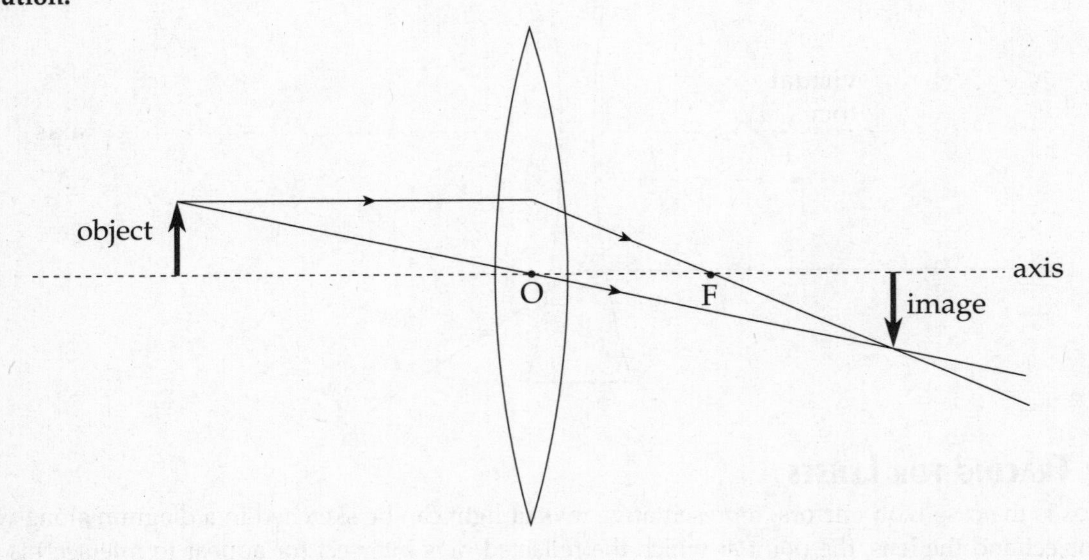

Example 16.19 The figure below shows a diverging lens and an object. Use a ray diagram to locate the image of the object.

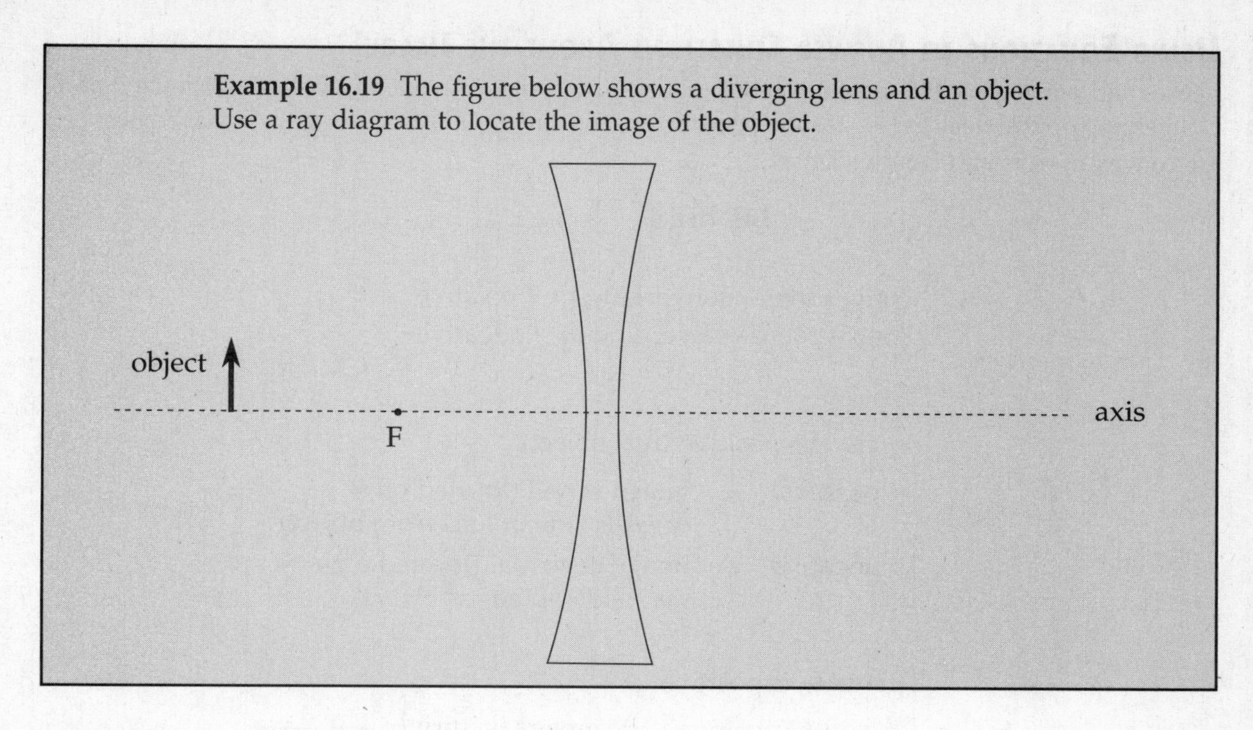

Solution.

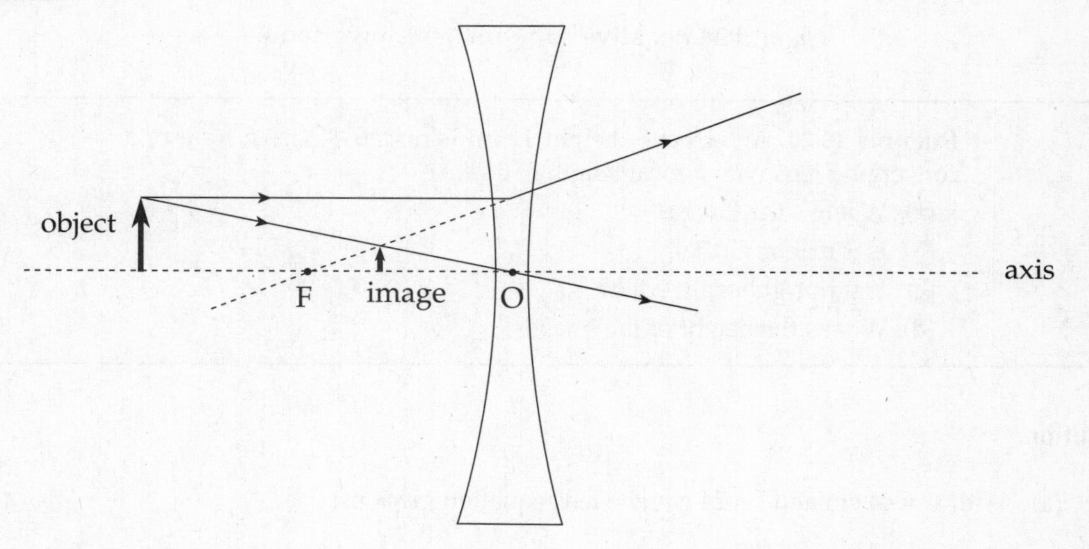

The nature of the image—that is, whether it's real or virtual—is determined by the side of the lens upon which the image is formed. If the image is formed on the side of the lens that's opposite the object, then the image is real, and if the image is formed on the same side of the lens as the object, then it's virtual. Another way of looking at this is: If you had to trace lines back to form an image, that image is virtual. Therefore, the image in Example 16.18 is real, while the image in Example 16.19 is virtual.

USING EQUATIONS TO ANSWER QUESTIONS ABOUT THE IMAGE

Lenses and mirrors use the same equations, notation, and sign conventions, with the following note. Converging optical devices ($+f$) are con<u>cave</u> mirrors and con<u>vex</u> lenses. Diverging optical devices ($-f$) are con<u>vex</u> mirrors and con<u>cave</u> lenses.

LENSES

convex lens—converging $\Leftrightarrow$ f positive
concave lens—diverging $\Leftrightarrow$ f negative

$$\frac{1}{s_o}+\frac{1}{s_i}=\frac{1}{f}$$

- s_o always positive (real object)
- s_i positive $\Rightarrow$ image is real (located on *opposite* side of lens from object)
- s_i negative $\Rightarrow$ image is virtual (located on *same* side of lens as object)

$$M=\frac{h_i}{h_o}=\frac{-s_i}{s_o}$$

- given h_o is positive
- h_i and M positive $\Rightarrow$ image is upright
- h_i and M negative $\Rightarrow$ image is inverted

Example 16.20 An object of height 11 cm is placed 44 cm in front of a converging lens with a focal length of 24 cm.

(a) Where's the image?
(b) Is it real or virtual?
(c) Is it upright or inverted?
(d) What's the height of the image?

Solution.

(a) With s_o = 44 cm and f = 24 cm, the lens equation gives us:

$$\frac{1}{s_o}+\frac{1}{s_i}=\frac{1}{f} \Rightarrow \frac{1}{44\text{ cm}}+\frac{1}{s_i}=\frac{1}{24\text{ cm}} \Rightarrow \frac{1}{s_i}=0.0189\text{ cm}^{-1} \Rightarrow s_i=53\text{ cm}$$

So, the image is located 53 cm from the lens, on the opposite side from the object.

(b) Because s_i is positive, the image is real.

(c) Real images are inverted.

(d) $$\frac{h_i}{h_o}=\frac{-s_i}{s_o}\rightarrow h_i=\frac{-s_i h_o}{s_o}$$

$$\rightarrow h_i=\frac{(-53\text{ cm})(11\text{ cm})}{44\text{ cm}}=-13\text{ cm}$$

The negative h_i reaffirms that we have an inverted image. These results are illustrated in Example 16.18.

Example 16.21 An object of height 11 cm is placed 48 cm in front of a diverging lens with a focal length of –24.5 cm.

(a) Where's the image?
(b) Is it real or virtual?
(c) Is it upright or inverted?
(d) What's the height of the image?

Solution.

(a) With $s_o = 48$ cm and $f = -24.5$ cm, the lens equation gives us:

$$\frac{1}{s_o}+\frac{1}{s_i}=\frac{1}{f} \Rightarrow \frac{1}{48\text{ cm}}+\frac{1}{s_i}=\frac{1}{-24.5\text{ cm}} \Rightarrow \frac{1}{s_i}=-0.0616\text{ cm}^{-1} \Rightarrow s_i=-16\text{ cm}$$

The image is 16 cm from the lens, on the same side as the object.

(b) Because s_i is negative, the image is virtual.

(c) Virtual images are upright.

(d) $$\frac{h_i}{h_o}=\frac{-s_i}{s_o} \rightarrow h_i=\frac{-s_i h_o}{s_o}$$

$$\rightarrow h_i=\frac{-(-16\text{ cm})(11\text{ cm})}{48\text{ cm}}=3.7\text{ cm}$$

These results are illustrated in Example 16.19.

Example 16.22 The **power** of a lens, P, is defined as $1/f$, where f is the lens's focal length (in meters). Lens power is expressed in **diopters** (abbreviated D), where 1 D = 1 m^{-1}. When an object is placed 10 cm from a converging lens, the real image formed is twice the height of the object. What's the power of this lens?

Solution. Since h_i, the height of the image, is twice h_o, the height of the object, the value of the magnification is either +2 or –2. To decide, we simply notice that the image is real; real images are inverted, so the magnification, m, must be negative. Since $m = -2$, we find that

$$-\frac{s_i}{s_o}=-2 \Rightarrow s_o=\frac{1}{2}s_i$$

Because $s_o = 10$ cm, it must be true that $s_i = 20$ cm, so the lens equation gives us:

$$\frac{1}{s_o}+\frac{1}{s_i}=\frac{1}{f} \Rightarrow \frac{1}{10\text{ cm}}+\frac{1}{20\text{ cm}}=\frac{1}{f} \Rightarrow f=\frac{20}{3}\text{ cm}$$

Expressing the focal length in meters and then taking the reciprocal gives the lens power in diopters:

$$f=\frac{20}{3}\text{ cm}\times\frac{1\text{ m}}{100\text{ cm}}=\frac{1}{15}\text{ m} \Rightarrow P=\frac{1}{f}=\frac{1}{\frac{1}{15}\text{ m}}=15\text{ D}$$

CHAPTER 16 REVIEW QUESTIONS

Solutions can be found in Chapter 18.

SECTION I: MULTIPLE CHOICE

1. What is the wavelength of an X-ray whose frequency is 1.0×10^{18} Hz?

   (A) 3.3×10^{-11} m
   (B) 3.0×10^{-10} m
   (C) 3.3×10^{-9} m
   (D) 3.0×10^{-8} m
   (E) 3.0×10^{26} m

2. In Young's double-slit interference experiment, what is the difference in path length of the light waves from the two slits at the center of the first bright fringe above the central maximum?

   (A) 0
   (B) $\frac{1}{4}\lambda$
   (C) $\frac{1}{2}\lambda$
   (D) λ
   (E) $\frac{3}{2}\lambda$

3. A beam of light in air is incident upon the smooth surface of a piece of flint glass, as shown:

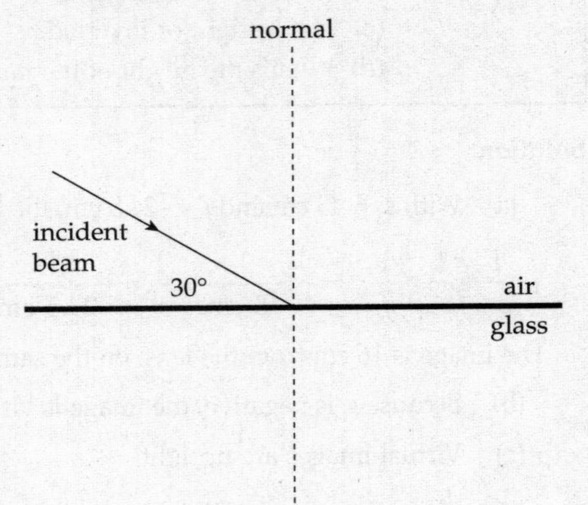

   If the reflected beam and refracted beam are perpendicular to each other, what is the index of refraction of the glass?

   (A) $\frac{1}{2}$
   (B) $\frac{1}{2}\sqrt{3}$
   (C) $\sqrt{3}$
   (D) 2
   (E) $2\sqrt{3}$

4. When green light (wavelength = 500 nm in air) travels through diamond (refractive index = 2.5), what is its wavelength?

   (A) 200 nm
   (B) 300 nm
   (C) 500 nm
   (D) 1000 nm
   (E) 1250 nm

5. A beam of light traveling in Medium 1 strikes the interface to another transparent medium, Medium 2. If the speed of light is less in Medium 2 than in Medium 1, the beam will

(A) refract toward the normal.
(B) refract away from the normal.
(C) undergo total internal reflection.
(D) have an angle of reflection smaller than the angle of incidence.
(E) have an angle of reflection greater than the angle of incidence.

6. If a clear liquid has a refractive index of 1.45 and a transparent solid has an index of 2.90 then, for total internal reflection to occur at the interface between these two media, which of the following must be true?

	incident beam originates in	at an angle of incidence greater than
(A)	the solid	30°
(B)	the liquid	30°
(C)	the solid	60°
(D)	the liquid	60°
(E)	Total internal reflection cannot occur.	

7. An object is placed 60 cm in front of a concave spherical mirror whose focal length is 40 cm. Which of the following best describes the image?

	Nature of image	Distance from mirror
(A)	Virtual	24 cm
(B)	Real	24 cm
(C)	Virtual	120 cm
(D)	Real	120 cm
(E)	Real	240 cm

8. An object is placed 60 cm from a spherical convex mirror. If the mirror forms a virtual image 20 cm from the mirror, what's the magnitude of the mirror's radius of curvature?

(A) 7.5 cm
(B) 15 cm
(C) 30 cm
(D) 60 cm
(E) 120 cm

9. The image created by a converging lens is projected onto a screen that's 60 cm from the lens. If the height of the image is 1/4 the height of the object, what's the focal length of the lens?

(A) 36 cm
(B) 45 cm
(C) 48 cm
(D) 72 cm
(E) 80 cm

10. Which of the following is true concerning a bi-concave lens? (A bi-concave lens has both surfaces concave.)

(A) Its focal length is positive.
(B) It cannot form real images.
(C) It cannot form virtual images.
(D) It can magnify objects.
(E) None of the above

SECTION II: FREE RESPONSE

1. Two trials of a double-slit interference experiment are set up as follows. The slit separation is $d = 0.50$ mm, and the distance to the screen, L, is 4.0 m.

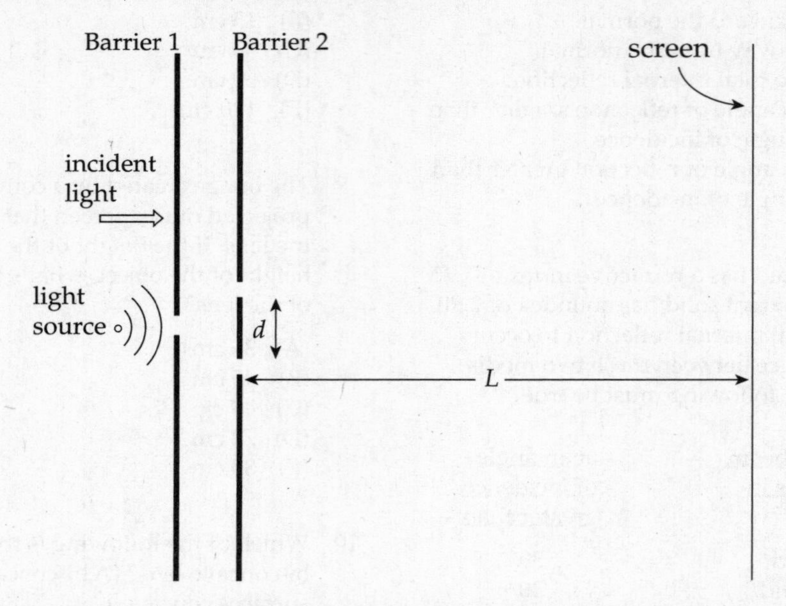

(a) What is the purpose of the first (single-slit) barrier? Why not use two light sources, one at each slit at the second barrier? Explain briefly.

In the first trial, white light is used.

(b) What is the vertical separation on the screen (in mm) between the first-order maxima for red light ($\lambda = 750$ nm) and violet light ($\lambda = 400$ nm)?

(c) Locate the nearest point to the central maximum where an intensity maximum for violet light ($\lambda = 400$ nm) coincides with an intensity maximum for orange-yellow light ($\lambda = 600$ nm).

In the second trial, the entire region between the double-slit barrier and the screen is filled with a large slab of glass of refractive index $n = 1.5$, and monochromatic green light ($\lambda = 500$ nm in air) is used.

(d) What is the separation between adjacent bright fringes on the screen?

2. In order to determine the criteria for constructive and destructive interference, the following rules are used:

   i) When light strikes the boundary to a medium with a higher refractive index than the incident medium, it undergoes a 180° phase change upon reflection (this is equivalent to a shift by one-half wavelength).

   ii) When light strikes the boundary to a medium with a lower refractive index than the incident medium, it undergoes no phase change upon reflection.

These rules can be applied to the two situations described below.

A thin soap film of thickness T, consisting of a mixture of water and soap (refractive index = 1.38), has air on both sides. Incident sunlight is reflected off the front face and the back face, causing interference.

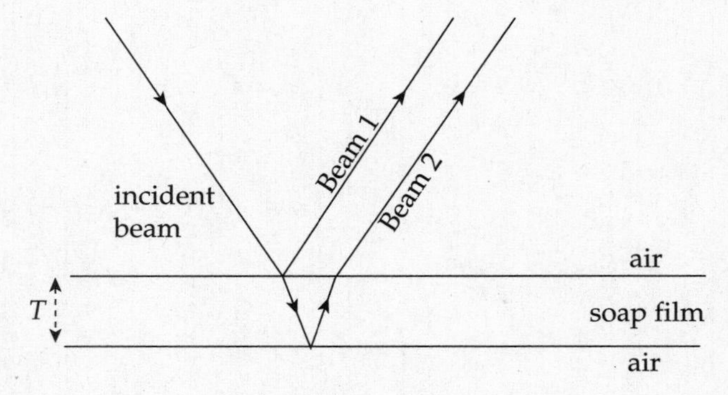

(a) Which beam, 1 or 2, suffers a 180° phase change upon reflection?

(b) Since the beams are out of phase, destructive interference will occur if the difference in their path lengths, $\Delta\ell \approx 2T$ for near-normal incidence, is equal to a whole number of wavelengths (wavelength as measured in the soap film). What is the criterion for *constructive* interference? Write your answer as an algebraic equation.

3. An object of height 5 cm is placed 40 cm in front of a spherical concave mirror. An image is formed 72 cm behind the mirror.

   (a) Is the image real or virtual?

   (b) Is the image upright or inverted?

   (c) What's the height of the image?

   (d) What is the mirror's radius of curvature?

   (e) In the figure below, sketch the mirror, labeling its vertex and focal point, and then construct a ray diagram (with a minimum of two rays) showing the formation of the image.

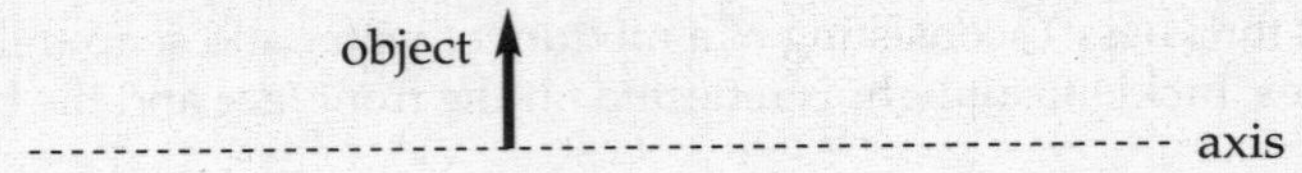

SUMMARY

- All electromagnetic waves travel at the speed of light ($c = 3 \times 10^8$ m/s). They obey the wave equation $v = f\lambda$. When monochromatic coherent light goes through double slits (or a diffraction grating) there will be constructive interference when the path difference is given by $PD = m\lambda$ or $d \sin\theta = m\lambda$ where $m = 0, 1, 2 \ldots$. There will be destructive interference for $PD = (m+\frac{1}{2})\lambda$ or $d \sin\theta = (m+\frac{1}{2})\lambda$. For small angles we can say $x_m = \frac{m\lambda L}{d}$, where x_m is the distance the mth maximum is located from the central maximum and shows where constructive interference occurs.

- Geometric optics always measures angles from the normal line. Topics include the fact that the angle of incidence is equal to the angle of reflection and that these angles lie in the same plane.

- Snell's Law states when light enters a new medium (material) it changes speeds and may change directions. This is stated in the formula

  $$n_1 \sin\theta_1 = n_2 \sin\theta_2$$

  where n is the index of refraction, which is a ratio of the speed of light in a vacuum to the speed of light in the substance ($n = \frac{c}{v}$). The index of refraction is always greater than 1 and has no units. When going from a high index of refraction to a lower index of refraction, light may have total internal reflection if the angle is larger than the critical angle ($\sin\theta_c = \frac{n_2}{n_1}$).

- For both thin lenses and curved mirrors you can use the thin lens equation: $\frac{1}{s_o} + \frac{1}{s_i} = \frac{1}{f}$

  Where f is positive for a convex lens or a concave mirror and f is negative for a concave lens or a convex mirror s_0 is always positive.

- For a lens, if s_i is negative the lens image is virtual and located on the same side of the lens as the object. If s_i is positive, the lens image is real and located on the opposite side of the lens from the object.

- For a mirror, if s_i is negative, the image is virtual and located on the opposite side of the mirror from the object. If s_i is positive, the image is real and located on the same side of the mirror as the object.

- The magnification of a lens is given by $M = \frac{h_i}{h_o} = \frac{-s_i}{s_o}$.

- A **diverging optical device** could be either a convex mirror or a concave lens. The thing these two have in common is that no matter where the object is located they form images that are virtual and upright. The sign of h_i will always be positive and the sign of s_i will always be negative. The sign of the magnification will be positive and its value will be less than one. The image will always be located between the focal length and the optical device.

- A **converging optical device** could be either a concave mirror or a convex lens. The thing these two have in common is that when s_o is *outside* the focal length they form images that are real and inverted. The sign of h_i will be negative and the sign of s_i will be positive. The sign of the magnification will be negative and its absolute value can be any number greater than zero. The other thing they have in common is that when s_o is *inside* the focal length they form images that are virtual and upright. The sign of h_i will be positive and the sign of s_i will be negative. The sign of the magnification will be positive and its value will be greater than one.

17

Atomic and Nuclear Physics

INTRODUCTION

The subject matter of the previous chapters was developed in the seventeenth, eighteenth, and nineteenth centuries, but as we delve into the physics of the very small, we enter the twentieth century.

PHOTONS AND THE PHOTOELECTRIC EFFECT

The particle-like nature of light was revealed and studied through the work of Max Planck in 1900, and later Albert Einstein (who won the 1921 Nobel Prize for work in this area.) Electromagnetic radiation is emitted and absorbed by matter as though it existed in individual bundles called **quanta**. A quantum of electromagnetic energy is known as a **photon**. Light behaves like a stream of photons, and this is illustrated by the **photoelectric effect**.

When a piece of metal is illuminated by electromagnetic radiation (specifically visible light, ultraviolet light, or X-rays), the energy absorbed by electrons near the surface of the metal can liberate them from their bound state, and these electrons can fly off. The released electrons are known as **photoelectrons**. In this case, the classical, wave-only theory of light would predict three results:

(1) There would be a significant time delay between the moment of illumination and the ejection of photoelectrons, as the electrons absorbed incident energy until their kinetic energy was sufficient to release them from the atoms' grip.

(2) Increasing the intensity of the light would cause the electrons to leave the metal surface with greater kinetic energy.

(3) Photoelectrons would be emitted regardless of the frequency of the incident energy, as long as the intensity was high enough.

Surprisingly, none of these predictions was observed. Photoelectrons were ejected within just a few billionths of a second after illumination, disproving prediction (1). Secondly, increasing the intensity of the light did not cause photoelectrons to leave the metal surface with greater kinetic energy. Although more electrons were ejected as the intensity was increased, there was a maximum photoelectron kinetic energy; prediction (2) was false. And, for each metal, there was a certain **threshold frequency**, f_0: If light of frequency lower than f_0 were used to illuminate the metal surface, *no* photoelectrons were ejected, regardless of how intense the incident radiation was; prediction (3) was also false. Clearly, something was wrong with the wave-only theory of light.

Einstein explained these observations by postulating that the energy of the incident electromagnetic wave was absorbed in individual bundles (photons). The energy of a photon is proportional to the frequency of the wave,

$$E = hf$$

where h is **Planck's constant** (about 6.63×10^{-34} J·s). A certain amount of energy had to be imparted to an electron on the metal surface in order to liberate it; this was known as the metal's **work function**, or ϕ. If an electron absorbed a photon whose energy E was greater than ϕ, it would leave the metal with a maximum kinetic energy equal to $E - \phi$. This process could occur *very* quickly, which accounts for the rapidity with which photoelectrons are produced after illumination.

Increasing the intensity of the incident energy means bombardment with more photons and results in the ejection of more photoelectrons—but since the energy of each incident photon is fixed by the equation $E = hf$, the value of K_{max} will still be $E - \phi$. This can be expressed as

$$K_{max} = hf - \phi$$

This accounts for the observation that disproved prediction (2).

Finally, if the incident energy had a frequency that was less than ϕ/h, the incident photons would each have an energy that was less than ϕ; this would not be enough energy to liberate electrons. Blasting the metal surface with more photons (that is, increasing the intensity of the incident beam) would also do nothing; none of the photons would have enough energy to eject electrons, so whether there were one or one million of them wouldn't make any difference. This accounts for the observation of a threshold frequency, which we now know is ϕ/h. This can be expressed as

$$f_o = \phi/h$$

Before we get to some examples, it's worthwhile to introduce a new unit for energy. The SI unit for energy is the joule, but it's too large to be convenient in the domains we're studying now. We'll use a much smaller unit, the **electronvolt** (abbreviated **eV**). The eV is equal to the energy gained (or lost) by an electron accelerated through a potential difference of one volt. Using the equation $\Delta U_E = qV$, we find that

$$1 \text{ eV} = (1 \text{ e})(1 \text{ V}) = (1.6 \times 10^{-19} \text{ C})(1 \text{ V}) = 1.6 \times 10^{-19} \text{ J}$$

In terms of electronvolts, the value of Planck's constant is 4.14×10^{-15} eV·s.

Example 17.1 The work function, ϕ, for aluminum is 4.08 eV.

(a) What is the threshold frequency required to produce photoelectrons from aluminum?

(b) Classify the electromagnetic radiation that can produce photoelectrons.

(c) If light of frequency $f = 4.00 \times 10^{15}$ Hz is used to illuminate a piece of aluminum,

(i) what is K_{max}, the maximum kinetic energy of ejected photoelectrons?

(ii) what's the maximum speed of the photoelectrons? (Electron mass = 9.11×10^{-31} kg.)

(d) If the light described in part (b) were increased by a factor of 2 in intensity, what would happen to the value of K_{max}?

Solution.

(a) We know from the statement of the question that, in order for a photon to be successful in liberating an electron from the surface of the aluminum, its energy cannot be less than 4.08 eV. Therefore, the minimum frequency of the incident light—the threshold frequency—must be

$$f_0 = \frac{\phi}{h} = \frac{4.08 \text{ eV}}{4.14 \times 10^{-15} \text{ eV} \cdot \text{s}} = 9.86 \times 10^{14} \text{ Hz}$$

(b) Based on the electromagnetic spectrum given in the previous chapter, the electromagnetic radiation used to produce photoelectrons from aluminum must be at least in the ultraviolet region of the EM spectrum.

(c) (i) The maximum kinetic energy of photoelectrons is found from the equation

$$\begin{aligned} K_{max} &= hf - \phi \\ &= (4.14 \times 10^{-15} \text{ eV} \cdot \text{s})(4.00 \times 10^{15} \text{ Hz}) - (4.08 \text{ eV}) \\ &= 12.5 \text{ eV} \end{aligned}$$

(ii) Using the above result and $K = \frac{1}{2}mv^2$, we can find v_{max}:

$$\begin{aligned} v_{max} &= \sqrt{\frac{2}{m_e} K_{max}} \\ &= \sqrt{\frac{2}{9.11 \times 10^{-31} \text{ kg}} \left(12.5 \text{ eV} \cdot \frac{1.6 \times 10^{-19} \text{ J}}{1 \text{ eV}} \right)} \\ &= 2.1 \times 10^6 \text{ m/s} \end{aligned}$$

(d) Nothing will happen. Increasing the intensity of the illuminating radiation will cause more photons to impinge on the metal surface, thereby ejecting more photoelectrons, but their maximum kinetic energy will remain the same. The only way to increase K_{max} would be to increase the frequency of the incident energy.

THE BOHR MODEL OF THE ATOM

In the years immediately following Rutherford's announcement of his nuclear model of the atom, a young physicist, Niels Bohr, would add an important piece to the atomic puzzle. Rutherford told us where the positive charge of the atom was located; Bohr would tell us about the electrons.

For fifty years it had been known that atoms in a gas discharge tube emit and absorb light only at specific wavelengths. The light from a glowing gas, passed through a prism to disperse the beam into its component wavelengths, produces patterns of sharp lines called **atomic spectra**. The visible wavelengths that appear in the emission spectrum of hydrogen had been summarized by the *Balmer formula*

$$\frac{1}{\lambda_n} = R\left(\frac{1}{2^2} - \frac{1}{n^2}\right)$$

where R is the *Rydberg constant* (about 1.1×10^7 m^{-1}). The formula worked—that is, it fit the observational data—but it had no theoretical basis. So, the question was, *why* do atoms emit (or absorb) radiation only at certain discrete wavelengths?

Bohr's model of the atom explains the spectroscopists' observations. Using the simplest atom, hydrogen (with only one electron), Bohr postulated that the electron would orbit the nucleus only at certain discrete radii. When the electron is in one of these special orbits, it does not radiate away energy (as the classical theory would predict). However, if the electron absorbs a certain amount of energy, it is **excited** to a higher orbit, one with a greater radius. After a short time in this excited state, it returns to a lower orbit, emitting a photon in the process. Since each allowed orbit—or **energy level**—has a specific radius (and corresponding energy), the photons emitted in each jump have only specific wavelengths.

Bohr found that the energy of each excited state was a particular fraction of the energy of the ground state. Specifically, the energy of the nth energy level in hydrogen is given by the formula

$$E_n = \frac{1}{n^2}E_1$$

where E_1 is the **ground-state energy**, equal to −13.6 eV. The energy E_1 (and, therefore, all the E_n) are negative. This corresponds to the fact that the electron is bound within the atom. Energy must be supplied to cause it to leave (**ionizing** the atom), bringing its energy to zero. For one-electron atoms, such as hydrogen, ionized helium ($Z = 2$), double-ionized lithium ($Z = 3$), etc., the value of the ground-state energy is also known as the **ionization energy**, the minimum amount of energy that must be supplied to release the atom's electron. The energy levels for these atoms are given by

$$E_n = \frac{Z^2}{n^2}(-13.6 \text{ eV})$$

where Z is the number of protons in the atom's nucleus.

When an excited electron drops from energy level $n = j$ to a lower one, $n = i$, the transition causes a photon of energy to be emitted, and the energy of the photon is the difference between the two energy levels:

$$E_{\text{emitted photon}} = |\Delta E| = E_j - E_i$$

The wavelength of this photon is

$$\lambda = \frac{c}{f} = \frac{c}{E_{\text{photon}} / h} = \frac{hc}{E_j - E_i}$$

Example 17.2

The first five energy levels of an atom are shown in the diagram below:

–3 eV __________ $n = 5$
–4 eV __________ $n = 4$
–7 eV __________ $n = 3$
–15 eV __________ $n = 2$
–62 eV __________ $n = 1$ ground state

(a) If the atom begins in the $n = 3$ level, what photon energies could be emitted as it returns to its ground state?
(b) What could happen if this atom, while in an undetermined energy state, were bombarded with a photon of energy 10 eV?

Solution.

(a) If the atom is in the $n = 3$ state, it could return to ground state by a transition from $3 \rightarrow 1$, or from $3 \rightarrow 2$ and then $2 \rightarrow 1$. The energy emitted in each of these transitions is simply the difference between the energies of the corresponding levels:

$$E_{3\rightarrow1} = E_3 - E_1 = (-7\text{ eV}) - (-62\text{ eV}) = 55\text{ eV}$$
$$E_{3\rightarrow2} = E_3 - E_2 = (-7\text{ eV}) - (-15\text{ eV}) = 8\text{ eV}$$
$$E_{2\rightarrow1} = E_2 - E_1 = (-15\text{ eV}) - (-62\text{ eV}) = 47\text{ eV}$$

(b) Since no two energy levels in this atom are separated by 10 eV, the atom could not absorb a 10 eV photon, and as a result, nothing would happen. This atom would be transparent to light of energy 10 eV.

Example 17.3

(a) How much energy must a ground-state electron in a hydrogen atom absorb to be excited to the $n = 4$ energy level?

(b) With the electron in the $n = 4$ level, what wavelengths are possible for the photon emitted when the electron drops to a lower energy level? In what regions of the EM spectrum do these photons lie?

Solution.

(a) The ground-state energy level ($n = 1$) for hydrogen is –13.6 eV, and using the equation $E_n = (1/n^2)E_1$, the $n = 4$ energy level (the third excited state) is

$$E_4 = \frac{1}{4^2}(-13.6\text{ eV}) = -0.85\text{ eV}$$

Therefore, in order for an electron to make the transition from E_1 to E_4, it must absorb energy in the amount $E_4 - E_1 = (-0.85\text{ eV}) - (-13.6\text{ eV}) = 12.8\text{ eV}$.

(b) An electron in the $n = 4$ energy level can make several different transitions: It can drop to $n = 3$, $n = 2$, or all the way down to the ground state, $n = 1$. The energies of the $n = 2$ and $n = 3$ levels are

$$E_2 = \frac{1}{2^2}(-13.6\text{ eV}) = -3.4\text{ eV} \quad \text{and} \quad E_3 = \frac{1}{3^2}(-13.6\text{ eV}) = -1.5\text{ eV}$$

The following diagram shows the electron dropping from $n = 4$ to $n = 3$:

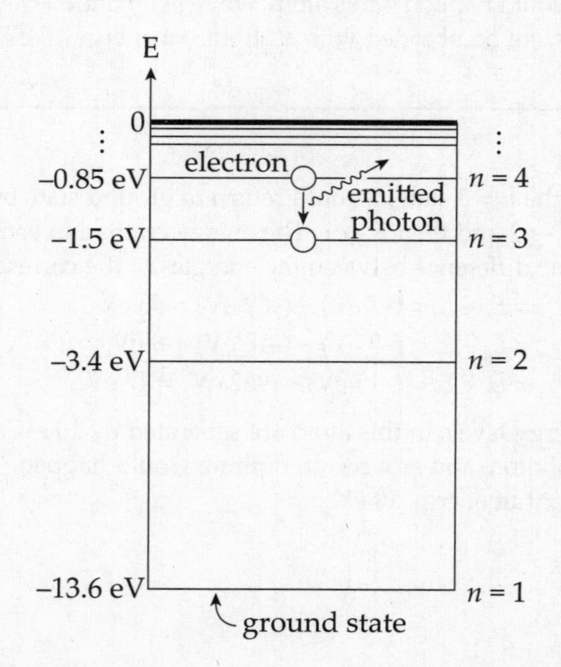

There are three possible values for the energy of the emitted photon, $E_{4\to3}$, $E_{4\to2}$, or $E_{4\to1}$:

$$E_{4\to3} = E_4 - E_3 = (-0.85 \text{ eV}) - (-1.5 \text{ eV}) = 0.65 \text{ eV}$$
$$E_{4\to2} = E_4 - E_2 = (-0.85 \text{ eV}) - (-3.4 \text{ eV}) = 2.55 \text{ eV}$$
$$E_{4\to1} = E_4 - E_1 = (-0.85 \text{ eV}) - (-13.6 \text{ eV}) = 12.8 \text{ eV}$$

From the equation $E = hf = hc/\lambda$, we get $\lambda = hc/E$, so

$$\lambda_{4\to3} = \frac{hc}{E_{4\to3}} = \frac{(4.14\times10^{-15} \text{ eV}\cdot\text{s})(3.00\times10^{8} \text{ m/s})}{0.65 \text{ eV}} = 1910 \text{ nm}$$
$$\lambda_{4\to2} = \frac{hc}{E_{4\to2}} = \frac{(4.14\times10^{-15} \text{ eV}\cdot\text{s})(3.00\times10^{8} \text{ m/s})}{2.55 \text{ eV}} = 487 \text{ nm}$$
$$\lambda_{4\to1} = \frac{hc}{E_{4\to1}} = \frac{(4.14\times10^{-15} \text{ eV}\cdot\text{s})(3.00\times10^{8} \text{ m/s})}{12.8 \text{ eV}} = 97 \text{ nm}$$

Note that $\lambda_{4\to2}$ is in the visible spectrum; this wavelength corresponds to the color blue-green; $\lambda_{4\to3}$ is an ultraviolet wavelength, and $\lambda_{4\to3}$ is infrared.

WAVE-PARTICLE DUALITY

Light and other electromagnetic waves exhibit wave-like characteristics through interference and diffraction. However, as we saw in the photoelectric effect, light also behaves as if its energy were granular, composed of particles. This is **wave-particle duality**: Electromagnetic radiation propagates like a wave but exchanges energy like a particle.

Since an electromagnetic wave can behave like a particle, can a particle of matter behave like a wave? In 1923, the French physicist Louis de Broglie proposed that the answer is "yes." His conjecture, which has since been supported by experiment, is that a particle of mass m and speed v—and thus with linear momentum $p = mv$—has an associated wavelength, which is called its **de Broglie wavelength**:

$$\lambda = \frac{h}{p}$$

Particles in motion can display wave characteristics and behave as if they had a wavelength.

Since the value of h is so small, ordinary macroscopic objects do not display wave-like behavior. For example, a baseball (mass = 0.15 kg) thrown at a speed of 40 m/s has a de Broglie wavelength of

$$\lambda = \frac{h}{p} = \frac{h}{mv} = \frac{6.63\times10^{-34} \text{ J}\cdot\text{s}}{(0.15 \text{ kg})(40 \text{ m/s})} = 1.1\times10^{-34} \text{ m}$$

This is much too small to measure. However, with subatomic particles, the wave nature is clearly evident. The 1937 Nobel prize was awarded for experiments that revealed that a stream of electrons exhibited diffraction patterns when scattered by crystals—a behavior that's characteristic of waves.

There's an interesting connection between the de Broglie wavelength for electrons and Bohr's model of quantized orbital radii (once tested on the AP Physics Exam!). Bohr postulated that the electron's orbital angular momentum, $m_e v_n r_n$, had to be an integral multiple of $\hbar = \frac{h}{2\pi}$. The equation

$$m_e v_n r_n = n\frac{h}{2\pi}$$

can be rewritten as follows:

$$2\pi r_n = n\frac{h}{m_e v_n} \quad \Rightarrow \quad 2\pi r_n = n\frac{h}{p_e} \quad \Rightarrow \quad 2\pi r_n = n\lambda_e$$

In this last form, the equation says that the circumference of the electron's orbit must be equal to a whole number of wavelengths in order for it to be stable, a restriction that should remind you of sustained standing waves.

Example 17.4 Electrons in a diffraction experiment are accelerated through a potential difference of 200 V. What is the de Broglie wavelength of these electrons?

Solution. By definition, the kinetic energy of these electrons is 200 eV. Since the relationship between linear momentum and kinetic energy is $p = \sqrt{2mK}$,

$$\lambda = \frac{h}{p} = \frac{h}{\sqrt{2mK}} = \frac{6.63\times10^{-34}\ \text{J}\cdot\text{s}}{\sqrt{2(9.11\times10^{-31}\ \text{kg})\left[200\ \text{eV}\cdot\frac{1.6\times10^{-19}\ \text{J}}{1\ \text{eV}}\right]}} = 8.7\times10^{-11}\ \text{m} = 0.087\ \text{nm}$$

This wavelength is characteristic of X-rays.

NUCLEAR PHYSICS

The nucleus of the atom is composed of particles called **protons** and **neutrons**, which are collectively called **nucleons**. The number of protons in a given nucleus is called the atom's **atomic number**, and is denoted Z, and the number of neutrons (the **neutron number**) is denoted N. The total number of nucleons, $Z + N$, is called the **mass number** (or **nucleon number**), and is denoted A. The number of protons in the nucleus of an atom defines the element. For example, the element chlorine (abbreviated Cl) is characterized by the fact that the nucleus of every chlorine atom contains 17 protons, so the atomic number of chlorine is 17; but, different chlorine atoms may contain different numbers of neutrons. In fact, about three-fourths of all naturally occurring chlorine atoms have 18 neutrons in their nuclei (mass number = 35), and most of the remaining one-fourth contain 20 neutrons (mass number = 37). Nuclei that contain the same numbers of protons but different numbers of neutrons are called **isotopes**. The notation for a **nuclide**—the term for a nucleus with specific numbers of protons and neutrons—is to write Z and A before the chemical symbol of the element:

$${}^{A}_{Z}\text{X}$$

The isotopes of chlorine mentioned above would be written as follows:

$${}^{35}_{17}\text{Cl} \quad \text{and} \quad {}^{37}_{17}\text{Cl}$$

Example 17.5 How many protons and neutrons are contained in the nuclide $^{63}_{29}\text{Cu}$?

Solution. The subscript (the atomic number, Z) gives the number of protons, which is 29. The superscript (the mass number, A) gives the total number of nucleons. Since $A = 63 = Z + N$, we find that $N = 63 - 29 = 34$.

Example 17.6 The element neon (abbreviated Ne, atomic number 10) has several isotopes. The most abundant isotope contains 10 neutrons, and two others contain 11 and 12. Write symbols for these three nuclides.

Solution. The mass numbers of these isotopes are 10 + 10 = 20, 10 + 11 = 21, and 10 + 12 = 22. So, we'd write them as follows:

$$^{20}_{10}\text{Ne}, \quad ^{21}_{10}\text{Ne}, \quad \text{and} \quad ^{22}_{10}\text{Ne}$$

Another common notation—which we also use—is to write the mass number after the name of the element. These three isotopes of neon would be written as neon-20, neon-21, and neon-22.

The Nuclear Force

Why wouldn't any nucleus that has more than one proton be unstable? After all, protons are positively charged and would therefore experience a repulsive Coulomb force from each other. Why don't these nuclei explode? And what holds neutrons—which have no electric charge—in the nucleus? These issues are resolved by the presence of another fundamental force, the **strong nuclear force**, which binds neutrons and protons together to form nuclei. Although the strength of the Coulomb force can be expressed by a simple mathematical formula (it's inversely proportional to the square of their separation), the nuclear force is much more complicated; no simple formula can be written for the strength of the nuclear force.

Binding Energy

The masses of the proton and neutron are listed below.

$$\text{proton: } m_p = 1.6726 \times 10^{-27} \text{ kg}$$

$$\text{neutron: } m_n = 1.6749 \times 10^{-27} \text{ kg}$$

Because these masses are so tiny, a much smaller mass unit is used. With the most abundant isotope of carbon (carbon-12) as a reference, the **atomic mass unit** (abbreviated **amu** or simply **u**) is defined as 1/12 the mass of a ^{12}C atom. The conversion between kg and u is $1 \text{ u} = 1.6605 \times 10^{-27}$ kg. In terms of atomic mass units,

$$m_p = 1.00728 \text{ u}$$

$$m_n = 1.00867 \text{ u}$$

Now consider the **deuteron**, the nucleus of **deuterium**, an isotope of hydrogen that contains 1 proton and 1 neutron. The mass of a deuteron is 2.01356 u, which is a little *less* than the sum of the individual masses of the proton and neutron. The difference between the mass of any bound nucleus and the sum of the masses of its constituent nucleons is called the **mass defect**, Δm. In the case of the deuteron (symbolized **d**), the mass defect is

$$\begin{aligned}\Delta m &= (m_p + m_n) - m_d \\ &= (1.00728\ \text{u} + 1.00867\ \text{u}) - (2.01356\ \text{u}) \\ &= 0.00239\ \text{u}\end{aligned}$$

What happened to this missing mass? It was converted to energy when the deuteron was formed. It also represents the amount of energy needed to break the deuteron into a separate proton and neutron. Since this tells us how strongly the nucleus is bound, it is called the **binding energy** of the nucleus. The conversion between mass and energy is given by Einstein's **mass-energy equivalence** equation, $E = mc^2$ (where c is the speed of light); the binding energy, E_B, is equal to the mass defect, Δm:

$$E_B = (\Delta m)c^2$$

Using $E = mc^2$, the energy equivalent of 1 atomic mass unit is

$$\begin{aligned}E &= (1.6605\times10^{-27}\ \text{kg})(2.9979\times10^{8}\ \text{m/s})^2 \\ &= 1.4924\times10^{-10}\ \text{J} \\ &= 1.4924\times10^{-10}\ \text{J}\cdot\frac{1\ \text{eV}}{1.6022\times10^{-19}\ \text{J}} \\ &= 9.31\times10^{8}\ \text{eV} \\ &= 931\ \text{MeV}\end{aligned}$$

In terms of electronvolts, then, the binding energy of the deuteron is

$$E_B\ (\text{deuteron}) = 0.00239\ \text{u}\times\frac{931\ \text{MeV}}{1\ \text{u}} = 2.23\ \text{MeV}$$

Since the deuteron contains 2 nucleons, the **binding-energy-per-nucleon** is

$$\frac{2.23\ \text{MeV}}{2\ \text{nucleons}} = 1.12\ \text{MeV/nucleon}$$

This is the lowest value of all nuclides. The highest, 8.8 MeV/nucleon, is for an isotope of nickel, ^{62}Ni. Typically, when nuclei smaller than nickel are fused to form a single nucleus, the binding energy per nucleon increases, which tells us that energy is released in the process. On the other hand, when nuclei *larger* than nickel are *split*, binding energy per nucleon again increases, releasing energy.

Example 17.7 What is the maximum wavelength of EM radiation that could be used to photodisintegrate a deuteron?

Solution. The binding energy of the deuteron is 2.23 MeV, so a photon would need to have at least this much energy to break the deuteron into a proton and neutron. Since $E = hf$ and $f = c/\lambda$,

$$E = \frac{hc}{\lambda} \Rightarrow \lambda_{max} = \frac{hc}{E_{min}} = \frac{(4.14\times10^{-15}\ \text{eV}\cdot\text{s})(3.00\times10^{8}\ \text{m/s})}{2.23\times10^{6}\ \text{eV}} = 5.57\times10^{-13}\ \text{m}$$

Example 17.8 The atomic mass of $^{27}_{13}\text{Al}$ is 26.9815 u. What is its nuclear binding energy per nucleon? (Mass of electron = 0.0005486 u.)

Solution. The nuclear mass of $^{27}_{13}\text{Al}$ is equal to its atomic mass minus the mass of its electrons. Since an aluminum atom has 13 protons, it must also have 13 electrons. So,

$$\begin{aligned}\text{nuclear mass of } ^{27}_{13}\text{Al} &= (\text{atomic mass of } ^{27}_{13}\text{Al}) - 13m_e \\ &= 26.9815 \text{ u} - 13(0.0005486 \text{ u}) \\ &= 26.9744 \text{ u}\end{aligned}$$

Now, the nucleus contains 13 protons and 27 – 13 = 14 neutrons, so the total mass of the individual nucleons is

$$\begin{aligned}M &= 13m_p + 14m_n \\ &= 13(1.00728 \text{ u}) + 14(1.00867 \text{ u}) \\ &= 27.2160 \text{ u}\end{aligned}$$

and, the mass defect of the aluminum nucleus is

$$\Delta m = M - m = 27.2160 \text{ u} - 26.9744 \text{ u} = 0.2416 \text{ u}$$

Converting this mass to energy, we can see that

$$E_B = 0.2416 \text{ u} \times \frac{931 \text{ MeV}}{1 \text{ u}} = 225 \text{ MeV}$$

so the binding energy per nucleon is

$$\frac{225 \text{ MeV}}{27} = 8.3 \text{ MeV/nucleon}$$

NUCLEAR REACTIONS

Natural radioactive decay provides one example of a nuclear reaction. Other examples of nuclear reactions include the bombardment of target nuclei with subatomic particles to artificially induce radioactivity, such as the emission of a particle or the splitting of the nucleus (this is **nuclear fission**), and the **nuclear fusion** of small nuclei at extremely high temperatures. In all cases of nuclear reactions that we'll study, nucleon number and charge must be conserved. In order to balance nuclear reactions, we write $^{1}_{1}\text{p}$ or $^{1}_{1}\text{H}$ for a proton and $^{1}_{0}\text{n}$ for a neutron. Gamma-ray photons can also be produced in nuclear reactions; they have no charge or nucleon number and are represented as $^{0}_{0}\gamma$.

Example 17.9 A mercury-198 nucleus is bombarded by a neutron, which causes a nuclear reaction:

$$^{1}_{0}\text{n} + ^{198}_{80}\text{Hg} \rightarrow ^{197}_{79}\text{Au} + ^{?}_{?}\text{X}$$

What's the unknown product particle, X?

Solution. In order to balance the superscripts, we must have $1 + 198 = 197 + A$, so $A = 2$, and the subscripts are balanced if $0 + 80 = 79 + Z$, so $Z = 1$:

$$^{1}_{0}\text{n} + ^{198}_{80}\text{Hg} \rightarrow ^{197}_{79}\text{Au} + ^{2}_{1}\text{X}$$

Therefore, X must be a deuteron, $^{2}_{1}\text{H}$ (or just d).

DISINTEGRATION ENERGY

Nuclear reactions not only produce new nuclei and other subatomic product particles, they also involve the absorption or emission of energy. Nuclear reactions must conserve total energy, so changes in mass are accompanied by changes in energy according to Einstein's equation

$$\Delta E = (\Delta m)c^2$$

A general nuclear reaction is written

$$\text{A} + \text{B} \rightarrow \text{C} + \text{D} + Q$$

where Q denotes the **disintegration energy**. If Q is positive, the reaction is **exothermic** (or **exoergic**) and the reaction can occur spontaneously; if Q is negative, the reaction is **endothermic** (or **endoergic**) and the reaction cannot occur spontaneously. The energy Q is calculated as follows:

$$Q = [(m_A + m_B) - (m_C + m_D)]c^2 = (\Delta m)c^2$$

For spontaneous reactions—ones that liberate energy—most of the energy is revealed as kinetic energy of the least massive product nuclei.

Example 17.10 The process that powers the Sun—and upon which all life on Earth is dependent—is the fusion reaction:

$$4\,^{1}_{1}\text{H} \rightarrow ^{4}_{2}\alpha + 2\,^{0}_{+1}\text{e} + 2\nu_e + \gamma$$

(a) Show that this reaction releases energy.
(b) How much energy is released per proton?

Use the fact that $m_\alpha = 4.0015$ u and ignore the mass of the electron-neutrino, ν_e.

Solution.

(a) We need to find the mass difference between the reactants and products:

$$\begin{aligned}\Delta m &= 4m_p - (m_\alpha + 2m_e) \\ &= 4(1.00728 \text{ u}) - [4.0015 \text{ u} + 2(0.0005486 \text{ u})] \\ &= 0.02652 \text{ u}\end{aligned}$$

Since Δm is positive, the reaction is exothermic: Energy is released.

(b) Converting the mass difference to energy gives

$$Q = 0.02652 \text{ u} \times \frac{931 \text{ MeV}}{1 \text{ u}} = 24.7 \text{ MeV}$$

Since four protons went into the reaction, the energy liberated per proton is

$$\frac{24.7 \text{ MeV}}{4 \text{ p}} = 6.2 \text{ MeV/proton}$$

Example 17.11 Can the following nuclear reaction occur spontaneously?

$${}^{4}_{2}\alpha + {}^{14}_{7}\text{N} \rightarrow {}^{17}_{8}\text{O} + {}^{1}_{1}\text{H}$$

(The mass of the nitrogen nucleus is 13.9992 u, and the mass of the oxygen nucleus is 16.9947 u.)

Solution. We first figure out the mass equivalent of the disintegration energy:

$$\begin{aligned}\Delta m &= (m_\alpha + m_N) - (m_O + m_p) \\ &= (4.0015 \text{ u} + 13.9992 \text{ u}) - (16.9947 \text{ u} + 1.00728 \text{ u}) \\ &= -0.00128 \text{ u}\end{aligned}$$

Since Δm is negative, this reaction is nonspontaneous; energy must be *supplied* in order for this reaction to proceed. But how much?

$$|Q| = 0.00128 \text{ u} \times \frac{931 \text{ MeV}}{1 \text{ u}} = 1.19 \text{ MeV}$$

CHAPTER 17 REVIEW QUESTIONS

Solutions can be found in Chapter 18.

Section I: Multiple Choice

1. What's the energy of a photon whose wavelength is 2.07 nm?

   (A) 60 eV
   (B) 600 eV
   (C) 960 eV
   (D) 6000 eV
   (E) 9600 eV

2. A metal whose work function is 6.0 eV is struck with light of frequency 7.2×10^{15} Hz. What is the maximum kinetic energy of photoelectrons ejected from the metal's surface?

   (A) 7 eV
   (B) 13 eV
   (C) 19 eV
   (D) 24 eV
   (E) No photoelectrons will be produced.

3. An atom with one electron has an ionization energy of 25 eV. How much energy will be released when the electron makes the transition from an excited energy level, where $E = -16$ eV, to the ground state?

   (A) 9 eV
   (B) 11 eV
   (C) 16 eV
   (D) 25 eV
   (E) 41 eV

4. The single electron in an atom has an energy of –40 eV when it's in the ground state, and the first excited state for the electron is at –10 eV. What will happen to this electron if the atom is struck by a stream of photons, each of energy 15 eV?

   (A) The electron will absorb the energy of one photon and become excited halfway to the first excited state, then quickly return to the ground state, without emitting a photon.
   (B) The electron will absorb the energy of one photon and become excited halfway to the first excited state, then quickly return to the ground state, emitting a 15 eV photon in the process.
   (C) The electron will absorb the energy of one photon and become excited halfway to the first excited state, then quickly absorb the energy of another photon to reach the first excited state.
   (D) The electron will absorb two photons and be excited to the first excited state.
   (E) Nothing will happen.

5. What is the de Broglie wavelength of a proton whose linear momentum has a magnitude of 3.3×10^{-23} kg·m/s?

   (A) 0.0002 nm
   (B) 0.002 nm
   (C) 0.02 nm
   (D) 0.2 nm
   (E) 2 nm

6. A partial energy-level diagram for an atom is shown below. What photon energies could this atom emit if it begins in the $n = 3$ state?

–3 eV	$n = 4$
–5 eV	$n = 3$
–8 eV	$n = 2$
–12 eV	$n = 1$ ground state

(A) 5 eV only
(B) 7 eV only
(C) 3 eV or 7 eV only
(D) 2 eV, 3 eV, or 7 eV
(E) 3 eV, 4 eV, or 7 eV

7. The ground-state energy level for He^+ is –54.4 eV. How much energy must the electron in the ground state of He^+ absorb in order to be excited to the next higher energy level?

(A) 13.6 eV
(B) 20.4 eV
(C) 27.2 eV
(D) 40.8 eV
(E) 68.0 eV

8. What would happen to the energy of a photon if its wavelength were reduced by a factor of 2?

(A) It would decrease by a factor of 4.
(B) It would decrease by a factor of 2.
(C) It would increase by a factor of 2.
(D) It would increase by a factor of 4.
(E) It would increase by a factor of $2h$.

9. In an exothermic nuclear reaction, the difference in mass between the reactants and the products is m, and the energy released is Q. In a separate exothermic nuclear reaction in which the mass difference between reactants and products is $m/4$, how much energy will be released?

(A) $Q/4$
(B) $Q/2$
(C) $(Q/4)c^2$
(D) $(Q/2)c^2$
(E) $4Q$

10. What's the missing particle in the following nuclear reaction?

$$^{2}_{1}\text{H} + ^{63}_{29}\text{Cu} \rightarrow ^{64}_{30}\text{Zn} + (?)$$

(A) proton
(B) neutron
(C) electron
(D) positron
(E) deuteron

11. What's the missing particle in the following nuclear reaction?

$$^{196}_{78}\text{Pt} + ^{1}_{0}\text{n} \rightarrow ^{197}_{78}\text{Pt} + (?)$$

(A) proton
(B) neutron
(C) electron
(D) positron
(E) gamma

SECTION II: FREE RESPONSE

1. The Bohr model of electron energy levels can be applied to any one-electron atom, such as doubly ionized lithium (Li^{2+}). The energy levels for the electron are given by the equation

$$E_n = \frac{Z^2}{n^2}(-13.6 \text{ eV})$$

where Z is the atomic number. The emission spectrum for Li^{2+} contains four spectral lines corresponding to the following wavelengths:

11.4 nm, 13.5 nm, 54.0 nm, 72.9 nm

(a) What's the value of Z for Li^{2+}?

(b) Identify which energy-level transitions give rise to the four wavelengths cited.

(c) Can the emission spectrum for Li^{2+} contain a line corresponding to a wavelength between 54.0 nm and 72.9 nm? If so, calculate its wavelength. If not, briefly explain.

(d) What is the next shortest wavelength in the emission spectrum (closest to, but shorter than, 11.4 nm)?

SUMMARY

- The energy available to liberate electrons near the surface of a metal (the photoelectric effect) is proportional to the frequency of the incident photon. This idea is expressed in $E = hf$ where h is Planck's constant. ($h = 6.63 \times 10^{-34}$J·s).

- The work function (ϕ) indicates the amount of energy needed to liberate the electron. There is a minimum frequency needed to liberate the electrons: $f_0 = \frac{\phi}{h}$. The kinetic energy of the emitted electron is $K_{max} = hf - \phi$.

- Particles in motion have wavelike properties: $\lambda = \frac{h}{p}$ where p is the momentum of the particle. Combining this with $E = hf$ and $c = f\lambda$ yields $E = pc$.

- Standard notation for an element is ${}^{A}_{Z}X$, where A is the mass number, Z is the number of protons in the nucleus, and $A = Z+N$, where N is the number of neutrons in the nucleus.

- A nuclear reaction produces new nuclei, other subatomic particles, and the absorption or emission of energy. The change in mass between the reactants and the products tells how much energy is released (exothermic or $+Q$) or how much energy is needed to produce the reaction (endothermic or $-Q$) in the general equation $A + B \rightarrow C + D + Q$, where A and B are reactants and C and D are products. This energy released or absorbed is given by $\Delta E = (\Delta m)c^2$.

18

Solutions to the Chapter Review Questions

CHAPTER 2 REVIEW QUESTIONS

SECTION I: MULTIPLE CHOICE

1. **A** Traveling once around a circular path means that the final position coincides with the initial position. Therefore, the displacement is zero. The average speed, which is *total* distance traveled divided by elapsed time, cannot be zero. Since the velocity changed (because its direction changed), there was a nonzero acceleration. Therefore, only Statement I is true.

2. **D** Section 1 represents a constant positive speed. Section 2 shows an object slowing down, moving in the positive direction. Section 3 represents an object speeding up in the negative direction. Section 4 demonstrates a constant positive speed, and section 5 represents the correct answer: slowing down moving in the negative direction.

3. **C** Statement I is false since a projectile experiencing only the constant acceleration due to gravity can travel in a parabolic trajectory. Statement II is true: Zero acceleration means no change in speed (or direction). Statement III is false: An object whose speed remains constant but whose velocity vector is changing direction is accelerating.

4. **C** The baseball is still under the influence of Earth's gravity. Its acceleration throughout the *entire* flight is constant, equal to g downward.

5. **A** Use Big Five #3 with $v_0 = 0$:

$$x = x_0 + v_0 t + \frac{1}{2}at^2 = \frac{1}{2}at^2 = \frac{1}{2}at^2 \quad \Rightarrow \quad t = \sqrt{\frac{2\Delta x}{a}} = \sqrt{\frac{2(200 \text{ m})}{5 \text{ m/s}^2}} = 9 \text{ s}$$

6. **D** Use Big Five #5 with $v_0 = 0$ (calling *down* the positive direction):

$$v^2 = v_0^2 + 2a(x - x_0) = 2a(x - x_0) \quad \Rightarrow \quad (x - x_0) = \frac{v^2}{2a} = \frac{v^2}{2g} = \frac{(30 \text{ m/s})^2}{2(10 \text{ m/s}^2)} = 45 \text{ m}$$

7. **C** Apply Big Five #3 to the vertical motion, calling *down* the positive direction:

$$\Delta y = v_{0y}t + \frac{1}{2}a_y t^2 = \frac{1}{2}a_y t^2 = \frac{1}{2}gt^2 \quad \Rightarrow \quad t = \sqrt{\frac{2\Delta y}{g}} = \sqrt{\frac{2(80 \text{ m})}{10 \text{ m/s}^2}} = 4 \text{ s}$$

Note that the stone's initial horizontal speed ($v_{0x} = 10$ m/s) is irrelevant.

8. **B** First we determine the time required for the ball to reach the top of its parabolic trajectory (which is the time required for the vertical velocity to drop to zero).

$$v_y \stackrel{\text{set}}{=} 0 \quad \Rightarrow \quad v_{0y} - gt = 0 \quad \Rightarrow \quad t = \frac{v_{0y}}{g}$$

The total flight time is equal to twice this value:

$$t_t = 2t = 2\frac{v_{0y}}{g} = 2\frac{v_0 \sin\theta_0}{g} = \frac{2(10 \text{ m/s})\sin 30°}{10 \text{ m/s}^2} = 1 \text{ s}$$

9. **C** After 4 seconds, the stone's vertical speed has changed by $\Delta v_y = a_y t = (10 \text{ m/s}^2)(4 \text{ s}) = 40 \text{ m/s}$. Since $v_{0y} = 0$, the value of v_y at $t = 4$ is 40 m/s. The horizontal speed does not change. Therefore, when the rock hits the water, its velocity has a horizontal component of 30 m/s and a vertical component of 40 m/s.

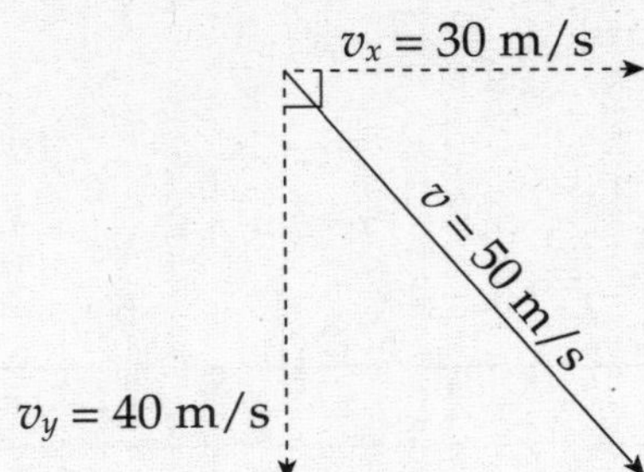

By the Pythagorean theorem, the magnitude of the total velocity, v, is 50 m/s.

10. **E** Since the acceleration of the projectile is always downward (because it's gravitational acceleration), the vertical speed decreases as the projectile rises and increases as the projectile falls. Statements (A), (B), (C), and (D) are all false.

Section II: Free Response

1. (a) At time $t = 1$ s, the car's velocity starts to decrease as the acceleration (which is the slope of the given v vs. t graph) changes from positive to negative.

(b) The average velocity between $t = 0$ and $t = 1$ s is $\frac{1}{2}(v_{t=0} + v_{t=1}) = \frac{1}{2}(0 + 20 \text{ m/s}) = 10 \text{ m/s}$, and the average velocity between $t = 1$ and $t = 5$ is $\frac{1}{2}(v_{t=1} + v_{t=5}) = \frac{1}{2}(20 \text{ m/s} + 0) = 10 \text{ m/s}$. The two average velocities are the same.

(c) The displacement is equal to the area bounded by the graph and the t axis, taking areas above the t axis as positive and those below as negative. In this case, the displacement from $t = 0$ to $t = 5$ s is equal to the area of the triangular region whose base is the segment along the t axis from $t = 0$ to $t = 5$ s:

$$\Delta x\ (t = 0 \text{ to } t = 5 \text{ s}) = \frac{1}{2} \times \text{base} \times \text{height} = \frac{1}{2}(5 \text{ s})(20 \text{ m/s}) = 50 \text{ m}$$

The displacement from $t = 5$ s to $t = 7$ s is equal to the negative of the area of the triangular region whose base is the segment along the t axis from $t = 5$ s to $t = 7$ s:

$$\Delta x\ (t = 5 \text{ s to } t = 7 \text{ s}) = -\frac{1}{2} \times \text{base} \times \text{height} = -\frac{1}{2}(2 \text{ s})(10 \text{ m/s}) = -10 \text{ m}$$

Therefore, the displacement from $t = 0$ to $t = 7$ s is

$$\Delta x\ (t = 0 \text{ to } t = 5 \text{ s}) + \Delta s\ (t = 5 \text{ s to } t = 7 \text{ s}) = 50 \text{ m} + (-10 \text{ m}) = 40 \text{ m}$$

(d) The acceleration is the slope of the v vs. t graph. The segment of the graph from $t = 0$ to $t = 1$ s has a slope of $a = \Delta v/\Delta t = (20 \text{ m/s} - 0)/(1 \text{ s} - 0) = 20 \text{ m/s}^2$, and the segment of the graph from $t = 1$ s to $t = 7$ s has a slope of $a = \Delta v/\Delta t = (-10 \text{ m/s} - 20 \text{ s})/(7 \text{ s} - 1 \text{ s}) = -5 \text{ m/s}^2$. Therefore, the graph of a vs. t is

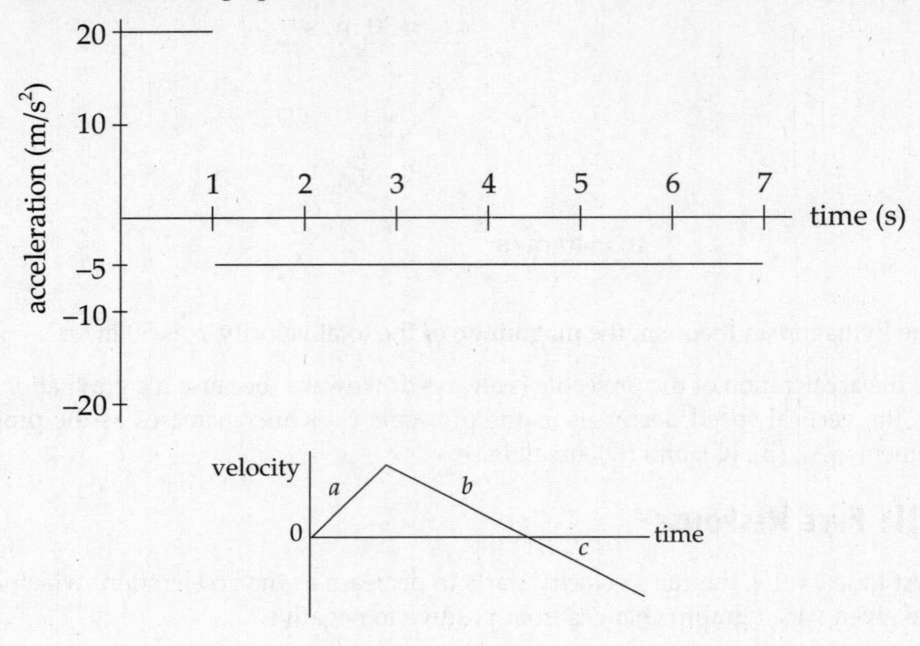

Section a tells us the object is speeding up in the positive direction. Section b tells us the object is slowing down, yet still moving in the positive direction. At five seconds, the object has stopped for an instant. Section c tells us the object is moving in the negative direction and speeding up. The corresponding position-vs.-time graph for each section would look like this:

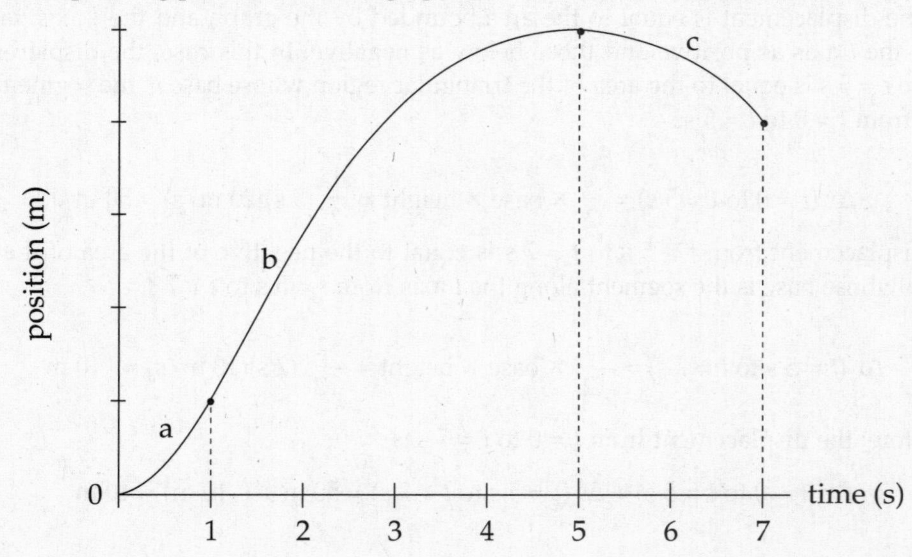

2. (a) The maximum height of the projectile occurs at the time at which its vertical velocity drops to zero:

$$v_y \stackrel{\text{set}}{=} 0 \;\Rightarrow\; v_{0y} - gt = 0 \;\Rightarrow\; t = \frac{v_{0y}}{g}$$

The vertical displacement of the projectile at this time is computed as follows:

$$\Delta y = v_{0y}t - \frac{1}{2}gt^2 \;\Rightarrow\; H = v_{0y}\frac{v_{0y}}{g} - \frac{1}{2}g\left(\frac{v_{0y}}{g}\right)^2 = \frac{v_{0y}^2}{2g} = \frac{v_0^2 \sin^2\theta_0}{2g}$$

(b) The total flight time is equal to twice the time computed in part (a):

$$t_t = 2t = 2\frac{v_{0y}}{g}$$

The horizontal displacement at this time gives the projectile's range:

$$\Delta x = v_{0x}t \;\Rightarrow\; R = v_{0x}t_t = \frac{v_{0x}\cdot 2v_{0y}}{g} = \frac{2v_0^2 \sin\theta_0 \cos\theta_0}{g} \text{ or } \frac{v_0^2 \sin 2\theta_0}{g}$$

(c) For any given value of v_0, the range,

$$\Delta x = v_{0x}t \;\Rightarrow\; R = v_{0x}t_t = \frac{v_{0x}\cdot 2v_{0y}}{g} = \frac{2v_0^2 \sin\theta_0 \cos\theta_0}{g} \text{ or } \frac{v_0^2 \sin 2\theta_0}{g}$$

will be maximized when $\sin 2\theta_0$ is maximized. This occurs when $2\theta_0 = 90°$, that is, when $\theta_0 = 45°$.

(d) Set the general expression for the projectile's vertical displacement equal to h and solve for the two values of t (assuming that $g = +10\text{ m/s}^2$):

$$v_{0y}t - \frac{1}{2}gt^2 \stackrel{\text{set}}{=} h \;\Rightarrow\; \frac{1}{2}gt^2 - v_{0y}t + h = 0$$

Applying the quadratic formula, we find that

$$t = \frac{v_{0y} \pm \sqrt{(-v_{0y})^2 - 4(\frac{1}{2}g)(h)}}{2(\frac{1}{2}g)} = \frac{v_{0y} \pm \sqrt{v_{0y}^2 - 2gh}}{g}$$

Therefore, the two times at which the projectile crosses the horizontal line at height h are

$$t_1 = \frac{v_{0y} - \sqrt{v_{0y}^2 - 2gh}}{g} \quad \text{and} \quad t_2 = \frac{v_{0y} + \sqrt{v_{0y}^2 - 2gh}}{g}$$

so the amount of time that elapses between these events is

$$\Delta t = t_2 - t_1 = \frac{2\sqrt{v_{0y}^2 - 2gh}}{g}$$

3. (a) The cannonball will certainly reach the wall (which is only 220 m away) since the ball's range is

$$R = \frac{v_0^2 \sin 2\theta_0}{g} = \frac{(50 \text{ m/s})^2 \sin 2(40°)}{9.8 \text{ m/s}^2} = 251 \text{ m}$$

We simply need to make sure that the cannonball's height is less than 30 m at the point where its horizontal displacement is 220 m (so that the ball actually hits the wall rather than flying over it). To do this, we find the time at which $x = 220$ m by first writing

$$x = v_{0x}t \quad \Rightarrow \quad t = \frac{x}{v_{0x}} = \frac{x}{v_0 \cos\theta_0} \qquad (1)$$

Thus, the cannonball's vertical position can be written in terms of its horizontal position as follows:

$$y = v_{0y}t - \tfrac{1}{2}gt^2 = v_0 \sin\theta_0 \frac{x}{v_0 \cos\theta_0} - \tfrac{1}{2}g\left(\frac{x}{v_0 \cos\theta_0}\right)^2$$

$$= x \tan\theta_0 - \frac{gx^2}{2v_0^2 \cos^2\theta_0} \qquad (2)$$

Substituting the known values for x, θ_0, g, and v_0, we get

$$y(\text{at } x = 220 \text{ m}) = (220 \text{ m})\tan 40° - \frac{(9.8 \text{ m/s}^2)(220 \text{ m})^2}{2(50 \text{ m/s})^2 \cos^2 40°}$$

$$= 23 \text{ m}$$

This is indeed less than 30 m, as desired.

(b) From Equation (1) derived in part (a),

$$t = \frac{x}{v_0 \cos\theta_0} = \frac{220 \text{ m}}{(50 \text{ m/s})\cos 40°} = 5.7 \text{ s}$$

(c) The height at which the cannonball strikes the wall was determined in part (a) to be 23 m.

CHAPTER 3 REVIEW QUESTIONS

SECTION I: MULTIPLE CHOICE

1. **B** Because the person is not accelerating, the net force he feels must be zero. Therefore, the magnitude of the upward normal force from the floor must balance that of the downward gravitational force. Although these two forces have equal magnitudes, they do not form an action/reaction pair because they both act on the same object (namely, the person). The forces in an action/reaction pair always act on different objects.

2. **D** First draw a free-body diagram:

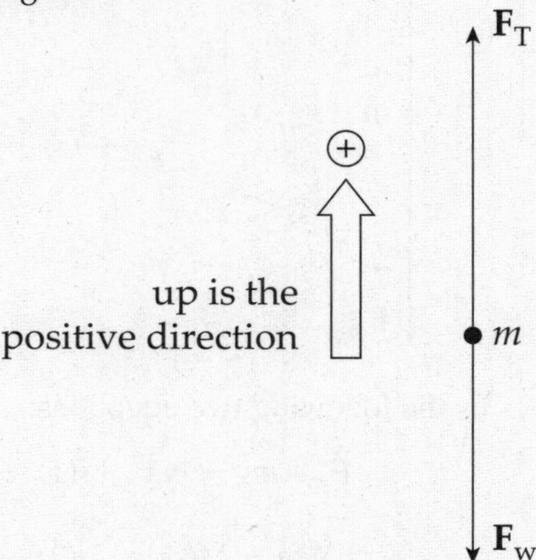

The person exerts a downward force on the scale, and the scale pushes up on the person with an equal (but opposite) force, $\mathbf{F}_N$. Thus, the scale reading is F_N, the magnitude of the normal force. Since $F_N - F_w = ma$, we have $F_N = F_w + ma = (800\text{ N}) + [800\text{ N}/(10\text{ m/s}^2)](5\text{ m/s}^2) = 1200\text{ N}$.

3. **A** The net force that the object feels on the inclined plane is $mg \sin \theta$, the component of the gravitational force that is parallel to the ramp. Since $\sin \theta = (5\text{ m})/(20\text{ m}) = 1/4$, we have $F_{net} = (2\text{ kg})(10\text{ N/kg})(1/4) = 5\text{ N}$.

4. **C** The net force on the block is $F - F_f = F - \mu_k F_N = F - \mu_k F_w = (18\text{ N}) - (0.4)(20\text{ N}) = 10\text{ N}$. Since $F_{net} = ma = (F_w/g)a$, we find that $10\text{ N} = [(20\text{ N})/(10\text{ m/s}^2)]a$, which gives $a = 5\text{ m/s}^2$.

5. **A** The force pulling the block down the ramp is $mg \sin \theta$, and the maximum force of static friction is $\mu_s F_N = \mu_s mg \cos \theta$. If $mg \sin \theta$ is greater than $\mu_s mg \cos \theta$, then there is a net force down the ramp, and the block will accelerate down. So, the question becomes, "Is $\sin \theta$ greater than $\mu_s \cos \theta$?" Since $\theta = 30°$ and $\mu_s = 0.5$, the answer is "yes."

6. **E** One way to attack this question is to notice that if the two masses happen to be equal, that is, if $M = m$, then the blocks won't accelerate (because their weights balance). The only expression given that becomes zero when $M = m$ is the one given in choice (E). If we draw a free-body diagram,

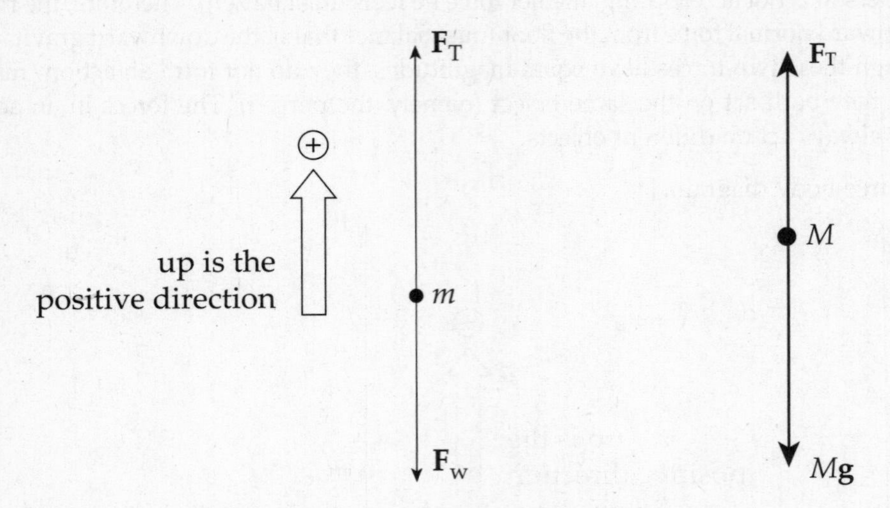

Newton's Second Law gives us the following two equations:

$$F_T - mg = ma \qquad (1)$$

$$F_T - Mg = M(-a) \qquad (2)$$

Subtracting these equations yields $Mg - mg = ma + Ma = (M + m)a$, so

$$a = \frac{Mg - mg}{M + m} = \frac{M - m}{M + m}g$$

7. **E** If $\mathbf{F}_{net} = \mathbf{0}$, then $\mathbf{a} = \mathbf{0}$. No acceleration means constant speed (possibly, but not necessarily, zero) with no change in direction. Therefore, statements (B), (C), and (D) are false, and statement (A) is not necessarily true.

8. **D** The horizontal motion across the frictionless tables is unaffected by (vertical) gravitational acceleration. It would take as much force to accelerate the block across the table on Earth as it would on the Moon. (If friction *were* taken into account, then the smaller weight of the block on the Moon would imply a smaller normal force by the table and hence a smaller frictional force. Less force would be needed on the Moon in this case.)

9. **D** The maximum force which static friction can exert on the crate is $\mu_s F_N = \mu_s F_w = \mu_s mg =$ (0.4)(100 kg)(10 N/kg) = 392 N. Since the force applied to the crate is only 344 N, static friction is able to apply that same magnitude of force on the crate, keeping it stationary. [Choice (B) is incorrect because the static friction force is *not* the reaction force to **F**; both **F** and $\mathbf{F}_{f\,(static)}$ act on the same object (the crate) and therefore cannot form an action/reaction pair.]

10. **A** With Crate #2 on top of Crate #1, the force pushing downward on the floor is greater, so the normal force exerted by the floor on Crate #1 is greater, which increases the friction force. Statements (B), (C), (D), and (E) are all false.

SECTION II: FREE RESPONSE

1. (a) The forces acting on the crate are $\mathbf{F}_T$ (the tension in the rope), $\mathbf{F}_w$ (the weight of the block), $\mathbf{F}_N$ (the normal force exerted by the floor), and $\mathbf{F}_f$ (the force of kinetic friction):

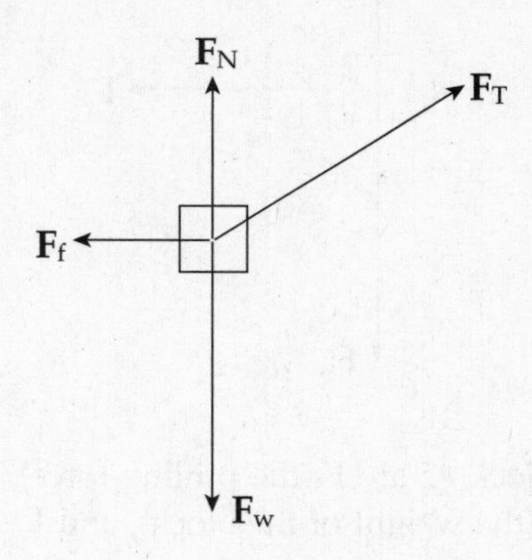

(b) First break $\mathbf{F}_T$ into its horizontal and vertical components:

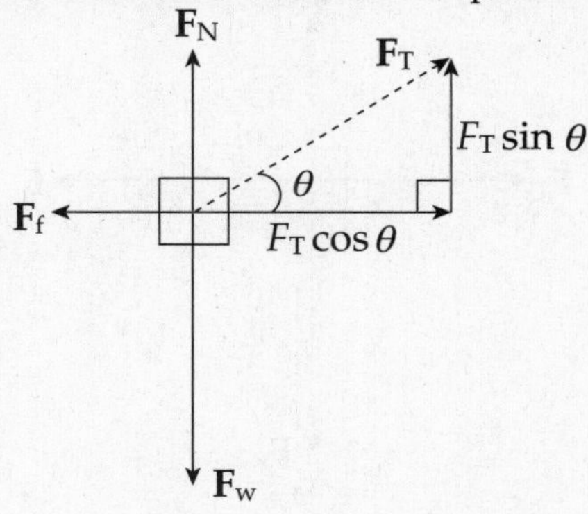

Since the net vertical force on the crate is zero, we have $F_N + F_T \sin\theta = F_w$, so $F_N = F_w - F_T \sin\theta = mg - F_T \sin\theta$.

(c) From part (b), we see that the net horizontal force acting on the crate is

$$F_T \cos\theta - F_f = F_T \cos\theta - \mu F_N = F_T \cos\theta - \mu(mg - F_T \sin\theta)$$

so the crate's horizontal acceleration across the floor is

$$a = \frac{F_{net}}{m} = \frac{F_T \cos\theta - \mu(mg - F_T \sin\theta)}{m}$$

2. (a) The forces acting on Block #1 are $\mathbf{F}_T$ (the tension in the string connecting it to Block #2), $\mathbf{F}_{w1}$ (the weight of the block), and $\mathbf{F}_{N1}$ (the normal force exerted by the tabletop):

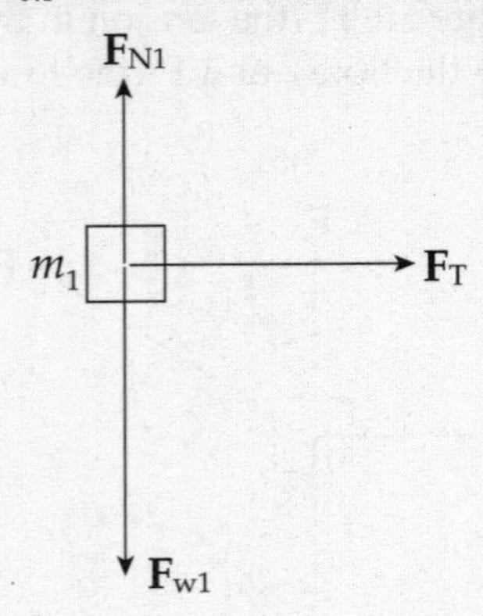

(b) The forces acting on Block #2 are **F** (the pulling force), $\mathbf{F}_T$ (the tension in the string connecting it to Block #1), $\mathbf{F}_{w2}$ (the weight of the block), and $\mathbf{F}_{N2}$ (the normal force exerted by the tabletop):

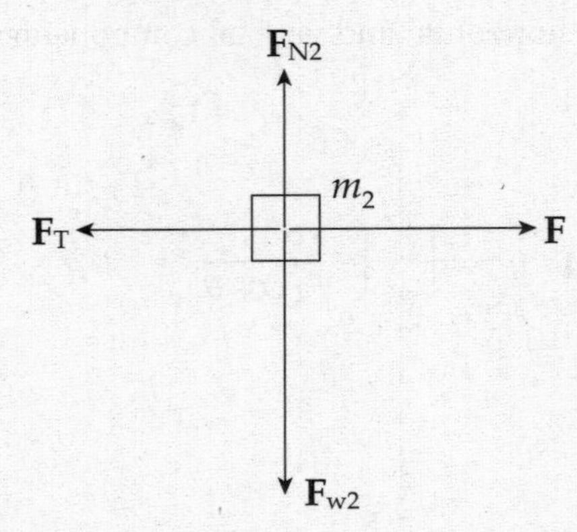

(c) Newton's Second Law applied to Block #2 yields $F - F_T = m_2a$, and applied to Block #1 yields $F_T = m_1a$. Adding these equations, we find that $F = (m_1 + m_2)a$, so

$$a = \frac{F}{m_1 + m_2}$$

(d) Substituting the result of part (c) into the equation $F_T = m_1a$, we get

$$F_T = m_1a = \frac{m_1}{m_1 + m_2}F$$

(e) (i) Since the force **F** must accelerate all three masses—m_1, m, and m_2—the common acceleration of all parts of the system is

$$a = \frac{F}{m_1 + m + m_2}$$

(e) (ii) Let $\mathbf{F}_{T1}$ denote the tension force in the connecting string acting on Block #1, and let $\mathbf{F}_{T2}$ denote the tension force in the connecting string acting on Block #2. Then, Newton's Second Law applied to Block #1 yields $F_{T1} = m_1a$ and applied to Block #2 yields $F - F_{T2} = m_2a$. Therefore, using the value for a computed above, we get

$$\begin{aligned} F_{T2} - F_{T1} &= (F - m_2a) - m_1a \\ &= F - (m_1 + m_2)a \\ &= F - (m_1 + m_2)\frac{F}{m_1 + m + m_2} \\ &= F\left(1 - \frac{m_1 + m_2}{m_1 + m + m_2}\right) \\ &= F\frac{m}{m_1 + m + m_2} \end{aligned}$$

3. (a) First draw free-body diagrams for the two boxes:

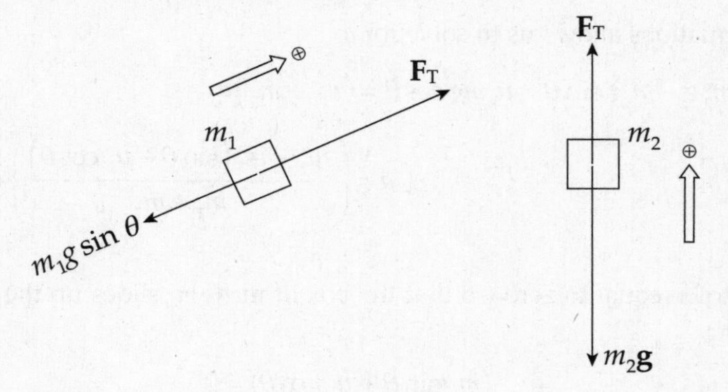

Applying Newton's Second Law to the boxes yields the following two equations:

$$F_T - m_1g \sin \theta = m_1a \qquad (1)$$

$$F_T - m_2g = m_2(-a) \qquad (2)$$

Subtracting the equations allows us to solve for a:

$$m_2g - m_1g \sin \theta = (m_1 + m_2)a$$

$$a = \frac{m_2 - m_1 \sin \theta}{m_1 + m_2} g$$

(i) For a to be positive, we must have $m_2 - m_1 \sin \theta > 0$, which implies that $\sin \theta < m_2/m_1$, or, equivalently, $\theta < \sin^{-1}(m_2/m_1)$.

(ii) For a to be zero, we must have $m_2 - m_1 \sin \theta = 0$, which implies that $\sin \theta = m_2/m_1$, or, equivalently, $\theta = \sin^{-1}(m_2/m_1)$.

(b) Including the force of kinetic friction, the force diagram for m_1 is

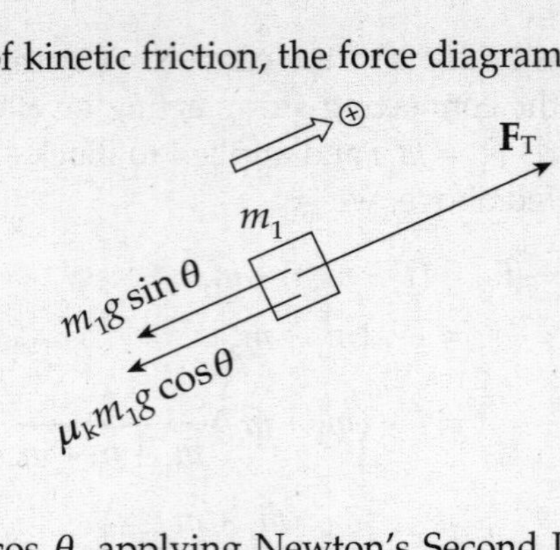

Since $F_f = \mu_k F_N = \mu_k m_1 g \cos\theta$, applying Newton's Second Law to the boxes yields these two equations:

$$F_T - m_1 g \sin\theta - \mu_k mg \cos\theta = m_1 a \qquad (1)$$

$$m_2 g - F_T = m_2 a \qquad (2)$$

Adding the equations allows us to solve for a:

$$m_2 g - m_1 g \sin\theta - \mu_k mg \cos\theta = (m_1 + m_2)a$$

$$a = \left(\frac{m_2 - m_1(\sin\theta + \mu_k \cos\theta)}{m_1 + m_2}\right)g$$

If we want a to be equal to zero (so that the box of mass m_1 slides up the ramp with constant velocity), then

$$m_2 - m_1(\sin\theta + \mu_k \cos\theta) = 0$$

$$\sin\theta + \mu_k \cos\theta = \frac{m_2}{m_1}$$

4. (a) The forces acting on the sky diver are $\mathbf{F}_r$, the force of air resistance (upward), and $\mathbf{F}_w$, the weight of the sky diver (downward):

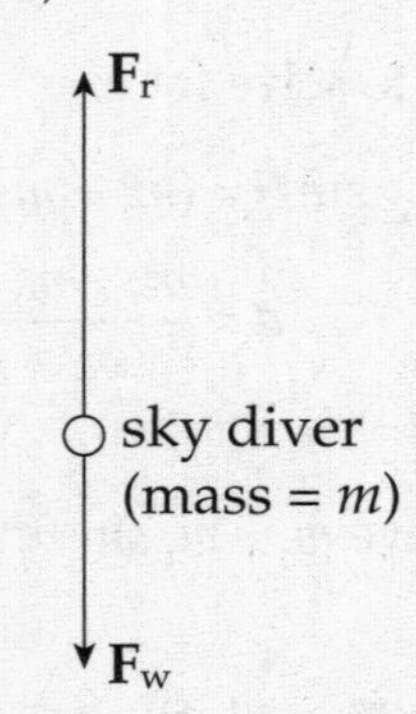

(b) Since $F_{net} = F_w - F_r = mg - kv$, the sky diver's acceleration is

$$a = \frac{F_{net}}{m} = \frac{mg - kv}{m}$$

(c) Terminal speed occurs when the sky diver's acceleration becomes zero, since then the descent velocity becomes constant. Setting the expression derived in part (b) equal to 0, we find the speed $v = v_t$ at which this occurs:

$$v = v_t \text{ when } a = 0 \quad \Rightarrow \quad \frac{mg - kv_t}{m} = 0 \quad \Rightarrow \quad v_t = \frac{mg}{k}$$

(d) The sky diver's descent speed is initially v_0 and the acceleration is (close to) g. However, once the parachute opens, the force of air resistance provides a large (speed-dependent) upward acceleration, causing her descent velocity to decrease. The slope of the v vs. t graph (the acceleration) is not constant but instead decreases to zero as her descent speed decreases from v_0 to v_t. Therefore, the graph is not linear.

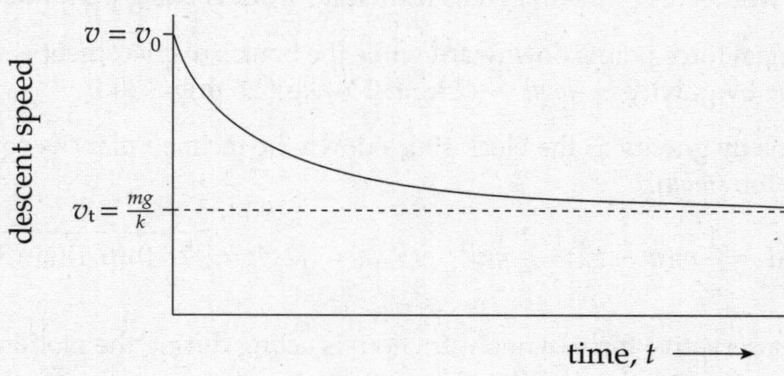

CHAPTER 4 REVIEW QUESTIONS

SECTION I: MULTIPLE CHOICE

1. **A** Since the force **F** is perpendicular to the displacement, the work it does is zero.

2. **B** By the work–energy theorem,

$$W = \Delta K = \frac{1}{2}m(v^2 - v_0^2) = \frac{1}{2}(4\text{ kg})[(6\text{ m/s})^2 - (3\text{ m/s})^2] = 54\text{ J}$$

3. **B** Since the box (mass m) falls through a vertical distance of h, its gravitational potential energy decreases by mgh. The length of the ramp is irrelevant here.

4. **C** Since the centripetal force always points along a radius toward the center of the circle, and the velocity of the object is always tangent to the circle (and thus perpendicular to the radius), the work done by the centripetal force is zero. Alternatively, since the object's speed remains constant, the work–energy theorem tells us that no work is being performed.

5. **A** The gravitational force points downward while the book's displacement is upward. Therefore, the work done by gravity is $-mgh = -(2\text{ kg})(10\text{ N/kg})(1.5\text{ m}) = -30\text{ J}$.

6. **D** The work done by gravity as the block slides down the inclined plane is equal to the potential energy at the top (mgh).

$$mgh = W = \Delta K = \frac{1}{2}m(v^2 - v_0^2) = \frac{1}{2}mv^2 \quad \Rightarrow \quad v = \sqrt{2gh} = \sqrt{2(10)(6.4)(\sin 30^\circ)} = 8\text{ m/s}$$

7. **D** Since a nonconservative force (namely, friction) is acting during the motion, we use the modified Conservation of Mechanical Energy equation.

m F $\Leftarrow U_i = mgh$
s h
$U_f = 0 \Rightarrow$

$$K_i + U_i + W_{\text{friction}} = K_i + U_f$$
$$0 + mgh - Fs = K_f + 0$$
$$mgh - Fs = K_f$$

8. **E** Apply Conservation of Mechanical Energy (including the negative work done by $\mathbf{F}_r$, the force of air resistance):

$$K_i + U_i + W_r = K_f + U_f$$
$$0 + mgh - F_r h = \frac{1}{2}mv^2 + 0$$
$$v = \sqrt{\frac{2h(mg - F_r)}{m}}$$
$$= \sqrt{\frac{2(40\text{ m})[(4\text{ kg})(10\text{ N/kg}) - 20\text{ N}]}{4\text{ kg}}}$$
$$= 20\text{ m/s}$$

9. **E** Because the rock has lost half of its gravitational potential energy, its kinetic energy at the halfway point is half of its kinetic energy at impact. Since K is proportional to v^2, if $K_{\text{at halfway point}}$ is equal to $\frac{1}{2}K_{\text{at impact}}$, then the rock's speed at the halfway point is $\sqrt{1/2} = 1/\sqrt{2}$ its speed at impact.

10. **D** Using the equation $P = Fv$, we find that $P = (200\text{ N})(2\text{ m/s}) = 400\text{ W}$.

SECTION II: FREE RESPONSE

1. (a) Applying Conservation of Energy,

$$K_A + U_A = K_{\text{at } H/2} + U_{\text{at } H/2}$$
$$0 + mgH = \frac{1}{2}mv^2 + mg(\frac{1}{2}H)$$
$$\frac{1}{2}mgH = \frac{1}{2}mv^2$$
$$v = \sqrt{gH}$$

(b) Applying Conservation of Energy again,

$$K_A + U_A = K_B + U_B$$
$$0 + mgH = \frac{1}{2}mv_B^2 + 0$$
$$v_B = \sqrt{2gH}$$

(c) By the work–energy theorem, we want the work done by friction to be equal (but opposite) to the kinetic energy of the box at Point B:

$$W = \Delta K = \frac{1}{2}m(v_C^2 - v_B^2) = -\frac{1}{2}mv_B^2 = -\frac{1}{2}m(\sqrt{2gH})^2 = -mgH$$

Therefore,

$$W = -mgH \;\Rightarrow\; -F_f x = -mgH \;\Rightarrow\; -\mu_k mgx = -mgH \;\Rightarrow\; \mu_k = H/x$$

(d) Apply Conservation of Energy (including the negative work done by friction as the box slides up the ramp from B to C):

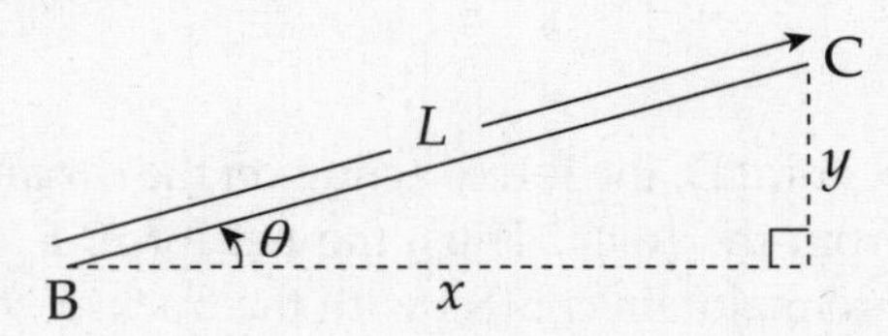

$$K_B + U_B + W_f = K_C + U_C$$

$$\frac{1}{2}m(\sqrt{2gH})^2 + 0 - F_f L = 0 + mgy$$

$$mgH + 0 - F_f L = 0 + mgy$$

$$mg(H - y) - (\mu_k mg\cos\theta)(L) = 0$$

$$\mu_k = \frac{H - y}{L\cos\theta} = \frac{H - y}{x}$$

(e) The result of part (b) reads $v_B = \sqrt{2gH}$. Therefore, by Conservation of Mechanical Energy (with the work done by the frictional force on the slide included), we get

$$K_A + U_A + W_f = K_B' + U_B$$

$$0 + mgH + W_f = \frac{1}{2}m\left(\frac{1}{2}v_B\right)^2 + 0$$

$$mgH + W_f = \frac{1}{2}m\left(\frac{1}{2}\sqrt{2gH}\right)^2$$

$$mgH + W_f = \frac{1}{4}mgH$$

$$W_f = -\frac{3}{4}mgH$$

2. (a) The centripetal acceleration of the car at Point C is given by the equation $a = v_C^2/r$, where v_C is the speed of the car at C. To find v_C^2, we apply Conservation of Energy:

$$K_A + U_A = K_C + U_C$$

$$0 + mgH = \frac{1}{2}mv_C^2 + mgr$$

$$mg(H - r) = \frac{1}{2}mv_C^2$$

$$v_C^2 = 2g(H - r)$$

Therefore,

$$a_c = \frac{v_C^2}{r} = \frac{2g(H - r)}{r}$$

(b) When the car reaches Point D, the forces acting on the car are its weight, $\mathbf{F}_w$, and the downward normal force, $\mathbf{F}_N$, from the track. Thus, the net force, $\mathbf{F}_w + \mathbf{F}_N$, provides the centripetal force. In order for the car to maintain contact with the track, $\mathbf{F}_N$ must not vanish. Therefore, the cut-off speed for ensuring that the car makes it safely around the track is the speed at which $\mathbf{F}_N$ just becomes zero; any greater speed would imply that the car would make it around. Thus,

$$F_w + F_N = m\frac{v^2}{r} \Rightarrow F_w + 0 = m\frac{v_{\text{cut-off}}^2}{r} \Rightarrow v_{\text{cut-off}} = \sqrt{\frac{rF_w}{m}} = \sqrt{gr}$$

(c) Using the cut-off speed calculated in part (c), we now apply Conservation of Mechanical Energy:

$$
\begin{aligned}
K_A + U_A &= K_D + U_D \\
0 + mgH &= \frac{1}{2}mv_{\text{cut-off}}^2 + mg(2r) \\
mgH &= \frac{1}{2}m(gr) + mg(2r) \\
&= \frac{5}{2}mgr \\
H &= \frac{5}{2}r
\end{aligned}
$$

(d) First, we calculate the car's kinetic energy at Point B; then, we determine the distance x the car must travel from B to F for the work done by friction to eliminate this kinetic energy. So, applying Conservation of Mechanical Energy, we find

$$
\begin{aligned}
K_A + U_A &= K_B + U_B \\
0 + mg(6r) &= \frac{1}{2}mv_B^2 + 0 \\
6mgr &= \frac{1}{2}mv_B^2
\end{aligned}
$$

Now, by the work–energy theorem,

$$
W = \Delta K = \frac{1}{2}mv_F^2 - \frac{1}{2}mv_B^2 = -\frac{1}{2}mv_B^2 \quad \Rightarrow \quad
\begin{aligned}
-F_f x &= -\frac{1}{2}mv_B^2 \\
-\mu mgx &= -6mgr \\
x &= \frac{6r}{\mu} = \frac{6r}{0.5} = 12r
\end{aligned}
$$

3. (a) Using conservation of energy $K_i + U_i = K_f + U_f$ and $v_i = 0$ this becomes $U_i = K_f + U_f$ or $K_f = U_i - U_f$. This is equivalent to $\frac{1}{2}mv^2 = mgh_i - mgh_f$, which simplifies to $\frac{1}{2}v^2 - gh_i = -gh_f$ or $h_f = h_i - \frac{v^2}{2g}$. Now we can fill in the table on the next page.

Time (s)	Velocity (m/s)	Height (m)
0.00	0.00	1.5
0.05	1.41	1.4
0.10	2.45	1.2
0.15	3.74	0.8
0.20	3.74	0.8
0.25	3.46	0.9
0.30	3.16	1
0.35	2.83	1.1
0.40	3.46	0.9
0.45	4.24	0.6
0.50	4.47	0.5

(b) The greatest acceleration would occur where there is the greatest change in velocity. This occurs between 0.00 and 0.05 seconds. The acceleration during that time interval is given by

$$a = \frac{v_f - v_i}{t_f - t_i} \Rightarrow \frac{1.41 - 0}{0.05 - 0.00} \text{ or } a = 2.8 \text{ m/s}^2.$$

(c) Changing the mass does not effect the time spent falling or the velocity of the object. Thus, a change in mass will not affect the results.

CHAPTER 5 REVIEW QUESTIONS

Section I: Multiple Choice

1. **C** The magnitude of the object's linear momentum is $p = mv$. If $p = 6$ kg · m/s and $m = 2$ kg, then $v = 3$ m/s. Therefore, the object's kinetic energy is $K = \frac{1}{2}mv^2 = \frac{1}{2}(2 \text{ kg})(3 \text{ m/s})^2 = 9$ J.

2. **C** The impulse delivered to the ball, $J = F\Delta t$, equals its change in momentum. Since the ball started from rest, we have

$$F\Delta t = mv \quad \Rightarrow \quad \Delta t = \frac{mv}{F} = \frac{(0.5 \text{ kg})(4 \text{ m/s})}{20 \text{ N}} = 0.1 \text{ s}$$

3. **E** The impulse delivered to the ball, $J = \bar{F}\Delta t$, equals its change in momentum. Thus,

$$\bar{F}\Delta t = \Delta p = p_f - p_i = m(v_f - v_i) \quad \Rightarrow \quad \bar{F} = \frac{m(v_f - v_i)}{\Delta t} = \frac{(2 \text{ kg})(8 \text{ m/s} - 4 \text{ m/s})}{0.5 \text{ s}} = 16 \text{ N}$$

4. **D** The impulse delivered to the ball is equal to its change in momentum. The momentum of the ball was $m\mathbf{v}$ before hitting the wall and $m(-\mathbf{v})$ after. Therefore, the change in momentum is $m(-\mathbf{v}) - m\mathbf{v} = -2m\mathbf{v}$, so the magnitude of the momentum change (and the impulse) is $2mv$.

5. **B** By definition of *perfectly inelastic*, the objects move off together with one common velocity, $\mathbf{v}'$, after the collision. By Conservation of Linear Momentum,

$$\begin{aligned} m_1\mathbf{v}_1 + m_2\mathbf{v}_2 &= (m_1 + m_2)\mathbf{v}' \\ \mathbf{v}' &= \frac{m_1v_1 + m_2v_2}{m_1 + m_2} \\ &= \frac{(3 \text{ kg})(2 \text{ m/s}) + (5 \text{ kg})(-2 \text{ m/s})}{3 \text{ kg} + 5 \text{ kg}} \\ &= -0.5 \text{ m/s} \end{aligned}$$

6. **D** First, apply Conservation of Linear Momentum to calculate the speed of the combined object after the (perfectly inelastic) collision:

$$\begin{aligned} m_1v_1 + m_2v_2 &= (m_1 + m_2)v' \\ v' &= \frac{m_1v_1 + m_2v_2}{m_1 + m_2} \\ &= \frac{m_1v_1 + (2m_1)(0)}{m_1 + 2m_1} \\ &= \frac{1}{3}v_1 \end{aligned}$$

Therefore, the ratio of the kinetic energy after the collision to the kinetic energy before the collision is

$$\frac{K'}{K} = \frac{\frac{1}{2}m'v'^2}{\frac{1}{2}m_1v_1^2} = \frac{\frac{1}{2}(m_1 + 2m_1)(\frac{1}{3}v_1)^2}{\frac{1}{2}m_1v_1^2} = \frac{1}{3}$$

7. **C** Total linear momentum is conserved in a collision during which the net external force is zero. If kinetic energy is lost, then by definition, the collision is not elastic.

8. **E** Because the two carts are initially at rest, the initial momentum is zero. Therefore, the final total momentum must be zero.

9. **D** The linear momentum of the bullet must have the same magnitude as the linear momentum of the block in order for their combined momentum after impact to be zero. The block has momentum MV to the left, so the bullet must have momentum MV to the right. Since the bullet's mass is m, its speed must be $v = MV/m$.

10. **C** In a perfectly inelastic collision, kinetic energy is never conserved; some of the initial kinetic energy is always lost to heat and some is converted to potential energy in the deformed shapes of the objects as they lock together.

SECTION II: FREE RESPONSE

1. (a) First draw a free-body diagram:

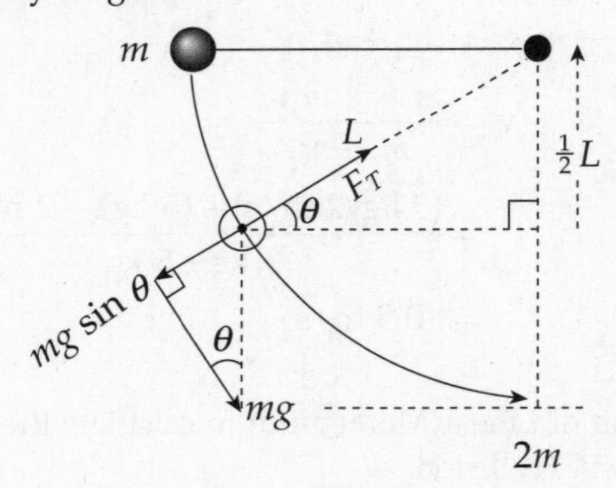

The net force toward the center of the steel ball's circular path provides the centripetal force. From the geometry of the diagram, we have

$$F_T - mg\sin\theta = \frac{mv^2}{L} \qquad (*)$$

In order to determine the value of mv^2, we use Conservation of Mechanical Energy:

$$K_i + U_i = K_f + U_f$$

$$0 + mgL = \frac{1}{2}mv^2 + mg(\frac{1}{2}L)$$

$$\frac{1}{2}mgL = \frac{1}{2}mv^2$$

$$mgL = mv^2$$

Substituting this result into Equation (*), we get

$$F_T - mg\sin\theta = \frac{mgL}{L}$$

$$F_T = mg(1+\sin\theta)$$

Now, from the free-body diagram we see that $\sin\theta = \frac{1}{2}L/L = \frac{1}{2}$, so

$$F_T = mg(1+\frac{1}{2}) = \frac{3}{2}mg$$

(b) Applying Conservation of Energy, we find the speed of the ball just before impact:

$$K_i + U_i = K_f + U_f$$

$$0 + mgL = \frac{1}{2}mv^2 + 0$$

$$v = \sqrt{2gL}$$

We can now use Conservation of Linear Momentum and the fact that kinetic energy is conserved to derive expressions for the speeds of the ball and block immediately after their collision. Since the collision is elastic, head-on, and the target object is at rest, this was done in Example 5.8. The velocity of the block after the collision is

$$v_2' = \frac{2m_1}{m_1+m_2}v_1 = \frac{2m}{m+4m}\sqrt{2gL} = \frac{2}{5}\sqrt{2gL}$$

(c) We can quote the result of Example 5.8 to find the velocity of the ball immediately after the collision:

$$v_1' = \frac{m_1-m_2}{m_1+m_2}v_1 = \frac{m-4m}{m+4m}\sqrt{2gL} = -\frac{3}{5}\sqrt{2gL}$$

Now, applying Conservation of Mechanical Energy, we find

$$K_i + U_i = K_f + U_f$$

$$\frac{1}{2}mv_1'^2 + 0 = 0 + mgh$$

$$h = \frac{v_1'^2}{2g} = \frac{\left(\frac{3}{5}\sqrt{2gL}\right)^2}{2g} = \frac{9}{25}L$$

2. (a) By Conservation of Linear Momentum, $mv = (m + M)v'$, so $v' = \dfrac{mv}{m+M}$

Now, by Conservation of Mechanical Energy,

$$K_i + U_i = K_f + U_f$$

$$\frac{1}{2}(m+M)v'^2 + 0 = 0 + (m+M)gy$$

$$\frac{1}{2}v'^2 = gy$$

$$\frac{1}{2}\left(\frac{mv}{m+M}\right)^2 = gy$$

$$v = \frac{m+M}{m}\sqrt{2gy}$$

(b) Use the result derived in part (a) to compute the kinetic energy of the block and bullet immediately after the collision:

$$K' = \frac{1}{2}(m+M)v'^2 = \frac{1}{2}(m+M)\left(\frac{mv}{m+M}\right)^2 = \frac{1}{2}\frac{m^2v^2}{m+M}$$

Since $K = \dfrac{1}{2}mv^2$, the difference is

$$\Delta K = K' - K = \frac{1}{2}\frac{m^2v^2}{m+M} - \frac{1}{2}mv^2$$

$$= \frac{1}{2}mv^2\left(\frac{m}{m+M} - 1\right)$$

$$= K\left(\frac{-M}{m+M}\right)$$

Therefore, the fraction of the bullet's original kinetic energy that was lost is $M/(m + M)$. This energy is manifested as heat (the bullet and block are warmer after the collision than before), and some was used to break the intermolecular bonds within the wooden block to allow the bullet to penetrate.

(c) From the geometry of the diagram,

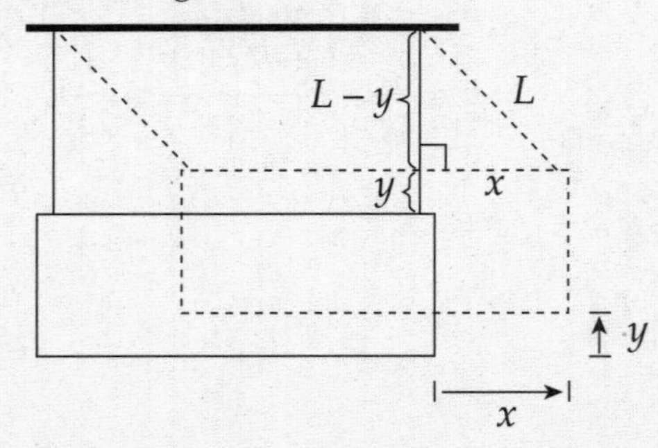

the Pythagorean theorem implies that $(L-y)^2+x^2=L^2$. Therefore,

$$L^2-2Ly+y^2+x^2=L^2 \quad \Rightarrow \quad y=\frac{x^2}{2L}$$

(where we have used the fact that y^2 is small enough to be neglected). Substituting this into the result of part (a), we derive the following equation for the speed of the bullet in terms of x and L instead of y:

$$v=\frac{m+M}{m}\sqrt{2gy}=\frac{m+M}{m}\sqrt{2g\frac{x^2}{2L}}=\frac{m+M}{m}x\sqrt{\frac{g}{L}}$$

(d) No; momentum is conserved only when the net external force on the system is zero (or at least negligible). In this case, the block and bullet feel a net nonzero force that causes it to slow down as it swings upward. Since its speed is decreasing as it swings upward, its linear momentum cannot remain constant.

CHAPTER 6 REVIEW QUESTIONS

Section I: Multiple Choice

1. **E** Neither the velocity nor the acceleration is constant because the direction of each of these vectors is always changing as the object moves along its circular path. And the net force on the object is not zero, because a centripetal force must be acting to provide the necessary centripetal acceleration to maintain the object's circular motion.

2. **B** When the bucket is at the lowest point in its vertical circle, it feels a tension force $\mathbf{F}_T$ upward and the gravitational force $\mathbf{F}_w$ downward. The net force toward the center of the circle, which is the centripetal force, is $F_T - F_w$. Thus,

$$F_T-F_w=m\frac{v^2}{r} \quad \Rightarrow \quad v=\sqrt{\frac{r(F_T-mg)}{m}}=\sqrt{\frac{(0.60\text{ m})[50\text{ N}-(3\text{ kg})(10\text{ N/kg})]}{3\text{ kg}}}=2\text{ m/s}$$

3. **C** When the bucket reaches the topmost point in its vertical circle, the forces acting on the bucket are its weight, $\mathbf{F}_w$, and the downward tension force, $\mathbf{F}_T$. The net force, $\mathbf{F}_w + \mathbf{F}_T$, provides the centripetal force. In order for the rope to avoid becoming slack, $\mathbf{F}_T$ must not vanish. Therefore, the cut-off speed for ensuring that the bucket makes it around the circle is the speed at which $\mathbf{F}_T$ just becomes zero; any greater speed would imply that the bucket would make it around. Thus,

$$F_w+F_T=m\frac{v^2}{r} \quad \Rightarrow \quad F_w+0=m\frac{v_{\text{cut-off}}^2}{r} \quad \Rightarrow \quad v_{\text{cut-off}}=\sqrt{\frac{rF_w}{m}}=\sqrt{gr}$$

$$=\sqrt{(10\text{ m/s}^2)(0.60\text{ m})}$$

$$=2.4\text{ m/s}$$

4. **D** Centripetal acceleration is given by the equation $a_c = v^2/r$. Since the object covers a distance of $2\pi r$ in 1 revolution, its speed is $2\pi r\ \text{s}^{-1}$. Therefore,

$$a_c = \frac{v^2}{r} = \frac{(2\pi r\ \text{s}^{-1})^2}{r} = 4\pi^2 r\ \text{s}^{-2}$$

5. **D** The torque is $\tau = rF = (0.20\ \text{m})(20\ \text{N}) = 4\ \text{N} \cdot \text{m}$.

6. **D** From the diagram,

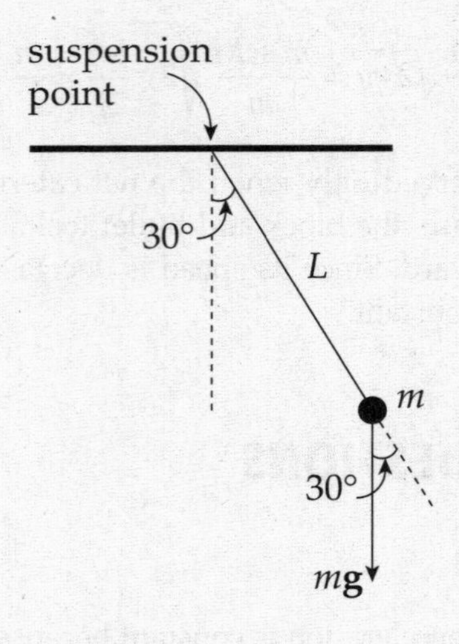

we calculate that

$$\begin{aligned}\tau = rF\sin\theta &= Lmg\sin\theta \\ &= (0.80\ \text{m})(0.50\ \text{kg})(10\ \text{N/kg})(\sin 30°) \\ &= 2.0\ \text{N}\cdot\text{m}\end{aligned}$$

7. **B** The stick will remain at rest in the horizontal position if the torques about the suspension point are balanced:

$r_1 = 50$ cm $r_2 = 30$ cm

$F_1 = m_1 g$

$F_2 = m_2 g$

$$\begin{aligned}\tau_{\text{CCW}} &= \tau_{\text{CW}} \\ r_1 F_1 &= r_2 F_2 \\ r_1 m_1 g &= r_2 m_2 g \\ m_2 &= \frac{r_1 m_1}{r_2} = \frac{(50\ \text{cm})(3\ \text{kg})}{30\ \text{cm}} = 5\ \text{kg}\end{aligned}$$

8. **A** Gravitational force obeys an inverse-square law: $F_{grav} \propto 1/r^2$. Therefore, if r increases by a factor of 2, then F_{grav} decreases by a factor of $2^2 = 4$.

9. **E** Mass is an intrinsic property of an object and does not change with location. This eliminates choices (A) and (C). If an object's height above the surface of the earth is equal to $2R_E$, then its distance from the center of the earth is $3R_E$. Thus, the object's distance from the earth's center increases by a factor of 3, so its weight decreases by a factor of $3^2 = 9$.

10. **C** The gravitational force that the Moon exerts on the planet is equal in magnitude to the gravitational force that the planet exerts on the Moon (Newton's Third Law).

11. **D** The gravitational acceleration at the surface of a planet of mass M and radius R is given by the equation $g = GM/R^2$. Therefore, for the dwarf planet Pluto:

$$g_{\text{Pluto}} = G\frac{M_{\text{Pluto}}}{R^2_{\text{Pluto}}} = G\frac{\frac{1}{500}M_{\text{Earth}}}{\left(\frac{1}{15}R_{\text{Earth}}\right)^2} = \frac{15^2}{500}\cdot G\frac{M_{\text{Earth}}}{R^2_{\text{Earth}}} = \frac{225}{500}(10\ g^{m/s^2}) = \frac{225}{50}\text{m/s}^2$$

12. **B** The gravitational pull by the earth provides the centripetal force on the satellite, so $GMm/R^2 = mv^2/R$. This gives $\frac{1}{2}mv^2 = GMm/2R$, so the kinetic energy K of the satellite is inversely proportional to R. Therefore if R increases by a factor of 2, then K decreases by a factor of 2.

13. **E** The gravitational pull by Jupiter provides the centripetal force on its moon:

$$\begin{aligned} G\frac{Mm}{R^2} &= \frac{mv^2}{R} \\ G\frac{M}{R} &= v^2 \\ &= \left(\frac{2\pi R}{T}\right)^2 \\ &= \frac{4\pi^2 R^2}{T^2} \\ M &= \frac{4\pi^2 R^3}{GT^2} \end{aligned}$$

14. **E** Let the object's distance from Body A be x; then its distance from Body B is $R - x$. In order for the object to feel no net gravitational force, the gravitational pull by A must balance the gravitational pull by B. Therefore, if we let M denote the mass of the object, then

$$G\frac{m_A M}{x^2} = G\frac{m_B M}{(R-x)^2}$$

$$\frac{m}{x^2} = \frac{4m}{(R-x)^2}$$

$$\frac{(R-x)^2}{x^2} = \frac{4m}{m}$$

$$\left(\frac{R-x}{x}\right)^2 = 4$$

$$\left(\frac{R}{x}-1\right)^2 = 4$$

$$\frac{R}{x}-1 = 2$$

$$x = R/3$$

15. **B** Because the planet is spinning clockwise and the velocity is tangent to the circle, the velocity must point down. The acceleration and force point toward the center of the circle.

16. **B** If there were no forces or balanced forces in and out the satellite would have a net force of zero. If the net force were zero, the satellite would continue in a straight line and not orbit the planet.

SECTION II: FREE RESPONSE

1. (a) Given a mass of 1 kg, weight $F_g = 5N$, and diameter = 8×10^6 m (this gives us $r = 4 \times 10^6$ m), we can fill in this equation:

$$F_G = \frac{Gm_1m_2}{r^2} \Rightarrow m_1 = \frac{F_G r^2}{Gm_2}$$

This becomes:

$$m_1 = \frac{(5N)(4\times10^6\,\text{m})^2}{\left(6.67\times10^{-11}\,\frac{\text{Nm}^2}{\text{kg}^2}\right)(1\text{kg})} = 1.2 \times 10^{24}\ \text{kg}$$

(b) $$g = \frac{Gm_1}{r^2} \Rightarrow g = \frac{\left(6.67\times10^{-11}\,\frac{\text{Nm}^2}{\text{kg}^2}\right)1.2\times10^{24}\,\text{kg}}{(4\times10^6\,\text{m})^2} = 5\ \text{m/s}^2$$

Note that we could have also observed that, because a 1 kg mass (which normally weighs 10 N on the surface of the earth) only weighed 5 N, gravity on this planet must be half the earth's gravity.

If you want to look at it in terms of g, g is 10 m/s^2 on Earth, so we can simply convert

$$5\ \text{m}/\text{s}^2\left(\frac{1g}{10\ \text{m}/\text{s}^2}\right) = 0.5g$$

(c) Density is given by mass per unit volume ($\rho = \frac{m}{V}$). We also use the equation for the volume of a sphere as $V = \frac{4}{3}\pi r^3$ to get

$$\rho = \frac{m}{\frac{4}{3}\pi r^3} \Rightarrow \rho = \frac{3m}{4\pi r^3} \Rightarrow \rho = \frac{3(1.2\times10^{24}\,\text{kg})}{4(3.14)(4\times10^6\,\text{m})^3} \text{ or}$$

$$\rho = 4480\frac{\text{kg}}{\text{m}^3}$$

Using Kepler's Third Law $\frac{T^2}{R^3} = \frac{4\pi^2}{Gm_p}$, where m_p is the weight of the uncharted planet, we can solve for the period.

$$T = \sqrt{\frac{4\pi^2R^3}{Gm_p}} \Rightarrow T = \sqrt{\frac{4\pi^2(6\times10^9\,\text{m})^3}{\left(6.67\times10^{-11}\,\frac{\text{Nm}^2}{\text{kg}^2}\right)(1.2\times10^{24}\,\text{kg})}} = 3.3 \times 10^8 \text{ seconds.}$$

2. (a) The centripetal acceleration is given by the equation $a_c = \frac{v^2}{R}$. We also know that for objects traveling in circles (Earth's orbit can be considered a circle), $v = \frac{2\pi R}{T}$. Substituting this v into the above equation for centripetal acceleration, we get $a_c = \frac{4\pi^2 R}{T^2}$. This becomes

$$a_c = \frac{4\pi^2 1.5\times10^{11}\,\text{m}}{(3.15\times10^7\,\text{s})^2} = 6.0 \times 10^{-3}\,\text{m/s}^2.$$

(b) The gravitational force is the force that keeps the earth traveling in a circle around the Sun. More specifically, $ma_c = (6 \times 10^{24}\text{ kg})6.0 \times 10^{-3}\,\text{m/s}^2$.

(c) Using Kepler's Third Law $\frac{T^2}{R^3} = \frac{4\pi^2}{GM_{sun}}$, we can solve for the mass of the planet:

$$M_{sun} = \frac{4\pi^2 R^3}{GT^2} \Rightarrow M_{sun} = \frac{4\pi^2(1.5\times10^{11}\,\text{m})^3}{\left(6.67\times10^{-11}\,\frac{\text{Nm}^2}{\text{kg}^2}\right)(3.15\times10^7\,\text{s})} = 2.0 \times 10^{30}\text{ kg}.$$

3. (a) The forces acting on a person standing against the cylinder wall are gravity ($\mathbf{F}_w$, downward), the normal force from the wall ($\mathbf{F}_N$, directed toward the center of the cylinder), and the force of static friction ($\mathbf{F}_f$, directed upward):

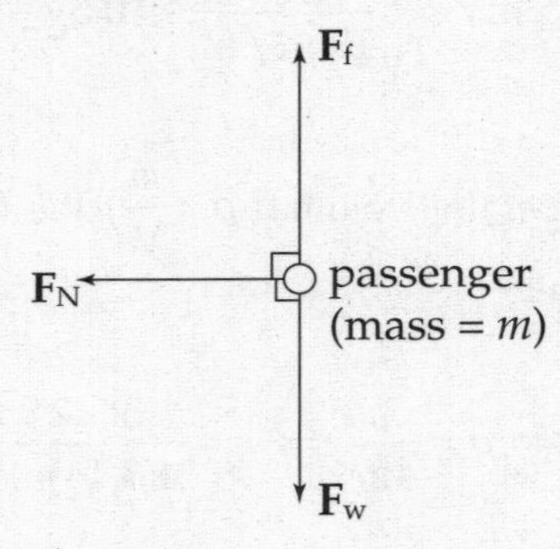

(b) In order to keep the passenger from sliding down the wall, the maximum force of static friction must be at least as great as the passenger's weight: $F_{f\,(max)} \geq mg$. Since $F_{f\,(max)} = \mu_s F_N$, this condition becomes

$$\mu_s F_N \geq mg$$

Now, consider the circular motion of the passenger. Neither $\mathbf{F}_f$ nor $\mathbf{F}_w$ has a component toward the center of the path, so the centripetal force is provided entirely by the normal force:

$$F_N = \frac{mv^2}{r}$$

Substituting this expression for F_N into the previous equation, we get

$$\mu_s \frac{mv^2}{r} \geq mg$$

$$\mu_s \geq \frac{gr}{v^2}$$

Therefore, the coefficient of static friction between the passenger and the wall of the cylinder must satisfy this condition in order to keep the passenger from sliding down.

(c) Since the mass m canceled out in deriving the expression for μ_s, the conditions are independent of mass. Thus, the inequality $\mu_s \geq gr/v^2$ holds for both the adult passenger of mass m and the child of mass $m/2$.

4. (a) The forces acting on the car are gravity ($\mathbf{F}_w$, downward), the normal force from the road ($\mathbf{F}_N$, upward), and the force of static friction ($\mathbf{F}_f$, directed toward the center of curvature of the road):

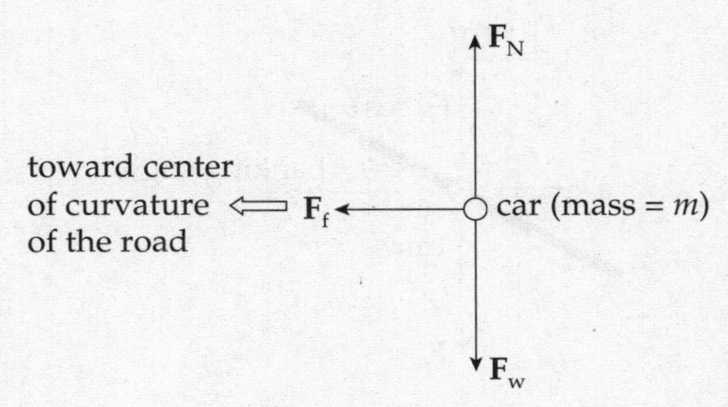

(b) The force of static friction [we assume static friction because we *don't* want the car to slide (that is, skid)] provides the necessary centripetal force:

$$F_f = \frac{mv^2}{r}$$

Therefore, to find the maximum speed at which static friction can continue to provide the necessary force, we write

$$F_{f\,(max)} = \frac{mv^2_{max}}{r}$$

$$\mu_s F_N = \frac{mv^2_{max}}{r}$$

$$\mu_s mg = \frac{mv^2_{max}}{r}$$

$$v_{max} = \sqrt{\mu_s gr}$$

(c) Ignoring friction, the forces acting on the car are gravity ($\mathbf{F}_w$, downward) and the normal force from the road ($\mathbf{F}_N$, which is now tilted toward the center of curvature of the road):

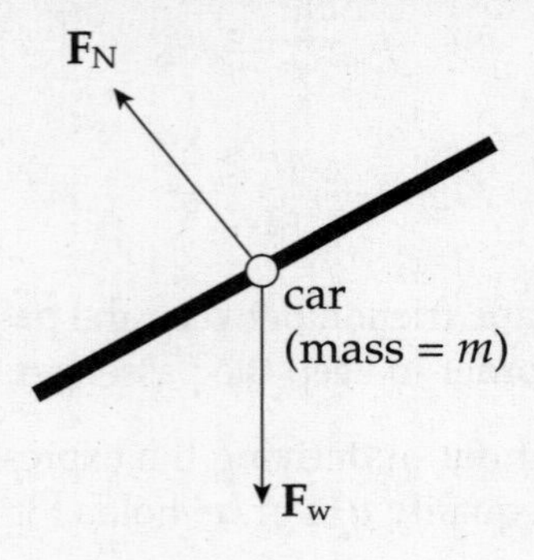

(d) Because of the banking of the turn, the normal force is tilted toward the center of curvature of the road. The component of $\mathbf{F}_N$ toward the center can provide the centripetal force, making reliance on friction unnecessary.

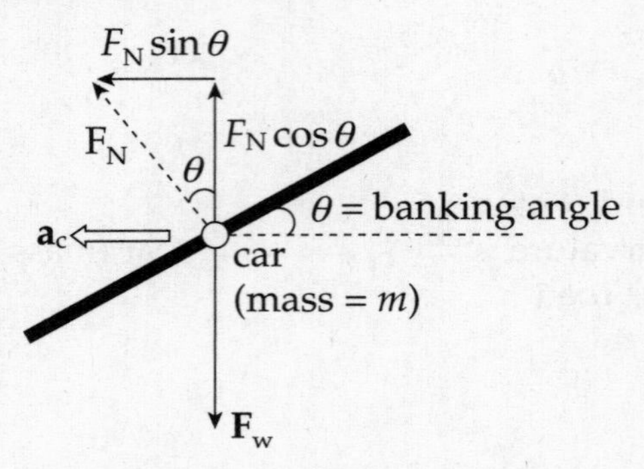

However, there's no vertical acceleration, so we can say that there is no net vertical force. Therefore, $F_N \cos\theta = F_w = mg$, so $F_N = mg/\cos\theta$. The component of F_N toward the center of curvature of the turn, $F_N \sin\theta$, provides the centripetal force:

$$F_N \sin\theta = \frac{mv^2}{r}$$

$$\frac{mg}{\cos\theta}\sin\theta = \frac{mv^2}{r}$$

$$g\tan\theta = \frac{v^2}{r}$$

$$\theta = \tan^{-1}\frac{v^2}{gr}$$

CHAPTER 7 REVIEW QUESTIONS

Section I: Multiple Choice

1. **D** The acceleration of a simple harmonic oscillator is not constant, since the restoring force—and, consequently, the acceleration—depends on position. Therefore Statement I is false. However, both Statements II and III are fundamental, defining characteristics of simple harmonic motion.

2. **C** The acceleration of the block has its maximum magnitude at the points where its displacement from equilibrium has the maximum magnitude (since $a = F/m = kx/m$). At the endpoints of the oscillation region, the potential energy is maximized and the kinetic energy (and hence the speed) is zero.

3. **E** By Conservation of Mechanical Energy, $K + U_S$ is a constant for the motion of the block. At the endpoints of the oscillation region, the block's displacement, x, is equal to $\pm A$. Since $K = 0$ here, all the energy is in the form of potential energy of the spring, $\frac{1}{2}kA^2$. Because $\frac{1}{2}kA^2$ gives the total energy at these positions, it also gives the total energy at any other position.

   Using the equation $U_S(x) = \frac{1}{2}kx^2$, we find that, at $x = \frac{1}{2}A$,

$$K + U_S = \frac{1}{2}kA^2$$

$$K + \frac{1}{2}k(\frac{1}{2}A)^2 = \frac{1}{2}kA^2$$

$$K = \frac{3}{8}kA^2$$

   Therefore,

$$K/E = \frac{3}{8}kA^2 \Big/ \frac{1}{2}kA^2 = \frac{3}{4}$$

4. **C** As we derived in Example 7.2, the maximum speed of the block is given by the equation $v_{max} = A\sqrt{k/m}$. Therefore, v_{max} is inversely proportional to $\sqrt{m}$. If m is increased by a factor of 2, then v_{max} will decrease by a factor of $\sqrt{2}$.

5. **D** The period of a spring–block simple harmonic oscillator is independent of the value of g. (Recall that $T = 2\pi\sqrt{m/k}$.) Therefore, the period will remain the same.

6. **D** The frequency of a spring–block simple harmonic oscillator is given by the equation $f = (1/2\pi)\sqrt{k/m}$. Squaring both sides of this equation, we get $f^2 = (k/4\pi^2)(1/m)$. Therefore, if f^2 is plotted vs. $(1/m)$, then the graph will be a straight line with slope $k/4\pi^2$. (Note: The slope of the line whose equation is $y = ax$ is a.)

7. **C** For small angular displacements, the period of a simple pendulum is essentially independent of amplitude.

SECTION II: FREE RESPONSE

1. (a) Since the spring is compressed to 3/4 of its natural length, the block's position relative to equilibrium is $x = -\frac{1}{4}L$. Therefore, from $F_S = -kx$, we find

$$a = \frac{F_S}{m} = \frac{-k(-\frac{1}{4}L)}{m} = \frac{kL}{4m}$$

(b) Let v_1 denote the velocity of Block 1 just before impact, and let v_1' and v_2' denote, respectively, the velocities of Block 1 and Block 2 immediately after impact. By conservation of linear momentum, we write $mv_1 = mv_1' + mv_2'$, or

$$v_1 = v_1' + v_2' \qquad (1)$$

The initial kinetic energy of Block 1 is $\frac{1}{2}mv_1^2$. If half is lost to heat, then $\frac{1}{4}mv_1^2$ is left to be shared by Block 1 and Block 2 after impact: $\frac{1}{4}mv_1^2 = \frac{1}{2}mv_1'^2 + \frac{1}{2}mv_2'^2$, or

$$v_1^2 = 2v_1'^2 + 2v_2'^2 \qquad (2)$$

Square Equation (1) and multiply by 2 to give

$$2v_1^2 = 2v_1'^2 + 4v_1'v_2' + 2v_2'^2 \qquad (1')$$

then subtract Equation (2) from Equation (1′):

$$v_1^2 = 4v_1'v_2' \qquad (3)$$

Square Equation (1) again,

$$v_1^2 = v_1'^2 + 2v_1'v_2' + v_2'^2$$

and substitute into this the result of Equation (3):

$$4v_1'v_2' = v_1'^2 + 2v_1'v_2' + v_2'^2$$
$$0 = v_1'^2 - 2v_1'v_2' + v_2'^2$$
$$0 = (v_1' - v_2')^2$$
$$v_1' = v_2' \qquad (4)$$

Thus, combining Equations (1) and (4), we find that

$$v_1' = v_2' = \frac{1}{2}v_1$$

(c) When Block 1 reaches its new amplitude position, A', all of its kinetic energy is converted to elastic potential energy of the spring. That is,

$$K_1' \rightarrow U_S' \quad \Rightarrow \quad \frac{1}{2}mv_1'^2 = \frac{1}{2}kA'^2$$

$$A'^2 = \frac{m}{k}v_1'^2$$

$$A'^2 = \frac{m}{k}(\frac{1}{2}v_1)^2$$

$$A'^2 = \frac{mv_1^2}{4k} \qquad (1)$$

But the original potential energy of the spring, $U_S = \frac{1}{2}k(-\frac{1}{4}L)^2$, gave K_1:

$$U_S \rightarrow K_1 \quad \Rightarrow \quad \frac{1}{2}k(-\frac{1}{4}L)^2 = \frac{1}{2}mv_1^2 \quad \Rightarrow \quad mv_1^2 = \frac{1}{16}kL^2 \qquad (2)$$

Substituting this result into Equation (1) gives

$$A'^2 = \frac{\frac{1}{16}kL^2}{4k} = \frac{L^2}{64} \quad \Rightarrow \quad A' = \frac{1}{8}L$$

(d) The period of a spring–block simple harmonic oscillator depends only on the spring constant k and the mass of the block. Since neither of these changes, the period will remain the same; that is, $T' = T_0$.

(e) As we showed in part (b), Block 2's velocity as it slides off the table is $\frac{1}{2}v_1$ (horizontally). The time required to drop the vertical distance H is found as follows (calling *down* the positive direction):

$$\Delta y = v_{0y}t + \frac{1}{2}gt^2 \quad \Rightarrow \quad H = \frac{1}{2}gt^2 \quad \Rightarrow \quad t = \sqrt{\frac{2H}{g}}$$

Therefore,

$$R = (\frac{1}{2}v_1)t = \frac{1}{2}v_1\sqrt{\frac{2H}{g}}$$

Now, from Equation (2) of part (c), $v_1 = \sqrt{kL^2/16m}$, so

$$R = \frac{1}{2}\sqrt{\frac{kL^2}{16m}}\sqrt{\frac{2H}{g}} = \frac{L}{8}\sqrt{\frac{2kH}{mg}}$$

2. (a) By conservation of linear momentum,

$$mv = (m+M)v' \quad \Rightarrow \quad v' = \frac{mv}{m+M}$$

(b) When the block is at its amplitude position (maximum compression of spring), the kinetic energy it (and the embedded bullet) had just after impact will become potential energy of the spring:

$$K' \rightarrow U_S$$

$$\frac{1}{2}(m+M)\left(\frac{mv}{m+M}\right)^2 = \frac{1}{2}kA^2$$

$$A = \frac{mv}{\sqrt{k(m+M)}}$$

(c) Since the mass on the spring is $m + M$, $f = \frac{1}{2\pi}\sqrt{\frac{k}{m+M}}$.

(d) The position of the block is given by the equation $x = A \sin(\omega t + \phi_0)$, where $\omega = 2\pi f$ and A is the amplitude. Since $x = 0$ at time $t = 0$, the initial phase, ϕ_0, is 0. From the results of parts (b) and (c), we have

$$x = A\sin(2\pi f t) = \frac{mv}{\sqrt{k(m+M)}}\sin\left(t\sqrt{\frac{k}{m+M}}\right)$$

3. (a) By Conservation of Mechanical Energy, $K + U = E$, so

$$\frac{1}{2}Mv^2 + \frac{1}{2}k(\frac{1}{2}A)^2 = \frac{1}{2}kA^2$$

$$\frac{1}{2}Mv^2 = \frac{3}{8}kA^2$$

$$v = A\sqrt{\frac{3k}{4M}}$$

(b) Since the clay ball delivers no horizontal linear momentum to the block, horizontal linear momentum is conserved. Thus,

$$Mv = (M+m)v'$$

$$v' = \frac{Mv}{M+m} = \frac{MA}{M+m}\sqrt{\frac{3k}{4M}} = \frac{A}{M+m}\sqrt{\frac{3kM}{4}}$$

(c) Applying the general equation for the period of a spring–block simple harmonic oscillator,

$$T = 2\pi\sqrt{\frac{M+m}{k}}$$

(d) The total energy of the oscillator after the clay hits is $\frac{1}{2}kA'^2$, where A' is the new amplitude. Just after the clay hits the block, the total energy is

$$K' + U_S = \frac{1}{2}(M+m)v'^2 + \frac{1}{2}k(\frac{1}{2}A)^2$$

Substitute for v' from part (b), set the resulting sum equal to $\frac{1}{2}kA'^2$, and solve for A'.

$$\frac{1}{2}(M+m)\left(\frac{A}{M+m}\sqrt{\frac{3kM}{4}}\right)^2+\frac{1}{2}k(\frac{1}{2}A)^2=\frac{1}{2}kA'^2$$

$$\frac{A^2\cdot 3kM}{8(M+m)}+\frac{1}{8}kA^2=\frac{1}{2}kA'^2$$

$$A^2\left(\frac{3M}{M+m}+1\right)=4A'^2$$

$$A'=\frac{1}{2}A\sqrt{\frac{3M}{M+m}+1}$$

(e) No, since the period depends only on the mass and the spring constant k.

(f) Yes. For example, if the clay had landed when the block was at $x = A$, the speed of the block would have been zero immediately before the collision and immediately after. No change in the block's speed would have meant no change in K, so no change in E, so no change in $A = \sqrt{2E/k}$.

CHAPTER 8 REVIEW QUESTIONS

Section I: Multiple Choice

1. **D** Since Point X is 5 m below the surface of the water, the pressure due to the water at X, P_X, is $\rho g h_X = \rho g(5\text{ m})$, where ρ is the density of water. Because Point Y is 4 m below the surface of the water, the pressure due to the water at Y, p_Y, is $\rho g(4\text{ m})$. Therefore,

$$\frac{P_X}{P_Y}=\frac{\rho g(5\text{ m})}{\rho g(4\text{ m})}=\frac{5}{4}\quad\Rightarrow\quad 4P_X=5P_Y$$

2. **C** Because the top of the box is at a depth of $D - z$ below the surface of the liquid, the pressure on the top of the box is

$$P=P_{\text{atm}}+\rho gh=P_{\text{atm}}+\rho g(D-z)$$

The area of the top of the box is $A = xy$, so the force on the top of the box is

$$F=PA=[P_{\text{atm}}+\rho g(D-z)]xy$$

3. **C** Let P_0 be the pressure of the gas at the surface of the liquid. Then the pressures at points Y and Z are, respectively,

$$P_Y=P_0+\rho g(1\text{ m})$$
$$P_Z=P_0+\rho g(3\text{ m})$$

Subtracting the first equation from the second, we get $P_Z - P_Y = \rho g(2\text{ m})$. Because we're given the values of P_Y and P_Z, we know that $P_Z - P_Y = 16{,}000$ Pa, so we can write

$$16{,}000\text{ Pa} = \rho g(2\text{ m}) \Rightarrow \rho g = 8000\ \tfrac{\text{Pa}}{\text{m}}$$

Substituting this into the equation for P_Y, we find that

$$P_Y = P_0 + \rho g(1\text{ m}) \Rightarrow P_0 = P_Y - \rho g(1\text{ m}) = (13{,}000\text{ Pa}) - (8000\ \tfrac{\text{Pa}}{\text{m}})(1\text{ m}) = 5000\text{ Pa}$$

4. **A** The density of the plastic cube is

$$\rho = \frac{m}{V} = \frac{m}{l^3} = \frac{100\text{ kg}}{(\frac{1}{2}\text{ m})^3} = 800\ \tfrac{\text{kg}}{\text{m}^3}$$

This is 4/5 the density of the water, so 4/5 of the cube's volume is submerged; this means that 1/5 of the cube's volume is above the surface of the water.

5. **B** The buoyant force on the styrofoam block is $F_{\text{buoy}} = \rho_L Vg$, and the weight of the block is $F_g = m_S g = \rho_S Vg$. Because $\rho_L > \rho_S$, the net force on the block is upward and has magnitude

$$F_{\text{net}} = F_{\text{buoy}} - F_g = (\rho_L - \rho_S)Vg$$

Therefore, by Newton's Second Law, we have

$$a = \frac{F_{\text{net}}}{m} = \frac{(\rho_L - \rho_S)Vg}{\rho_S V} = \left(\frac{\rho_L}{\rho_S} - 1\right)g$$

6. **B** The buoyant force acting on the ball is

$$F_{\text{buoy}} = \rho_{\text{water}} V_{\text{sub}} g = \rho_{\text{water}} Vg = (1000\ \tfrac{\text{kg}}{\text{m}^3})(5 \times 10^{-3}\text{ m}^3)(10\ \tfrac{\text{N}}{\text{kg}}) = 50\text{ N}$$

The weight of the ball is

$$F_g = \rho_{\text{ball}} Vg = (400\ \tfrac{\text{kg}}{\text{m}^3})(5 \times 10^{-3}\text{ m}^3)(10\ \tfrac{\text{N}}{\text{kg}}) = 20\text{ N}$$

Because the upward force on the ball, F_{buoy}, balances the total downward force, $F_g + F_T$, the tension in the string is

$$F_T = F_{\text{buoy}} - F_g = 50\text{ N} - 20\text{ N} = 30\text{ N}$$

7. **A** If the object weighs 100 N less when completely submerged in water, the buoyant force must be 100 N; therefore

$$F_{\text{buoy}} = \rho_{\text{water}} V_{\text{sub}} g = \rho_{\text{water}} Vg = 100\text{ N} \Rightarrow V = \frac{100\text{ N}}{\rho_{\text{water}} g} = \frac{100\text{ N}}{(1000\ \frac{\text{kg}}{\text{m}^3})(10\ \frac{\text{N}}{\text{kg}})} = 10^{-2}\text{ m}^3$$

Now that we know the volume of the object, we can figure out its weight:

$$F_g = mg = \rho_{\text{object}} Vg = (2000\ \tfrac{\text{kg}}{\text{m}^3})(10^{-2}\ \text{m}^3)(10\ \tfrac{\text{N}}{\text{kg}}) = 200\ \text{N}$$

8. **A** The cross-sectional diameter at Y is 3 times the cross-sectional diameter at X, so the cross-sectional area at Y is $3^2 = 9$ times that at X. The Continuity Equation tells us that the flow speed, v, is inversely proportional to the cross-sectional area, A. So, if A is 9 times greater at Point Y than it is at X, then the flow speed at Y is 1/9 the flow speed at X; that is, $v_Y = (1/9)v_X = (1/9)(6\ \text{m/s}) = 2/3\ \text{m/s}$.

9. **E** Each side of the rectangle at the bottom of the conduit is 1/4 the length of the corresponding side at the top. Therefore, the cross-sectional area at the bottom is $(1/4)^2 = 1/16$ the cross-sectional area at the top. The Continuity Equation tells us that the flow speed, v, is inversely proportional to the cross-sectional area, A. So, if A at the bottom is 1/16 the value of A at the top, then the flow speed at the bottom is 16 times the flow speed at the top.

10. **D** We'll apply Bernoulli's Equation to a point at the pump (Point 1) and at the nozzle (the exit point, Point 2). We'll choose the level of Point 1 as the horizontal reference level; this makes $y_1 = 0$ and $y_2 = 1$ m. Now, because the cross-sectional diameter decreases by a factor of 10 between Points 1 and 2, the cross-sectional area decreases by a factor of $10^2 = 100$, so flow speed must increase by a factor of 100; that is, $v_2 = 100v_1 = 100(0.4\ \text{m/s}) = 40\ \text{m/s}$. Because the pressure at Point 2 is P_{atm}, Bernoulli's Equation becomes

$$P_1 + \frac{1}{2}\rho v_1^2 = P_{\text{atm}} + \rho g y_2 + \frac{1}{2}\rho v_2^2$$

Therefore,

$$\begin{aligned}
P_1 - P_{\text{atm}} &= \rho g y_2 + \frac{1}{2}\rho v_2^2 - \frac{1}{2}\rho v_1^2 \\
&= (1000\ \tfrac{\text{kg}}{\text{m}^3})(10\text{m/s}^2)(1\ \text{m}) + \frac{1}{2}(1000\ \tfrac{\text{kg}}{\text{m}^3})(40\ \text{m/s})^2 - \frac{1}{2}(1000\ \tfrac{\text{kg}}{\text{m}^3})(0.4\ \text{m/s})^2 \\
&\approx (1000\ \tfrac{\text{kg}}{\text{m}^3})(10\text{m/s}^2)(1\ \text{m}) + \frac{1}{2}(1000\ \tfrac{\text{kg}}{\text{m}^3})(40\ \text{m/s})^2 \\
&= (10{,}000\ \text{Pa}) + (800{,}000\ \text{Pa}) \\
&= 810{,}000\ \text{Pa} \\
&= 810\ \text{kPa}
\end{aligned}$$

SECTION II: FREE RESPONSE

1. (a) The pressure at the top surface of the block is $P_{\text{top}} = P_{\text{atm}} + \rho_L gh$. Since the area of the top of the block is $A = xy$, the force on the top of the block has magnitude

$$F_{\text{top}} = P_{\text{bottom}} A = (P_{\text{atm}} + \rho_L gh)xy$$

The pressure at the bottom of the block is $P_{\text{bottom}} = P_{\text{atm}} + \rho_L g(h+z)$. Since the area of the bottom face of the block is also $A = xy$, the force on the bottom surface of the block has magnitude

$$F_{\text{bottom}} = P_{\text{bottom}} A = \left[P_{\text{atm}} + \rho_L g(h+z)\right]xy$$

These forces are sketched below:

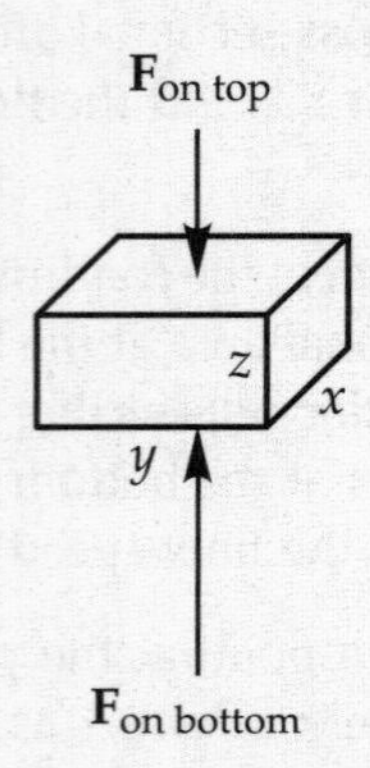

(b) Each of the other four faces of the block (left and right, front and back) is at an average depth of $h + \frac{1}{2}z$, so the average pressure on each of these four sides is

$$\bar{P}_{\text{sides}} = P_{\text{atm}} + \rho_L g(h + \frac{1}{2}z)$$

The left and right faces each have area $A = xz$, so the magnitude of the average force on this pair of faces is

$$\bar{F}_{\text{left and right}} = \bar{P}_{\text{sides}} A = [P_{\text{atm}} + \rho_L g(h + \frac{1}{2}z)]xz$$

The front and back faces each have area $A = yz$, so the magnitude of the average force on this pair of faces is

$$\bar{F}_{\text{front and back}} = \bar{P}_{\text{sides}} A = [P_{\text{atm}} + \rho_L g(h + \frac{1}{2}z)]yz$$

These forces are sketched below:

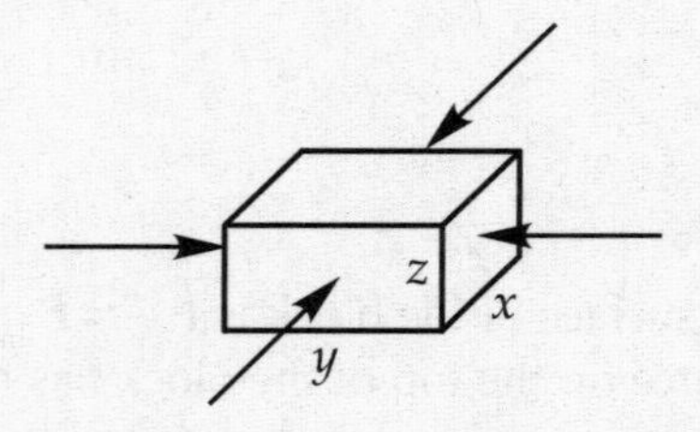

(c) The four forces sketched in part (b) add up to zero, so the total force on the block due to the pressure is the sum of $F_{\text{on top}}$ and $F_{\text{on bottom}}$; because $F_{\text{on bottom}} > F_{\text{on top}}$, this total force points upward and its magnitude is

$$F_{\text{on bottom}} - F_{\text{on top}} = [P_{\text{atm}} + \rho_L g(h+z)]xy - (P_{\text{atm}} + \rho_L gh)xy = \rho_L gxyz$$

(d) By Archimedes' Principle, the buoyant force on the block is upward and has magnitude

$$F_{buoy} = \rho_L V_{sub} g = \rho_L V g = \rho_L xyzg$$

This is the same as the result we found in part (c).

(e) The weight of the block is

$$F_g = mg = \rho_B V g = \rho_B xyzg$$

If F_T is the tension in the string, then the total upward force on the block, $F_T + F_{buoy}$, must balance the downward force, F_g; that is, $F_T + F_{buoy} = F_g$, so

$$F_T = F_g - F_{buoy} = \rho_B xyzg - \rho_L xyzg = xyzg(\rho_B - \rho_L)$$

2. (a) See the section on Torricelli's theorem for the derivation of the efflux speed from the hole; applying Bernoulli's Equation to a point on the surface of the water in the tank (Point 1) and a point at the hole (Point 2), the assumption that $v_1 \approx 0$ leads to the result

$$v = \sqrt{2gh}$$

(b) The initial velocity of the water, as it emerges from the hole, is horizontal. Since there's no initial vertical velocity, the time t required to drop the distance $y = D - h$ to the ground is found as follows:

$$y = \frac{1}{2}gt^2 \quad \Rightarrow \quad D - h = \frac{1}{2}gt^2 \quad \Rightarrow \quad t = \sqrt{\frac{2(D-h)}{g}}$$

Therefore, the horizontal distance the water travels is

$$x = v_x t = \sqrt{2gh} \cdot \sqrt{\frac{2(D-h)}{g}} = 2\sqrt{h(D-h)}$$

(c) The second hole would be at a depth of $h/2$ below the surface of the water, so the horizontal distance it travels—from the edge of the tank to the point where it hits the ground—is given by the same formula we found in part (b) except it will have $h/2$ in place of h; that is,

$$x_2 = 2\sqrt{\frac{1}{2}h(D - \frac{1}{2}h)}$$

If both streams land at the same point, then the value of x from part (b) is the same as x_2:

$$\begin{aligned}
2\sqrt{h(D-h)} &= 2\sqrt{\frac{1}{2}h(D - \frac{1}{2}h)} \\
h(D-h) &= \frac{1}{2}h(D - \frac{1}{2}h) \\
D - h &= \frac{1}{2}(D - \frac{1}{2}h) \\
4D - 4h &= 2D - h \\
-3h &= -2D \\
h &= \frac{2}{3}D
\end{aligned}$$

(d) Once again, we apply Bernoulli's Equation to a point on the surface of the water in the tank (Point 1) and to a point at the hole (Point 2). We'll choose the ground level as our horizontal reference level; then $y_1 = D$ and $y_2 = D - h$. If v_1 is the flow speed of Point 1—that is, the speed with which the water level in the tank drops—and v_2 is the efflux speed from the hole, then, by the Continuity Equation, $A_1v_1 = A_2v_2$, where A_1 and A_2 are the cross-sectional areas at Points 1 and 2, respectively. Therefore, $v_1 = (A_2/A_1)v_2$. Bernoulli's Equation then becomes

$$P_1 + \rho g D + \frac{1}{2}\rho v_1^2 = P_2 + \rho g(D-h) + \frac{1}{2}\rho v_2^2$$

Since $P_1 = P_2 = P_{atm}$, these terms cancel out; and substituting $v_1 = (A_2/A_1)v_2$, we have

$$\rho g D + \frac{1}{2}\rho\left(\tfrac{A_2}{A_1}v_2\right)^2 = \rho g(D-h) + \frac{1}{2}\rho v_2^2$$

$$\frac{1}{2}\rho v_2^2\left[\left(\tfrac{A_2}{A_1}\right)^2 - 1\right] = -\rho g h$$

$$v_2 = \sqrt{\frac{2gh}{1-\left(\tfrac{A_2}{A_1}\right)^2}}$$

Now, since $A_1 = \pi R^2$ and $A_2 = \pi r^2$, this final equation can be written as

$$v_2 = \sqrt{\frac{2gh}{1-\left(\tfrac{r}{R}\right)^4}}$$

[Note that if $r << R$, then $(r/R)^4 \approx 0$, and the equation above reduces to $v_2 = \sqrt{2gh}$, as in part (a).]

3. (a) Point X is at a depth of h_1 below Point 1, where the pressure is P_1. Therefore, the hydrostatic pressure at X is $P_X = P_1 + \rho_F g h_1$.

(b) Point Y is at a depth of $h_2 + d$ below Point 2, where the pressure is P_2. The column of static fluid above Point Y contains fluid of density of depth h_2 and fluid of density ρ_F of depth d. Therefore, the hydrostatic pressure at Y is $P_Y = P_2 + \rho_F g h_2 + \rho_V g d$.

(c) First, notice that Points 1 and 2 are at the same horizontal level; therefore, the heights y_1 and y_2 are the same, and the terms $\rho_F g y_1$ and $\rho_F g y_2$ will cancel out of the equation. Bernoulli's Equation then becomes

$$P_1 + \frac{1}{2}\rho_F v_1^2 = P_2 + \frac{1}{2}\rho_F v_2^2$$

By the Continuity Equation, we have $A_1v_1 = A_2v_2$, so $v_1 = (A_2/A_1)v_2$. Therefore,

$$\begin{aligned} P_1 - P_2 &= \frac{1}{2}\rho_F v_2^2 - \frac{1}{2}\rho_F v_1^2 \\ &= \frac{1}{2}\rho_F v_2^2 - \frac{1}{2}\rho_F\left(\frac{A_2}{A_1}v_2\right)^2 \\ &= \frac{1}{2}\rho_F v_2^2\left[1-\left(\frac{A_2}{A_1}\right)^2\right] \end{aligned}$$

(d) In parts (a) and (b) above, we found that $P_X = P_1 + \rho_F gh_1$ and $P_Y = P_2 + \rho_F gh_2 + \rho_v gd$. Since $P_X = P_Y$, we have

$$P_1 + \rho_F gh_1 = P_2 + \rho_F gh_2 + \rho_V gd$$

so

$$\begin{aligned} P_1 - P_2 &= \rho_F g(h_2 - h_1) + \rho_V gd \\ &= \rho_F g(-d) + \rho_V gd \\ &= (\rho_V - \rho_F)gd \end{aligned}$$

(e) In parts (c) and (d), we found two expressions for $P_1 - P_2$. Setting them equal to each other gives

$$\begin{aligned} \frac{1}{2}\rho_F v_2^2\left[1-\left(\frac{A_2}{A_1}\right)^2\right] &= (\rho_V - \rho_F)gd \\ v_2^2 &= \frac{\rho_V - \rho_F}{\rho_F}\cdot\frac{2gd}{1-\left(\frac{A_2}{A_1}\right)^2} \\ v_2 &= \sqrt{\frac{2gd\left(\frac{\rho_V}{\rho_F}-1\right)}{1-\left(\frac{A_2}{A_1}\right)^2}} \end{aligned}$$

The flow rate in the pipe is

$$f = A_2v_2 = A_2\sqrt{\frac{2gd\left(\frac{\rho_V}{\rho_F}-1\right)}{1-\left(\frac{A_2}{A_1}\right)^2}}$$

Since

$$f = A_2\sqrt{\frac{2g\left(\frac{\rho_V}{\rho_F}-1\right)}{1-\left(\frac{A_2}{A_1}\right)^2}}\cdot\sqrt{d}$$

we see that f is proportional to $\sqrt{d}$, as desired.

CHAPTER 9 REVIEW QUESTIONS

SECTION I: MULTIPLE CHOICE

1. **C** Because the average kinetic energy of a molecule of gas is directly proportional to the temperature of the sample, the fact that the gases are at the same temperature—since they're in the same container at thermal equilibrium—tells us that the molecules have the same average kinetic energy. The ratio of their kinetic energies is therefore equal to 1.

2. **E** From the Ideal Gas Law, we know that $P = nRT/V$. If both T and V are doubled, the ratio T/V remains unchanged, so P remains unchanged.

3. **A** The work done on the gas during a thermodynamic process is equal to the area of the region in the P–V diagram above the V-axis and below the path the system takes from its initial state to its final state. Since the area below path 1 is the greatest, the work done on the gas during the transformation along path 1 is the greatest.

4. **C** During an isothermal change, ΔU is always zero.

5. **B** Because the gas is confined, n remains constant, and because we're told the volume is fixed, V remains constant as well. Since R is a universal constant, the Ideal Gas Law, $PV = nRT$, tells us that P and T are proportional. Therefore, if T increases by a factor of 2, then so does P.

6. **E** Neither choice (A) nor (B) can be correct. Using $PV = nRT$, both containers have the same V, n is the same, P is the same, and R is a universal constant. Therefore, T must be the same for both samples. Choice (C) is also wrong, since R is a universal constant. The kinetic theory of gases predicts that the rms speed of the gas molecules in a sample of molar mass M and temperature T is

$$v_{rms} = \sqrt{\frac{3RT}{M}}$$

Hydrogen has a smaller molar mass than does helium, so v_{rms} for hydrogen must be greater than v_{rms} for helium (because both samples are at the same T).

7. **A** By convention, work done *on* the gas sample is designated as positive, so in the First Law of Thermodynamics, $\Delta U = Q + W$, we must write $W = +320$ J. Therefore, $Q = \Delta U - W = 560 \text{ J} - C\ 320 \text{ J} = +240$ J. Positive Q denotes heat *in*.

8. **C** No work is done during the step from state a to state b because the volume doesn't change. Therefore, the work done from a to c is equal to the work done from b to c. Since the pressure remains constant (this step is isobaric), we find that

$$W = -P\Delta V = -(3.0 \times 10^5 \text{ Pa})[(10 - 25) \times 10^{-3} \text{ m}^3] = 4500 \text{ J}$$

9. **D** Statement (A) is wrong because "no heat exchanged between the gas and its surroundings" is the definition of *adiabatic*, not *isothermal*. Statement (B) cannot be correct since the step described in the question is isothermal; by definition, the temperature does not change. This also eliminates statement (E) and supports statement (D). Statement (C) is false, because although the heat absorbed is converted completely to work ($\Delta U = 0$ since $\Delta T = 0$, so, by the First Law of Thermodynamics, $W = -Q_H$), the Second Law of Thermodynamics is not violated. If the sample

could be brought back to its initial state *and* have a 100% conversion of heat to work, *that* would violate the Second Law. The Second Law states that heat cannot be completely converted to work with no other change taking place. In this case, there are changes taking place: The pressure decreases and the volume increases.

10. **D** Be careful! Make sure that you always use absolute temperatures (in kelvins) when computing the Carnot efficiency, *not* temperatures in degrees Celsius. Since T_H = 800°C = 1073 K and T_C = 200°C = 473 K, the Carnot efficiency is

$$e_C = \frac{T_H - T_C}{T_H} = \frac{1073 - 473}{1073} = \frac{600}{1073}$$

This is a little less than 60%, which best matches 56%.

Section II: Free Response

1. (a) First, let's calculate ΔU_{acb}. Using path *acb*, the question tells us that Q = +70 J and W = –30 J (W is negative here because it is the *system* that does the work). The First Law, $\Delta U = Q + W$, then tells us that ΔU_{acb} = +40 J. Because $\Delta U_{a\to b}$ does not depend on the path taken from *a* to *b*, we must have ΔU_{ab} = +40 J, and $\Delta U_{ba} = -\Delta U_{ab}$ = –40 J. Thus, –40 J = $Q_{ba} + W_{ba}$, where Q_{ba} and W_{ba} are the values along the curved path from *b* to *a*. Since Q_{ba} = –60 J, it follows that W_{ba} = +20 J. Therefore, the surroundings do 20 J of work on the system.

(b) Again, using the fact that $\Delta U_{a\to b}$ does not depend on the path taken from *a* to *b*, we know that ΔU_{adb} = +40 J, as computed above. Writing $\Delta U_{adb} = QW_{adb} + W_{adb}$, if W_{adb}= –10 J, it follows that Q_{adb} = +50 J. That is, the system absorbs 50 J of heat.

(c) For the process *db*, there is no change in volume, so W_{db} = 0. Therefore, $\Delta U_{db} = Q_{db} + W_{db} = Q_{db}$. Now, since U_{ab} = +40 J, the fact that U_a = 0 J implies that U_b = 40 J, so $\Delta U_{db} = U_b - U_d$ = 40 J – 30 J = 10 J. Thus, Q_{db} = 10 J. Now let's consider the process *ad*. Since $W_{adb} = W_{ad} + W_{db} = W_{ad} + 0 = W_{ad}$, the fact that W_{adb} = –10 J [computed in part (c)] tells us that W_{ad} = –10 J. Because $\Delta U_{ad} = U_d - U_a$ = 30 J, it follows from $\Delta U_{ad} = Q_{ad} + W_{ad}$ that $Q_{ad} = \Delta U_{ad} - W_{ad}$ = 30 J – (–10 J) = 40 J.

(d) The process *adbca* is cyclic, so ΔU is zero. Because this cyclic process is traversed *counter-clockwise* in the *P–V* diagram, we know that W is *positive*. Then, since $\Delta U = Q + W$, it follows that Q must be negative.

2. (a) (i) Use the Ideal Gas Law:

$$T_a = \frac{P_a V_a}{nR} = \frac{(2.4\times10^5 \text{ Pa})(12\times10^{-3} \text{ m}^3)}{(0.4 \text{ mol})(8.31 \text{ J/mol}\cdot\text{K})} = 870 \text{ K}$$

(a) (ii) Since state *b* is on the isotherm with state *a*, the temperature of state *b* must also be 870 K.

(a) (iii) Use the Ideal Gas Law:

$$T_c = \frac{P_c V_c}{nR} = \frac{(0.6\times10^5 \text{ Pa})(12\times10^{-3} \text{ m}^3)}{(0.4 \text{ mol})(8.31 \text{ J/mol}\cdot\text{K})} = 220 \text{ K}$$

(b) (i) Since step *ab* takes place along an isotherm, the temperature of the gas does not change, so neither does the internal energy; $\Delta U_{ab} = 0$.

(b) (ii) By the First Law of Thermodynamics, $\Delta U_{bc} = Q_{bc} + W_{bc}$. Since step *bc* is isobaric (constant pressure), we have

$$Q_{bc} = nC_P \Delta T_{bc} = (0.4 \text{ mol})(29.1 \text{ J/mol} \cdot \text{K})(220 \text{ K} - 870 \text{ K}) = -7600 \text{ J}$$

Next,

$$W_{bc} = -P\Delta V_{bc} = -(0.6 \times 10^5 \ Pa)[(12 - 48) \times 10^{-3} \ m^3] = 2200 \text{ J}$$

Therefore,

$$\Delta U_{bc} = Q_{bc} + W_{bc} = (-7600 \text{ J}) + 2200 \text{ J} = -5400 \text{ J}$$

(b) (iii) By the First Law of Thermodynamics, $\Delta U_{ca} = Q_{ca} + W_{ca}$. Since step *ca* is isochoric (constant volume), we have

$$Q_{ca} = nC_V \Delta T_{ca} = (0.4 \text{ mol})(20.8 \text{ J/mol} \cdot \text{K})(870 \text{ K} - 220 \text{ K}) = 5400 \text{ J}$$

and $W_{ca} = 0$. Therefore, $\Delta U_{ca} = Q_{ca} + W_{ca} = (5400 \text{ J}) + 0 = 5400 \text{ J}$.

Note that part (iii) could have been answered as follows:

$$\begin{aligned} \Delta U_{aa} &= \Delta U_{ab} + \Delta U_{bc} + \Delta U_{ca} \\ 0 &= 0 + \Delta U_{bc} + \Delta U_{ca} \\ \Delta U_{ca} &= -\Delta U_{bc} \\ &= -(-5400 \text{ J}) \\ &= 5400 \text{ J} \end{aligned}$$

(c) Using the equation given, we find that

$$\begin{aligned} W_{ab} &= -nRT \cdot \ln \frac{V_b}{V_a} = -(0.4 \text{ mol})(8.31 \text{ J/mol} \cdot \text{K})(870 \text{ K}) \cdot \ln \frac{48 \times 10^{-3} \text{ m}^3}{12 \times 10^{-3} \text{ m}^3} \\ &= -4000 \text{ J} \end{aligned}$$

(d) The total work done over the cycle is equal to the sum of the values of the work done over each step:

$$\begin{aligned} W_{\text{cycle}} &= W_{ab} + W_{bc} + W_{ca} \\ &= W_{ab} + W_{bc} \\ &= (-4000 \text{ J}) + (2200 \text{ J}) \\ &= -1800 \text{ J} \end{aligned}$$

(e) (i) and (ii) By the First Law of Thermodynamics,

$$\begin{aligned} \Delta U_{ab} &= Q_{ab} + W_{ab} \\ 0 &= Q_{ab} + W_{ab} \\ Q_{ab} &= -W_{ab} \\ &= 4000 \text{ J} \quad [\text{from part (c)}] \end{aligned}$$

Since Q is positive, this represents heat *absorbed* by the gas.

(f) The maximum possible efficiency is the efficiency of a Carnot engine:

$$e_C = 1 - \frac{T_C}{T_H} = 1 - \frac{220 \text{ K}}{870 \text{ K}} = 75\%$$

CHAPTER 10 REVIEW QUESTIONS

Section I: Multiple Choice

1. **D** Electrostatic force obeys an inverse-square law: $F_E \propto 1/r^2$. Therefore, if r increases by a factor of 3, then F_E decreases by a factor of $3^2 = 9$.

2. **C** The strength of the electric force is given by kq^2/r^2, and the strength of the gravitational force is Gm^2/r^2. Since both of these quantities have r^2 in the denominator, we simply need to compare the numerical values of kq^2 and Gm^2. There's no contest: Since

$$kq^2 = (9 \times 10^9 \text{ N·m}^2/\text{C}^2)(1 \text{ C})^2 = 9 \times 10^9 \text{ N·m}^2$$

and

$$Gm^2 = (6.7 \times 10^{-11} \text{ N·m}^2/\text{kg}^2)(1 \text{ kg})^2 = 6.7 \times 10^{-11} \text{ N·m}^2$$

we see that $kq^2 >> Gm^2$, so F_E is much stronger than F_G.

3. **C** If the net electric force on the center charge is zero, the electrical repulsion by the $+2q$ charge must balance the electrical repulsion by the $+3q$ charge:

$$\frac{1}{4\pi\varepsilon_0}\frac{(2q)(q)}{x^2} = \frac{1}{4\pi\varepsilon_0}\frac{(3q)(q)}{y^2} \Rightarrow \frac{2}{x^2} = \frac{3}{y^2} \Rightarrow \frac{y^2}{x^2} = \frac{3}{2} \Rightarrow \frac{y}{x} = \sqrt{\frac{3}{2}}$$

4. **E** Since P is equidistant from the two charges, and the magnitudes of the charges are identical, the strength of the electric field at P due to $+Q$ is the same as the strength of the electric field at P due to $-Q$. The electric field vector at P due to $+Q$ points away from $+Q$, and the electric field vector at P due to $-Q$ points toward $-Q$. Since these vectors point in the same direction, the net electric field at P is (E to the right) + (E to the right) = ($2E$ to the right).

5. **D** The acceleration of the small sphere is

$$a = \frac{F_E}{m} = \frac{1}{4\pi\varepsilon_0}\frac{Qq}{mr^2}$$

As r increases (that is, as the small sphere is pushed away), a decreases. However, since a is always positive, the small sphere's speed, v, is always increasing.

6. **B** Since $\mathbf{F}_E$ (on q) = $q\mathbf{E}$, it must be true that $\mathbf{F}_E$ (on $-2q$) = $-2q\mathbf{E} = -2\mathbf{F}_E$.

7. **D** All excess electric charge on a conductor resides on the outer surface.

SECTION II: FREE RESPONSE

1. (a) From the figure below, we have $F_{1\text{-}2} = F_1/\cos 45°$.

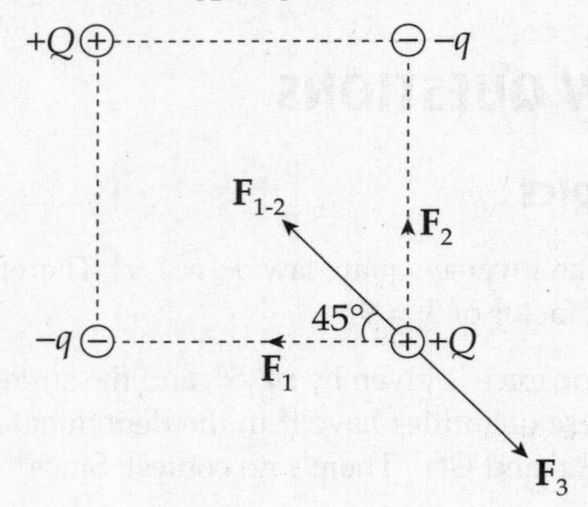

Since the net force on $+Q$ is zero, we want $F_{1\text{-}2} = F_3$. If s is the length of each side of the square, then:

$$F_{1-2} = F_3 \quad \Rightarrow \quad \frac{F_1}{\cos 45°} = F_3 \quad \Rightarrow \quad \frac{1}{\cos 45°}\frac{1}{4\pi\varepsilon_0}\frac{Qq}{s^2} = \frac{1}{4\pi\varepsilon_0}\frac{Q^2}{(s\sqrt{2})^2}$$

$$\sqrt{2}\cdot q = \frac{Q}{2}$$

$$q = \frac{Q}{2\sqrt{2}}$$

(b) No. If $q = Q/2\sqrt{2}$, as found in part (a), then the net force on $-q$ is not zero.

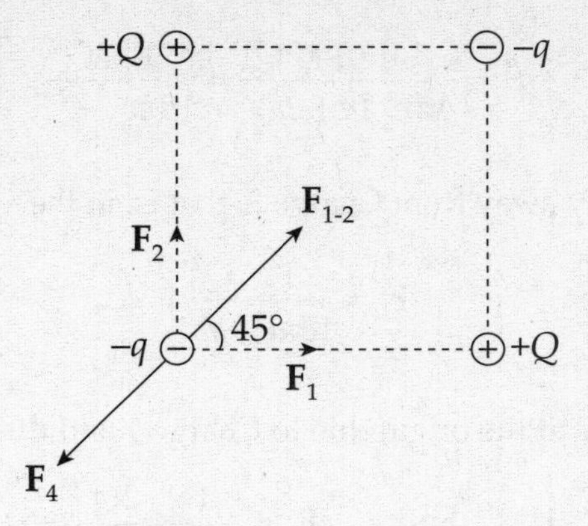

This is because $F_{1\text{-}2} \neq F_4$, as the following calculations show:

$$F_{1-2} = \frac{F_1}{\cos 45°} = \sqrt{2}\frac{1}{4\pi\varepsilon_0}\frac{Qq}{s^2} = \sqrt{2}\frac{1}{4\pi\varepsilon_0}\frac{Q\frac{Q}{2\sqrt{2}}}{s^2} = \frac{1}{8\pi\varepsilon_0}\frac{Q^2}{s^2}$$

but

$$F_4 = \frac{1}{4\pi\varepsilon_0}\frac{q^2}{(s\sqrt{2})^2} = \frac{1}{4\pi\varepsilon_0}\frac{\left(\frac{Q}{2\sqrt{2}}\right)^2}{(s\sqrt{2})^2} = \frac{1}{64\pi\varepsilon_0}\frac{Q^2}{s^2}$$

(c) By symmetry, $E_1 = E_2$ and $E_3 = E_4$, so the net electric field at the center of the square is zero:

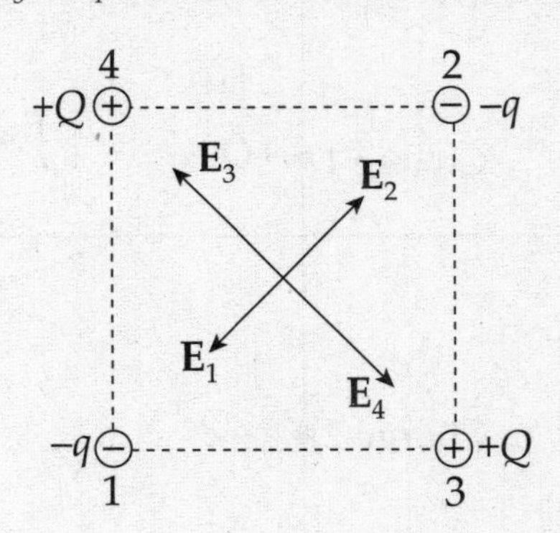

2. (a) The magnitude of the electric force on Charge 1 is

$$F_1 = \frac{1}{4\pi\varepsilon_0}\frac{(Q)(2Q)}{(a+2a)^2} = \frac{1}{18\pi\varepsilon_0}\frac{Q^2}{a^2}$$

The direction of $\mathbf{F}_1$ is directly away from Charge 2; that is, in the $+y$ direction, so

$$\mathbf{F}_1 = \frac{1}{18\pi\varepsilon_0}\frac{Q^2}{a^2}\hat{\mathbf{j}}$$

(b) The electric field vectors at the origin due to Charge 1 and due to Charge 2 are

$$\mathbf{E}_1 = \frac{1}{4\pi\varepsilon_0}\frac{Q}{a^2}(-\hat{\mathbf{j}}) \quad \text{and} \quad \mathbf{E}_2 = \frac{1}{4\pi\varepsilon_0}\frac{2Q}{(2a)^2}(+\hat{\mathbf{j}}) = \frac{1}{8\pi\varepsilon_0}\frac{Q}{a^2}(+\hat{\mathbf{j}})$$

Therefore, the net electric field at the origin is

$$\mathbf{E} = \mathbf{E}_1 + \mathbf{E}_2 = \frac{1}{4\pi\varepsilon_0}\frac{Q}{a^2}(-\hat{\mathbf{j}}) + \frac{1}{8\pi\varepsilon_0}\frac{Q}{a^2}(+\hat{\mathbf{j}}) = \frac{1}{8\pi\varepsilon_0}\frac{Q}{a^2}(-\hat{\mathbf{j}})$$

(c) No. The only point on the x axis where the individual electric field vectors due to each of the two charges point in exactly opposite directions is the origin (0,0). But at that point, the two vectors are not equal and thus do not cancel.

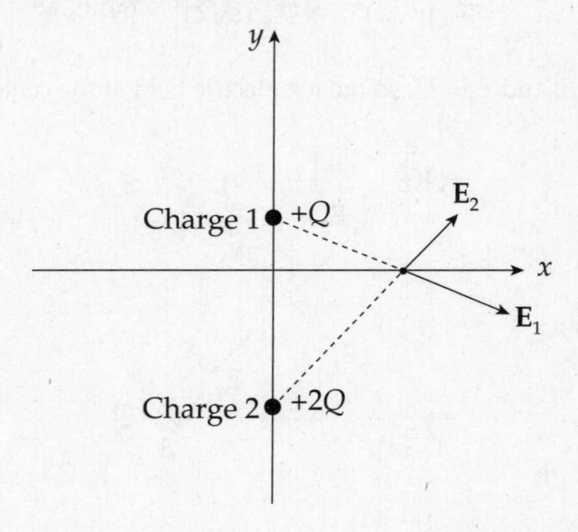

Therefore, at no point on the x axis could the total electric field be zero.

(d) Yes. There will be a Point P on the y axis between the two charges,

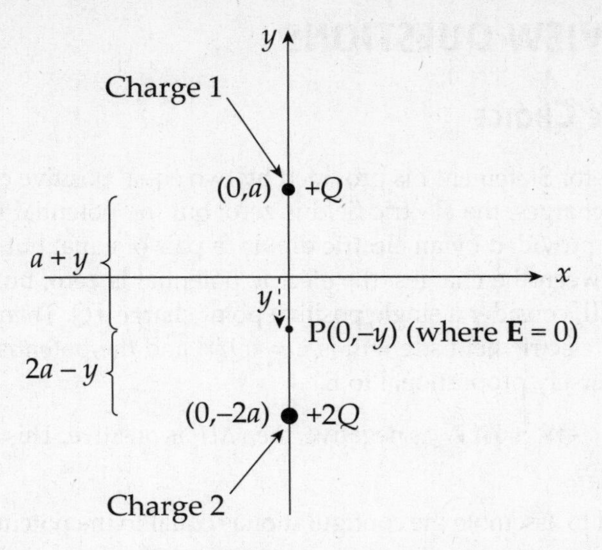

where the electric fields due to the individual charges will cancel each other out.

$$E_1 = E_2$$

$$\frac{1}{4\pi\varepsilon_0}\frac{Q}{(a+y)^2} = \frac{1}{4\pi\varepsilon_0}\frac{2Q}{(2a-y)^2}$$

$$\frac{1}{(a+y)^2} = \frac{2}{(2a-y)^2}$$

$$(2a-y)^2 = 2(a+y)^2$$

$$y^2 + 8ay - 2a^2 = 0$$

$$y = \frac{-8a \pm \sqrt{(8a)^2 - 4(-2a^2)}}{2}$$

$$= (-4 \pm 3\sqrt{2})a$$

Disregarding the value $y = (-4 - 3\sqrt{2})a$ (because it would place the point P below Charge 2 on the y axis, where the electric field vectors do not point in opposite directions), we have that $\mathbf{E} = \mathbf{0}$ at the point P = (0,–y) = (0, $(4 - 3\sqrt{2})a$).

(e) Use the result of part (b) with Newton's Second Law:

$$\mathbf{a} = \frac{\mathbf{F}}{m} = \frac{-q\mathbf{E}}{m} = \frac{-q}{m}\left[\frac{1}{8\pi\varepsilon_0}\frac{Q}{a^2}(-\hat{\mathbf{j}})\right] = \frac{1}{8\pi\varepsilon_0}\frac{qQ}{ma^2}(+\hat{\mathbf{j}})$$

CHAPTER 11 REVIEW QUESTIONS

SECTION I: MULTIPLE CHOICE

1. **E** A counterexample for Statement I is provided by two equal positive charges; at the point midway between the charges, the electric field is zero, but the potential is not. A counterexample for Statement II is provided by an electric dipole (a pair of equal but opposite charges); at the point midway between the charges, the electric potential is zero, but the electric field is not. As for Statement III, consider a single positive point charge $+Q$. Then at a distance r from this source charge, the electric field strength is $E = kQ/r^2$ and the potential is $V = kQ/r$. Thus, $V = rE$, so V is not inversely proportional to E.

2. **C** By definition, $\Delta U_E = -W_E$, so if W_E is negative, then ΔU_E is positive. This implies that the potential energy, U_E, increases.

3. **B** The work required to assemble the configuration is equal to the potential energy of the system:

$$W = U_E = \frac{1}{4\pi\varepsilon_0}\sum_{i<j}\frac{q_i q_j}{r_{ij}}$$

$$= \frac{1}{4\pi\varepsilon_0}\left(\frac{q_1q_2}{r_{12}} + \frac{q_1q_3}{r_{13}} + \frac{q_1q_4}{r_{14}} + \frac{q_2q_3}{r_{23}} + \frac{q_2q_4}{r_{24}} + \frac{q_3q_4}{r_{34}}\right)$$

$$= \frac{1}{4\pi\varepsilon_0}\left(\frac{q^2}{s} + \frac{q^2}{s\sqrt{2}} + \frac{q^2}{s} + \frac{q^2}{s} + \frac{q^2}{s\sqrt{2}} + \frac{q^2}{s}\right)$$

$$= \frac{1}{4\pi\varepsilon_0}\frac{q^2}{s}(4+\sqrt{2})$$

4. **B** Use the definition $\Delta V = -W_E/q$. If an electric field accelerates a negative charge doing positive work on it, then $W_E > 0$. If $q < 0$, then $-W_E/q$ is positive. Therefore, ΔV is positive, which implies that V increases.

5. **E** By definition,

$$V_{A \to B} = \Delta U_E/q, \text{ so } V_B - V_A = \Delta U_E/q$$

6. **C** Because **E** is uniform, the potential varies linearly with distance from either plate ($\Delta V = Ed$). Since Points 2 and 4 are at the same distance from the plates, they lie on the same equipotential. (The equipotentials in this case are planes parallel to the capacitor plates.)

7. **B** By definition, $W_E = -q\Delta V$, which gives

$$W_E = -q(V_B - V_A) = -(-0.05\text{ C})(100\text{ V} - 200\text{ V}) = -5\text{ J}$$

Note that neither the length of the segment AB nor that of the curved path from A to B is relevant.

8. **D**

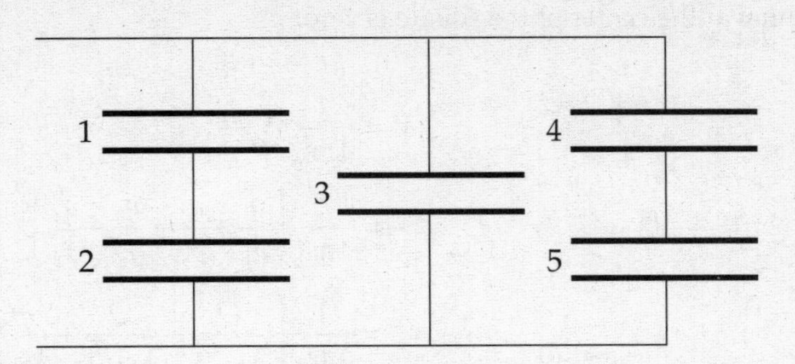

Capacitors 1 and 2 are in series, so their equivalent capacitance is $C_{1\text{-}2} = C/2$. (This is obtained from the equation $1/C_{1\text{-}2} = 1/C_1 + 1/C_2 = 1/C + 1/C = 2/C$.) Capacitors 4 and 5 are also in series, so their equivalent capacitance is $C_{4\text{-}5} = C/2$. The capacitances $C_{1\text{-}2}$, C_3, and $C_{4\text{-}5}$ are in parallel, so the overall equivalent capacitance is $(C/2) + C + (C/2) = 2C$.

Section II: Free Response

1. (a) Labeling the four charges as given in the diagram, we get

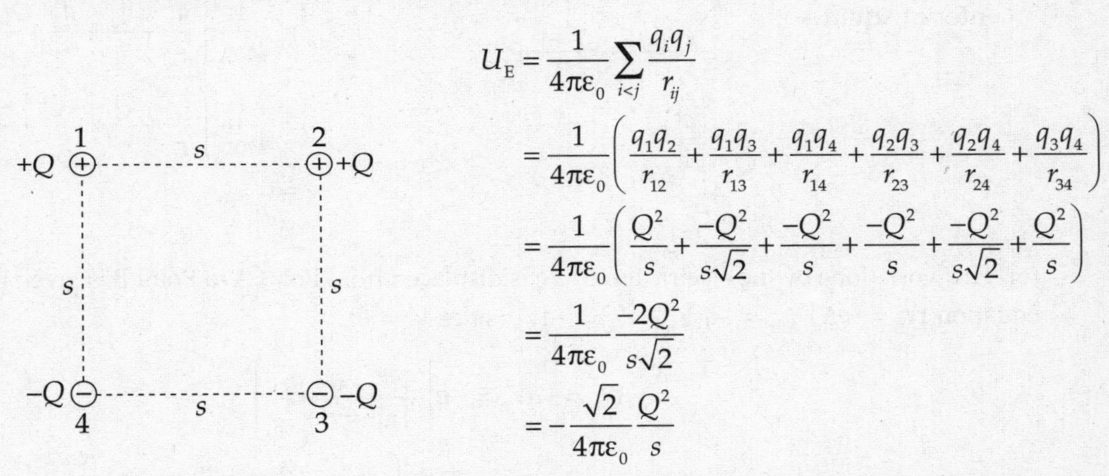

$$U_E = \frac{1}{4\pi\varepsilon_0}\sum_{i<j}\frac{q_i q_j}{r_{ij}}$$

$$= \frac{1}{4\pi\varepsilon_0}\left(\frac{q_1q_2}{r_{12}} + \frac{q_1q_3}{r_{13}} + \frac{q_1q_4}{r_{14}} + \frac{q_2q_3}{r_{23}} + \frac{q_2q_4}{r_{24}} + \frac{q_3q_4}{r_{34}}\right)$$

$$= \frac{1}{4\pi\varepsilon_0}\left(\frac{Q^2}{s} + \frac{-Q^2}{s\sqrt{2}} + \frac{-Q^2}{s} + \frac{-Q^2}{s} + \frac{-Q^2}{s\sqrt{2}} + \frac{Q^2}{s}\right)$$

$$= \frac{1}{4\pi\varepsilon_0}\frac{-2Q^2}{s\sqrt{2}}$$

$$= -\frac{\sqrt{2}}{4\pi\varepsilon_0}\frac{Q^2}{s}$$

(b) Let $\mathbf{E}_i$ denote the electric field at the center of the square due to Charge i. Then by symmetry, $\mathbf{E}_1 = \mathbf{E}_3$, $\mathbf{E}_2 = \mathbf{E}_4$, and $E_1 = E_2 = E_3 = E_4$. The horizontal components of the four individual field vectors cancel, leaving only a downward-pointing electric field of magnitude $E_{\text{total}} = 4E_1 \cos 45°$:

$$E_{\text{total}} = 4E_1 \cos 45°$$

$$= 4\frac{1}{4\pi\varepsilon_0}\frac{Q}{\left(\frac{1}{2}s\sqrt{2}\right)^2}\cdot\frac{\sqrt{2}}{2}$$

$$= \frac{\sqrt{2}}{\pi\varepsilon_0}\frac{Q}{s^2}$$

(c) The potential at the center of the square is zero:

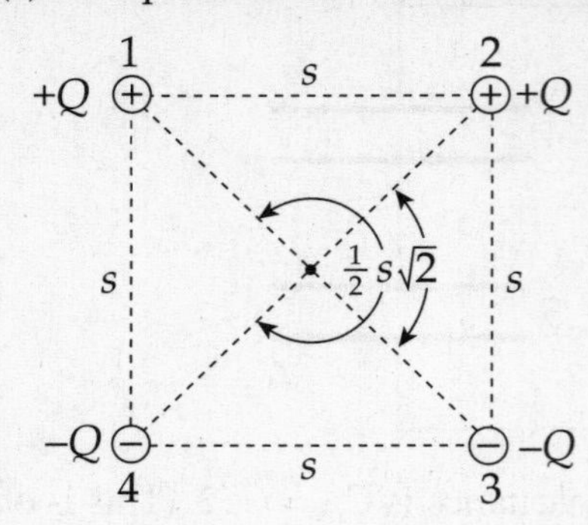

$$V = \frac{1}{4\pi\varepsilon_0}\sum_i \frac{q_i}{r_i}$$

$$= \frac{1}{4\pi\varepsilon_0}\left(\frac{q_1}{r_1} + \frac{q_2}{r_2} + \frac{q_3}{r_3} + \frac{q_4}{r_4}\right)$$

$$= \frac{1}{4\pi\varepsilon_0}\left(\frac{Q}{\frac{1}{2}s\sqrt{2}} + \frac{Q}{\frac{1}{2}s\sqrt{2}} + \frac{-Q}{\frac{1}{2}s\sqrt{2}} + \frac{-Q}{\frac{1}{2}s\sqrt{2}}\right)$$

$$= 0$$

(d) At every point on the center horizontal line shown, $r_1 = r_4$ and $r_2 = r_3$, so V will equal zero (just as it does at the center of the square):

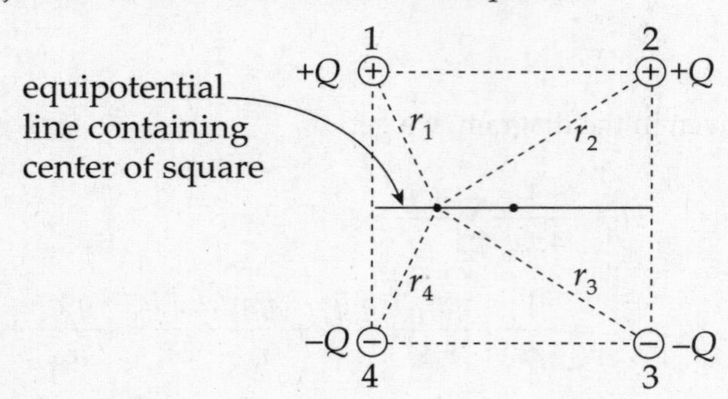

$$V = \frac{1}{4\pi\varepsilon_0}\sum_i \frac{q_i}{r_i}$$

$$= \frac{1}{4\pi\varepsilon_0}\left(\frac{q_1}{r_1} + \frac{q_2}{r_2} + \frac{q_3}{r_3} + \frac{q_4}{r_4}\right)$$

$$= \frac{1}{4\pi\varepsilon_0}\left(\frac{Q}{r_1} + \frac{Q}{r_2} + \frac{-Q}{r_3} + \frac{-Q}{r_4}\right)$$

$$= 0$$

(e) The work done by the electric force as q is displaced from Point A to Point B is given by the equation $W_E = -q\Delta V_{A\to B} = -q(V_B - V_A) = -qV_B$ (since $V_A = 0$).

$$W_E = -qV_B = -q\left[\frac{1}{4\pi\varepsilon_0}\sum_i \frac{q_i}{r_{i\to B}}\right]$$

$$= \frac{-q}{4\pi\varepsilon_0}\left(\frac{q_1}{r_{1\to B}} + \frac{q_2}{r_{2\to B}} + \frac{q_3}{r_{3\to B}} + \frac{q_4}{r_{4\to B}}\right)$$

$$= \frac{-q}{4\pi\varepsilon_0}\left(\frac{Q}{\frac{1}{2}s} + \frac{Q}{\frac{1}{2}s} + \frac{-Q}{\sqrt{s^2 + (\frac{1}{2}s)^2}} + \frac{-Q}{\sqrt{s^2 + (\frac{1}{2}s)^2}}\right)$$

$$= \frac{-q}{4\pi\varepsilon_0}\left(\frac{4Q}{s} + \frac{-2Q}{\frac{\sqrt{5}}{2}s}\right)$$

$$= \frac{-qQ}{\pi\varepsilon_0 s}\left(1 - \frac{1}{\sqrt{5}}\right)$$

2. (a) The capacitance is $C = \varepsilon_0 A/d$. Since the plates are rectangular, the area A is equal to Lw, so $C = \varepsilon_0 Lw/d$.

(b) and (c) Since the electron is attracted upward, the top plate must be the positive plate:

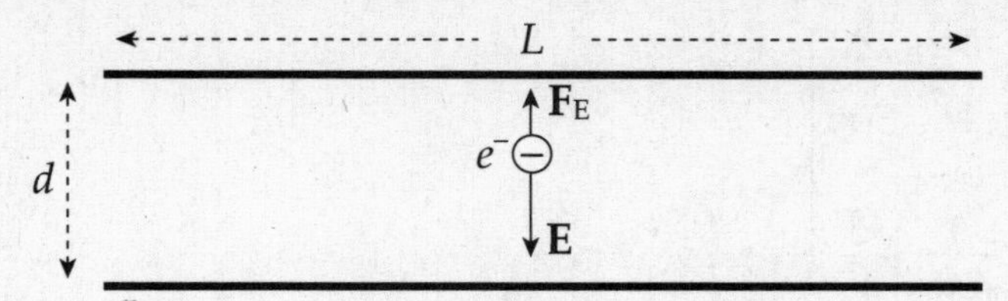

(d) The acceleration of the electron is $a = F_E/m = qE/m = eE/m$, vertically upward. Therefore, applying Big Five #3 for vertical motion, $\Delta y = v_{0y}t + \frac{1}{2}a_y t^2$, we get

$$\Delta y = \frac{1}{2}\frac{eE}{m}t^2 \quad (1)$$

To find t, notice that $v_{0x} = v_0$ remains constant (because there is no horizontal acceleration). Therefore, the time necessary for the electron to travel the horizontal distance L is $t = L/v_0$. In this time, Δy is $d/2$, so Equation (1) becomes

$$\frac{d}{2} = \frac{1}{2}\frac{eE}{m}\left(\frac{L}{v_0}\right)^2 \quad \Rightarrow \quad E = \frac{dmv_0^2}{eL^2}$$

(e) Substituting the result of part (d) into the equation $\Delta V = Ed$ gives

$$\Delta V = \frac{d^2mv_0^2}{eL^2}$$

Since $Q = C\Delta V$ (by definition), the result of part (a) now gives

$$Q = \frac{\varepsilon_0 Lw}{d} \cdot \frac{d^2mv_0^2}{eL^2} = \frac{\varepsilon_0 wdmv_0^2}{eL}$$

(f) Applying the equation $U_E = \frac{1}{2}C(\Delta V)^2$, we get

$$U_E = \frac{1}{2} \cdot \frac{\varepsilon_0 Lw}{d} \cdot \left(\frac{d^2mv_0^2}{eL^2}\right)^2 = \frac{\varepsilon_0 wd^3m^2v_0^4}{2e^2L^3}$$

3. (a) Outside the sphere, the sphere behaves as if all the charge were concentrated at the center. Inside the sphere, the electrostatic field is zero:

$$E(r) = \begin{cases} 0 & (r < a) \\ \dfrac{1}{4\pi\varepsilon_0}\dfrac{Q}{r^2} & (r > a) \end{cases}$$

(b) On the surface and outside the sphere, the electric potential is $(1/4\pi\varepsilon_0)(Q/r)$. Within the sphere, V is constant (because $E = 0$) and equal to the value on the surface. Therefore,

$$V(r) = \begin{cases} \dfrac{1}{4\pi\varepsilon_0}\dfrac{Q}{a} & (r \le a) \\ \dfrac{1}{4\pi\varepsilon_0}\dfrac{Q}{r} & (r > a) \end{cases}$$

(c) See diagrams:

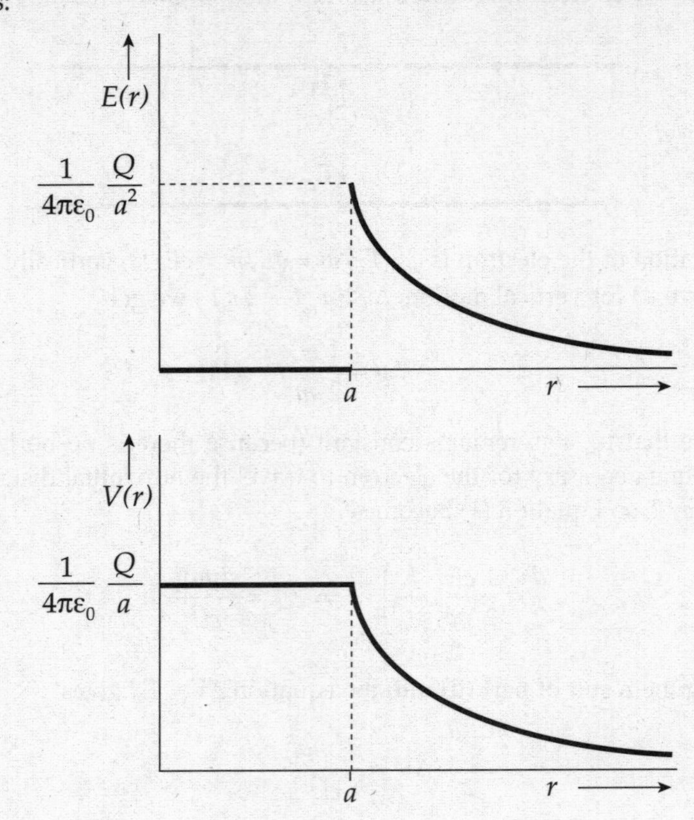

CHAPTER 12 REVIEW QUESTIONS

SECTION I: MULTIPLE CHOICE

1. **A** Let ρ_S denote the resistivity of silver and let A_S denote the cross-sectional area of the silver wire. Then

$$R_B = \frac{\rho_B L}{A_B} = \frac{(5\rho_S)L}{4^2 A_S} = \frac{5}{16}\frac{\rho_S L}{A_S} = \frac{5}{16}R_S$$

2. **D** The equation $I = V/R$ implies that increasing V by a factor of 2 will cause I to increase by a factor of 2.

3. **C** Use the equation $P = V^2/R$:

$$P = \frac{V^2}{R} \quad \Rightarrow \quad R = \frac{V^2}{P} = \frac{(120\text{ V})^2}{60\text{ W}} = 240\ \Omega$$

4. **B** The current through the circuit is

$$I = \frac{\varepsilon}{r+R} = \frac{40 \text{ V}}{(5\ \Omega)+(15\ \Omega)} = 2 \text{ A}$$

Therefore, the voltage drop across R is $V = IR = (2 \text{ A})(15\ \Omega) = 30$ V.

5. **E** The 12 Ω and 4 Ω resistors are in parallel and are equivalent to a single 3 Ω resistor, because 1/(12 Ω) + 1/(4 Ω) = 1/(3 Ω). This 3 Ω resistor is in series with the top 3 Ω resistor, giving an equivalent resistance in the top branch of 3 + 3 = 6 Ω. Finally, this 6 Ω resistor is in parallel with the bottom 3 Ω resistor, giving an overall equivalent resistance of 2 Ω, because 1/(6 Ω) + 1/(3 Ω) = 1/(2 Ω).

6. **D** If each of the identical bulbs has resistance R, then the current through each bulb is ε/R. This is unchanged if the middle branch is taken out of the parallel circuit. (What *will* change is the total amount of current provided by the battery.)

7. **B** The three parallel resistors are equivalent to a single 2 Ω resistor, because 1/(8 Ω) + 1/(4 Ω) + 1/(8 Ω) = 1/(2 Ω). This 2 Ω resistance is in series with the given 2 Ω resistor, so their equivalent resistance is 2 + 2 = 4 Ω. Therefore, three times as much current will flow through this equivalent 4 Ω resistance in the top branch as through the parallel 12 Ω resistor in the bottom branch, which implies that the current through the bottom branch is 3 A, and the current through the top branch is 9 A. The voltage drop across the 12 Ω resistor is therefore $V = IR = (3 \text{ A})(12\ \Omega) = 36$ V.

8. **E** Since points a and b are grounded, they're at the same potential (call it zero).

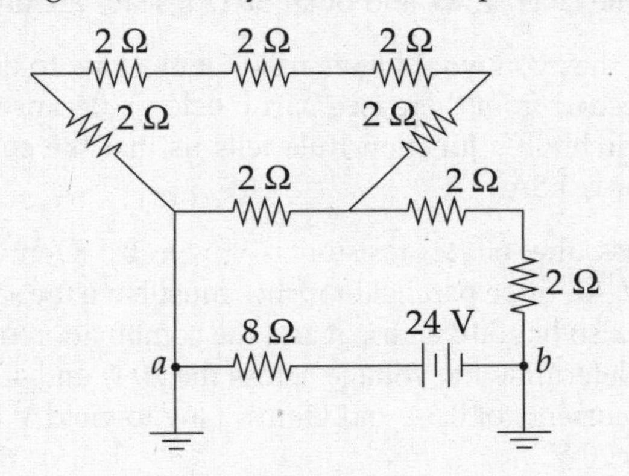

Traveling from b to a across the battery, the potential increases by 24 V, so it must decrease by 24 V across the 8 Ω resistor as we reach point a. Thus, $I = V/R = (24 \text{ V})/(8\ \Omega) = 3$ A.

9. **D** The equation $P = I^2R$ gives

$$P = (0.5 \text{ A})^2(100\ \Omega) = 25 \text{ W} = 25 \text{ J/s}$$

Therefore, in 20 s, the energy dissipated as heat is

$$E = Pt = (25 \text{ J/s})(20 \text{ s}) = 500 \text{ J}$$

SECTION II: FREE RESPONSE

1. (a) The two parallel branches, the one containing the 40 Ω resistor and the other a total of 120 Ω, is equivalent to a single 30 Ω resistance. This 30 Ω resistance is in series with the three 10 Ω resistors, giving an overall equivalent circuit resistance of 10 + 10 + 30 + 10 = 60 Ω. Therefore, the current supplied by the battery is $I = V/R$ = (120V)/(60) = 2 A, so it must supply energy at a rate of $P = IV$ = (2 A)(120 V) = 240 W.

   (b) Since three times as much current will flow through the 40 Ω resistor as through the branch containing 120 Ω of resistance, the current through the 40 Ω resistor must be 1.5 A.

   (c) (i) $V_a - V_b = IR_{20} + IR_{100}$ = (0.5 A)(20 Ω) + (0.5 A)(100 Ω) = 60 V.

   (ii) Point a is at the higher potential (current flows from high to low potential).

   (d) Because energy is equal to power multiplied by time, we get

$$E = Pt = I^2Rt = (0.5\text{ A})^2(100\ \Omega)(10\text{ s}) = 250\text{ J}$$

   (e) Using the equation $R = \rho L/A$, with $A = \pi r^2$, we find

$$R = \frac{\rho L}{\pi r^2} \quad \Rightarrow \quad r = \sqrt{\frac{\rho L}{\pi R}} = \sqrt{\frac{(0.45\ \Omega \cdot \text{m})(0.04\text{ m})}{\pi(100\ \Omega)}} = 0.0076\text{ m} = 7.6\text{ mm}$$

2. (a) There are many ways to solve this problem. If you notice that each of the three branches of the parallel section is 60 Ω, then they must all have the same current flowing through them. The currents through the 20 Ω, 40 Ω, and other 60 Ω resistor are all 0.5A.

   If you had not noticed this, you would have used Ohm's Law to determine the voltage across the resistors and proceeded from there (see part b below). Because 0.5A goes through each of the three pathways, Kirchhoff's Junction Rule tells us that the current that must have come though the 10Ω resistor is 1.5A.

   (b) The voltage across the 60 Ω resistor is given by Ohm's Law. Because $V = IR$, V = (0.5 A)(60 Ω) = 30V. All three parallel branches must have the same voltage across them, so the other 60 Ω resistor also has 30V across it and the combination of the 20 Ω and 60 Ω resistor must also be 30 V. To determine the voltage across the 20 Ω and 40 Ω resistor you can rely on the previously solved currents of 0.5A and Ohm's Law to yield $V = IR$ = (0.5 A)(20 Ω) = 10 V and V = (0.5A)(40 Ω) = 20 V.

   You also could have used the ratio of the resistors. That is, we know the two voltages must sum to 30 V and the voltage drop across the 40 Ω must be twice the amount across the 20 Ω. The voltage across the 10 Ω resistor can be found using Ohm's Law: V = (1.5 A)(10 Ω) = 15 V.

(c) The equivalent resistance of the circuit can be solved either by adding the resistances or by using Ohm's Law.

If you want to add resistances, start by summing the 20 Ω and 40 Ω resistors to get 60 Ω. Then add the three parallel branches using $\frac{1}{R_P} = \sum_i \frac{1}{R_i}$ or $\frac{1}{R_P} = \sum \frac{1}{60\ \Omega} + \frac{1}{60\ \Omega} + \frac{1}{60\ \Omega}$, which becomes 20 Ω. Then adding this section in series to the 10 Ω resistor to get $R_p = R_1 + R_2 = 10\ \Omega + 20\ \Omega = 30\ \Omega$.

You could have also realized that the total voltage drop across the battery is 45 V (15 V across the 10 Ω resistor and 30 V across the parallel branch). Using Ohm's Law again $R_{eq} = \frac{V_B}{I_B}$ or $R_{eq} = \frac{45V}{1.5A} = 30\ \Omega$.

CHAPTER 13 REVIEW QUESTIONS

SECTION I: MULTIPLE CHOICE

1. **D** Statement I is false: The magnetic field lines due to a current-carrying wire encircle the wire in closed loops. Statement II is also false: Since the magnetic force is always perpendicular to the charged particle's velocity vector, it cannot do work on the charged particle; therefore, it cannot change the particle's kinetic energy. Statement III, however, is true: If the charged particle's velocity is parallel (or antiparallel) to the magnetic field lines, then the particle will feel no magnetic force.

2. **C** The magnitude of the magnetic force is $F_B = qvB$, so the acceleration of the particle has magnitude

$$a = \frac{F_B}{m} = \frac{qvB}{m} = \frac{(4.0 \times 10^{-9}\ \text{C})(3 \times 10^4\ \text{m/s})(0.1\ \text{T})}{2 \times 10^{-4}\ \text{kg}} = 0.06\ \text{m/s}^2$$

3. **D** By the right-hand rule, the direction of **v** × **B** is into the plane of the page. Since the particle carries a negative charge, the magnetic force it feels will be out of the page.

4. **D** Since $\mathbf{F}_B$ is always perpendicular to **v**, **v** cannot be upward or downward in the plane of the page; this eliminates choices (B) and (C). The velocity vector also cannot be to the right (choice (A)), since then **v** would be antiparallel to **B**, and $\mathbf{F}_B$ would be zero. Because the charge is positive, the direction of $\mathbf{F}_B$ will be the same as the direction of **v** × **B**. In order for **v** × **B** to be downward in the plane of the page, the right-hand rule implies that **v** must be out of the plane of the page.

5. **A** The magnetic force provides the centripetal force on the charged particle. Therefore,

$$qvB = \frac{mv^2}{r} \Rightarrow qB = \frac{mv}{r} \Rightarrow mv = qBr \Rightarrow p = qBr$$

6. **D** The strength of the magnetic field at a distance r from a long, straight wire carrying a current I is given by the equation $B = (\mu_0/2\pi)(I/r)$. Therefore,

$$\frac{\mu_0}{2\pi}\frac{I}{r} = \frac{(4\pi \times 10^{-7}\ \text{T}\cdot\text{m/A})}{2\pi}\frac{10\ \text{A}}{0.02\ \text{m}} = 1\times 10^{-4}\ \text{T}$$

7. **D** By Newton's Third Law, neither choice (A) nor choice (B) can be correct. Also, as we learned in Example 13.9, if two parallel wires carry current in the same direction, the magnetic force between them is attractive; this eliminates choices (C) and (E). Therefore, the answer must be (D). The strength of the magnetic field at a distance r from a long, straight wire carrying a current I_1 is given by the equation $B_1 = (\mu_0/2\pi)(I_1/r)$. The magnetic force on a wire of length ℓ carrying a current I through a magnetic field **B** is $I(\ell \times \mathbf{B})$, so the force on Wire #2 (F_{B2}) due to the magnetic field of Wire #1 (B_1) is

$$F_{B2} = I_2 \ell B_1 = I_2 \ell \frac{\mu_0}{2\pi}\frac{I_1}{r}$$

which implies

$$\frac{F_{B2}}{\ell} = \frac{\mu_0}{2\pi}\frac{I_1 I_2}{r} = \frac{(4\pi \times 10^{-7}\ \text{N/A}^2)}{2\pi}\frac{(5\ \text{A})(10\ \text{A})}{0.01\ \text{m}} = 0.001\ \text{N/m}$$

8. **E** The strength of the magnetic field at a distance r from a long, straight wire carrying a current I is given by the equation $B = (\mu_0/2\pi)(I/r)$. Therefore, the strength of the magnetic field at Point P due to either wire is $B = (\mu_0/2\pi)(I/\frac{1}{2}d)$. By the right-hand rule, the direction of the magnetic field at P due to the top wire is into the plane of the page and the direction of the magnetic field at P due to the bottom wire is out of the plane of the page. Since the two magnetic field vectors at P have the same magnitude and opposite directions, the net magnetic field at Point P is zero.

9. **C** Use the right-hand rule for wires. If you point your thumb to the right and wrap your fingers along the wire, you will note that the magnetic field goes into the page when you are below the wire and comes out of the page above the wire. This allows us to eliminate choices (A) and (B). Because $B = \frac{\mu_o I}{2\pi r}$, the closer we are to the wire the stronger the magnetic field. Choice (C) is closer, so it is the correct answer. We could ignore choice (E), because point E is in line with the wire, so there is no radial distance and B becomes undefined there.

10. **D** Magnetic fields point from north to south. Therefore, the magnetic field between the two magnets is toward the right of the page. Use the right-hand rule. Because the B field is to the right and the charges through the wire flow to the bottom of the page, the force must be out of the page.

SECTION II: FREE RESPONSE

1. (a) The acceleration of an ion of charge q is equal to F_E/m. The electric force is equal to qE, where $E = V/d$. Therefore, $a = qV/(dm)$.

   (b) Using $a = qV/(dm)$ and the equation $v^2 = v_0^2 + 2ad = 2ad$, we get

$$v^2 = 2\frac{qV}{dm}d \quad \Rightarrow \quad v = \sqrt{\frac{2qV}{m}}$$

   As an alternate solution, notice that the change in the electrical potential energy of the ion from the source S to the entrance to the magnetic-field region is equal to qV; this is equal to the gain in the particle's kinetic energy.

   Therefore,

$$qV = \tfrac{1}{2}mv^2 \quad \Rightarrow \quad v = \sqrt{\frac{2qV}{m}}$$

   (c) (i) and (ii) Use the right-hand rule. Since **v** points to the right and **B** is into the plane of the page, the direction of **v** × **B** is upward. Therefore, the magnetic force on a positively charged particle (cation) will be upward, and the magnetic force on a negatively charged particle (anion) will be downward. The magnetic force provides the centripetal force that causes the ion to travel in a circular path. Therefore, a cation would follow Path 1 and an anion would follow Path 2.

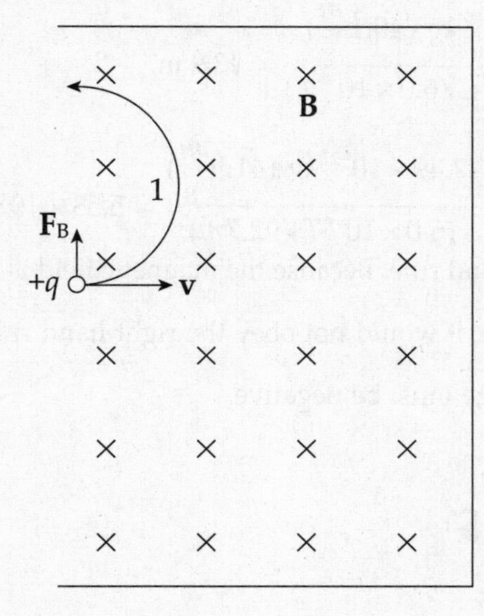

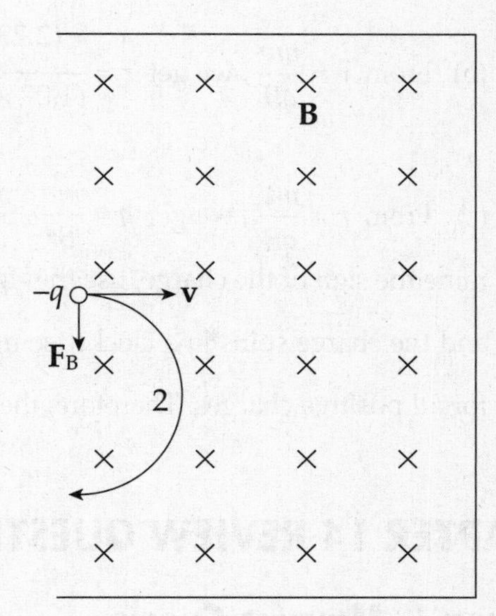

(d) Since the magnetic force on the ion provides the centripetal force,

$$qvB = \frac{mv^2}{r} \Rightarrow qvB = \frac{mv^2}{\frac{1}{2}y} \Rightarrow m = \frac{qBy}{2v}$$

Now, by the result of part (b),

$$m = \frac{qBy}{2\sqrt{\frac{2qV}{m}}} \Rightarrow m^2 = \frac{q^2B^2y^2}{\frac{8qV}{m}} \Rightarrow m^2 = \frac{mq^2B^2y^2}{8qV} \Rightarrow m = \frac{qB^2y^2}{8V}$$

(e) Since the magnetic force cannot change the speed of a charged particle, the time required for the ion to hit the photographic plate is equal to the distance traveled (the length of the semicircle) divided by the speed computed in part (b):

$$t = \frac{s}{v} = \frac{\pi \cdot \frac{1}{2}y}{\sqrt{\frac{2qV}{m}}} = \frac{1}{2}\pi y\sqrt{\frac{m}{2qV}}$$

(f) Since the magnetic force $\mathbf{F}_B$ is always perpendicular to a charged particle's velocity vector $\mathbf{v}$, it can do no work on the particle. Thus, the answer is zero.

2. (a) Because a photon has no charge, it will experience no force and therefore travel in a straight line.

(b) From $r = \frac{mv}{qB}$, we get $r = \frac{(2.23 \times 10^{-27} kg)(40.1\frac{m}{s})}{(1.07 \times 10^{-19} C)(6.0 \times 10^{-8} T)}$ = 13.9 m.

(c) From $r = \frac{mv}{qB}$, we get $q = \frac{mv}{Br}$. So $q = \frac{(7.49 \times 10^{-27} kg)(41.5\frac{m}{s})}{(6.0 \times 10^{-8} T)(92.7 m)}$ = 5.58×10^{-20}C. To determine the sign of the charge, use the right-hand rule. Because the magnetic field is into the page and the charge spins in a clockwise manner, it would not obey the right-hand rule that works for all positive charges. Therefore, the charge must be negative.

CHAPTER 14 REVIEW QUESTIONS

Section I: Multiple Choice

1. **E** Since $\mathbf{v}$ is upward and $\mathbf{B}$ is out of the page, the direction of $\mathbf{v} \times \mathbf{B}$ is to the right. Therefore, free electrons in the wire will be pushed to the left, leaving an excess of positive charge at the right. Therefore, the potential at Point b will be higher than at Point a, by $\mathcal{E} = vBL$ (motional emf).

2. **A** As shown in Example 14.3, the magnitude of the emf induced between the ends of the rod is $\mathcal{E} = BLv = (0.5\text{ T})(0.2\text{ m})(3\text{ m/s}) = 0.3\text{ V}$. Since the resistance is 10 Ω, the current induced will be $I = V/R = (0.3\text{ V})/(10\ \Omega) = 0.03\text{ A}$. To determine the direction of the current, we can note that since positive charges in the rod are moving to the left and the magnetic field points into the plane of the page, the right-hand rule tells us that the magnetic force, $q\mathbf{v} \times \mathbf{B}$, points downward. Since the resulting force on the positive charges in the rod is downward, so is the direction of the induced current.

3. **A** The magnetic field through the loop is $B = \mu_0 nI$. Since its area is $A = \pi r^2$, the magnetic flux through the loop is $\Phi_B = BA = (\mu_0 nI)(\pi r^2)$. If the current changes (with $\Delta I/\Delta t = -a$), then the magnetic flux through the loop changes, which, by Faraday's Law, implies that an emf (and a current) will be induced. We get

$$\mathcal{E} = -\frac{\Delta\Phi_B}{\Delta t} = -\frac{\Delta(\mu_0 nI \cdot \pi r^2)}{\Delta t} = -(\mu_0 n\pi r^2)\frac{\Delta I}{\Delta t} = -\mu_0 \pi n r^2(-a) = \mu_0 \pi n r^2 a$$

Since the magnetic flux into the page is decreasing, the direction of the induced current will be clockwise (opposing a *decreasing into-the-page flux* means that the induced current will create more into-the-page flux).

4. **C** By definition, magnetic field lines emerge from the north pole and enter at the south pole. Therefore, as the north pole is moved upward through the loop, the upward magnetic flux increases. To oppose an increasing upward flux, the direction of the induced current will be clockwise (as seen from above) to generate some downward magnetic flux. Now, as the south pole moves away from the center of the loop, there is a decreasing upward magnetic flux, so the direction of the induced current will be counterclockwise.

5. **E** Since the current in the straight wire is steady, there is no change in the magnetic field, no change in magnetic flux, and, therefore, no induced emf or current.

Section II: Free Response

1. (a) $\varepsilon = B\ell v \Rightarrow (2\text{T})(0.4\text{m})(1\text{ m/s})$ or 0.8V.

(b) $V = IR$ becomes $I = \frac{V}{R}$ or $I = \frac{0.8V}{20\Omega} = 0.04$ A. The direction is given by Lenz's Law. Current flows in order to oppose the change in the magnetic flux. Because there is suddenly a new flux "in," current flows to produce an outward flux. This would be in the counterclockwise direction.

(c) $\varepsilon = B\ell v \Rightarrow (2\text{T})(0.2\text{m})(1\text{ m/s})$ or 0.4 V

(d) In order to get the 0.8 V you would need $v = \frac{\varepsilon}{B\ell} \Rightarrow \frac{0.8\text{V}}{(2\text{T})(0.2\text{m})}$ or $v = 2.0$ m/s. You also could have noticed that if you cut the length in half, you have to compensate by doubling the speed.

CHAPTER 15 REVIEW QUESTIONS

Section I: Multiple Choice

1. **D** From the equation $\lambda f = v$, we find that

$$\lambda = \frac{v}{f} = \frac{10 \text{ m/s}}{5 \text{ Hz}} = 2 \text{ m}$$

2. **C** The speed of a transverse traveling wave on a stretched rope is given by the equation $v = \sqrt{F_T / \mu}$. Therefore,

$$v = \sqrt{\frac{F_T}{m/L}} = \sqrt{\frac{80 \text{ N}}{(1 \text{ kg})/(5 \text{ m})}} = \sqrt{400 \text{ m}^2/\text{s}^2} = 20 \text{ m/s}$$

3. **D** The time interval from a point moving from its maximum displacement above $y = 0$ (equilibrium) to its maximum displacement below equilibrium is equal to one-half the period of the wave. In this case,

$$T = \frac{1}{f} = \frac{\lambda}{v} = \frac{8 \text{ m}}{2 \text{ m/s}} = 4 \text{ s}$$

so the desired time is $\frac{1}{2}$(4 s) = 2 s.

4. **D** The standard equation for a transverse traveling wave has the form $y = A \sin(\omega t \pm \kappa x)$, where ω is the angular frequency and κ is the angular wave number. Here, $\kappa = \frac{1}{2}\pi \text{ cm}^{-1}$. Since, by definition, $\kappa = 2\pi/\lambda$, we have

$$\lambda = \frac{2\pi}{\kappa} = \frac{2\pi}{\frac{1}{2}\pi \text{ cm}^{-1}} = 4 \text{ cm}$$

5. **E** The distance between successive nodes is always equal to $\frac{1}{2}\lambda$. If a standing wave on a string fixed at both ends has a total of 4 nodes, the string must have a length L equal to $3(\frac{1}{2}\lambda)$. If $L = 6$ m, then λ must equal 4 m.

6. **B** We found in the previous question that $\lambda = 4$ m. Since $v = 40$ m/s, the frequency of this standing wave must be

$$f = \frac{v}{\lambda} = \frac{40 \text{ m/s}}{4 \text{ m}} = 10 \text{ Hz}$$

7. **A** In general, sound travels faster through solids than through gases. Therefore, when the wave enters the air from the metal rod, its speed will decrease. The frequency, however, will not change. Since $v = \lambda f$ must always be satisfied, a decrease in v implies a decrease in λ.

8. **A** The distance from S_2 to P is 5 m (it's the hypotenuse of 3-4-5 triangle), and the distance from S_1 to P is 4 m. The difference between the path lengths to Point P is 1 m, which is half the wavelength. Therefore, the sound waves are always exactly out of phase when they reach Point P from the two speakers, causing destructive interference there. By contrast, since Point Q is equidistant from the two speakers, the sound waves will always arrive in phase at Q, interfering constructively. Since there's destructive interference at P and constructive interference at Q, the amplitude at P will be less than at Q.

9. **E** The intensity (power per unit area) is proportional to $1/r^2$, where r is the distance between the source and the detector. If r increases by a factor of 10, the intensity decreases by a factor of 100. Because the decibel scale is logarithmic, if the intensity decreases by a factor of $100 = 10^2$, the decibel level decreases by $10 \log (10^2) = 20$ dB.

10. **B** An air column (such as an organ pipe) with one closed end resonates at frequencies given by the equation $f_n = nv/(4L)$ for odd integers n. The fundamental frequency corresponds, by definition, to $n = 1$. Therefore,

$$f_1 = \frac{v}{4L} = \frac{340 \text{ m/s}}{4(0.17 \text{ m})} = 500 \text{ Hz}$$

11. **B** The speed of the chirp is

$$v = \lambda f = (8.75 \times 10^{-3} \text{ m})(40 \times 10^3 \text{ Hz}) = 350 \text{ m/s}$$

If the distance from the bat to the tree is d, then the wave travels a total distance of $d + d = 2d$ (round-trip distance). If T is the time for this round-trip, then

$$2d = vT \Rightarrow d = \frac{vT}{2} = \frac{(350 \text{ m/s})(0.4 \text{ s})}{2} = 70 \text{ m}$$

12. **A** Since the car is traveling away from the stationary detector, the observed frequency will be lower than the source frequency. This eliminates choices B and C. Using the Doppler effect equation, we find that

$$f_D = \frac{v}{v + v_S} \cdot f_S = \frac{340 \text{ m/s}}{(340 + 20) \text{ m/s}} \cdot (600 \text{ Hz}) = \frac{34}{36}(600 \text{ Hz})$$

SECTION II: FREE RESPONSE

1. (a) The speed of a transverse traveling wave on a stretched rope is given by the equation $v = \sqrt{F_T / \mu}$. Therefore,

$$v^2 = \frac{F_T}{\mu} \Rightarrow F_T = \mu v^2 = (0.4 \text{ kg/m})(12 \text{ m/s})^2 = 58 \text{ N}$$

(b) Use the fundamental equation $\lambda f = v$:

$$f = \frac{v}{\lambda} = \frac{12 \text{ m/s}}{2 \text{ m}} = 6 \text{ Hz}$$

(c) (i) Because higher harmonic numbers correspond to shorter wavelengths, the harmonic number of the 3.2 m standing wave must be higher than that of the 4 m standing wave. We're told that these harmonic numbers are consecutive integers, so if n is the harmonic number of the 4 m standing wave, then $n + 1$ is the harmonic number of the 3.2 m wave. Therefore,

$$\frac{2L}{n} = 4 \text{ m} \quad \text{and} \quad \frac{2L}{n+1} = 3.2 \text{ m}$$

The first equation says that $2L = 4n$, and the second one says that $2L = 3.2(n + 1)$. Therefore, $4n$ must equal $3.2(n + 1)$; solving this equation gives $n = 4$. Substituting this into either one of the displayed equations then gives $L = 8$ m.

(c) (ii) Because $\mu = 0.4$ kg/m, the mass of the rope must be

$$m = \mu L = (0.4 \text{ kg/m})(8 \text{ m}) = 3.2 \text{ kg}$$

(d) We determined this in the solution to part (c) (i). The 4 m standing wave has harmonic number $n = 4$.

(e)

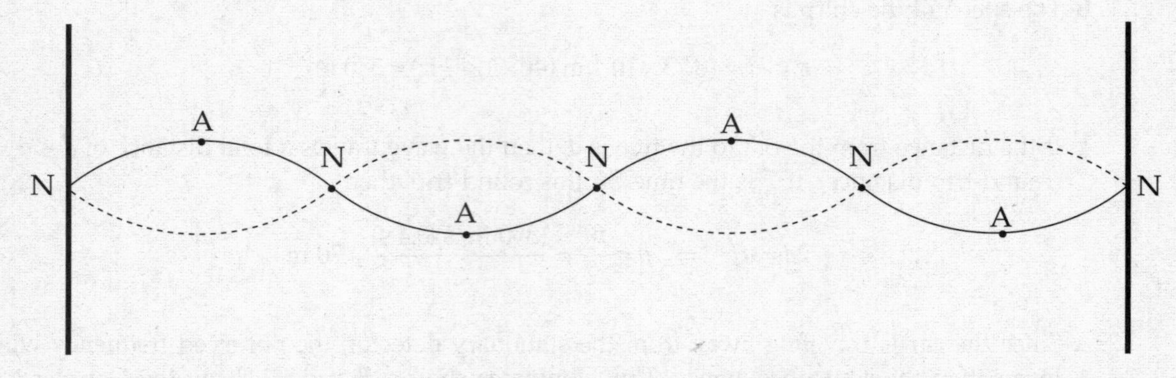

2. (a) For a stationary observer and a source moving away, the Doppler effect equation predicts that

$$f_D = \frac{v}{v + v_C} \cdot f_C = \frac{345 \text{ m/s}}{(345 + 50) \text{ m/s}}(500 \text{ Hz}) = 437 \text{ Hz}$$

(b) Yes. If the source is stationary and the observer moves away, the Doppler effect equation now gives

$$f_D = \frac{v - v_D}{v} \cdot f_C = \frac{(345 - 50) \text{ m/s}}{345 \text{ m/s}}(500 \text{ Hz}) = 428 \text{ Hz}$$

(c) With $\theta = 60°$, the given equation becomes

$$f' = \frac{v}{v - v_C \cos\theta} \cdot f = \frac{345 \text{ m/s}}{345 \text{ m/s} - (50 \text{ m/s})\cos 60°}(500 \text{ Hz}) = 539 \text{ Hz}$$

Note that the observed frequency is higher than the source frequency, which we expect since the car is traveling toward the students.

(d) With $\theta = 120°$, the given equation becomes

$$f' = \frac{v}{v - v_C \cos\theta} \cdot f = \frac{345 \text{ m/s}}{345 \text{ m/s} - (50 \text{ m/s})\cos 120°}(500 \text{ Hz}) = 466 \text{ Hz}$$

Note that the observed frequency is now lower than the source frequency, which we expect since the car is traveling away from the students.

(e) When the car is far to the left of the students' position, θ is very small (approaches 0°). Therefore, for large negative x, the observed frequency should approach

$$f' = \frac{v}{v - v_C \cos\theta} \cdot f = \frac{345 \text{ m/s}}{345 \text{ m/s} - (50 \text{ m/s})\cos 0°}(500 \text{ Hz}) = 585 \text{ Hz}$$

When the car is far to the right of the students' position, θ approaches 180°. Therefore, for large positive x, the observed frequency should approach

$$f' = \frac{v}{v - v_C \cos\theta} \cdot f = \frac{345 \text{ m/s}}{345 \text{ m/s} - (50 \text{ m/s})\cos 180°}(500 \text{ Hz}) = 437 \text{ Hz}$$

When the car is directly in front of the students—that is, when $x = 0$ (and $\theta = 90°$)—the equation given predicts that f' will equal 500 Hz. The graph of f' versus x should therefore have the following general shape:

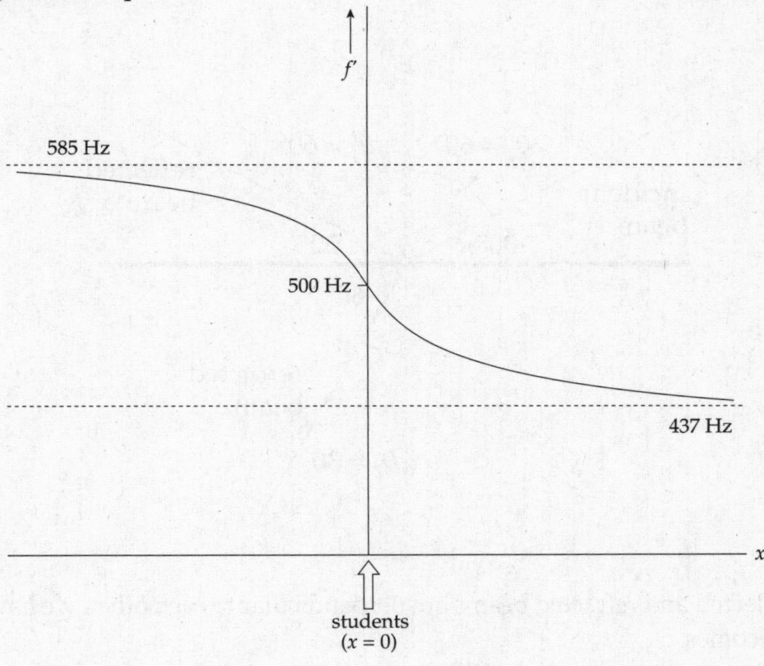

CHAPTER 16 REVIEW QUESTIONS

SECTION I: MULTIPLE CHOICE

1. **B** From the equation $\lambda f = c$, we find that

$$\lambda = \frac{c}{f} = \frac{3.0 \times 10^{8} \text{ m/s}}{1.0 \times 10^{18} \text{ Hz}} = 3.0 \times 10^{-10} \text{ m}$$

2. **D** Since the fringe is bright, the waves must interfere constructively. This implies that the difference in path lengths must be a whole number times the wavelength, eliminating choices (B), (C), and (E). The central maximum is equidistant from the two slits, so $\Delta\ell = 0$ there. At the first bright fringe above the central maximum, we have $\Delta\ell = \lambda$.

3. **C** First, eliminate choices (A) and (B): The index of refraction is never smaller than 1. Refer to the following diagram:

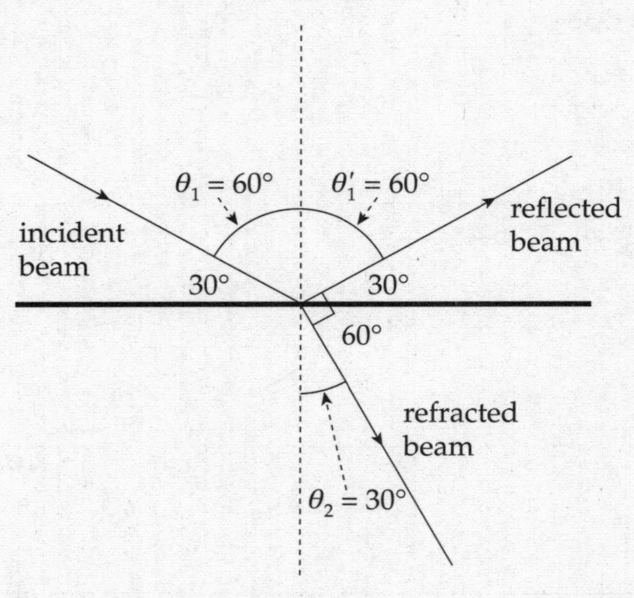

Since the reflected and refracted beams are perpendicular to each other, we have $\theta_2 = 30°$. Snell's Law then becomes

$$n_1 \sin\theta_1 = n_2 \sin\theta_2$$
$$1 \cdot \sin 60° = n_2 \sin 30°$$
$$\tfrac{1}{2}\sqrt{3} = n_2 \cdot \tfrac{1}{2}$$
$$\sqrt{3} = n_2$$

4. **A** The frequency is unchanged, but because the speed of light in diamond is less than in air, the wavelength of the light in diamond is shorter than its wavelength in air:

$$\lambda_{\text{in diamond}} = \frac{1}{f} v_{\text{in diamond}} = \frac{1}{f}\frac{c}{n_{\text{diamond}}} = \frac{1}{n_{\text{diamond}}}\lambda_{\text{in air}} = \frac{500 \text{ nm}}{2.5} = 200 \text{ nm}$$

5. **A** If the speed of light is less in Medium 2 than in Medium 1, then Medium 2 must have the higher index of refraction; that is, $n_2 > n_1$. Snell's Law then implies that $\theta_2 < \theta_1$: The beam will refract toward the normal upon transmission into Medium 2.

6. **A** The critical angle for total internal reflection is computed as follows:

$$n_1 \sin\theta_c = n_2 \sin(90°)$$

$$\sin\theta_c = \frac{n_2}{n_1} = \frac{1.45}{2.90} = \frac{1}{2} \Rightarrow \theta_c = 30°$$

Total internal reflection can happen only if the incident beam originates in the medium with the higher index of refraction and strikes the interface of the other medium at an angle of incidence greater than the critical angle.

7. **D** If $s_o = 60$ cm and $f = 40$ cm, the mirror equation tells us that

$$\frac{1}{s_o} + \frac{1}{s_i} = \frac{1}{f} \Rightarrow \frac{1}{60\text{ cm}} + \frac{1}{s_i} = \frac{1}{40\text{ cm}} \Rightarrow \frac{1}{s_i} = \frac{1}{120\text{ cm}} \Rightarrow s_i = 120\text{ cm}$$

And, since s_i is positive, the image is real.

8. **D** Because the image is virtual, we must write the image distance, s_i, as a negative quantity: $s_i = -20$ cm. Now, with $s_o = 60$ cm, the mirror equation gives

$$\frac{1}{s_o} + \frac{1}{s_i} = \frac{1}{f} \Rightarrow \frac{1}{60\text{ cm}} + \frac{1}{-20\text{ cm}} = \frac{1}{f} \Rightarrow \frac{1}{f} = \frac{1}{-30\text{ cm}} \Rightarrow f = -30\text{ cm}$$

The focal length is half the radius of curvature, so

$$f = \frac{R}{2} \Rightarrow R = 2f = 2(-30\text{ cm}) = -60\text{ cm} \Rightarrow |R| = 60\text{ cm}$$

9. **C** Since the image is projected onto a screen, it must be real, and therefore inverted. The magnification must be negative, so

$$M = -\frac{1}{4} \Rightarrow -\frac{s_i}{s_o} = -\frac{1}{4} \Rightarrow s_o = 4s_i$$

Because $s_i = 60$ cm, the object distance, s_o, must be 240 cm. Therefore,

$$\frac{1}{s_o} + \frac{1}{s_i} = \frac{1}{f} \Rightarrow \frac{1}{240\text{ cm}} + \frac{1}{60\text{ cm}} = \frac{1}{f} \Rightarrow \frac{1}{f} = \frac{1}{48\text{ cm}} \Rightarrow f = 48\text{ cm}$$

10. **B** A bi-concave lens is a diverging lens. Diverging lenses (like convex mirrors) have negative focal lengths and therefore cannot form real images. (Note that statement (D) is false; diverging lenses and convex mirrors always form diminished, virtual images, as you can verify using the mirror and magnification equations.)

Section II: Free Response

1. (a) Interference effects can be observed only if the light is coherent. Using two independent light sources at the slits in Barrier 2 would generate incoherent light waves. The set-up shown guarantees that the light reaching the two slits will be coherent.

(b) Maxima are located at positions given by the equation $x_m = m\lambda L/d$, where m is an integer ($y_0 = 0$ is the central maximum). The first-order maximum for red light occurs at

$$x_{1,\text{ red}} = \frac{1 \cdot \lambda_{\text{red}} L}{d} = \frac{1 \cdot (750 \times 10^{-9}\text{ m})(4.0\text{ m})}{0.50 \times 10^{-3}\text{ m}} = 0.006\text{ m} = 6.0\text{ mm}$$

and the first-order maximum for violet light occurs at

$$x_{1,\text{ violet}} = \frac{1 \cdot \lambda_{\text{violet}} L}{d} = \frac{1 \cdot (400 \times 10^{-9}\text{ m})(4.0\text{ m})}{0.50 \times 10^{-3}\text{ m}} = 0.0032\text{ m} = 3.2\text{ mm}$$

Therefore, the vertical separation of these maxima on the screen is $\Delta x = 6.0\text{ mm} - 3.2\text{ mm} = 2.8\text{ mm}$.

(c) Maxima are located at positions given by the equation $x_m = m\lambda L/d$, where m is an integer. We therefore want to solve the equation

$$x_{m,\text{ violet}} = x_{n,\text{ orange-yellow}}$$

$$\frac{m\lambda_{\text{violet}} L}{d} = \frac{n\lambda_{\text{orange-yellow}} L}{d}$$

$$m\lambda_{\text{violet}} = n\lambda_{\text{orange-yellow}}$$

$$m(400\text{ nm}) = n(600\text{ nm})$$

The smallest integers that satisfy this equation are $m = 3$ and $n = 2$. That is, the third-order maximum for violet light coincides with the second-order maximum for orange-yellow light. The position on the screen (relative to the central maximum at $x = 0$) of these maxima is

$$x_{3,\text{ violet}} = \frac{m\lambda_{\text{violet}} L}{d} = \frac{3(400 \times 10^{-9}\text{ m})(4.0\text{ m})}{0.50 \times 10^{-3}\text{ m}} = 0.0096\text{ m} = 9.6\text{ mm}$$

(d) Within the glass, the wavelength is reduced by a factor of $n = 1.5$ from the wavelength in air. Therefore, the difference in path lengths, $d \sin \theta$, must be equal to $m(\lambda/n)$ in order for constructive interference to occur. This implies that the maxima are located at positions given by $x_m = m\lambda L/(nd)$. The distance between adjacent bright fringes is therefore

$$x_{m+1} - x_m = \frac{(m+1)\lambda L}{nd} - \frac{m\lambda L}{nd} = \frac{\lambda L}{nd} = \frac{(500 \times 10^{-9}\text{ m})(4.0\text{ m})}{1.5(0.50 \times 10^{-3}\text{ m})} = 0.0027\text{ m} = 2.7\text{ mm}$$

2. (a) Beam 1 undergoes a 180° phase change when it reflects off the soap film. The refracted portion of the incident beam does not undergo a phase change upon reflection at the film/air boundary, because it strikes the boundary to a medium whose index is lower; therefore, Beam 2 does not suffer a 180° phase change.

(b) We're told that the criterion for destructive interference in this case is $\Delta\ell = 2T = m(\lambda/n)$, where n is the refractive index of the soap film and m is a whole number. Therefore, the criterion for *constructive* interference must be

$$\Delta\ell = 2T = (m + \tfrac{1}{2})\frac{\lambda}{n}$$

Alternatively, $\Delta\ell = 2T = \frac{k}{2} \bullet \frac{\lambda}{n}$ where k is an *odd* whole number.

3. (a) Since the image is formed behind the mirror, it is virtual.

(b) Virtual images are upright.

(c) Writing $s_i = -72$ cm (negative because the image is virtual), the magnification is

$$M = -\frac{s_i}{s_o} = -\frac{-72 \text{ cm}}{40 \text{ cm}} = 1.8$$

so the height of the image is $h_i = |m| \cdot h_o = 1.8(5 \text{ cm}) = 9$ cm.

(d) The mirror equation can be used to solve for the mirror's focal length:

$$\frac{1}{s_o} + \frac{1}{s_i} = \frac{1}{f} \quad \Rightarrow \quad \frac{1}{40 \text{ cm}} + \frac{1}{-72 \text{ cm}} = \frac{1}{f} \quad \Rightarrow \quad \frac{1}{90 \text{ cm}} = \frac{1}{f} \quad \Rightarrow \quad f = 90 \text{ cm}$$

Therefore, the mirror's radius of curvature is $R = 2f = 2(90 \text{ cm}) = 180$ cm.

(e)

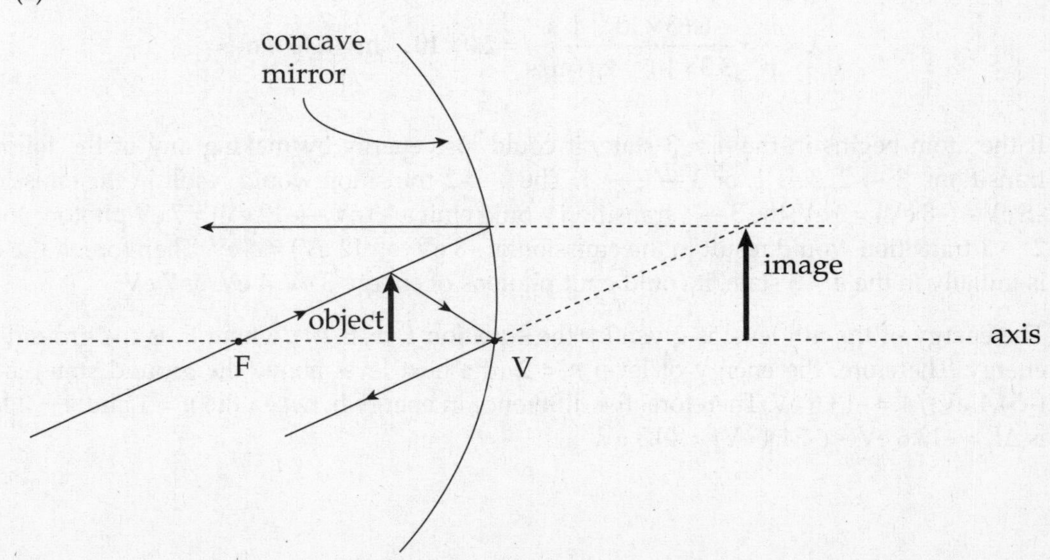

CHAPTER 17 REVIEW QUESTIONS

SECTION I: MULTIPLE CHOICE

1. **B** Combining the equation $E = hf$ with $f = c/\lambda$ gives us

$$E = \frac{hc}{\lambda} = \frac{(4.14 \times 10^{-15} \text{ eV} \cdot \text{s})(3.00 \times 10^{8} \text{ m/s})}{2.07 \times 10^{-9} \text{ m}} = 600 \text{ eV}$$

2. **D** The energy of the incident photons is

$$E = hf = (4.14 \times 10^{-15} \text{ eV} \cdot \text{s})(7.2 \times 10^{15} \text{ Hz}) = 30 \text{ eV}$$

Since $E > \phi$, photoelectrons will be produced, with maximum kinetic energy

$$K_{max} = E - \phi = 30 \text{ eV} - 6 \text{ eV} = 24 \text{ eV}$$

3. **A** If the atom's ionization energy is 25 eV, then the electron's ground-state energy must be –25 eV. Making a transition from the –16 eV energy level to the ground state will cause the emission of a photon of energy

$$\Delta E = (-16 \text{ eV}) - (-25\text{eV}) = 9 \text{ eV}$$

4. **E** The gap between the ground-state and the first excited state is

$$-10 \text{ eV} - (-40 \text{ eV}) = 30 \text{ eV}$$

Therefore, the electron must absorb the energy of a 30 eV photon (at least) in order to move even to the first excited state. Since the incident photons have only 15 eV of energy, the electron will be unaffected.

5. **C** The de Broglie wavelength of a particle whose momentum is p is $\lambda = h/p$. For this proton, we find that

$$\lambda = \frac{h}{p} = \frac{6.63 \times 10^{-34} \text{ J} \cdot \text{s}}{3.3 \times 10^{-23} \text{ kg} \cdot \text{m/s}} = 2.0 \times 10^{-11} \text{ m} = 0.02 \text{ nm}$$

6. **E** If the atom begins in the $n = 3$ state, it could lose energy by making any of the following transitions: $3 \to 2$, $3 \to 1$, or $3 \to 2 \to 1$. The $3 \to 2$ transition would result in the emission of –5 eV – (–8 eV) = 3 eV; the $3 \to 1$ transition would emit a –5 eV – (–12 eV) = 7 eV photon; and the $2 \to 1$ transition would result in the emission of –8 eV – (–12 eV) = 4 eV. Therefore, if the atom is initially in the $n = 3$ state, it could emit photons of energy 3 eV, 4 eV, or 7 eV.

7. **D** The energy of the nth level is given by the equation $E_n = E_1/n^2$, where E_1 is the ground-state energy. Therefore, the energy of level $n = 2$ (the next level above the ground state) is $E_2 = (-54.4 \text{ eV})/4 = -13.6$ eV. Therefore, the difference in energy between the $n = 1$ and $n = 2$ levels is $\Delta E = -13.6 \text{ eV} - (-54.4 \text{ eV}) = 40.8$ eV.

8. **C** The energy of a photon is given by the equation $E = hf$, or equivalently by $E = hc/\lambda$. Therefore, E is inversely proportional to λ. If λ decreases by a factor of 2, then E will increase by a factor of 2.

9. **A** The equation that relates the mass difference m and the disintegration energy Q is Einstein's mass–energy equivalence formula, $Q = mc^2$. Because c^2 is a constant, we see that Q is proportional to m. Therefore, if m decreases by a factor of 4, then so will Q.

10. **B** In order to balance the mass number (the superscripts), we must have $2 + 63 = 64 + A$, so $A = 1$. In order to balance the charge (the subscripts), we need $1 + 29 = 30 + Z$, so $Z = 0$. A particle with a mass number of 1 and no charge is a neutron, ${}^{1}_{0}\text{n}$.

11. **E** In order to balance the mass number (the superscripts), we must have $196 + 1 = 197 + A$, so $A = 0$. In order to balance the charge (the subscripts), we need $78 + 0 = 78 + Z$, so $Z = 0$. The only particle listed that has zero mass number and zero charge is a gamma-ray photon, ${}^{0}_{0}\gamma$.

SECTION II: FREE RESPONSE

1. (a) If ionizing lithium twice leaves the atom with just one electron, then it must have originally had three electrons. Since neutral atoms have the same number of protons as electrons, Z must be 3 for lithium.

   (b) First, use the formula given for E_n to determine the values of the first few electron energy levels:

$$E_1 = \frac{3^2}{1^2}(-13.6 \text{ eV}) = -122.4 \text{ eV}$$

$$E_2 = \frac{3^2}{2^2}(-13.6 \text{ eV}) = -30.6 \text{ eV}$$

$$E_3 = \frac{3^2}{3^2}(-13.6 \text{ eV}) = -13.6 \text{ eV}$$

$$E_4 = \frac{3^2}{4^2}(-13.6 \text{ eV}) = -7.65 \text{ eV}$$

Differences between electron energy levels equal the energies of the emitted photons. Since $E = hf = hc/\lambda$, we have $\lambda = hc/E$. Since the E's are in electronvolts and the wavelengths in nanometers, it is particularly helpful to note that

$$hc = (4.14 \times 10^{-15} \text{ eV} \cdot \text{s})(3.00 \times 10^8 \times 10^9 \text{ nm/s}) = 1240 \text{ eV} \cdot \text{nm}$$

The goal here is to match the energy-level differences with the given wavelengths, using the equation $\lambda = (1240 \text{ eV·nm})/E$.

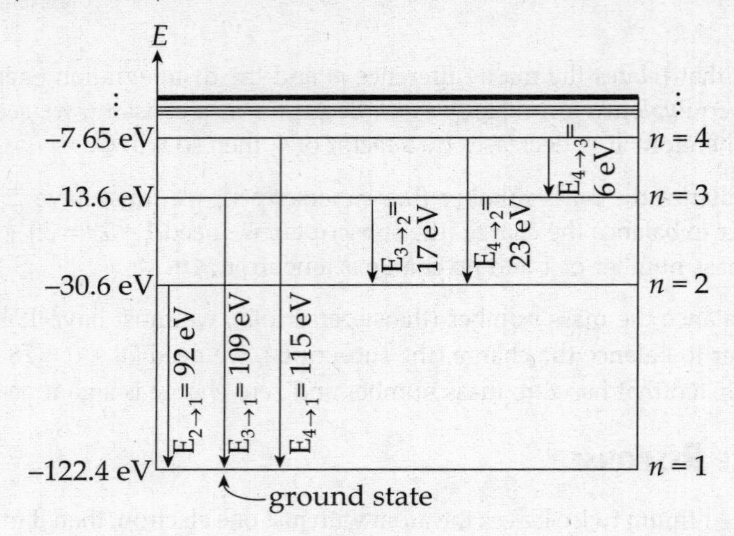

Using this diagram, the four wavelengths given correspond to the following photon energies and energy-level transitions:

$$\lambda = 11.4 \text{ nm} \quad \leftrightarrow \quad E = \frac{hc}{\lambda} = \frac{1240 \text{ eV} \cdot \text{nm}}{11.4 \text{ nm}} = 109 \text{ eV} \quad \leftrightarrow \quad (3 \rightarrow 1 \text{ transition})$$

$$\lambda = 13.5 \text{ nm} \quad \leftrightarrow \quad E = \frac{hc}{\lambda} = \frac{1240 \text{ eV} \cdot \text{nm}}{13.5 \text{ nm}} = 92 \text{ eV} \quad \leftrightarrow \quad (2 \rightarrow 1 \text{ transition})$$

$$\lambda = 54.0 \text{ nm} \quad \leftrightarrow \quad E = \frac{hc}{\lambda} = \frac{1240 \text{ eV} \cdot \text{nm}}{54.0 \text{ nm}} = 23 \text{ eV} \quad \leftrightarrow \quad (4 \rightarrow 2 \text{ transition})$$

$$\lambda = 72.9 \text{ nm} \quad \leftrightarrow \quad E = \frac{hc}{\lambda} = \frac{1240 \text{ eV} \cdot \text{nm}}{72.9 \text{ nm}} = 17 \text{ eV} \quad \leftrightarrow \quad (3 \rightarrow 2 \text{ transition})$$

(c) No. In order for a spectral line to have a wavelength between 54.0 nm and 72.9 nm, there must be an energy-level transition with an energy between 17 eV and 23 eV. As the diagram above shows, no such intermediate transition is possible.

(d) The 11.4 nm line corresponds to a photon energy of 109 eV. Therefore, the next shortest wavelength must correspond to the next highest possible photon energy. According to the diagram, the next highest available transition energy after 109 eV is 115 eV. This wavelength of the photon emitted in this $(4 \rightarrow 1)$ transition is

$$\lambda = \frac{hc}{E} = \frac{1240 \text{ eV} \cdot \text{nm}}{115 \text{ eV}} = 10.8 \text{ nm}$$

20

AP Physics B Practice Exam 1: Answers and Explanations

SECTION I

1. **D** If the acceleration is not zero, then the velocity is changing. Since linear momentum (Item II) is the product of mass and velocity, it cannot remain constant if the velocity is changing. However, the object's speed (Item I) could be constant (consider, for example, uniform circular motion), so its kinetic energy (Item III), the scalar $K = \frac{1}{2}mv^2$, could be constant as well.

2. **E** The gravitational force between the objects is given by

$$F = G\frac{m_1 m_2}{r^2}$$

where m_1 and m_2 are the masses of the objects and r is the distance between their centers. If both masses are doubled and r is doubled, then the gravitational force will be

$$F' = G\frac{(2m_1)(2m_2)}{(2r)^2} = G\frac{4m_1 m_2}{4r^2} = G\frac{m_1 m_2}{r^2} = F$$

That is, the force remains the same.

3. **D** Since the particle travels in a circular path with constant kinetic energy (which implies constant speed), the net force on the particle is the centripetal force:

$$F = \frac{mv^2}{r} = \frac{2(\frac{1}{2}mv^2)}{r} = \frac{2K}{r} = \frac{2(4\text{ J})}{0.2\text{ m}} = 40\text{ N}$$

4. **E** Energy is a scalar, not a vector, quantity. Displacement, velocity, acceleration, and linear momentum (the product of mass and velocity) are all vector quantities.

5. **D** The acceleration of the block is $a = g \sin \theta$, where θ is the incline angle, and the distance it must travel—the length of the incline—is $h/\sin \theta = h/\sin 30° = 2h$. Using the equation $s = \frac{1}{2}at^2$, we find that

$$t = \sqrt{\frac{2s}{a}} = \sqrt{\frac{2(2h)}{g\sin\theta}} = \sqrt{\frac{4(5\text{ m})}{(10\text{ m/s}^2)\sin 30°}} = 2\text{ s}$$

6. **B** The easiest way to answer this question is to use Conservation of Energy:

$$mgh = \frac{1}{2}mv^2 \quad \Rightarrow \quad v = \sqrt{2gh} = \sqrt{2(10\text{ m/s}^2)(5\text{ m})} = 10\text{ m/s}$$

7. **A** Since the normal force is perpendicular to the object's displacement, it does no work.

8. **B** The gravitational force provides the centripetal force on each satellite, so

$$\frac{GMm}{r^2} = \frac{mv^2}{r} \quad \Rightarrow \quad v = \sqrt{\frac{GM}{r}}$$

Therefore,

$$\frac{v_2}{v_1} = \frac{\sqrt{GM/r_2}}{\sqrt{GM/r_1}} = \sqrt{\frac{r_1}{r_2}} \quad \Rightarrow \quad v_2 = v_1\sqrt{\frac{r_1}{r_2}}$$

9. **C** The capacitance of a parallel-plate capacitor is $C = \kappa\varepsilon_0 A/d$ where κ is the dielectric constant, A is the area of each plate, and d is their separation distance. Decreasing d will cause C to increase.

10. **A** The period is the reciprocal of the frequency: $T = 1/f = 1/(2.5\text{ Hz}) = 0.4\text{ s}$.

11. **B** One way to answer this question is to invent some numbers that satisfy the conditions of the question and observe what happens. Let's choose $f = 1$ m, and take $s_{o1} = 3$ m and $s_{o2} = 2$ m (notice that $f < s_{o2} < s_{o1}$, which means that the object is moving closer to the focal point). Using the mirror equation, we find that

$$\frac{1}{s_{o1}} + \frac{1}{s_{i1}} = \frac{1}{f} \Rightarrow \frac{1}{3\text{ m}} + \frac{1}{s_{i1}} = \frac{1}{1\text{ m}} \Rightarrow s_{i1} = \frac{3}{2}\text{ m}$$

$$\frac{1}{s_{o2}} + \frac{1}{s_{i2}} = \frac{1}{f} \Rightarrow \frac{1}{2\text{ m}} + \frac{1}{s_{i2}} = \frac{1}{1\text{ m}} \Rightarrow s_{i2} = 2\text{ m}$$

Since $s_{i2} > s_{i1}$, the image is moving away from the mirror (eliminating choices (C) and (D)). Now for the magnifications:

$$M_1 = -\frac{s_{i1}}{s_{o1}} = -\frac{\frac{3}{2}\text{ m}}{3\text{ m}} = -\frac{1}{2}$$

$$M_2 = -\frac{s_{i2}}{s_{o2}} = -\frac{2\text{ m}}{2\text{ m}} = -1$$

Since $|M_2| > |M_1|$, the image gets taller. (You can also use ray diagrams.)

12. **B** The energies emitted during electron transitions are equal to the differences between the allowed energy levels. Choice (A), 17 eV, is equal to the energy emitted when the electron drops from the –21 eV level to the –38 eV level. Choice (C), 64 eV, is equal to the energy emitted when the electron drops from the –21 eV level to the –85 eV level. Choice (D), 255 eV, is equal to the energy emitted when the electron drops from the –85 eV level to the –340 eV level. And choice (E), 302 eV, is equal to the energy emitted when the electron drops from the –38 eV level to the –340 eV level. However, no electron transition in this atom could give rise to a 42 eV photon.

13. **A** Because the expansion is isothermal, no change in temperature occurs, which implies that no change in the internal energy of the gas takes place. Since $\Delta U = Q + W$ (the first law of thermodynamics), the fact that $\Delta U = 0$ implies that $Q = -W$. Since $W = -150$ J, it must be true that $Q = +150$ J.

14. **C** The frequency does not change, so the wavelength must (because the wave speed changes):

$$\lambda_{\text{in brass}} = \frac{v_{\text{in brass}}}{f} = \frac{10v_{\text{in air}}}{f} = 10\lambda_{\text{in air}}$$

15. **C** The net force on the two-block system is $3mg - F_f$, where F_f is the magnitude of the frictional force acting on the small block. Since $F_f = \mu mg$, we find that

$$a = \frac{F_{\text{net}}}{\text{total mass}} = \frac{3mg - (0.2)mg}{m + 3m} = \frac{(2.8)mg}{4m} = (0.7)g$$

16. **D** By Newton's Third Law, the force on the heavier person is equal but opposite to the force on the lighter person, eliminating choices (A) and (B). Since the lighter person has $\frac{1}{2}$ the mass of the heavier person and acceleration is inversely proportional to mass, the magnitude of the lighter person's acceleration will be twice that of the heavier person.

17. **E** The mass of the object is $m = w/g = (6 \text{ N})/(10 \frac{\text{N}}{\text{kg}}) = 0.6$ kg, so the density of the object is $\rho = m/V = (0.6 \text{ kg})/(2 \times 10^{-3} \text{ m}^3) = 300 \text{ kg/m}^3$. Since this is 30% of the density of water ($\rho_{\text{water}} = 1000 \text{ kg/m}^3$), Archimedes' Principle tells us that 30% of the floating object's volume will be submerged. This leaves 70% above the surface of the water.

18. **A** If the mass of the $+q$ charge is m, then its acceleration is

$$a = \frac{F_E}{m} = \frac{1}{4\pi\varepsilon_0} \frac{Qq}{mr^2}$$

The graph in (A) best depicts an inverse-square relationship between a and r.

19. **D** When the block comes to rest, the spring force, kx, balances the gravitational force, mg. Therefore,

$$kx = mg \quad \Rightarrow \quad x = \frac{mg}{k} = \frac{(4.0 \text{ kg})(10 \text{ N/kg})}{800 \text{ N/m}} = 0.05 \text{ m} = 5.0 \text{ cm}$$

20. **E** Since the crate is being lifted with constant velocity, the magnitude of the force exerted by the crane must also be constant and equal to the magnitude of the gravitational force on the crate. Power can then be found from the equation

$$P = Fv = (5000 \text{ N})(4 \text{ m/s}) = 20{,}000 \text{ W}$$

21. **C** When the particle enters the magnetic field, the magnetic force provides the centripetal force to cause the particle to execute uniform circular motion:

$$|q|vB = \frac{mv^2}{r} \quad \Rightarrow \quad r = \frac{mv}{|q|B}$$

Since v and B are the same for all the particles, the largest r is found by maximizing the ratio $m/|q|$. The value of $m/|q|$ for an alpha particle is about twice that for a proton and thousands of times greater than that of an electron or positron.

22. **B** The value of g near the surface of a planet depends on the planet's mass and radius:

$$mg = \frac{GMm}{r^2} \quad \Rightarrow \quad g = \frac{GM}{r^2}$$

Therefore, calling this Planet X, we find that

$$g_X = \frac{GM_X}{r_X^2} = \frac{G(2M_{\text{Earth}})}{(2r_{\text{Earth}})^2} = \frac{1}{2}\frac{GM_{\text{Earth}}}{r_{\text{Earth}}^2} = \frac{1}{2}g_{\text{Earth}}$$

Since g is half as much on the planet as it is on Earth, the astronaut's weight (mg) will be half as much as well.

23. **E** The maximum net force on the object occurs when all four forces act in the same direction, giving $F_{net} = 4F = 4(10 \text{ N}) = 40$ N, and a resulting acceleration of $a = F_{net}/m = (40 \text{ N})/(5 \text{ kg}) = 8 \text{ m/s}^2$. These four forces could not give the object an acceleration greater than this.

24. **A** The Carnot efficiency is equal to $(T_H - T_C)/T_H$, where the temperatures are expressed in *kelvins*. Since 127°C = 127 + 273 = 400 K and 27°C = 27 + 273 = 300 K, we get

$$e = \frac{T_H - T_C}{T_H} = \frac{400 \text{ K} - 300 \text{ K}}{400 \text{ K}} = \frac{1}{4} = 25\%$$

25. **C** The nth harmonic frequency is equal to n times the fundamental frequency, f_1. Therefore,

$$\frac{f_6}{f_3} = \frac{6f_1}{3f_1} = 2$$

26. **B** We use the equation for the power dissipated by a resistor, $P = IV$:

$$I = \frac{P}{V} = \frac{60 \text{ W}}{120 \text{ V}} = 0.5 \text{ A}$$

27. **B** A point charge placed at point A, B, or C would experience no net force only if the electric field at that point were zero. Let's label the four charges 1, 2, 3, and 4, starting at the upper right and counting counterclockwise. The diagram below shows the individual electric field vectors at points A, B, and C due to the four charges.

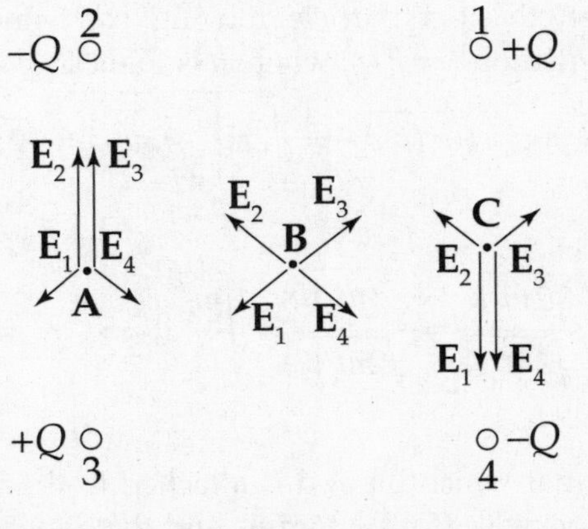

Only at point B do the four electric field vectors cancel each other to give a sum of zero, so only at point B would a charge feel no net force.

28. **E** The electric potential is a scalar, given by the formula $V = kQ/r$, where $k = 1/4\pi\varepsilon_0$ is Coulomb's constant and r is the distance from the charge Q. In this case, the total potential at any point is the sum of the four potentials produced by the four source charges. Here's the computation of the potential at each of the points A, B, and C, where a denotes the length of each side of the square in the diagram. Note that all of them equal zero.

$$V_\text{A} = V_\text{A1} + V_\text{A2} + V_\text{A3} + V_\text{A4} = k\frac{+Q}{\frac{\sqrt{5}}{2}a} + k\frac{-Q}{\frac{1}{2}a} + k\frac{+Q}{\frac{1}{2}a} + k\frac{-Q}{\frac{\sqrt{5}}{2}a} = 0$$

$$V_\text{B} = V_\text{B1} + V_\text{B2} + V_\text{B3} + V_\text{B4} = k\frac{+Q}{\frac{\sqrt{2}}{2}a} + k\frac{-Q}{\frac{\sqrt{2}}{2}a} + k\frac{+Q}{\frac{\sqrt{2}}{2}a} + k\frac{-Q}{\frac{\sqrt{2}}{2}a} = 0$$

$$V_\text{C} = V_\text{C1} + V_\text{C2} + V_\text{C3} + V_\text{C4} = k\frac{+Q}{\frac{1}{2}a} + k\frac{-Q}{\frac{\sqrt{5}}{2}a} + k\frac{+Q}{\frac{\sqrt{5}}{2}a} + k\frac{-Q}{\frac{1}{2}a} = 0$$

29. **D** The buoyant force is given by $F_\text{B} = \rho_\text{air} V_\text{sub} g$. Because the entire balloon is surrounded by air, V_sub is equal to V, the full volume of the balloon. Therefore, the buoyant force is proportional to the balloon's volume. Since the volume of a sphere is given by $V = \frac{4}{3}\pi r^3$, if r is doubled, V will increase by a factor of $2^3 = 8$. Thus, F_B increases by a factor of 8.

30. **C** When a wave enters a new medium, its frequency does not change, but its wave speed does. Since $\lambda f = v$, the change in wave speed implies a change in wavelength also.

31. **D** Since the magnetic flux out of the page is decreasing, the induced current will oppose this change (as always), attempting to create more magnetic flux out of the page. In order to do this, the current must circulate in the counterclockwise direction (remember the right-hand rule). As B decreases uniformly (that is, while $\Delta B/\Delta t$ is negative and constant), the induced emf,

$$\varepsilon = -\frac{\Delta\Phi_\text{B}}{\Delta t} = -A\frac{\Delta B}{\Delta t}$$

is nonzero and constant, which implies that the induced current, $I = \varepsilon/R$, is also nonzero and constant.

32. **A** The de Broglie wavelength of a particle moving with linear momentum of magnitude $p = mv$ is given by the equation $\lambda = h/p$, where h is Planck's constant. Because

$$p = mv = \sqrt{m(mv^2)} = \sqrt{2m\left(\frac{1}{2}mv^2\right)} = \sqrt{2mK}$$

we find that

$$\frac{\lambda_\text{e}}{\lambda_\text{p}} = \frac{h/\sqrt{2m_\text{e}K}}{h/\sqrt{2m_\text{p}K}} = \frac{\sqrt{2m_\text{p}K}}{\sqrt{2m_\text{e}K}} = \sqrt{\frac{m_\text{p}}{m_\text{e}}} \Rightarrow \lambda_\text{e} = \lambda_\text{p}\sqrt{\frac{m_\text{p}}{m_\text{e}}}$$

33. **D** With respect to the point at which the rod is attached to the vertical wall, the tension in the string exerts a counterclockwise (CCW) torque, and the gravitational force—which acts at the

rod's center of mass—exerts a clockwise (CW) torque. If the rod is in equilibrium, these torques must balance. Letting L denote the length of the rod, this gives

$$\tau_{CCW} = \tau_{CW}$$

$$F_T L \sin\theta = (\frac{1}{2}L)(mg)\sin\theta$$

$$F_T = \frac{1}{2}mg$$

34. **A** Statement (B) is false, because a solid must *absorb* thermal energy in order to melt. Statement (C) is false since it generally requires much more energy to break the intermolecular bonds of a liquid to change its state to vapor than to loosen the intermolecular bonds of a solid to change its state to liquid. And statements (D) and (E) are false: While a substance undergoes a phase change, its temperature remains constant. The only true statement is (A).

35. **C** Relative to the central maximum, the locations of the bright fringes on the screen are given by the expression $m(\lambda L/d)$, where λ is the wavelength of the light used, L is the distance to the screen, d is the separation of the slits, and m is an integer. The width of a fringe is therefore $(m + 1)(\lambda L/d) - m(\lambda L/d) = \lambda L/d$. One way to increase $\lambda L/d$ is to decrease d.

36. **B** Nuclear reactions must conserve mass number and charge, so the missing particle must be a proton:

$$^{13}_{6}\text{C} + ^{1}_{1}\text{p} \rightarrow ^{13}_{7}\text{N} + ^{1}_{0}\text{n}$$

37. **D** Resistors b, c, and d are equivalent to a single 1 Ω resistor, since

$$\frac{1}{3\ \Omega} + \frac{1}{3\ \Omega} + \frac{1}{3\ \Omega} = \frac{1}{1\ \Omega}$$

Therefore, the total resistance in the circuit is (3 Ω) + (1 Ω) + (3 Ω) + (3 Ω) = 10 Ω. From $V = IR$, we find that V = (3 A)(10 Ω) = 30 V.

38. **E** The power dissipated by a resistor is I^2R. Since the current through Resistor b is $\frac{1}{3}$ of the current through Resistor e, we find that

$$\frac{P_b}{P_e} = \frac{I_b^2 R}{I_e^2 R} = \frac{I_b^2}{I_e^2} = \frac{\left(\frac{1}{3}I_e\right)^2}{I_e^2} = \frac{1}{9} \quad \Rightarrow \quad P_e = 9P_b$$

39. **C** If both the source and detector travel in the same direction and at the same speed, there will be no relative motion and hence no Doppler shift.

40. **D** The SI unit for power, P, is the watt, where 1 watt = 1 joule per second. One joule is equal to 1 newton·meter, and 1 newton = 1 kg·m/s². Therefore,

$$[P] = \frac{\text{J}}{\text{s}} = \frac{\text{N} \cdot \text{m}}{\text{s}} = \frac{\frac{\text{kg} \cdot \text{m}}{\text{s}^2} \cdot \text{m}}{\text{s}} = \frac{\text{kg} \cdot \text{m}^2}{\text{s}^3} \quad \Rightarrow \quad [P] = \frac{\text{ML}^2}{\text{T}^3}$$

41. **A** In order for total internal reflection to occur, the beam must be incident in the medium with the higher index of refraction and strike the interface at an angle of incidence greater than the critical angle. Since $v_1 < v_2$, the refractive index of Medium 1, $n_1 = c/v_1$, must be *greater* than the index of Medium 2 ($n_2 = c/v_2$), and the critical angle is $\theta_c = \sin^{-1}(n_2/n_1) = \sin^{-1}(v_1/v_2)$.

42. **A** Because the initial height of the object is comparable to the radius of the Moon, we cannot simply use *mgh* for its initial potential energy. Instead, we must use the more general expression $U = -GMm/r$, where r is the distance from the center of the Moon. Note that since $h = R$, the object's initial distance from the Moon's center is $h + R = R + R = 2R$. Conservation of Energy then gives

$$K_i + U_i = K_f + U_f$$
$$0 - \frac{GMm}{2R} = \frac{1}{2}mv^2 - \frac{GMm}{R}$$
$$\frac{GMm}{2R} = \frac{1}{2}mv^2$$
$$v = \sqrt{\frac{GM}{R}}$$

43. **C** Since the magnetic force is always perpendicular to the object's velocity, it does zero work on any charged particle. Zero work means zero change in kinetic energy, so the speed remains the same. Remember: The magnetic force can only change the direction of a charged particle's velocity, not its speed.

44. **E** The intensity of a wave is equal to the power delivered per unit area. By increasing the distance from the source by a factor of 2, the area is increased by a factor of $2^2 = 4$. Since the intensity decreases by a factor that is less than 10, the sound level (in decibels) drops by an amount that is less than 10. The answer must be E.

45. **E** By definition, an electric dipole consists of a pair of charges that are equal in magnitude but opposite in sign. The electric field at the point midway between the charges is equal to the sum of the electric fields due to each charge alone:

⊕ • $\mathbf{E}_1$ ⇒ $\mathbf{E}_2$ ⊖

$\frac{1}{2}d$ $\frac{1}{2}d$

Because the individual electric field vectors point in the same direction, their magnitudes add to give the total electric field:

$$E_{\text{total}} = 2E = 2\frac{kQ}{r^2} = 2\frac{kQ}{\left(\frac{1}{2}d\right)^2} = 8\frac{kQ}{d^2} = 8\cdot\frac{(9\times10^9\ \text{N}\cdot\text{m}^2/\text{C}^2)(4.0\times10^{-9}\ \text{C})}{(2.0\times10^{-2}\ \text{m})^2}$$
$$= 7.2\times10^5\ \text{N/C}$$

46. **C** By Newton's Third Law, both vehicles experience the same magnitude of force and, therefore, the same impulse; so Statement I is false. Invoking Newton's Second Law, in the form *impulse = change in momentum*, we see that Statement II is therefore also false. However, since the car has a smaller mass than the truck, its acceleration will be greater in magnitude than that of the truck, so Statement III is true.

47. **B** The projectile reaches its maximum height when its vertical velocity drops to zero. Since $v_y = v_{0y} - gt$, this occurs at time $t = v_{0y}/g$ after launch. The vertical displacement at this time is

$$\Delta y = v_{0y}t - \frac{1}{2}gt^2 = v_{0y}\frac{v_{0y}}{g} - \frac{1}{2}g\left(\frac{v_{0y}}{g}\right)^2 = \frac{v_{0y}^2}{2g} = \frac{(a\sin\beta)^2}{2g}$$

48. **A** At the instant a simple harmonic oscillator passes through equilibrium, the restoring force is zero. Therefore, the tangential acceleration of the pendulum bob is also zero.

49. **D** The angular momentum, L, of a particle is equal to rmv, so we must first determine v. Since the earth's gravitational pull provides the centripetal force on the Moon,

$$\frac{GMm}{R^2} = \frac{mv^2}{R} \quad \Rightarrow \quad v = \sqrt{\frac{GM}{R}}$$

Therefore,

$$L = rmv = Rm\sqrt{\frac{GM}{R}} = m\sqrt{GMR}$$

50. **E** If the photons of the incident light have insufficient energy to liberate electrons from the metal's surface, then simply increasing the number of these weak photons (that is, increasing the intensity of the light) will do nothing. In order to produce photoelectrons, each photon of the incident light must have an energy at least as great as the work function of the metal.

51. **D** The force per unit length on each wire is $(\mu_0/2\pi)I_1I_2/r$, where r is the separation distance. Since $I_1 = I_2 = I$ and $r = a$, the force per unit length on each wire is $(\mu_0/2\pi)I^2/a$. Remember that when the currents are in opposite directions, the force between the wires is repulsive.

52. **D** The distance from the mirror to the image is equal to the distance from the mirror to the object. Therefore, the distance from the object to the image is 100 cm + 100 cm = 200 cm.

53. **B** Use the Ideal Gas Law:

$$P' = \frac{nRT'}{V'} = \frac{nR(2T)}{8V} = \frac{1}{4}\frac{nRT}{V} = \frac{1}{4}P$$

54. **E** The nuclide $^{68}_{30}\text{Zn}$ has 30 protons and 68 – 30 = 38 neutrons, so the excess number of neutrons over protons is 38 – 30 = 8. Of the choices given, only the nuclide $^{78}_{35}\text{Br}$ shares this property; it has 35 protons and 78 – 35 = 43 neutrons, giving an excess of neutrons over protons of 43 – 35 = 8.

55. **B** The charge on each plate has magnitude $Q = CV = (2 \times 10^{-3}\text{ F})(5\text{ V}) = 0.01\text{ C}$.

56. **C** Use the work-energy theorem:

$$W = \Delta K = \frac{1}{2}m(v_f^2 - v_i^2) = \frac{1}{2}(2 \text{ kg})[(4 \text{ m/s})^2 - (2 \text{ m/s})^2] = 12 \text{ J}$$

57. **E** Since the process is cyclic, $\Delta U = 0$, neither choice (A) nor (B) can be true. Since the work done during an isochoric step is always zero (because there is no change in volume), choice (C) is incorrect also. Choice (D) is not true; for the cycle described—whether it's clockwise or counterclockwise in a *P–V* diagram, the total work *W* will be negative. But choice (E) must be true; since $\Delta U = 0$, the First Law of Thermodynamics says that $W = -Q$.

58. **C** Draw a free-body diagram:

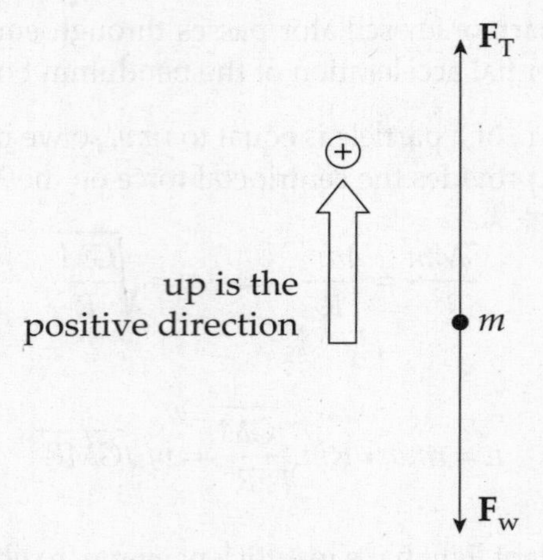

The net force on the object is $F - F_W$, so

$$\begin{aligned} F - F_W &= ma \\ F &= F_W + ma \\ &= F_W + \frac{F_W}{g}a \\ &= (50 \text{ N}) + \frac{50 \text{ N}}{10 \text{ m/s}^2}(10 \text{ m/s}^2) \\ &= 100 \text{ N} \end{aligned}$$

59. **D** The impulse delivered to the ball is equal to the ball's change in momentum, and the presence of the mitt does not change this; choices (A) and (C) are eliminated. The extra area decreases the *pressure* on the catcher's hand, but not the force, eliminating choice (B). The padding of the mitt causes the time during which the ball comes to a stop to increase, which decreases the magnitude of the ball's acceleration (since $a = \Delta v/\Delta t$). A decrease in the magnitude of the acceleration means a decrease in the magnitude of the force (since $a = F/m$).

60. **C** The Continuity Equation, Av = constant, tells us that v, the flow speed, is inversely proportional to A, the cross-sectional area of the pipe. Since A is proportional to d^2, where d is the diameter of the pipe, we can say that v is inversely proportional to d^2. Now, if d increases by a factor of 3/2 (from 2 cm to 3 cm), then v decreases by a factor of $(3/2)^2 = 9/4$. Thus, the flow speed at the point where the pipe is 3 cm in diameter is (4/9)(18 m/s) = 8 m/s.

61. **A** A bi-convex lens is a converging lens, so it has a positive focal length. Since $f = R/2$, we have f = (20 cm)/2 = 10 cm. The lens equation then gives

$$\frac{1}{s_o}+\frac{1}{s_i}=\frac{1}{f} \quad\Rightarrow\quad \frac{1}{30\text{ cm}}+\frac{1}{s_i}=\frac{1}{10\text{ cm}} \quad\Rightarrow\quad s_i = 15\text{ cm}$$

Because s_i is positive, the image is real, and real images are inverted. This eliminates choices (B), (C), and (D). The magnification is

$$M=-\frac{s_i}{s_o}=-\frac{15\text{ cm}}{30\text{ cm}}=-\frac{1}{2}$$

so the height of the image is

$$h_i=|M|\cdot h_o=\frac{1}{2}(2\text{ cm})=1\text{ cm}$$

62. **E** If Lenz's Law were violated, then energy conservation would also be violated. The other principles listed as choices are unrelated to Lenz's law.

63. **D** Because no electrostatic field can exist within the body of a conductor, a charge of –1 mC will appear on the wall of the cavity to "guard" the +1 mC charge of the plastic ball. Since the sphere has a charge of +1 mC, a net charge of (+1 mC) – (–1 mC) = +2 mC will appear on the outer surface of the copper sphere.

64. **A** The resistance of a wire is given by the formula $R = \rho L/A$, where ρ is the resistivity of the wire, L is its length, and A is its cross-sectional area. Since both wires are made of the same material, the value of ρ is the same for both wires, so we only need to look at the ratio L/A. Because the diameter of Wire Y is twice the diameter of Wire X, the cross-sectional area of Wire Y is 4 times the cross-sectional area of Wire X. Since

$$\frac{L_Y}{A_Y}=\frac{\frac{1}{2}L_X}{4A_X}=\frac{1}{8}\frac{L_X}{A_X}$$

we see that the resistance of wire Y is 1/8 the resistance of wire X.

65. **C** The work required to charge the capacitor is equal to the resulting increase in the stored electrical potential energy:

$$U=\frac{1}{2}CV^2=\frac{1}{2}(10\times10^{-6}\text{ F})(100\text{ V})^2=0.05\text{ J}$$

66. **A** According to the Bernoulli effect, the flow speed will be lowest, and thus the pressure will be greatest, in the section of the pipe with the greatest diameter.

67. **A** Choices B, C, D, and E all change the magnetic flux through the loop containing the light bulb, thus inducing an emf and a current. However, just swinging the handle over as described in choice A changes neither the area presented to the magnetic field lines from the bottom coil nor the density of the field lines at the position of the loop. No change in magnetic flux means no induced emf and no induced current.

68. **E** Gamma rays and X-rays are very high-energy, short-wavelength radiations. Ultraviolet light has a higher energy and shorter wavelength than the visible spectrum. Within the visible spectrum, the colors are—in order of increasing frequency—ROYGBV, so orange light has a lower frequency (and thus longer wavelength) than blue light.

69. **A** If these waves interfere completely destructively, the displacement of the resultant wave will be 6 cm – 4 cm = 2 cm. If they interfere completely constructively, then the displacement of the resultant wave will be 6 cm + 4 cm = 10 cm. In general, then, the displacement of the resultant wave will be no less than 2 cm and no greater than 10 cm.

70. **C** Since $\mathbf{F}_{net} = \mathbf{F}_1 + \mathbf{F}_2 = \mathbf{0}$, the bar cannot accelerate translationally, so (B) is false. The net torque does *not* need to be zero, as the following diagram shows (eliminating choices (A) and (D)):

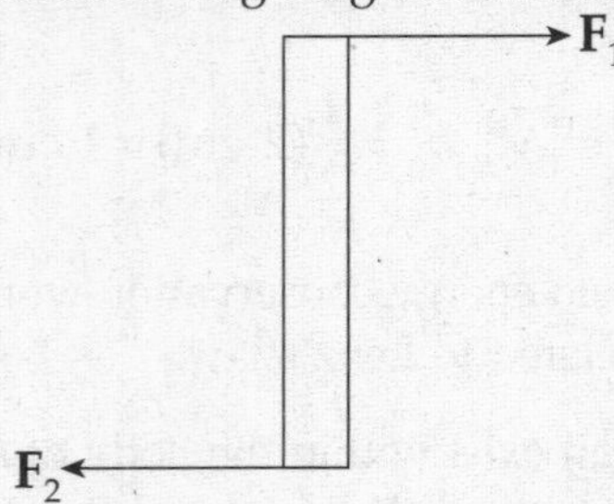

However, since $\mathbf{F}_2 = -\mathbf{F}_1$, choice (C) is true; one possible illustration of this is given below:

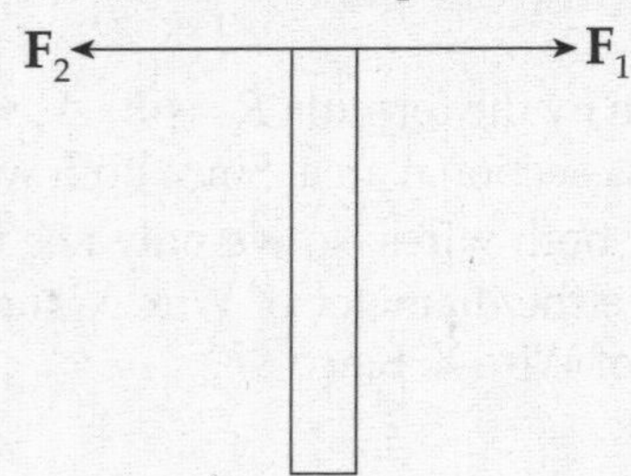

SECTION II

1. (a) The elastic potential energy of the spring turns into kinetic energy of the ball, which then becomes gravitational potential energy of the ball. Let the zero for gravitational potential energy be the initial position of the ball on the compressed spring (i.e., 8 cm below ground level). Then

$$K_i + U_i = K_f + U_f$$

$$0 + (0 + \frac{1}{2}kx^2) = 0 + [mg(h + 0.08 \text{ m}) + 0]$$

$$h = \frac{\frac{1}{2}kx^2}{mg} - 0.08 \text{ m}$$

$$= \frac{\frac{1}{2}(500 \text{ N/m})(0.08 \text{ m})^2}{(0.15 \text{ kg})(10 \text{ N/kg})} - 0.08 \text{ m}$$

$$= 0.99 \text{ m}$$

(b) (i) Use the impulse–momentum theorem (Newton's Second Law):

$$\begin{aligned} J &= \Delta p \\ F_{\text{avg}}\Delta t &= m\Delta v \\ F_{\text{avg}} &= m\frac{\Delta v}{\Delta t} \\ &= (0.15\text{ kg})\frac{30\text{ m/s}}{0.004\text{ s}} \\ &= 1100\text{ N} \end{aligned}$$

(b) (ii) The ball is in flight until its vertical displacement is $-h$ (calling *up* the positive direction). Therefore,

$$\Delta y = v_{0y}t - \frac{1}{2}gt^2 = -\frac{1}{2}gt^2 \quad \Rightarrow \quad t = \sqrt{\frac{-2\Delta y}{g}} = \sqrt{\frac{2h}{g}} = \sqrt{\frac{2(1.0\text{ m})}{10\text{ m/s}^2}} = 0.45\text{ s}$$

(b) (iii) The ball's horizontal displacement is $v_{0x}t$, where t is the time calculated in part (b) (ii):

$$\Delta x = v_{0x}t = (30\text{ m/s})(0.45\text{ s}) = 13.5\text{ m}$$

(c) The frequency of this simple harmonic oscillator is

$$f = \frac{1}{2\pi}\sqrt{\frac{k}{m}} = \frac{1}{2\pi}\sqrt{\frac{500\text{ N/m}}{0.15\text{ kg}}} = 9.2\text{ Hz}$$

2. (a) Since the total resistance in the circuit is $r + R$, Ohm's law gives

$$I = \frac{\mathcal{E}}{r+R}$$

(b) The power dissipated as heat by resistor R is $P = I^2R$; multiplying this by the time, t, gives Q, the heat dissipated:

$$Q = Pt = I^2Rt = \frac{\mathcal{E}^2Rt}{(r+R)^2}$$

(c) (i) Use the equation $Q = mc\Delta T$ with Q as given in part (b):

$$\begin{aligned} Q &= mc\Delta T \\ \frac{\mathcal{E}^2Rt}{(r+R)^2} &= mc(T - T_i) \\ T &= \frac{\mathcal{E}^2R}{(r+R)^2 mc}t + T_i \end{aligned}$$

(c) (ii) The graph of T vs. t is a straight line with slope $\mathcal{E}^2R/[(r+R)^2 mc]$:

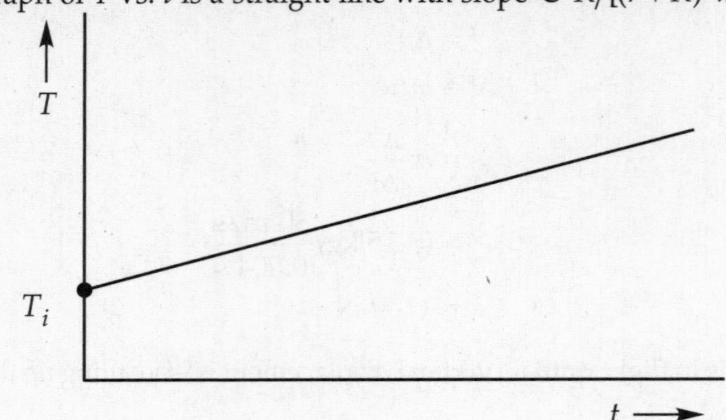

(d) The movement of the blades of the stirrer through the water transfers mechanical work to thermal energy. The faster the stirrer rotates (i.e., the more rotational kinetic energy it has), the more heat will be transferred to the water and the faster the temperature will rise. [In fact, given enough time, the rotating stirrer could bring the water to boil all by itself (assuming that heat losses to the beaker, the lid, etc. could be kept low)].

3. (a) Magnetic forces alone can only change the direction of a charged particle's velocity, not its speed. This is because $\mathbf{F}_B = q(v \times \mathbf{B})$ is always perpendicular to v, the particle's velocity, so it can do no work and thus cause no change in kinetic energy. The electric field is required to increase the particle's speed.

(b) At the moment shown below, v is upward in the plane of the page, and **F** must point toward the center of the circular path.

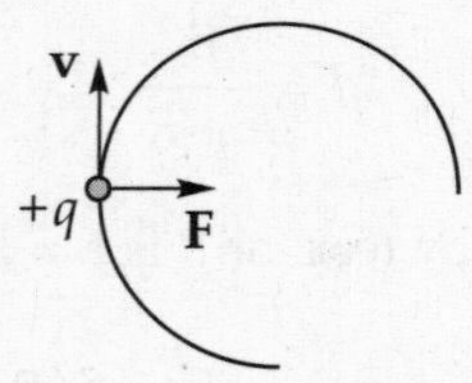

In order for $\mathbf{F}_B = q(v \times \mathbf{B})$ to be consistent with this, the right-hand rule implies that **B** must point out of the plane of the page (because q is positive).

(c) The time to complete one revolution, T, is equal to $2\pi r/v$, where r is the radius of the path and v is the speed. Since the magnetic force provides the centripetal force,

$$qvB = \frac{mv^2}{r} \quad \Rightarrow \quad qB = \frac{mv}{r} \quad \Rightarrow \quad \frac{r}{v} = \frac{m}{qB} \quad \Rightarrow \quad T = \frac{2\pi r}{v} = \frac{2\pi m}{qB} = \frac{2\pi m}{eB}$$

Therefore, T depends only on e, m, and B.

(d) Since T is the time for one revolution, the number of revolutions per unit time is the reciprocal of T, $eB/(2\pi m)$.

(e) The description given with the problem states that "the voltage must be alternated twice during each revolution of the particle." Therefore, the frequency of the alternating voltage must be twice the value found in part (d), $2eB/(2\pi m) = eB/(\pi m)$.

(f) From part (c), we know that

$$evB = \frac{mv^2}{r} \Rightarrow v = \frac{eBr}{m}$$

Therefore,

$$K_{max} = \frac{1}{2}mv_{max}^2 = \frac{1}{2}m\left(\frac{eBr_{max}}{m}\right)^2 = \frac{1}{2}m\left(\frac{eBR}{m}\right)^2 = \frac{e^2B^2R^2}{2m}$$

4. (a) First draw free-body diagrams for the blocks:

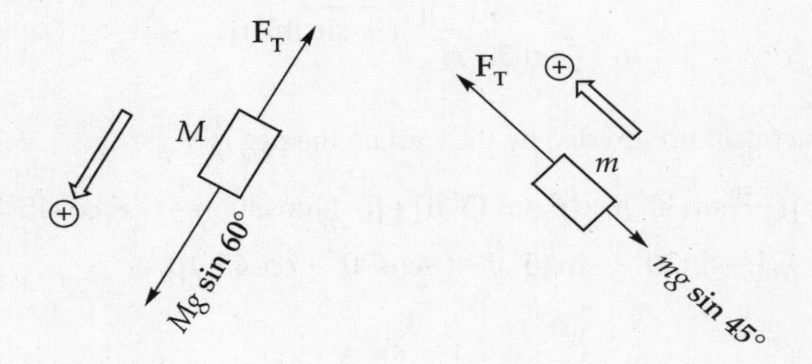

Now, apply Newton's Second Law:

$$Mg \sin 60° - F_T = Ma$$

add equations

$$F_T - mg \sin 45° = ma$$

$$Mg \sin 60° - mg \sin 45° = Ma + ma$$

$$a = \frac{M \sin 60° - m \sin 45°}{M + m} g$$

$$= \frac{(8 \text{ kg}) \sin 60° - (2 \text{ kg}) \sin 45°}{8 \text{ kg} + 2 \text{ kg}} (10 \text{ m/s}^2)$$

$$= 5.5 \text{ m/s}^2$$

(b) Using the value for a determined above, find F_T; we can use either of the two original equations from part (a):

$$\begin{aligned} F_T &= mg\sin 45^\circ + ma \\ &= (2\text{ kg})(10\text{ m/s}^2)\sin 45^\circ + (2\text{ kg})(5.5\text{ m/s}^2) \\ &= 25.1\text{ N} \end{aligned}$$

From the following diagram,

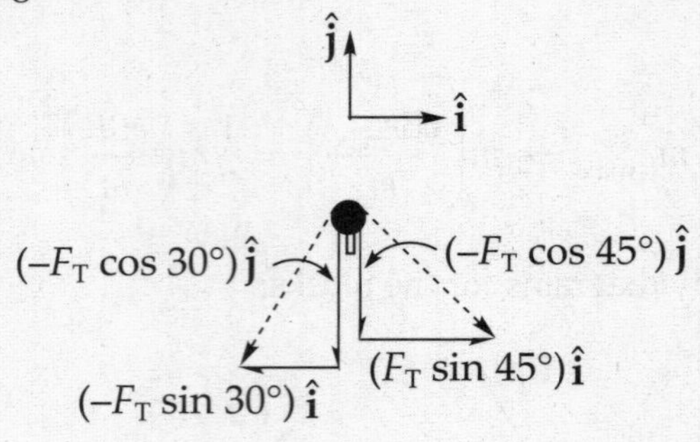

we see that the total force exerted by the cord on the peg is

$$\begin{aligned} \mathbf{F} &= [(-F_T\sin30^\circ)\hat{\mathbf{i}} + (F_T\sin45^\circ)\hat{\mathbf{i}}] + [(-F_T\cos30^\circ)\hat{\mathbf{j}} + (-F_T\cos45^\circ)\hat{\mathbf{j}}] \\ &= F_T[(-\sin30^\circ + \sin45^\circ)\hat{\mathbf{i}} + (-\cos30^\circ - \cos45^\circ)\hat{\mathbf{j}}] \end{aligned}$$

so

$$\begin{aligned} F &= F_T\sqrt{(-\sin 30^\circ + \sin 45^\circ)^2 + (-\cos 30^\circ - \cos 45^\circ)^2} \\ &= (25.1\text{ N})\sqrt{2.518} \\ &= 40\text{ N} \end{aligned}$$

(c) Refer to the following diagram:

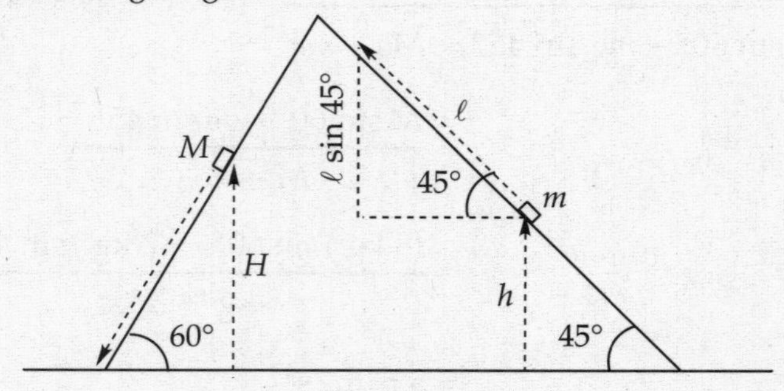

We have the constant acceleration of 5.5 m/s² from part (a), so we may use $v^2 = v_0{}^2 - 2a(x - x_0)$.

We know that $v_0 = 0$ because it starts from rest. The distance M moves is given by: $(x - x_0) = \dfrac{H}{\sin 60^\circ}$.

We now have:

$$v^2 = 0^2 + 2(5.5\ \text{m/s}^2)\left(\frac{1.5m}{\sin 60^\circ}\right) = 4.4\ \text{m/s}$$

(d) First draw free-body diagrams for the blocks:

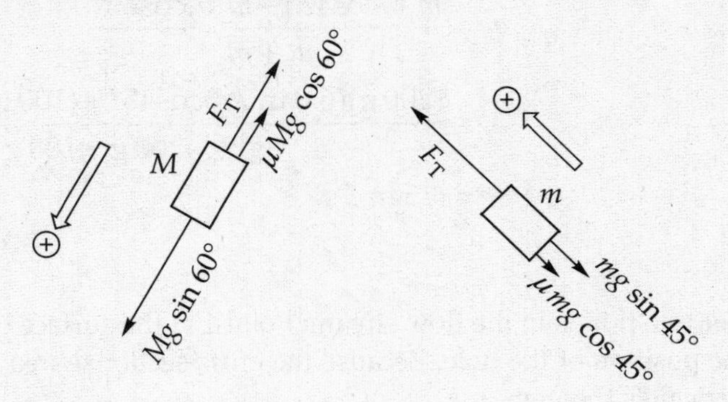

Now apply Newton's Second Law:

$$Mg \sin 60° - F_T - \mu Mg \cos 60° = Ma$$

add equations

$$F_T - mg \sin 45° - \mu mg \cos 45° = ma$$

$$Mg \sin 60° - mg \sin 45° - \mu Mg \cos 60° - \mu mg \cos 45° = Ma + ma$$

This gives

$$a = \frac{M(\sin 60^\circ - \mu \cos 60^\circ) - m(\sin 45^\circ + \mu \cos 45^\circ)}{M + m} g$$

$$= \frac{(8\ \text{kg})(\sin 60^\circ - 0.2 \cos 60^\circ) - (2\ \text{kg})(\sin 45^\circ + 0.2 \cos 45^\circ)}{8\ \text{kg} + 2\ \text{kg}}(10\ \text{m/s}^2)$$

$$= 4.4\ \text{m/s}^2$$

5. (a) The diagram given with the question shows that after the clay balls collide, they move in the $-y$ direction, which means that the horizontal components of their linear momenta cancel. In other words, $\mathbf{p}_{1x} + \mathbf{p}_{2x} = \mathbf{0}$:

$$m_1 \mathbf{v}_{1x} + m_2 \mathbf{v}_{2x} = \mathbf{0}$$

$$(m_1 v_1 \sin 45^\circ)\hat{\mathbf{i}} + (m_2 v_2 \sin 60^\circ)(-\hat{\mathbf{i}}) = \mathbf{0}$$

$$v_2 = \frac{m_1 v_1 \sin 45^\circ}{m_2 \sin 60^\circ}$$

$$= \frac{(200\ \text{g})(6.0\ \text{m/s}) \sin 45^\circ}{(100\ \text{g}) \sin 60^\circ}$$

$$= 9.8\ \text{m/s}$$

(b) Apply Conservation of Linear Momentum in the y direction:

$$m_1\mathbf{v}_{1y} + m_2\mathbf{v}_{2y} = (m_1 + m_2)\mathbf{v}'_y$$
$$(m_1v_1\cos45°)(-\hat{\mathbf{j}}) + (m_2v_2\cos60°)(-\hat{\mathbf{j}}) = (m_1 + m_2)\mathbf{v}'_y$$
$$v'_y = \frac{m_1v_1\cos45° + m_2v_2\cos60°}{m_1 + m_2}$$
$$= \frac{(200\text{ g})(6.0\text{ m/s})\cos45° + (100\text{ g})(9.8\text{ m/s})\cos60°}{200\text{ g} + 100\text{ g}}$$
$$= 4.5\text{ m/s}$$

6. (a) We'll consider two points in the flow stream: Point 1 at the surface of the water in the tank and Point 2 at the position of the hole. Because the cross-sectional area of the tank is so large, we can apply Torricelli's theorem,

$$v = \sqrt{2g(y_1 - y_2)}$$

where y_1 is the vertical distance from the ground to Point 1 and y_2 is the vertical distance from the ground to Point 2. We conclude that the efflux speed of the water is

$$v = \sqrt{2(10\text{ m/s}^2)(50\text{m} - 35\text{m})} = \sqrt{2(10\text{ m/s}^2)(15\text{m})} = 17.3\text{m/s}$$

(b) The flow rate is equal to the cross-sectional area times the flow speed: $f = Av$. Using the result of part (a), we first find that

$$A = \frac{f}{v} = \frac{3\times10^{-4}\ \frac{\text{m}^3}{\text{min}} \cdot \frac{1\text{ min}}{60\text{ s}}}{17.3\text{ m/s}} = 2.89\times10^{-7}\text{ m}^2$$

Since the area of the hole is given by $A = \pi r^2$, we find that the radius of the hole is

$$r = \sqrt{\frac{A}{\pi}} = \sqrt{\frac{2.89\times10^{-7}\text{ m}^2}{\pi}} = 3.03\times10^{-4}\text{ m}$$

Therefore, the diameter of the hole is twice this value: $d = 2r = 6.06\times10^{-4}$ m.

(c) The horizontal distance, x, will be equal to vt, where v is the efflux speed we found in part (a) and t is the time it takes for water to reach the ground. Because the stream of water has zero vertical speed when it exits the hole, we can use the equation $y = \frac{1}{2}at^2$, with $a = g$, to find t. This gives

$$t = \sqrt{\frac{2y}{g}}$$

Since the vertical distance the water drops after exiting the hole is $y = 35$ m, the horizontal distance we're looking for is

$$x = v\sqrt{\frac{2y}{g}} = (17.3\text{ m/s})\sqrt{\frac{2(35\text{ m})}{10\text{ m/s}^2}} = 45.8\text{ m}$$

HOW TO SCORE PRACTICE TEST 1

SECTION I: MULTIPLE-CHOICE

__________ × 1.3043 = __________

Number of Correct (out of 70) — Weighted Section I Score (Do not round)

SECTION II: FREE RESPONSE

(See if you can find a teacher or classmate to score your essays using the guidelines in Chapter 3.)

Question 1 __________ (out of 15) × 1.0000 = __________ (Do not round)

Question 2 __________ (out of 15) × 1.0000 = __________ (Do not round)

Question 3 __________ (out of 15) × 1.0000 = __________ (Do not round)

Question 4 __________ (out of 15) × 1.0000 = __________ (Do not round)

Question 5 __________ (out of 10) × 1.5000 = __________ (Do not round)

Question 6 __________ (out of 10) × 1.5000 = __________ (Do not round)

Sum = __________

Weighted Section II Score (Do not round)

AP Score Conversion Chart Physics B Exam

Composite Score Range	AP Score
112–180	5
85–111	4
57–84	3
40–56	2
0–39	1

COMPOSITE SCORE

__________ + __________ = __________

Weighted Section I Score + Weighted Section II Score = Composite Score (Round to nearest whole number)

21

The Princeton Review AP Physics B Practice Exam 2

AP® Physics B Exam

SECTION I: Multiple-Choice Questions

DO NOT OPEN THIS BOOKLET UNTIL YOU ARE TOLD TO DO SO.

At a Glance

Total Time
90 minutes
Number of Questions
70
Percent of Total Grade
50%
Writing Instrument
Pen required

Instructions

Section I of this examination contains 70 multiple-choice questions. Fill in only the ovals for numbers 1 through 70 on your answer sheet.

CALCULATORS MAY NOT BE USED IN THIS PART OF THE EXAMINATION.

Indicate all of your answers to the multiple-choice questions on the answer sheet. No credit will be given for anything written in this exam booklet, but you may use the booklet for notes or scratch work. After you have decided which of the suggested answers is best, completely fill in the corresponding oval on the answer sheet. Give only one answer to each question. If you change an answer, be sure that the previous mark is erased completely. Here is a sample question and answer.

Sample Question	Sample Answer
Chicago is a (A) state (B) city (C) country (D) continent (E) village	Ⓐ ● Ⓒ Ⓓ Ⓔ

Use your time effectively, working as quickly as you can without losing accuracy. Do not spend too much time on any one question. Go on to other questions and come back to the ones you have not answered if you have time. It is not expected that everyone will know the answers to all the multiple-choice questions.

About Guessing

Many candidates wonder whether or not to guess the answers to questions about which they are not certain. Multiple choice scores are based on the number of questions answered correctly. Points are not deducted for incorrect answers, and no points are awarded for unanswered questions. Because points are not deducted for incorrect answers, you are encouraged to answer all multiple-choice questions. On any questions you do not know the answer to, you should eliminate as many choices as you can, and then select the best answer among the remaining choices.

GO ON TO THE NEXT PAGE.

Advanced Placement Examination

PHYSICS B

SECTION I

TABLE OF INFORMATION FOR 2011

CONSTANTS AND CONVERSION FACTORS

1 unified atomic mass unit,	$1\ u = 1.66 \times 10^{-27}$ kg
	$= 931\ \text{MeV}/c^2$
Proton mass,	$m_p = 1.67 \times 10^{-27}$ kg
Neutron mass,	$m_n = 1.67 \times 10^{-27}$ kg
Electron mass,	$m_e = 9.11 \times 10^{-31}$ kg
Electron charge magitude,	$e = 1.60 \times 10^{-19}$ C
Avogadro's number,	$N_0 = 6.02 \times 10^{23}\ \text{mol}^{-1}$
Universal gas constant,	$R = 8.31$ J/(mol•K)
Boltzmann's constant,	$k_B = 1.38 \times 10^{-23}$ J/K
Speed of light,	$c = 3.00 \times 10^8$ m/s
Planck's constant,	$h = 6.63 \times 10^{-34}$ J•s
	$= 4.14 \times 10^{-15}$ eV•s
	$hc = 1.99 \times 10^{-25}$ J•m
	$= 1.24 \times 10^3$ eV•nm
Vacuum permittivity,	$\epsilon_0 = 8.85 \times 10^{-12}\ \text{C}^2/\text{N}\cdot\text{m}^2$
Coulomb's law constant,	$k = 1/4\pi\epsilon_0 = 9.0 \times 10^9\ \text{N}\cdot\text{m}^2/\text{C}^2$
Vacuum permeability,	$\mu_0 = 4\pi \times 10^{-7}$ (T•m)/A
Magnetic constant,	$k' = \mu_0/4\pi = 10^{-7}$ (T•m)/A
Universal gravitational constant,	$G = 6.67 \times 10^{-11}\ \text{m}^3/\text{kg}\cdot\text{s}^2$
Acceleration due to gravity at Earth's surface,	$g = 9.8\ \text{m/s}^2$
1 atmosphere pressure,	$1\ \text{atm} = 1.0 \times 10^5\ \text{N/m}^2$
	$= 1.0 \times 10^5$ Pa
1 electron volt,	$1\ \text{eV} = 1.60 \times 10^{-19}$ J

UNITS

Name	Symbol
meter	m
kilogram	kg
second	s
ampere	A
kelvin	K
mole	mol
hertz	Hz
newton	N
pascal	Pa
joule	J
watt	W
coulomb	C
volt	V
ohm	Ω
henry	H
farad	F
tesla	T
degree Celsius	°C
electron-volt	eV

PREFIXES

Factor	Prefix	Symbol
10^9	giga	G
10^6	mega	M
10^3	kilo	k
10^{-2}	centi	c
10^{-3}	milli	m
10^{-6}	micro	μ
10^{-9}	nano	n
10^{-12}	pico	p

VALUES OF TRIGONOMETRIC FUNCTIONS FOR COMMON ANGLES

θ	$\sin\theta$	$\cos\theta$	$\tan\theta$
0°	0	1	0
30°	1/2	$\sqrt{3}/2$	$\sqrt{3}/3$
37°	3/5	4/5	3/4
45°	$\sqrt{2}/2$	$\sqrt{2}/2$	1
53°	4/5	3/5	4/3
60°	$\sqrt{3}/2$	1/2	$\sqrt{3}$
90°	1	0	∞

The following conventions are used in this examination.

I. Unless otherwise stated, the frame of reference of any problem is assumed to be inertial.

II. The direction of any electric current is the direction of flow of positive charge (conventional current).

III. For any isolated electric charge, the electric potential is defined as zero at an infinite distance from the charge.

IV. For mechanics and thermodynamics equations, W represents the work done on a system.

GO ON TO THE NEXT PAGE.

PHYSICS B

SECTION I

Directions: Each of the questions or incomplete statements below is followed by five suggested answers or completions. Select the one that is best in each case and then fill in the corresponding oval on the answer sheet.

Note: To simplify calculations, you may use $g = 10\ m/s^2$ in all problems.

1. Which of the following must be true if the net force acting on an object is zero?

    I. The object is at rest
    II. The object has a constant velocity
    III. No forces act on the object

    (A) I only
    (B) II only
    (C) III only
    (D) Either I or III, but not II
    (E) None of these must be true

2. An elevator has a mass of 500 kg. It may safely carry an additional mass of 1000 kg and accelerate upward at $2\ m/s^2$. What is the tension in the cable under these maximum conditions?

    (A) 4,900 N
    (B) 9,800 N
    (C) 11,700 N
    (D) 14,700 N
    (E) 17,700 N

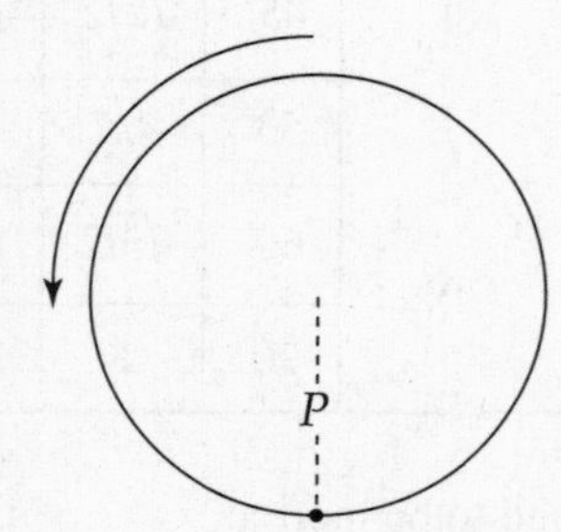

3. Here is the top view of a stopper tied to a string and spun in a horizontal circle. Which of the following would match with an object moving at point P in a counter-clockwise direction?

    (A) **v**↓ **a**↑ $\mathbf{F}_{net}$↑
    (B) **v**↓ **a**↓ $\mathbf{F}_{net}$↓
    (C) **v**→ **a**→ $\mathbf{F}_{net}$→
    (D) **v**→ **a**↓ $\mathbf{F}_{net}$↓
    (E) **v**→ **a**↑ $\mathbf{F}_{net}$↑

4. A ball is thrown with a velocity of 20 m/s at an angle of 30 degrees above the horizon. When the ball reaches its highest point, which of the following is true?

    (A) It has zero velocity.
    (B) It has zero acceleration.
    (C) It has minimum potential energy
    (D) It has minimum kinetic energy.
    (E) Two of the above

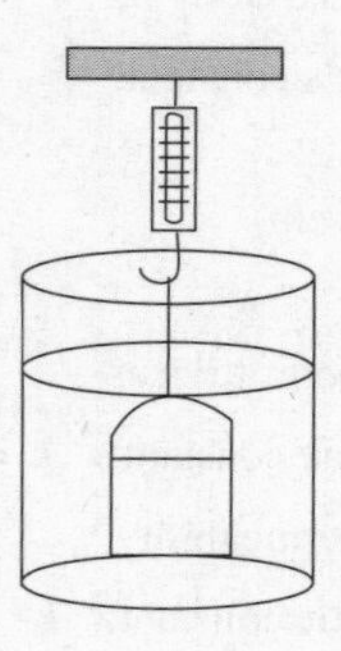

5. A 0.50 kg mass hangs from a spring scale. It is placed a container whose water level is initially at 1000 ml. The water level rises to 1200 ml when the mass is placed in the container as shown. What is the reading on the spring scale when the object is in the water?

    (A) 2.0 N
    (B) 3.0N
    (C) 5.0 N
    (D) 7.0 N
    (E) 12 N

GO ON TO THE NEXT PAGE.

Questions 6–7 refer to the following figure:

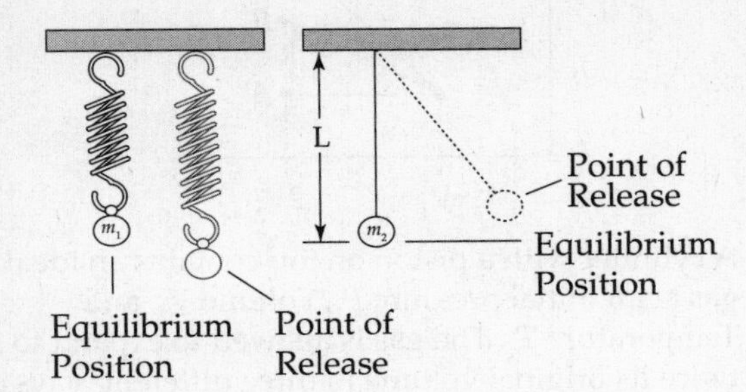

A mass on a spring and a simple pendulum undergo simple harmonic motion. There is no friction present as the mass on the spring, m_1, is displaced downward and released from rest or as the mass on the string, m_2, is displaced with a small sideways amplitude and released from rest.

6. Which of the following statements are true?

   I. Mechanical energy is constant.
   II. The momentum of the mass is constant.
   III. The period of the mass is constant.

   (A) I only
   (B) II only
   (C) III only
   (D) I and III, but not II
   (E) None of these must be true

7. If both masses have the same frequency of oscillation, what is the expression for the length of the string?

   (A) $\frac{m_1}{gk}$

   (B) $\frac{m_1 k}{g}$

   (C) $\frac{m_1 g}{k}$

   (D) $\frac{g}{m_1 k}$

   (E) $\frac{k}{m_1 g}$

8. A simple pendulum oscillates back and forth at a very small angle and undergoes simple harmonic motion with period T. The angle is halved, the mass hanging is reduced by a factor of three, and the length of the string is cut by a factor of four. What is the period of this new pendulum?

   (A) $4T$

   (B) $2T$

   (C) T

   (D) $\frac{1}{2}T$

   (E) $\frac{1}{4}T$

9. A frictionless cart of mass $2M$ moves at a velocity V_0. It collides and sticks to another frictionless cart of mass M initially at rest. What is the final velocity of the two carts?

   (A) V_0

   (B) $\frac{V_0}{2}$

   (C) $\frac{3}{2}V_0$

   (D) $\frac{2}{3}V_0$

   (E) $\frac{1}{3}V_0$

10. A planet in a science fiction film is twice the radius of the earth and yet has only half the earth's mass. Gravity on that planet would be about

   (A) $\frac{1}{8}g$

   (B) $\frac{1}{4}g$

   (C) $1g$

   (D) $4g$

   (E) $8g$

GO ON TO THE NEXT PAGE.

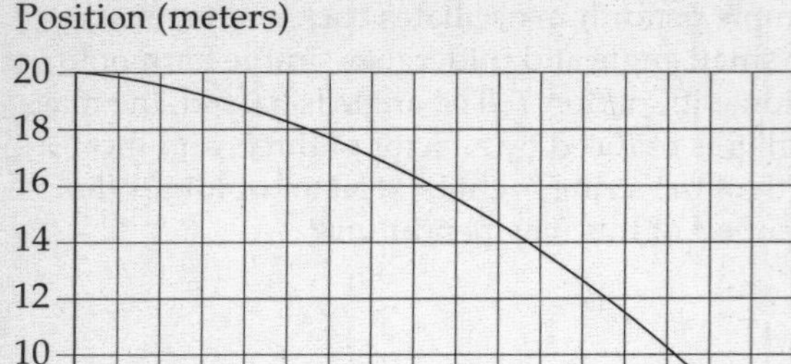
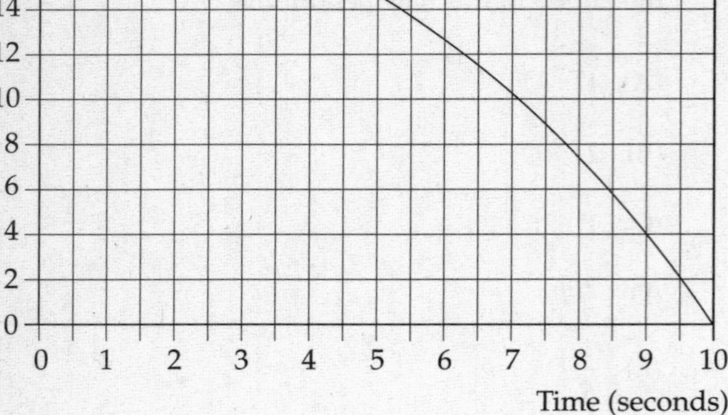

11. The instantaneous velocity at 9 seconds is closest to

(A) 4/9 m/s
(B) –4/9 m/s
(C) 7/4 m/s
(D) –7/4 m/s
(E) 2 m/s

12. A 100 g mass of Al has a specific heat of 0.27 cal/g°C. It is placed in an insulated container that contains 500 g of water which has a specific heat of 1 cal/g°C. They are allowed to come into thermal equilibrium. Consider the following statements:

I. The heat gained by the Al is equal to the heat lost by the water.
II. The temperature charge of the Al is equal to the temperature change of the water.
III. When in thermal equilibrium the two objects contain the same internal energy.

(A) I only
(B) II only
(C) III only
(D) Both I and II, but not III
(E) Both I and III, but not II

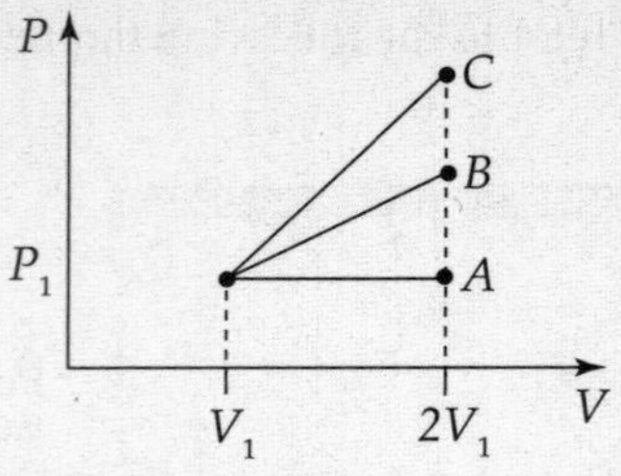

13. A cylinder with a piston on top contains an ideal gas at an initial pressure P_1, volume V_1, and Temperature T_1. The gas is allowed to expand to twice its original volume in three different ways as shown. The mechanical work done by the gas is

(A) greatest for path *A*
(B) greatest for path *B*
(C) greatest for path *C*
(D) equal for all paths
(E) can not be determined without knowing if isobaric, isothermal, or adiabatic

14. A charged particle moves through a magnetic field and experiences a force *F*. New particles are sent into the same magnetic field. If the new particles have twice the charge, twice the mass, and twice the velocity, the new force would be

(A) $4F$
(B) $2F$
(C) F
(D) $\frac{1}{2}F$
(E) $\frac{1}{4}F$

15. There is a negatively charged hollow sphere. Which of the following statements can be said of the sphere?

I. The electric field on the inside is zero.
II. For some distance $r > r_{\text{sphere}}$ you can consider the charge to be concentrated at the center to the sphere.
III. The electric field points inward toward the center of the sphere.

(A) I only
(B) II only
(C) III only
(D) I, II, and III
(E) Both I and III, but not II

GO ON TO THE NEXT PAGE.

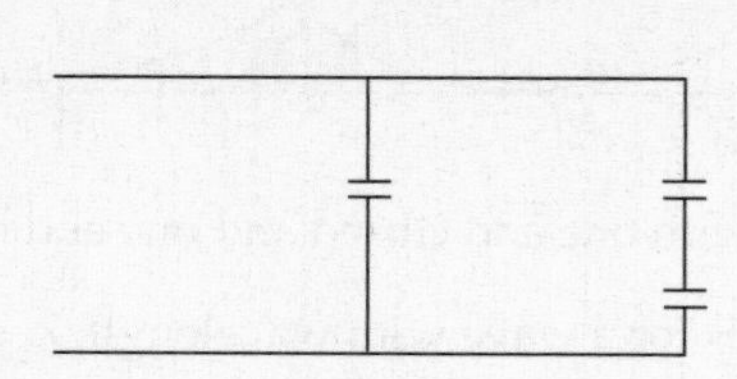

16. Each capacitor has a capacitance of C. What is the overall capacitance of the above circuit?

(A) 3C
(B) 3/2 C
(C) 2/3 C
(D) 1/3 C
(E) Cannot be determined with the voltage and charge

17. You have a wire of length L, radius r, and resistance R. You need to obtain half that resistance using the same material and changing only one factor. You could

(A) use half the length
(B) use twice the length
(C) use half the radius
(D) use twice the radius
(E) use twice the mass

18. An electric motor has a label on it that reads: Input: 120V AC, 1.0 Amps, 60 Hz – Efficiency – 75%. At what constant speed can the motor lift up a 6 kg mass?

(A) 0.5 m/s
(B) 1.0 m/s
(C) 1.5 m/s
(D) 2.0 m/s
(E) 2.5 m/s

Questions 19–20 refer to the following figure:

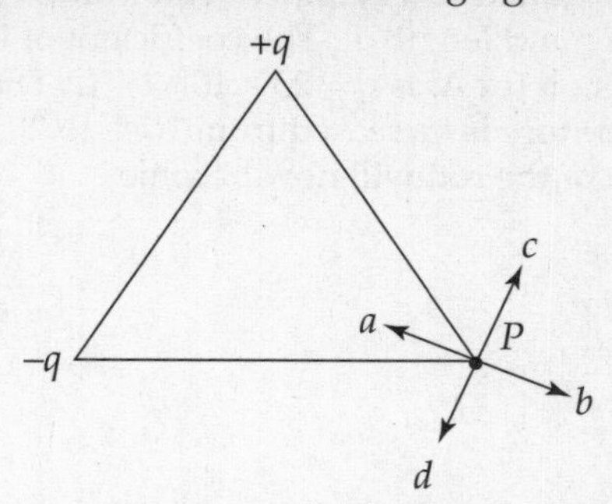

19. Two charges are placed as shown at the vertices of an equilateral triangle. What is the direction of the electric field at point P?

(A) a
(B) b
(C) c
(D) d
(E) the field cancels to zero at this point

20. If I placed a charge $+q$ at point P, the electric field at point P would

(A) increase by 1/3
(B) increase by 1/2
(C) decrease by 1/3
(D) decrease by 1/2
(E) remain the same

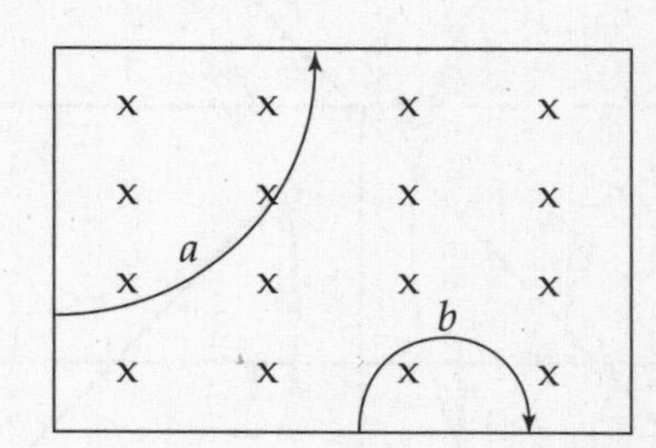

21. A machine shoots either a proton, neutron, or electron into a magnetic field at various locations. The paths of two particles are shown above. Assume they are far enough apart so that they do not interact. What can you say about the paths that represent each particle?

(A) a is the proton and b is the electron.
(B) b is the proton and a is the electron.
(C) Either may be a neutron.
(D) You cannot make any conclusions without knowing the velocities.
(E) Two of the above are true.

GO ON TO THE NEXT PAGE.

22. Heat is added to a cylindrical aluminum rod of radius r and length ℓ. The coefficient of linear expansion for Al is $\alpha = 25 \times 10^{-6}\ 1/°C$, The rod's temperature is increased from 10°C to 20°C. The length of the rod will now become

(A) 2ℓ
(B) 4ℓ
(C) ℓ^2
(D) $\frac{1}{2}\ \ell$
(E) remain just about ℓ

23. A rigid, solid container of constant volume holds an ideal gas of volume V_1 and temperature T_1 and pressure P_1. The temperature is increased in an isochoric process. Which of the following is NOT true?

(A) The average speed of the molecules increases.
(B) The pressure increases.
(C) The kinetic energy of the system increases.
(D) The volume increases.
(E) The number of collisions between molecules and the container increases.

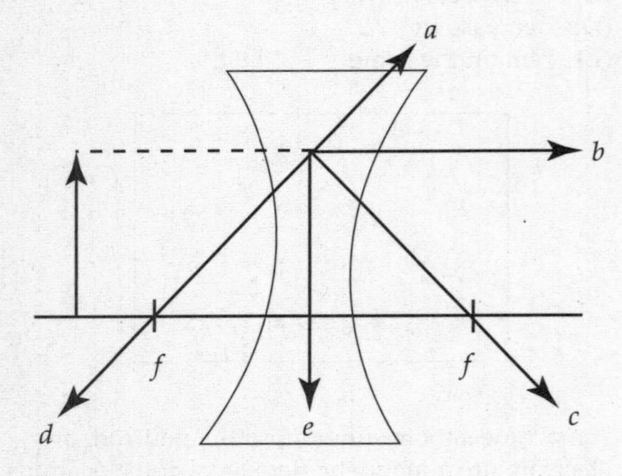

24. A ray of light hits an object and travels parallel to the principal axis as shown by the dotted line. Which line shows the correct continuation of the ray after it hits the concave lens?

(A) *a*
(B) *b*
(C) *c*
(D) *d*
(E) *e*

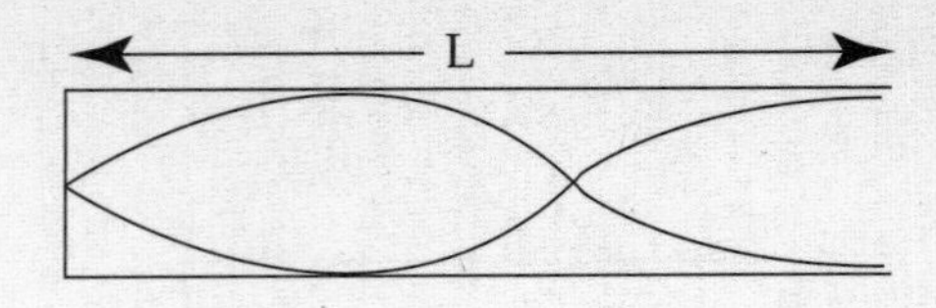

25. A tube with one end closed and one end open resonates for a wave with wavelength λ_a as shown. The next shorter wavelength at which resonance will occur is λ_b. The ratio of these two wavelengths $\frac{\lambda_a}{\lambda_b}$ is

(A) 1/4
(B) 1/3
(C) 3/5
(D) 5/3
(E) 3/1

26. Which of the following statements is true for a diverging lens?

I. It can only form virtual images.
II. The magnification is always less than 1.
III. It can form real images if the object is beyond the focal length and virtual images if the object is inside the focal length.

(A) I only
(B) II only
(C) III only
(D) II and III, only
(E) I and II only

27. Ultraviolet light has a wavelength of about 6×10^{-8} m. What is the frequency of this light?

(A) 5×10^{15} Hz
(B) 0.5 Hz
(C) 2 Hz
(D) 20 Hz
(E) 2×10^{-16} Hz

28. Which of the following statements is true for both sound and light waves?

I. They can both be polarized.
II. They can both diffract.
III. They can both refract.

(A) I only
(B) II only
(C) III only
(D) I and II only
(E) II and III only

GO ON TO THE NEXT PAGE.

Questions 29–30

A spring stretches 4 cm when a 2.0 kg mass is hung from it. The mass is lifted up four cm and dropped. The following graph represents the net force acting on a spring as it falls 8 cm from top to bottom.

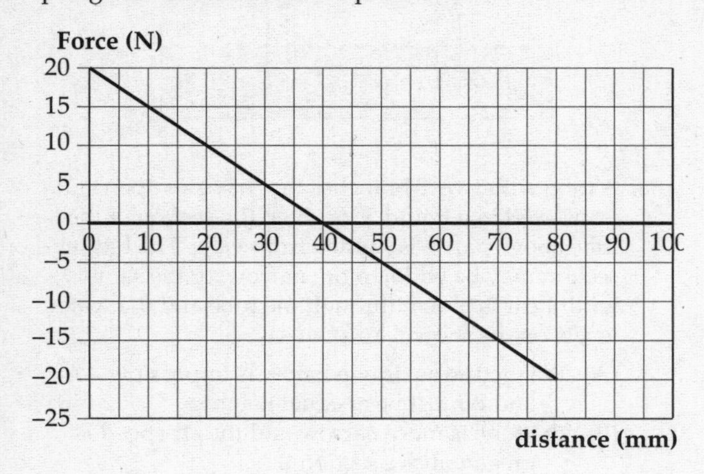

Net Force vs. Distance

29. What is the net work done as the mass falls from 0 to 8 cm?
    (A) 160 J
    (B) 80 J
    (C) 20
    (D) 0 J
    (E) –5 J

30. What is the maximum velocity of the mass?
    (A) 0.2 m/s
    (B) 0.4
    (C) 0.8 m/s
    (D) 1.41 m/s
    (E) 4 m/s

31. A padded catcher's mitt slows down a 150 g ball from 20 m/s to rest in 0.01 seconds? What force does it exert on the ball?
    (A) 0.3 N
    (B) 33 N
    (C) 75 N
    (D) 133 N
    (E) 300 N

Questions 32–33

A 100 g mass is hung at the 100 cm mark of a uniform meter stick. You notice you can balance the meter stick on your finger if you place your finger at the 75 cm mark.

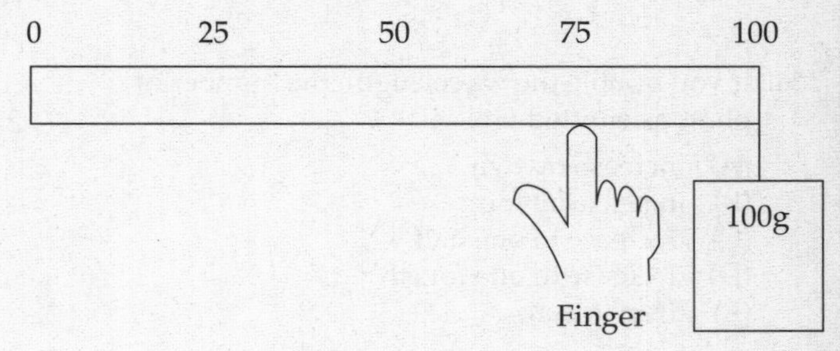

32. What is the torque the 100 g exerts on the meter stick taking the position of the finger to be the fulcrum?
    (A) 100 N·cm
    (B) 75 N·cm
    (C) 50 N·cm
    (D) 25 N·cm
    (E) 0 N·cm

33. What is the mass of the meter stick?
    (A) 200 g
    (B) 100 g
    (C) 75 g
    (D) 50 g
    (E) 25 g

GO ON TO THE NEXT PAGE.

Questions 34–35

Photons are emitted only when incoming radiation of wavelengths less than 2×10^{-6} m illuminates a metallic surface.

34. If you double the wavelength, the number of photons emitted will
    (A) increase by two
    (B) increase by four
    (C) decrease to one half
    (D) decrease to one fourth
    (E) drop to zero

35. The cutoff frequency is
    (A) 2.0×10^{-6} Hz
    (B) 1.5×10^{14} Hz
    (C) 6.7×10^{-15} Hz
    (D) 4.0×10^{12} Hz
    (E) 600 Hz

36. Which of the following is an isotope of Carbon $^{12}_{6}C$ with 7 neutrons in it?
    (A) $^{12}_{7}C$
    (B) $^{13}_{6}C$
    (C) $^{6}_{12}C$
    (D) $^{12}_{6}C$
    (E) $^{12}_{13}C$

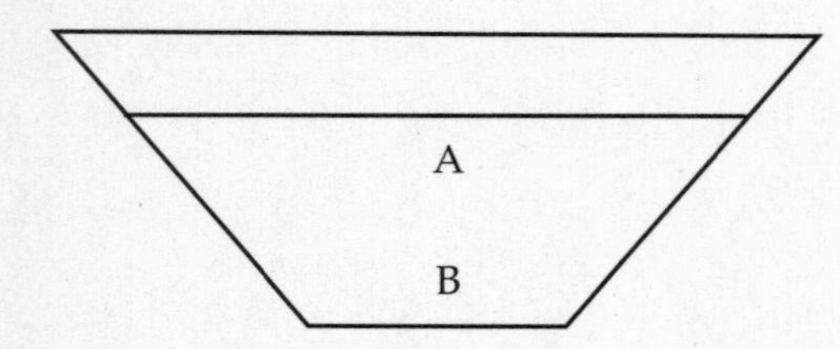

37. A cup of water is shaped as shown. You can say
    (A) the pressure is greater at point A due to the greater horizontal area
    (B) the pressure is greater at point A due to the smaller depth below the surface
    (C) the pressure is greater at point B due to the smaller horizontal area
    (D) the pressure is greater at point B due to the greater depth below the surface
    (E) two of the above are correct

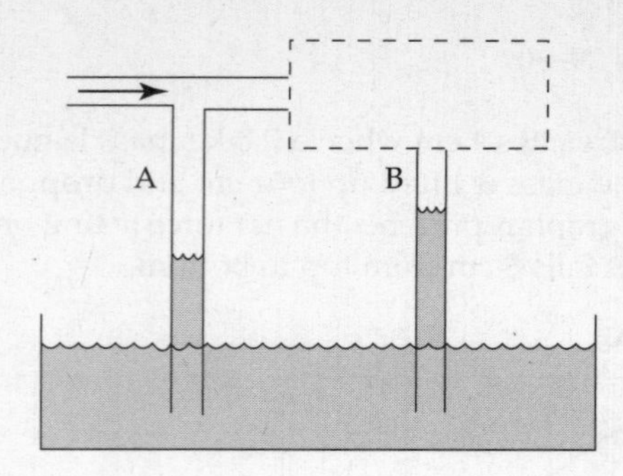

38. A tube with two T branches that have an open end is inserted in a liquid. However, the section of the tube above part B is hidden from view. The hidden section may be wider, more narrow, or the same width. Air is blown through the tube and the water levels rise as shown. You can say
    (A) the picture as drawn below is impossible—A and B must be at equal heights
    (B) the tube is more narrow and the air speed is greater above section B
    (C) the tube is more narrow and the air speed is less above section B
    (D) the tube is wider and the air speed is greater above section B
    (E) the tube is wider and the air speed is less above section B

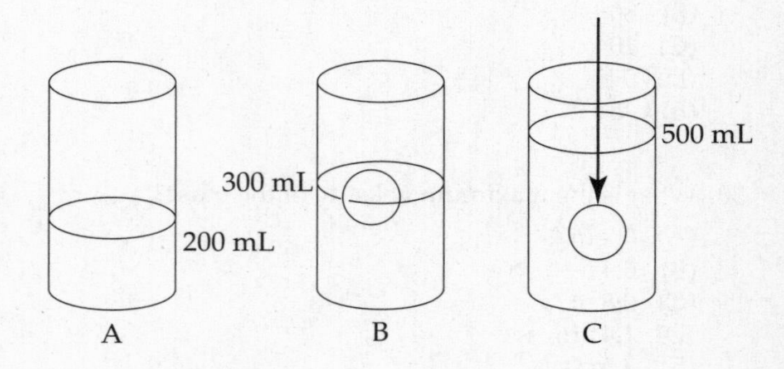

39. You have a cylinder with water in it as shown above in picture A. In picture B you place a ball in that floats. In picture C you physically push the ball under the water to just below the surface. What is the density of the ball?
    (A) 0.33 g/mL
    (B) 3 g/mL
    (C) 2 g/mL
    (D) 0.5 g/mL
    (E) 0.6 g/mL

GO ON TO THE NEXT PAGE.

40. Two students on ice skates are initially at rest. Student G weighs 150 pounds. Student K weighs 200 pounds. When they push off against each other the ratio of the magnitudes of velocities of student G to student K, $\frac{v_G}{v_K}$ is
    (A) 16/9
    (B) 4/3
    (C) 1/1
    (D) 3/4
    (E) 9/16

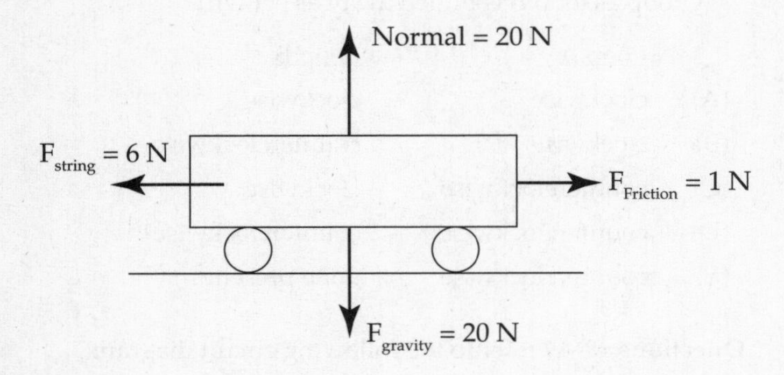

41. What is the acceleration of the cart shown?
    (A) 0 m/s/s
    (B) 2.5 m/s/s
    (C) 3 m/s/s
    (D) 9.8 m/s/s
    (E) 10 m/s/s

42. Knowing universal gravitation $F_G = \frac{Gm_1m_2}{r^2}$ and the circular motion equation $F_C = \frac{4\pi^2 m_1 r}{T^2}$ we can derive Kepler's third law for planets that travel in a basically circular orbit as
    (A) $\frac{T^2}{r^3} = \frac{4\pi^2 G}{m_2}$
    (B) $\frac{T^2}{r^3} = \frac{Gm_2}{4\pi^2}$
    (C) $\frac{T^2}{r^3} = \frac{4\pi^2}{Gm_2}$
    (D) $\frac{T^2}{r^3} = \frac{4\pi^2 m_2}{G}$
    (E) $\frac{T^2}{r^3} = \frac{4Gm_2}{\pi^2}$

43. A mass hanging from a spring and a simple pendulum have the same period, T. They are placed in an elevator that accelerates upward at 10 m/s². What can you say about the each of the periods on the elevator?
    (A) They both increase.
    (B) They both decrease.
    (C) They both remain the same.
    (D) The period of the mass on the spring remains the same but the period of the pendulum decreases.
    (E) The period of the pendulum remains the same but the period of the mass on the spring decreases.

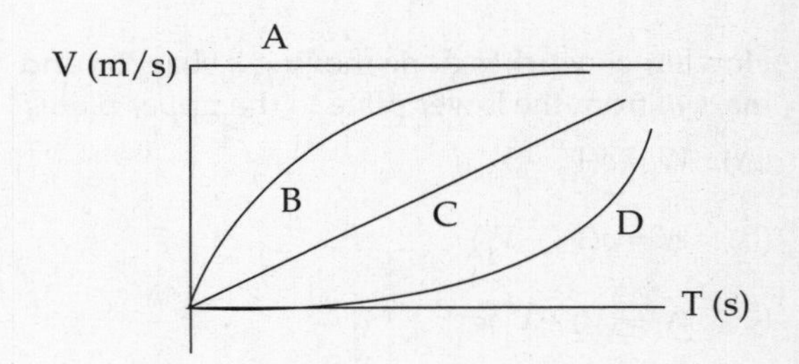

44. In deep space a rocket ship starts off at rest. It fires off its engines, using up fuel. The thruster exerts a constant force, but the mass decreases as it fires gas backwards. Which line would best represent the velocity of the rocket?
    (A) A
    (B) B
    (C) C
    (D) D
    (E) Cannot be determined

GO ON TO THE NEXT PAGE.

Questions 45–46

Two large flat parallel plates are separated by a distance of d. The lower plate is at a potential of V_1 with respect to the ground. The upper plate is at a potential of V_2 with respect to the ground.

45. How much work is done moving a charge q and mass m from the lower plate to the upper plates?
    (A) $W = q(V_2 - V_1)d$
    (B) $W = q(V_2 - V_1)$
    (C) $W = (V_2 - V_1)d$
    (D) $W = mq(V_2 - V_1)d$
    (E) $W = \frac{(V_2 - V_1)d}{q}$

46. The particle on the top plate is now released from rest. The velocity it has as it strikes the bottom plate is given by
    (A) $\sqrt{\frac{2m(V_2 - V_1)d}{q}}$
    (B) $\sqrt{\frac{(V_2 - V_1)}{2qmd}}$
    (C) $\sqrt{\frac{2q(V_2 - V_1)d}{m}}$
    (D) $\sqrt{\frac{2mq}{(V_2 - V_1)d}}$
    (E) $\sqrt{2qm(V_2 - V_1)d}$

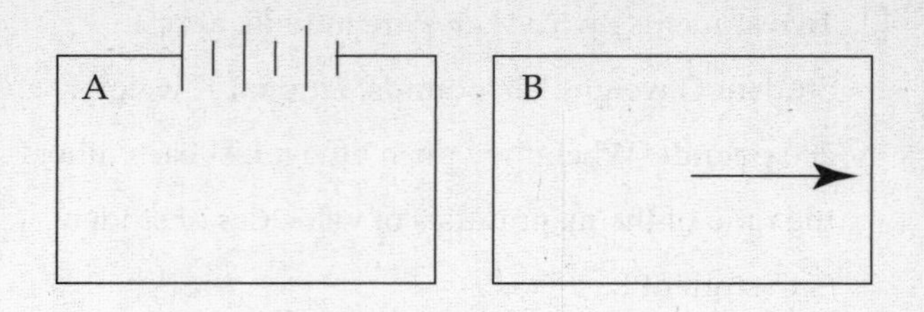

47. Two loops of conducting wire are near each other as shown. Loop A has a battery attached as indicated. Loop B is now pulled away from loop A. What can we say about the direction of the current in each loop as loop B is pulled away as shown?

	Loop A	Loop B
(A)	clockwise	clockwise
(B)	clockwise	counterclockwise
(C)	counterclockwise	clockwise
(D)	counterclockwise	counterclockwise
(E)	counterclockwise	none present

Questions 48–49 refer to the following circuit diagram.

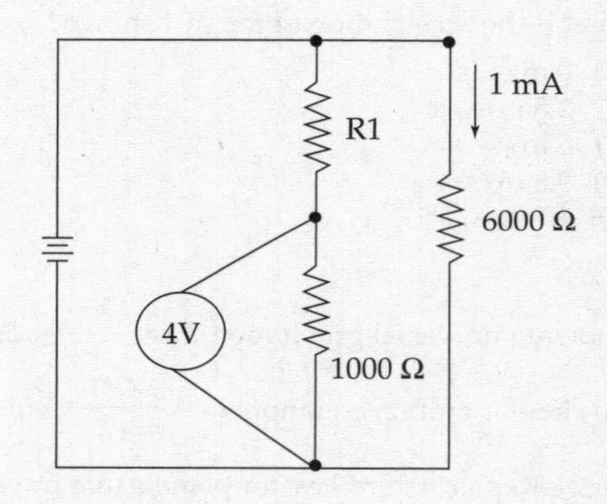

48. What is the voltage across R1?
    (A) 2 V
    (B) 4 V
    (C) 6 V
    (D) 8 V
    (E) 12 V

GO ON TO THE NEXT PAGE.

49. What is the current coming out of the battery?

(A) 1 mA
(B) 2 mA
(C) 3 mA
(D) 4 mA
(E) 5 mA

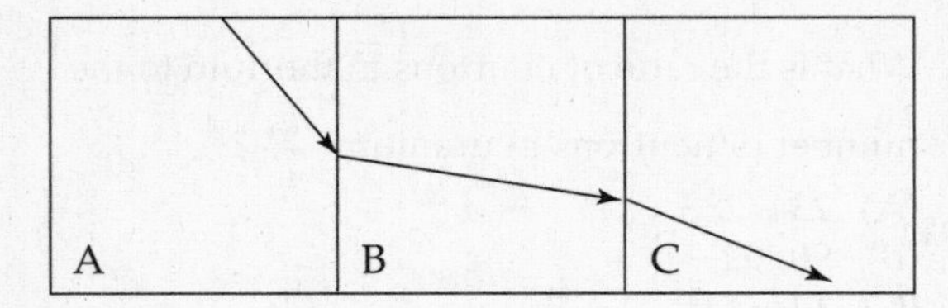

50. A ray of light travels though three different regions (mediums) as shown. List the three regions in order of increasing index of refraction.

(A) ABC
(B) CBA
(C) ACB
(D) BCA
(E) CAB

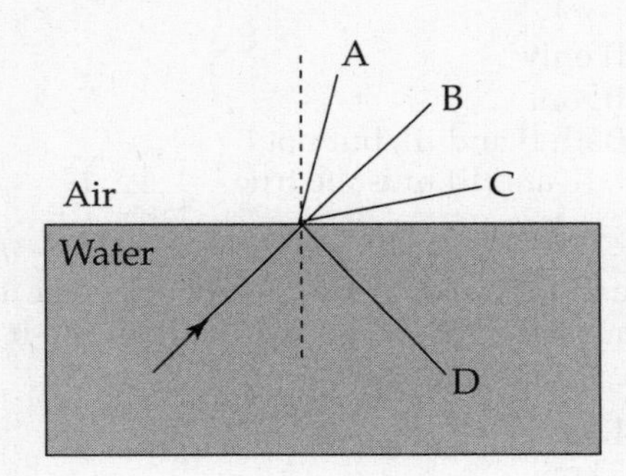

51. A beam of light goes from water to air. Depending on the actual angle that the light strikes the surface, which of the following are possible outcomes?

(A) A only
(B) B only
(C) C only
(D) A or D
(E) C or D

52. An alarm whose frequency is 400 Hz is dropped out of a third-floor window. The student who drops it measures the frequency with a very sensitive oscilloscope. The measured frequency

(A) appears higher than 400 Hz and the frequency increases as it falls
(B) appears higher than 400 Hz and the frequency decreases as it falls
(C) appears lower than 400 Hz and the frequency decreases as it falls
(D) appears lower than 400 Hz and the frequency increases as it falls
(E) remains constant throughout the fall

53. When a light wave in air hits a thin film ($n_{film}>n_{air}$), part of the wave is transmitted into the film and part is reflected. At the bottom of the thin film part of the wave is transmitted into air and part is reflected back up. What can you say about the reflected wave's phase at the top layer and bottom layers?

(A) Only the top reflection is inverted.
(B) Only the bottom reflection is inverted.
(C) Neither reflection is inverted.
(D) Both reflections are inverted.
(E) Only the transmitted waves are inverted.

54. An object is placed 10 cm in front of a diverging mirror. What is the focal length of the mirror if image appears 2 cm behind the mirror?

(A) –3/5 cm
(B) –5/3 cm
(C) –2/5 cm
(D) –5/2 cm
(E) –1/12 cm

55. Which of the following statements are true?

I. The focal length of a lens depends on the material the lens is made out of.
II. The focal length of the lens depends on the light intensity.
III. The focal length of the lens depends on the shape of the lens.

(A) I only
(B) II only
(C) III only
(D) I, II, and III
(E) Both I and III, but not II

GO ON TO THE NEXT PAGE.

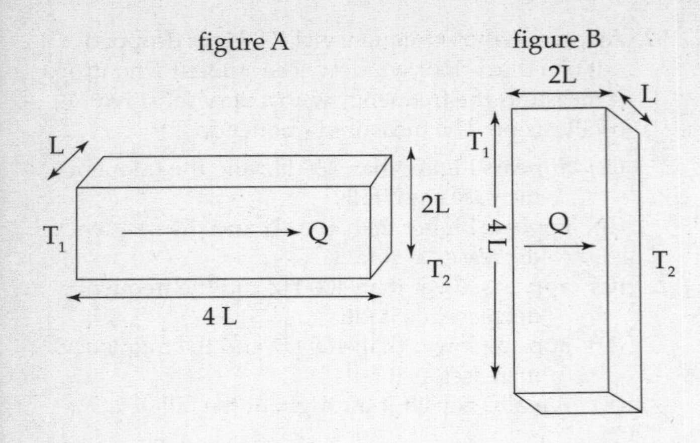

56. Heat flows through a block of iron as shown in figure A at a rate of H. The temperatures on each side of the block are T_1 and T_2 respectively. If the block is rotated 90º as shown in figure B, the rate at which heat is transferred is now

(A) $\frac{1}{4}H$

(B) $\frac{1}{2}H$

(C) H

(D) 2H

(E) 4H

57. The maximum efficiency of a heat engine which operates between a temperature of 27ºC and 227ºC is

(A) 12%
(B) 20%
(C) 40%
(D) 60%
(E) 75%

Questions 58–59

Uranium-238 decays into thorium-234 and an alpha particle as illustrated below.

$$^{238}_{92}U \rightarrow \, ^{234}_{90}Th + \, ^{4}_{2}\alpha$$

58. What is the ratio of neutrons in thorium to the number of neutrons in uranium $\frac{n_{Th}}{n_U}$?

(A) 234/238
(B) 90/92
(C) 144/146
(D) 1/1
(E) 324/330

59. Which of the following statements are true for the above reaction?

I. Charge is conserved.
II. The number of nucleons is conserved.
III. Mass–energy is conserved.

(A) I only
(B) II only
(C) III only
(D) Both II and III, but not I
(E) I, II, and III must be true

60. A generator can provide 2000 W of power for 30 seconds before it shuts off. How high can it lift a 50 kg mass in this time?

(A) 12 m
(B) 33 m
(C) 120 m
(D) 330 m
(E) 660 m

GO ON TO THE NEXT PAGE.

61. A projectile is launched with an initial velocity of $v_x = 5$ m/s and $v_{yi} = 20$ m/s. Which of the following is true about the projectile at the top of the motion?

	v_x (m/s)	v_y(m/s)	a (m/s²)
(A)	0	0	0
(B)	0	20	–10
(C)	5	0	0
(D)	5	20	–10
(E)	5	0	–10

62. The kinetic energy of the projectile launched at an angle is best represented by

(A)

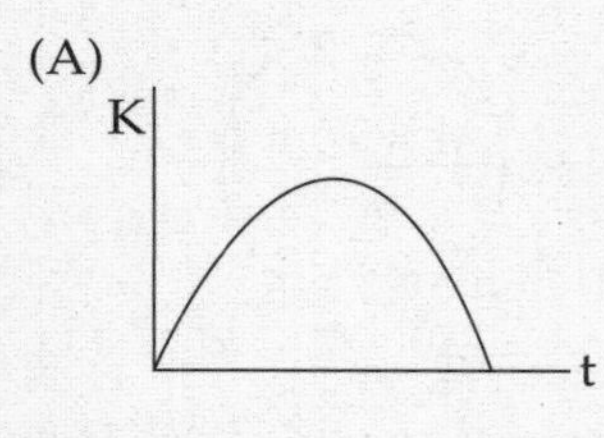

(B)

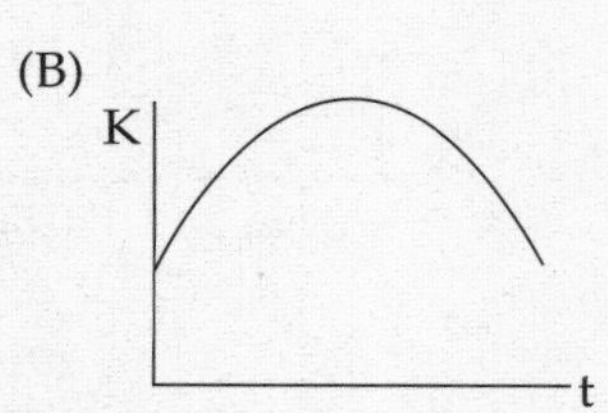

(C)

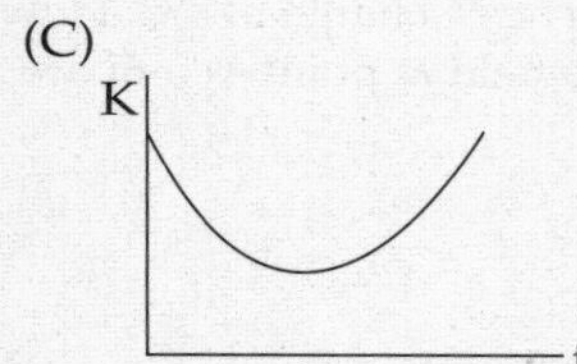

(D)

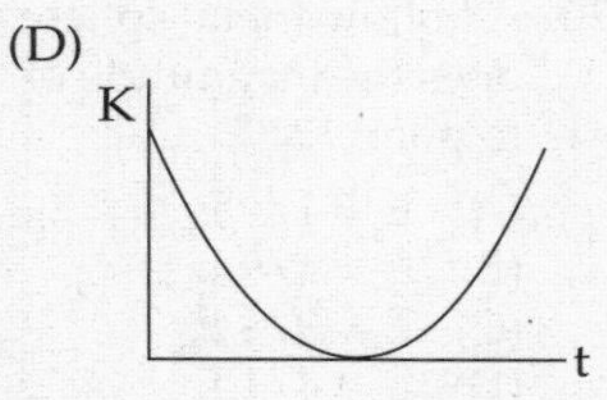

(E)

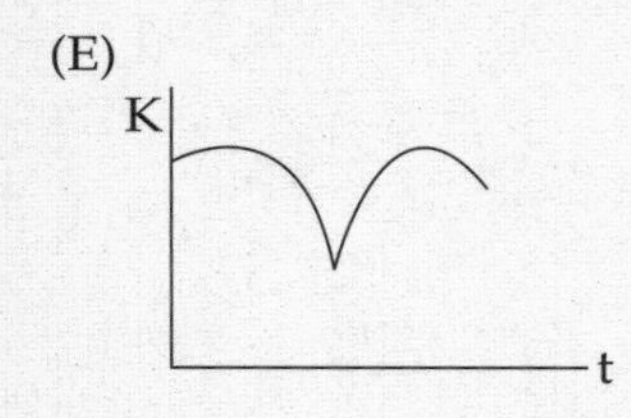

63. A small cart of mass M is initially at rest. It collides elastically with a large cart of mass 4M and velocity v. The large cart loses half its kinetic energy to the little cart. The little cart now has a velocity of

(A) 1.41v
(B) v
(C) 2v
(D) 4v
(E) 8v

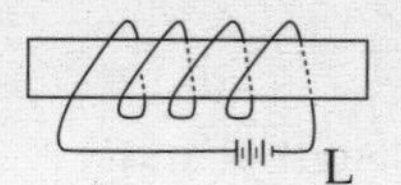

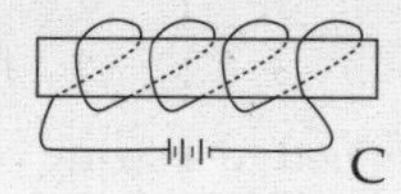

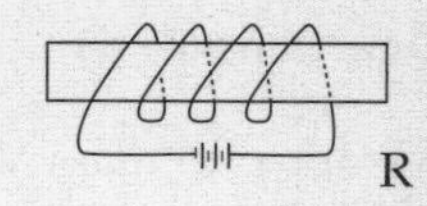

64. Three solenoids are arranged next to each other as shown above. What can we say about the solenoids interactions?

(A) L and C attract; C and R repel.
(B) L and C attract; C and R attract.
(C) L and C repel; C and R repel.
(D) L and C repel; C and R attract.
(E) L and R repel; L and C attract.

65. A particle of mass M and charge q and velocity v is directed toward a uniform electric field of strength E and travels a distance d. How far does the particle travel if the original velocity is doubled and the mass is cut in half?

(A) $4d$

(B) $2d$

(C) d

(D) $\frac{1}{2}d$

(E) $\frac{1}{4}d$

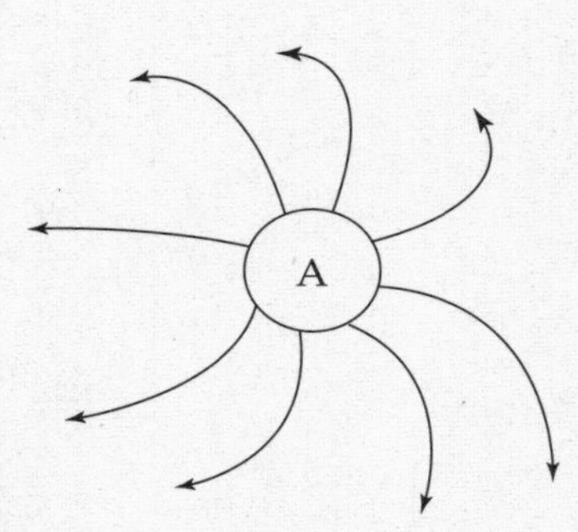

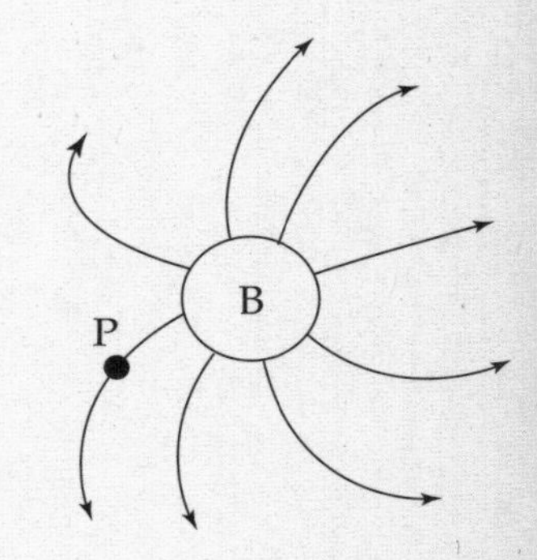

66. The diagram above shows the electric field lines around a region of space due to two charges A and B. Which of the following statements is true?

(A) The field is uniform.
(B) Charge A is positive and charge B is negative.
(C) Both charges are negative.
(D) A positive charge starting at rest from point P would move toward B.
(E) The equipotential lines are perpendicular to the field lines.

GO ON TO THE NEXT PAGE.

Questions 67–68

A stopper of mass M is tied to a string of length R and swings around horizontally in uniform circular motion with a velocity of v when acted on by a force of F.

67. What is the period of the stopper?

(A) 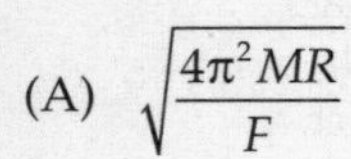 $\sqrt{\frac{4\pi^2 MR}{F}}$

(B) $\sqrt{\frac{MR}{4\pi^2 F}}$

(C) $\sqrt{\frac{4\pi^2}{FMR}}$

(D) $\sqrt{\frac{4\pi^2 R}{MF}}$

(E) $\sqrt{\frac{MR}{4\pi^2 F}}$

68. What is the power developed by the force?

(A) FR
(B) FV
(C) FV/R
(D) mv^2
(E) 0

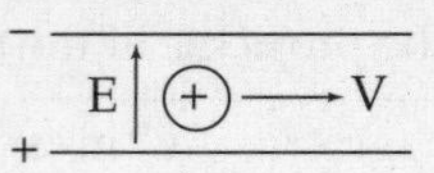

69. A particle travels through a uniform electric field E = 5 N/C as shown. What is the size and direction of the magnetic field needed to have a 0.2 C charge with mass 0.005 kg pass through undeflected at a velocity of 4 m/s?

(A) 20 T out of the page
(B) 20 T into the page
(C) 1.2 T out of the page
(D) 1.2 T into the page
(E) 0.8 T out of the page

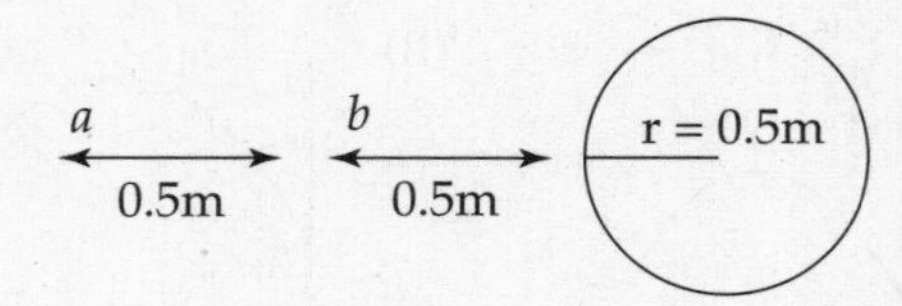

70. A hollow metal sphere carries a charge of 3 C. How does the magnitude of the field at point a compare to point b?

(A) $E_a = 1/4\ E_b$
(B) $E_a = 4/9\ E_b$
(C) $E_a = 9/4\ E_b$
(D) $E_a = 4/1\ E_b$
(E) $E_a = 3/2\ E_b$

END OF SECTION I

PHYSICS B

SECTION II

Free-Response Questions

Time—90 minutes

6 required questions. Questions 1–4 are worth 15 points each. Questions 5–6 are worth 10 points each.

Percent of total grade—50

General Instructions

Use a separate piece of paper to answer these questions.

Show your work. Be sure to write CLEARLY and LEGIBLY. If you make an error, you may save time by crossing it out rather than trying to erase it.

GO ON TO THE NEXT PAGE.

ADVANCED PLACEMENT PHYSICS B EQUATIONS FOR 2011

NEWTONIAN MECHANICS

$v = v_0 + at$

$x = x_0 + v_0 t + \frac{1}{2}at^2$

$v^2 = v_0^2 + 2a(x - x_0)$

$\Sigma \mathbf{F} = \mathbf{F}_{net} = m\mathbf{a}$

$F_{fric} \leq \mu N$

$a_c = \frac{v^2}{r}$

$\tau = rF \sin\theta$

$\mathbf{p} = m\mathbf{v}$

$\mathbf{J} = \mathbf{F}\Delta t = \Delta \mathbf{p}$

$K = \frac{1}{2}mv^2$

$\Delta U_g = mgh$

$W = F\Delta r \cos\theta$

$P_{avg} = \frac{W}{\Delta t}$

$P = Fv\cos\theta$

$\mathbf{F}_s = -k\mathbf{x}$

$U_s = \frac{1}{2}kx^2$

$T_s = 2\pi\sqrt{\frac{m}{k}}$

$T_p = 2\pi\sqrt{\frac{\ell}{g}}$

$T = \frac{1}{f}$

$F_G = -\frac{Gm_1m_2}{r^2}$

$U_G = -\frac{Gm_1m_2}{r}$

a = acceleration
F = force
f = frequency
h = height
J = impulse
K = kinetic energy
k = spring constant
ℓ = length
m = mass
N = normal force
P = power
p = momentum
r = radius or distance
T = period
t = time
U = potential energy
v = velocity or speed
W = work done on a system
x = position
μ = coefficient of friction
θ = angle
τ = torque

ELECTRICITY AND MAGNETISM

$F = \frac{1}{4\pi\epsilon_0}\frac{q_1q_2}{r^2}$

$\mathbf{E} = \frac{\mathbf{F}}{q}$

$U_E = qV = \frac{1}{4\pi\epsilon_0}\frac{q_1q_2}{r}$

$E_{avg} = -\frac{V}{d}$

$V = \frac{1}{4\pi\epsilon_0}\sum_i \frac{q_i}{r_i}$

$C = \frac{Q}{V}$

$C = \frac{\epsilon_0 A}{d}$

$U_c = \frac{1}{2}QV = \frac{1}{2}CV^2$

$I_{avg} = \frac{\Delta Q}{\Delta t}$

$R = \frac{\rho\ell}{A}$

$V = IR$

$P = IV$

$C_p = \sum_i C_i$

$\frac{1}{C_s} = \sum_i \frac{1}{C_i}$

$R_s = \sum_i R_i$

$\frac{1}{R_p} = \sum_i \frac{1}{R_i}$

$F_B = qvB\sin\theta$

$F_B = BI\ell\sin\theta$

$B = \frac{\mu_0}{2\pi}\frac{I}{r}$

$\phi_m = BA\cos\theta$

$\mathcal{E}_{avg} = -\frac{\Delta\phi_m}{\Delta t}$

$\mathcal{E} = B\ell v$

A = area
B = magnetic field
C = capacitance
d = distance
E = electric field
$\mathcal{E}$ = emf
F = force
I = current
ℓ = length
P = power
Q = charge
q = point charge
R = resistance
r = distance
t = time
U = potential (stored) energy
V = electric potential or potential difference
v = velocity or speed
ρ = resistivity
θ = angle
ϕ_m = magnetic flux

GO ON TO THE NEXT PAGE.

ADVANCED PLACEMENT PHYSICS EQUATIONS FOR 2011

FLUID MECHANICS AND THERMAL PHYSICS

$P = P_0 + \rho g h$

$F_{buoy} = \rho V g$

$A_1 v_1 = A_2 v_2$

$P + \rho g y + \frac{1}{2}\rho v^2 = \text{const.}$

$\Delta \ell = \alpha \ell_0 \Delta T$

$H = \frac{kA\Delta T}{L}$

$P = \frac{F}{A}$

$PV = nRT = Nk_B T$

$K_{avg} = \frac{3}{2}k_B T$

$v_{rms} = \sqrt{\frac{3RT}{M}} = \sqrt{\frac{3k_B T}{\mu}}$

$W = -P\Delta V$

$\Delta U = Q + W$

$e = \left|\frac{W}{Q_H}\right|$

$e_c = \frac{T_H - T_C}{T_H}$

A = area
e = efficiency
F = force
h = depth
H = rate of heat transfer
k = thermal conductivity
K_{avg} = average molecular kinetic energy
ℓ = length
L = thickness
M = molar mass
n = number of moles
N = number of molecules
P = pressure
Q = heat transferred to a system
T = temperature
U = internal energy
V = volume
v = velocity or speed
v_{rms} = root-mean-square velocity
W = work done on a system
y = height
α = coefficient of linear expansion
μ = mass of molecule
ρ = density

ATOMIC AND NUCLEAR PHYSICS

$E = hf = pc$

$K_{max} = hf - \phi$

$\lambda = \frac{h}{p}$

$\Delta E = (\Delta m)c^2$

E = energy
f = frequency
K = kinetic energy
m = mass
p = momentum
λ = wavelength
ϕ = work function

WAVES AND OPTICS

$v = f\lambda$

$n = \frac{c}{v}$

$n_1 \sin\theta_1 = n_2 \sin\theta_2$

$\sin\theta_c = \frac{n_2}{n_1}$

$\frac{1}{s_i} + \frac{1}{s_0} = \frac{1}{f}$

$M = \frac{h_i}{h_0} = -\frac{s_i}{s_0}$

$f = \frac{R}{2}$

$d\sin\theta = m\lambda$

$x_m \sim \frac{m\lambda L}{d}$

d = separation
f = frequency or focal length
h = height
L = distance
M = magnification
m = an integer
n = index of refraction
R = radius of curvature
s = distance
v = speed
x = position
λ = wavelength
θ = angle

GEOMETRY AND TRIGONOMETRY

Rectangle
$A = bh$

Triangle
$A = \frac{1}{2}bh$

Circle
$A = \pi r^2$
$C = 2\pi r$

Parallelepiped
$V = \ell wh$

Cylinder

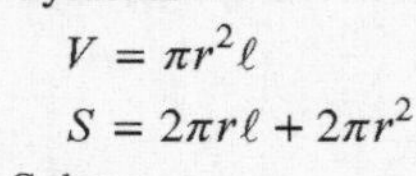

$V = \pi r^2 \ell$
$S = 2\pi r\ell + 2\pi r^2$

Sphere

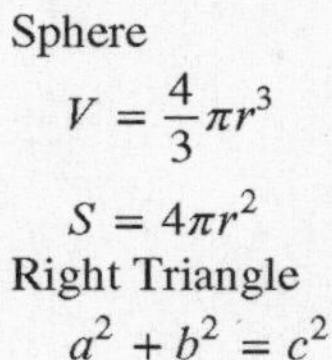

$V = \frac{4}{3}\pi r^3$
$S = 4\pi r^2$

Right Triangle
$a^2 + b^2 = c^2$

$\sin\theta = \frac{a}{c}$

$\cos\theta = \frac{b}{c}$

$\tan\theta = \frac{a}{b}$

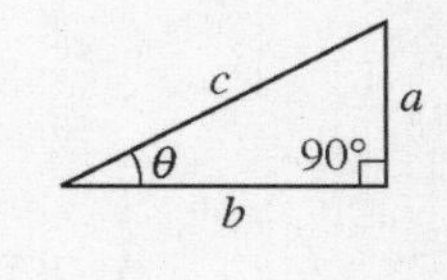

A = area
C = circumference
V = volume
S = surface area
b = base
h = height
ℓ = length
w = width
r = radius

GO ON TO THE NEXT PAGE.

PHYSICS B

SECTION II

Time—90 minutes

6 Questions

Directions: Answer all six questions. The suggested time is 17 minutes for answering each of questions 1–4, and about 11 minutes for answering each of questions 5–6. The parts within a question may not have equal weight.

1. (15 points)

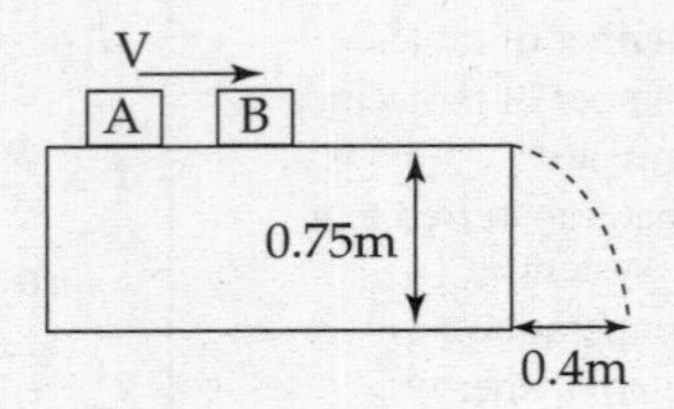

Block A has a mass of 0.5 kg and is sliding to the right across a frictionless table at 1.5 m/s. It collides with block B of unknown mass which is initially at rest and sticks to it. Together the two blocks slide off the table and hit the floor 0.25 meters from the end of the table. The table is 0.75 meters high.

(a) What was the speed of the blocks when they left the table?

(b) What is the mass of the block B?

(c) How much kinetic energy was lost in this collision?

GO ON TO THE NEXT PAGE.

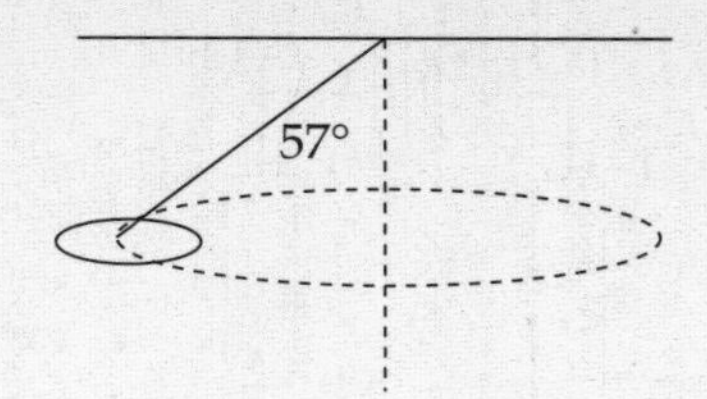

2. (15 points)

A conical pendulum is hanging from a string that is 2.2 meters long. It makes a horizontal circle. The mass of the ball at the end of the string is 0.5 kg.

(a) Below, make a free-body diagram for the ball at the point shown in the above illustration. Label each force with an appropriate letter.

(b) Write out Newton's second law in both the X and Y direction in terms used in your above free-body diagram.

(c) Calculate the centripetal acceleration from your free-body diagram.

(d) What is the radius of the circle that the ball is traveling in?

(e) What is the speed of the ball?

GO ON TO THE NEXT PAGE.

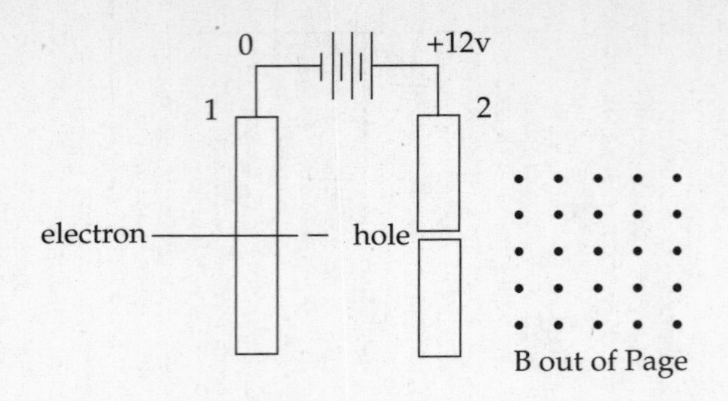

3. (15 points)

Two tests are run. In both trials you may ignore the effect of gravity.

Test 1: There are two large parallel plates separated by a distance $d = 0.5$ m with a potential difference of 0.12 V across them. There is a uniform magnetic field B pointing perpendicularly out of the paper of strength 0.002T starting to the right of plate 2. An electron is released from rest at plate 1 as shown above. It passes through a hole in plate 2 and enters the magnetic field and only experiences forces due to the magnetic field.

Test 2: The same set-up is run with the following two exceptions.

The battery is switched so that plate 1 becomes positive and plate 2 becomes negative.

A proton is used instead of an electron.

(a) Compare the force acting on each charge.

(b) Compare the speed of the proton as it emerges from the whole to the speed of the electron as it emerges from the hole.

(c) The charges will curve around once they are in the magnetic field. Make a sketch of each path. How far away will the two charges be when they strike the plate?

(d) Compare the time it takes for the electron to return to the plate to the time it takes the proton to return to the plate.

GO ON TO THE NEXT PAGE.

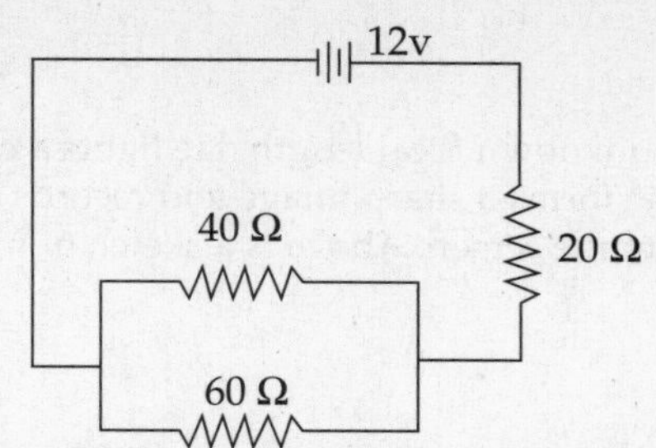

4. (15 points)

The circuit shown is built and all three resisters are placed in 1000 mL of water initially at a temperature of 10°C in an insulated container for 15 minutes. The specific heat of water is 1 cal/g°C and there are 4.186 J for every cal.

(a) What is the equivalent resistance of the circuit?

(b) What is the power?

(c) What is the energy transferred in the 15 minutes?

(d) What is the temperature change of the water?

GO ON TO THE NEXT PAGE.

5. (10 points)

A student has a convex lens of unknown focal length. He lights a candle in a darkened room and uses the lens and moves a screen until he forms a sharp image and records the distance from the candle to the lens and the distance from the lens to the screen. Above is a sketch of his set-up and his data.

s_o (cm)	s_i (cm)
15	61.5
20	29.8
30	19.3
40	18.0
50	15.6
60	15.3
70	14.7
80	13.9
90	13.7
100	13.7
110	13.4
120	13.2

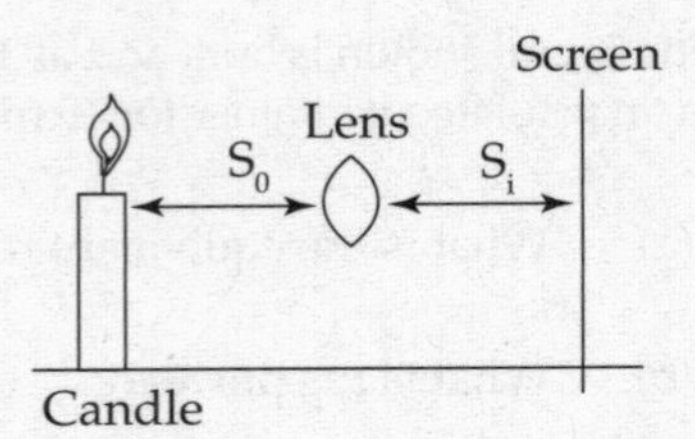

(a) What type of image is formed and what orientation is it?

(b) Where is the image formed if the candle is placed 6 cm from the lens?

(c) Use ray tracing to make a sketch when the object is 6 cm from the lens.

(d) What effect would it have if a lens with the same focal length was used but with only half the diameter?

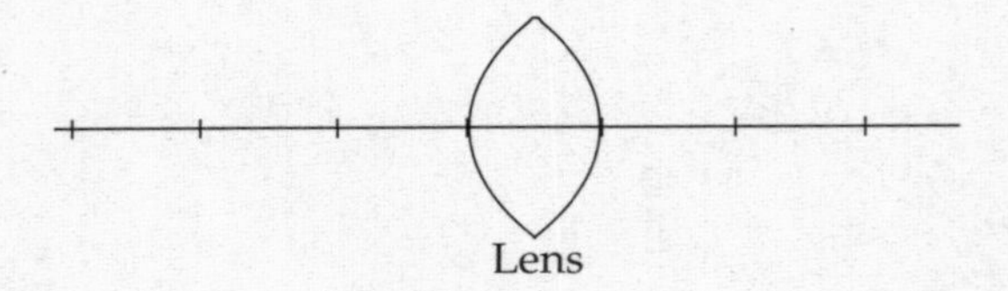

6. (10 points)

A laser whose wavelength is $\lambda_a = 4.7 \times 10^{-7}$ meter shines through a diffraction grating. The light hits a screen L = 2 m away and the third order maximum is detected 0.42 m on either side of the central maximum.

(a) What is the separation between slits?

(b) What is the path difference from the diffraction grating to the third maximum measured in meters?

(c) A new laser whose wavelength is $\lambda_b = 6.3 \times 10^{-7}$ meter shines though the slits and both its third maximums are labeled. What is the distance from the third order maximum using λ_a and the closest third maximum using λ_b?

STOP

END OF EXAM

22

AP Physics B Practice Exam 2: Answers and Explanations

SECTION I

1. **B** When the net force acting on an object is zero there is no acceleration, therefore no CHANGE in velocity. I. is not correct because the object could be at a constant nonzero velocity. III. is not correct because you could have two forces acting on the object in opposite directions that cancel each other out.

2. **E** $\sum F_y = ma_y \rightarrow T - mg = ma$ or $T = ma + mg$. This means $T = (1500 \text{ kg})(2 \text{ m/s}^2) + (1500 \text{ kg})(10 \text{ m/s}^2) = 18{,}000 \text{ N}$

3. **E** The velocity is tangent to the circle in the direction of motion. Both the acceleration and force point toward the center of the circle.

4. **D** The velocity is least at the top so the kinetic energy is least at the top. The other answers are incorrect because at the top the ball has a horizontal component of velocity so the velocity can not be zero, there is a vertical acceleration of -10 m/s^2 throughout the trip, and the potential energy is maximum at the top.

5. **B** The volume of water displaced is 200 mL (1,200 mL – 1,000 mL). Because the density of water is 1 g/mL, we can find the mass by $m = \rho V \rightarrow m - 1\frac{\text{g}}{\text{mL}}(200\text{mL}) = 200\text{g}$ of water. 200 g of water weighs 2 N (from $F_g = mg \rightarrow F_g = 0.2\text{kg}(10 \text{ m/s}^2)$). Archimedes' principle states the weight of the water displaced is equal to the buoyant force. Therefore there is an upward force of 2 N acting on the mass. A 0.50 kg mass weighs 5 N. The spring scale will read 3 N (5 N – 2N).

6. **D** Mechanical energy is conserved when there is no heat loss due to friction. The period of objects undergoing simple harmonic motion is constant. The velocity is continually changing so momentum of the mass is not conserved.

7. **C** $f_{spring} = \frac{1}{2\pi}\sqrt{\frac{k}{m_1}}$ and $f_{pendulum} = \frac{1}{2\pi}\sqrt{\frac{g}{\ell}}$. Setting these two equal to each other:
$\frac{1}{2\pi}\sqrt{\frac{k}{m_\partial}} = \frac{1}{2\pi}\sqrt{\frac{g}{\ell}}$ which becomes $\sqrt{\frac{k}{m_1}} = \sqrt{\frac{g}{\ell}}$ or $\frac{k}{m_1} = \frac{g}{\ell}$ or $\ell = \frac{m_1 g}{k}$

8. **D** $T_{pendulum} = 2\pi\sqrt{\frac{\ell}{g}}$ Changing the mass or amplitude has no effect on the period. The square root of one fourth the length reduces the period by a factor of 2.

9. **D** Using conservation of momentum, $m_1v_{1i} + m_2v_{2i} = (m_1 + m_2)v_f$. This becomes $2MV_0 + M(0) = (2M + M)v_f$. Solving for v_f you get (D).

10. **A** Gravity is given by $g = \frac{Gm_{planet}}{r^2}$. Doubling the radius yields one-fourth gravity. Reducing the mass cuts the gravity in half. One-fourth of one-half is one-eighth.

11. **D** The slope of the tangent line gives the instantaneous velocity. You could also use Process of Elimination. Recognize the answer must be negative, so you can eliminate (A), (C), and (E). The point in question is at (9,4). Dividing the y coordinate by the x coordinate is not the same as the obtaining the slope and is there to distract you.

12. **A** Statement I is correct because by conservation of energy, the heat lost by one object must be gained by the other. The temperature change depends on the equation $\Delta Q = mc\Delta T$. Because the water has both more mass and a higher specific heat, there will be a much lower temperature change of the water than the Al. Similarly, the internal energy of the object is given by $\Delta Q = mc\Delta T$. Although at the same temperature, the water again has both more mass and a higher specific heat and therefore more internal energy.

13. **C** The work done by a gas is determined by the area under a P vs. V graph. Because path C has the greatest area it does the greatest amount of work.

14. **A** The magnitude of the force is given by $F = qvB$. Doubling the charge doubles the force, doubling the velocity doubles the force, although doubling the mass will reduce the acceleration, it will not affect the force.

15. **D** All statements are true.

16. **B** First examine the two capacitors in series. The capacitors in series follow the rule $\frac{1}{C_S} = \sum_i \frac{1}{C_i}$. For two capacitors of equal value in parallel, the total capacitance is reduced by two. Once the series section is taken care of, the two branches in parallel can be determined by $C_P = \sum_i C_i$. This means the two branches can simply add. 1C+ 1/2 C = 3/2 C

17. **A** The resistance is given by $R = \frac{\rho \ell}{A}$. Using the same material means ρ is constant. Cutting the length in half will reduce the resistance by a factor of 2. Cutting the radius in half or doubling the radius will change the resistance by a factor of 1/4 (or 4) because $A = \pi r^2$. The only correct answer is (A).

18. **C** $P = IV$ means the electrical power is P = (1A)(120V) = 120 W. If this motor is only 75% efficient we get (0.75)(120 W) = 90 W of power output. Also $P = Fv$. Lifting a 6 kg mass upward at a constant velocity requires a 60 N force (by $F_g = mg \rightarrow (6\text{kg})(10\text{ m/s}^2)$). Therefore v = P/F or v = 90W/60N = 1/.5 m/s.

19. **D** The two electric fields vectors are shown. The resultant can be determined as shown below.

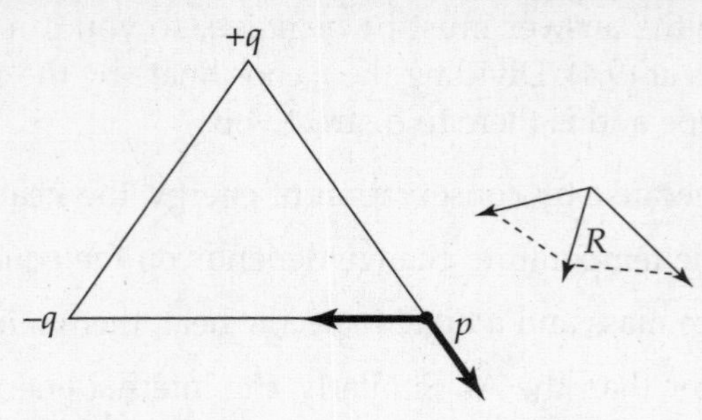

20. **E** Charges do not create an electric field at the location of the charge itself. Therefore the electric field at point p is not changed by the addition of a charge at point p.

21. **A** Path a follows the right-hand rule so must be positively charged. Path b is opposite the right hand rule so must be negatively charged. Neutrons would follow a straight line path, so it is impossible for either a or b to be a neutron.

22. **E** The change in length of a rod is given by $\Delta \ell = \alpha \ell_o$ T. Because α is a very small number, the length of a rod will be very much constant for a small temperature change. Formally, $\Delta \ell = \alpha \ell_o \Delta T$ becomes $\Delta \ell = (25 \times 10^{-6}\ 1/°C)(\ell_o)(10°C) = 2.5 \times 10^{-6} \ell_o$. $\ell = \ell_o + \Delta \ell$ or $\ell_o + 2.5 \times 10^{-6} \ell_o$. Again, this is just about ℓ.

23. **D** Isochoric means constant volume. Also the average speed of the molecules is given by $v_{rms} = \sqrt{\frac{3RT}{M}}$ so increasing T increases the speed and therefore the number of collisions. From the ideal gas law $PV = nRT$ so keeping a constant volume means increasing T also increases P. $K_{ave} = \frac{3}{2} k_b T$ tells us the kinetic energy increases as the temperature does.

24. **A** The rules for ray-tracing diagrams state that a line parallel to the principle axis bends away from the focal point as shown.

25. **D** Resonance occurs at $\lambda = \frac{4L}{n}$ for odd integers. This means at $4L, \frac{4}{3}L, \frac{4}{5}L$,.... We are shown a picture where a wave is in a tube of length L. This means $\lambda_a = \frac{4L}{3}$. The next shorter wavelength is $\lambda_b = \frac{4L}{5}$. The ratio of these is $\frac{\lambda_a}{\lambda_b} = \frac{\frac{4L}{3}}{\frac{4L}{5}}$ or 5/3.

26. **E** Statements I and II are true for a diverging (concave) lens. Statement III is true for a converging lens, not a diverging lens.

27. **A** $v = f\lambda$ so $\frac{v}{\lambda} = f$. $f = \frac{3 \times 10^8 \text{ m/s}}{6 \times 10^{-8} \text{ m}} \rightarrow 0.5 \times 10^{16} \text{ Hz}$

28. **E** Because sound is a longitudinal wave it cannot be polarized. Both longitudinal waves (sound) and transverse waves (light) can diffract (bending and spreading as it passes through an opening). Both can also refract (change speeds as the wave enters a new material).

29. **D** The positive area above the curve cancels the negative area below the curve to have zero work done. Another way of examining this problem is recognizing the velocity at the top and bottom of the bounce is zero. With no change in velocity there is no change in kinetic energy and therefore no net work done.

30. **E** The work done is given by the area under the curve. The maximum velocity occurs when there is a maximum positive area or from 0 to 4 cm. This area is given by $A = 1/2\ bh$ = ½ (0.04 m)·(20 N) = 0.4 N·m. The work done also gives the change in kinetic energy so we can say K = ½ mv^2 = 0.4 N·m. This becomes ½ (0.2kg)(v^2) = 0.4 N·m which becomes $0.1v^2$ = 0.4 N·m or v = 2 m/s.

31. **E** $Ft = m\Delta v \rightarrow F = \frac{m\Delta v}{t} \rightarrow \frac{(0.15\text{kg})(20\text{m/s})}{0.01\text{s}} \rightarrow 300\text{N}$

32. **D** 100 g weighs just about 1 N ($F_g = mg \rightarrow F_g = 0.1\text{kg}(10\text{ m/s}^2)$) and $\tau = dF \rightarrow (25\text{cm})(1\text{N})$ = 25N·cm

33. **B** For the bar to balance $\tau_{ccw} = \tau_{cw}$ or $d_L F_L = d_R F_R$. Furthermore the uniform stick can have its mass considered at the 50 cm mark which is 25 cm from the finger (fulcrum). This means $d_L F_L = d_R F_R$ becomes (25 cm)F_L = (25cm)(1N) or the stick weighs 1N which corresponds to a mass of 100g.

34. **E** If you double the wavelength you reduce the frequency by a factor of two. As soon as you are below the threshold frequency no photons are emitted.

35. **B** Given $v = f\lambda$ we know $f = \frac{v}{\lambda} \rightarrow \frac{3\times10^8\text{m/s}}{2\times10^{-6}\text{m}}$

36. **B** The lower number must remain 6 for carbon because the number of protons is equivalent to the atomic number. The upper left number is the sum of the protons and neutrons (6 + 7 = 13).

37. **D** The pressure in a non-moving fluid depends strictly on the depth.

38. **B** Because the fluid is higher in column B, the air speed must be greater above column B. This must be due to a narrower tube.

39. **A** To determine the density you need both the mass and the volume. From the picture, when the ball floats it displaces 100 mL of water (300mL – 200mL). This is equivalent to 100g of water from $m = \rho V$ and the density of water is 1g/mL. This weighs just about 1N (from $F_g = mg \rightarrow F_g = 0.1\text{kg}(10\text{ m/s}^2)$). When an object floats the buoyant force is equal to the weight of the object so the object weighs 1N which means the mass of the object is 100g. When an object is submerged the object displaces an amount of water equal to the object's volume. Therefore the volume of the object is 300 mL (500mL – 200mL). The density is now given by $\rho = \frac{m}{V} \rightarrow \frac{100\text{g}}{300\text{mL}} \rightarrow 0.33\frac{\text{g}}{\text{mL}}$

40. **B** Conservation of momentum tells us for a closed system $p_i = p_f$. Because the two students start at rest, the initial momentum is zero.

$0 = m_G v_G + m_K v_K \rightarrow -m_G v_G = m_K v_K \rightarrow \frac{v_G}{v_K} = \frac{m_K}{-m_G} \rightarrow \frac{v_G}{v_K} = \frac{200\text{lbs}}{-150\text{lbs}}$. You might have been told by your physics teachers always to use SI units and have an initial reaction to convert to kg. However, you know that mass is proportional to the weight and any conversion would appear in both equations and just cancel out. You can save some time by not converting (essential when trying to solve 70 problems in 90 minutes). The ratio is –4/3. The negative simply tells a direction and can be ignored.

41. **B** From the picture you can see there is no acceleration in the y direction because the upward and downward forces cancel. We can also determine the mass of the object because $F_G = mg \rightarrow 20\text{N} = m(10\text{ m/s}^2) \rightarrow m = 2.0\text{kg}$. In the x direction we have $\sum F_x = ma_x$ This becomes $F_s - F_F = ma_x \rightarrow a_x = \frac{F_s - F_F}{m} \rightarrow a_x = \frac{6\text{N} - 1\text{N}}{2.0\text{kg}} \rightarrow a_x = 2.5\text{ m/s}$.

42. **C** The centripetal force is provided by gravity. Setting $F_g = F_c$ we get $\frac{Gm_1m_2}{r^2} = \frac{4\pi^2 m_1 r}{T^2}$. Rearranging terms yields $\frac{T^2}{r^3} = \frac{4\pi^2}{Gm_2}$.

43. **D** By accelerating upward at 10 m/s^2 you are effectively experiencing 2 g's of acceleration. The period of a mass hanging from a spring does not depend on gravity, the period of a pendulum does. $T = 2\pi\sqrt{\frac{\ell}{g}}$ so as g increases, T decreases.

44. **D** $a = \frac{F}{m}$. If F is constant but mass decreases the acceleration will increase. As the acceleration increases the slope of a velocity-vs.-time graph will also increase.

45. **A** W = Fd. The force needed to push the charge from V_1 to V_2 is q V.

46. **C** The work done causes a change in kinetic energy. Because the particle starts from rest $W = K$. This becomes $q(V_2 - V_1)d = \frac{1}{2}mv^2 \rightarrow \frac{2q(V_2 - V_1)d}{m} = v^2 \rightarrow \sqrt{\frac{2q(V_2 - V_1)d}{m}} = v$

47. **C** Conventional current flows out of the positive side of the battery (the longer (left) side on the battery schematic). Therefore the current in A is CCW. The right-side section of wire in A produces a magnetic field on the right of the loop that is into the page as given by the right-hand rule. Therefore the flux in the B loop is into the page. As loop B moves away from loop A, the flux into the page decreases. From Lenz's Law a current will be produced in B in order to oppose this decrease in flux. This would require a CW current as given by the right-hand rule.

48. **A** The voltage across the 6000 Ω resister is given by V=IR or (0.001A)(6000 Ω) = 6V. In a parallel circuit the voltage across both branches is the same and so the total voltage drop across R1 and the 1000 Ω resistor must also be 6V. The voltage across R1 and the 1000 Ω will add to 6V because they are in series. Because there is 4V across the 1000 Ω resister, there must be 2V.

49. **E** The current through the 1000 Ω resistor is given by I = V/R or 4V/1000 Ω = 0.004 A. The current of the two branches will sum up to the current flowing out of the battery in a parallel circuit. I_B = 0.004A + 0.001A = 0.005A or 5 mA.

50. **C** The larger the angle when measured from the normal, the smaller the index of refraction. Because the angle in *A* > angle in *C* > angle in *B*, $n_A < n_C < n_B$.

51. **E** Water has a higher index of refraction than air, so if light refracts through the water the light will bend away from the normal—so C is possible. Another possibility is D, total internal reflection when the angle in water is greater than the critical angle.

52. **C** The velocity of the alarm is away from the detector so the frequency will lower due to the Doppler effect. As the speed increases away from the detector, due to the downward acceleration of gravity, the frequency will get even lower.

53. **A** Reflections are inverted when the new medium has a higher index of refraction. Reflections are not inverted when the new medium has a lower index of refraction. Because $n_{film} > n_{air}$ there is an inversion on the top, but not the bottom.

54. **D** The thin lens equation is given by $\frac{1}{s_i}+\frac{1}{s_o}=\frac{1}{f}$. Images formed behind a mirror have a negative in front of them. $\frac{1}{-2\text{cm}}+\frac{1}{10\text{cm}}=\frac{1}{f}\rightarrow\frac{-5}{10\text{cm}}+\frac{1}{10\text{cm}}=\frac{1}{f}\rightarrow\frac{-4}{10\text{cm}}=\frac{1}{f}$ or $f = -5/2$ cm. This answer makes sense because the focal length of a diverging mirror is negative.

55. **E** The intensity of the light does not affect the focal length.

56. **E** The rate at which heat is transferred is given by $H=\frac{kA\Delta T}{L}$. The material does not change, nor does the change in temperature, so *k* and ΔT won't affect the answer. In figure *a* the area is (L)(2L) or 2L². The length is 4L. This means the initial rate is $H=\frac{k2L^2\Delta T}{4L}\rightarrow H=\frac{kL\Delta T}{2}$. In figure *b* the area is given by (L)(4L) or 4L² and the length is 2L. The heat flow for *b* is $H=\frac{k4L^2\Delta T}{2L}\rightarrow H=2kL\Delta T$. Figure *b* has four times the rate as figure *a*.

57. **C** The maximum efficiency of a heat engine is given by $e=\frac{T_H-T_C}{T_H}$ using the Kelvin temperature scale. This means you have to add 273 to each answer to get 300K and 500K. The efficiency becomes $e=\frac{500\text{K}-300\text{K}}{500\text{K}}$ or 40%.

58. **C** The number of neutrons in an element can be found by taking the atomic mass minus the atomic number. For Thorium this becomes 234 – 90 = 144 and for Uranium this becomes 238 – 92 = 146. The ratio is therefore 144/146.

59. **E** All conservation rules are obeyed. Charge is conserved (92 – 90 + 2). The number of nucleons is conserved (92 protons + 146 neutrons) = (90 protons + 144 neutrons) + (2 protons + 2 neutrons). Mass–energy is conserved (238 = 234 + 4).

60. **C** The power is equal to $P = \frac{W}{t} \rightarrow P = \frac{mg\Delta h}{t} \rightarrow \Delta h = \frac{Pt}{mg} \rightarrow \Delta h = \frac{(2000\text{W})(30\text{s})}{(50\text{kg})(10\text{m}/\text{s}^2)} \rightarrow 120$ m

61. **E** The horizontal component of a projectile is constant (5 m/s). The vertical velocity at the top is zero. The acceleration due to gravity is constant at –10 m/s².

62. **C** As the time goes by the projectile rises and falls. Therefore the gravitational potential energy must increase and then decrease. Using conservation of energy, the kinetic energy must decrease and then increase. This lets us eliminate choices (A) and (B). The parabolic shape of a projectile lets us eliminate (E). Because there is some horizontal velocity at the top the kinetic energy can not be zero and (D) can be eliminated.

63. **A** The small cart starts with no kinetic energy because it is at rest. The large cart starts with $K = \frac{1}{2}(4M)v^2 \rightarrow K = 2Mv^2$. If it gives half its kinetic energy to the small cart it transfers $K = Mv^2$. The equation for the kinetic energy of the small cart is now $Mv^2 = \frac{1}{2}(M)v_f^{\,2} \rightarrow 2v^2 = v_f^{\,2} \rightarrow v_f = \sqrt{2v^2}$ or $1.41v$.

64. **D** The right-hand rule lets us determine that the solenoids will behave like magnets as shown.

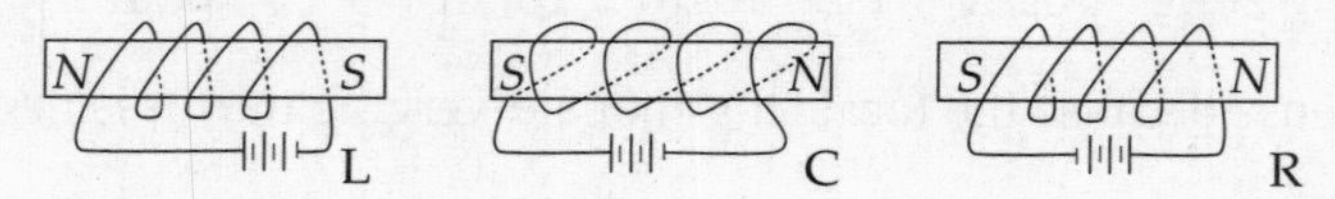

Because south poles of magnets repel south poles, and they attract north poles of magnets, we choose (D).

65. **B** The work done moving a particle against an electric field is given by $W = qEd$. Work is also equal to the change in kinetic energy. Because the final velocity is zero we can say $\frac{1}{2}mv^2 = qEd \rightarrow d = \frac{mv^2}{2qE}$. Cutting the mass in half cuts the distance in half but doubling the velocity makes the distance 4 times bigger. These factors combines to give the answer, (B).

66. **E** The field lines are neither evenly spaced nor parallel so the field is not uniform. The field lines point away from A and B and so both charges are positive and a positive charge would move in the direction of the field line away from B. (E) is true.

67. **A** Knowing the equations for circular motion $F = \frac{Mv^2}{R}$ and $v = \frac{2\pi R}{T}$ we can substitute in to get

$$F = \frac{M\left(\frac{2\pi R}{T}\right)^2}{R} \rightarrow F = \frac{4\pi^2 MR}{T^2} \rightarrow \text{T} = \sqrt{\frac{4\pi^2 MR}{F}}$$

68. **E** There is no work done on the stopper because the force applied is perpendicular to the direction of velocity and the kinetic energy does not change. Therefore, there is no power developed.

69. **C** The force acting on a particle moving through an electric field is given by $F = qE$. The force acting on a charge moving through a magnetic field is equal to $F = qvB$. Setting these two forces equal to each other we get $qE = qvB$ or $E = vB$. Therefore $B = \frac{E}{v} \rightarrow \frac{5\frac{\text{N}}{\text{C}}}{4\frac{\text{m}}{\text{s}}} \rightarrow 1.2\text{T}$. To determine the direction we know the force acting on the particle must be downward to counteract the upward push due to the electric field. Using the right-hand rule, we know the magnetic field must point out of the page.

70. **B** The electric field is given by $E = \frac{kq}{r^2}$ where r is measured from the center of the sphere to the point in question. $E_a = \frac{kq}{\left(\frac{3}{2}\right)^2} \rightarrow \frac{kq}{\frac{9}{4}} \rightarrow \frac{4kq}{9}$

$E_b = \frac{kq}{1^2} \rightarrow kq$. Substituting E_b for the kq in the first equation becomes $E_a = \frac{4E_b}{9}$.

SECTION II

1. (a) The height of the table and the distance the two blocks move horizontally can help us determine the speed of the blocks the moment before they leave the table. The time falling only depends upon the height of the table.

$$\Delta y = v_{yo}t + \frac{1}{2}at^2$$

We know that the blocks leave the table horizontally so that there is no initial y velocity and this simplifies to

$$\Delta y = \frac{1}{2}at^2 \rightarrow$$

$$t = \sqrt{\frac{2\Delta y}{a}}$$

Substituting in this becomes

$$t = \sqrt{\frac{2 \times 0.75\text{m}}{9.8\text{m}/\text{s}^2}}$$

$$t = 0.39\text{ s}$$

Knowing v_x is constant, the time falling and the horizontal distance, we can get

$$v_x = \frac{x}{t} \text{ or}$$

$$v_x = \frac{0.4\text{m}}{0.39\text{s}} \text{ or}$$

$$v_x = 1.0\frac{\text{m}}{\text{s}}$$

(b) We now know the final velocity, the instant the blocks leave the table. We can treat the problem like a momentum problem.

$$m_A v_{ai} + m_B v_{Bi} = m_A v_{Af} + m_B v_{Bf}$$

Because we are looking for the mass of B we can rearrange this to become:

$$+m_B v_{Bi} - m_B v_{Bf} = m_A v_{Af} - m_A v_{ai}$$

$$+m_B (v_{Bi} - v_{Bf}) = m_A v_{Af} - m_A v_{ai}$$

$$+m_B = \frac{m_A v_{Af} - m_A v_{ai}}{(v_{Bi} - v_{Bf})}$$

Substituting in the known numbers:

$$+m_B = \frac{(0.5\text{kg})(1\frac{\text{m}}{\text{s}}) - (0.5)(1.5\frac{\text{m}}{\text{s}})}{(0 - 1\frac{\text{m}}{\text{s}})}$$

Or $m_B = 0.25\text{kg}$

To determine the energy loss we simply have to compare the initial and final kinetic energies. Block B is initially at rest and so has no kinetic energy. Block A initially has a kinetic energy given by:

$K_i = \frac{1}{2} m_A v_{Ai}^2$. When you substitute in you get $K_i = \frac{1}{2}(0.5\text{kg})(1.5\frac{\text{m}}{\text{s}})^2 = 0.56$ J.

Because the two blocks hit and stick we can consider them together.

$$K_f = \frac{1}{2} m_{(A+B)} v_f^{\,2}$$

$$K_i = \frac{1}{2}(0.5\text{kg} + 0.25)(1\frac{\text{m}}{\text{s}})^2$$

or $K_f = 0.375$ J

The energy lost can be correctly expressed either of two ways—as a difference or a percentage. Difference: Subtract the total final energy from the total initial energy.

$$K_{lost} = K_f - K_i$$

$$= 0.56 \text{ J} - 0.375 \text{ J} = 0.185 \text{ J}$$

Percentage.

$$K_{lost} = \frac{K_f - K_i}{K_i}$$

$$K_{lost} = \frac{0.56\text{ J} - 0.375\text{ J}}{0.56\text{ J}}$$

$$= 0.33 \text{ or } 33\% \text{ lost.}$$

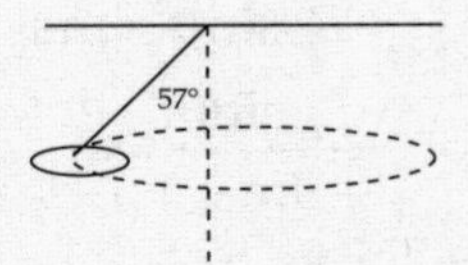

2. (a) A free-body diagram would include only the tension in the string and the force of gravity as shown below. Because it makes a horizontal circle take some care to draw the direction of the force represented by tension along the path of the string.

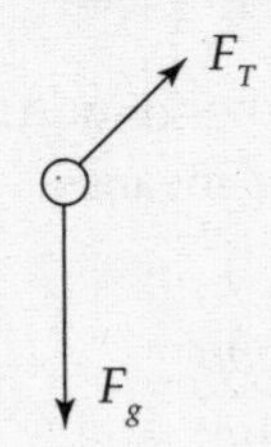

(b) Where θ is the angle between F_T and

$$\sum F_x = ma_x$$
$$F_T \sin\theta = ma_x$$

And

$$\sum F_y = ma_y$$
$$F_T \cos\theta - F_g = ma_y$$

(c) The centripetal force is the net force in the x direction. However, we need to use some information from the y direction. Because the conical pendulum travels in a horizontal circle there is no acceleration in the y direction and so Newton's Second Law in the y direction becomes

$$F_T \cos\theta - F_g = 0$$

or

$$F_T \cos\theta = F_g$$

$$F_T = \frac{F_g}{\cos\theta}$$

$$F_T = \frac{0.5\text{kg}(10\,\text{m}/\text{s}^2)}{\cos 57°}$$

$$F_T = 9.2\text{N}$$

Knowing this we can solve for the centripetal acceleration. It is the same as the acceleration in the x direction.

$$F_T \sin\theta = ma_c$$

$$\frac{F_T \sin\theta}{\text{m}} = a_c$$

$$\frac{9.2\text{N} \sin 57°}{0.5\text{kg}} = a_c$$

$$15.4\frac{\text{m}}{\text{s}^2}$$

(d) The radius the ball travels in can be found using some geometry. The length of the string is 2.2 m at an angle of 57 degrees. This means

$$\sin\theta = \frac{opp}{hyp}$$

$$\sin 57° = \frac{r}{2.2\text{m}}$$

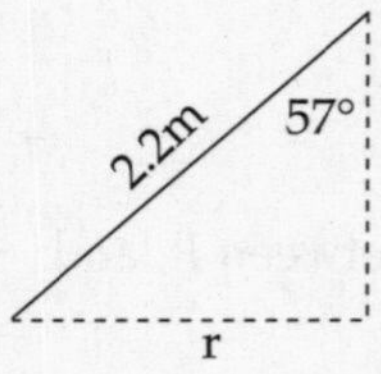

2.2m
57°
r

$$r = 1.84\text{m}$$

(e) The ball will travel at a speed of

$$a_c = \frac{v^2}{r}$$
$$v^2 = a_c r$$
$$v = \sqrt{a_c r}$$
$$v = \sqrt{15.4\frac{\text{m}}{\text{s}^2} \times 1.84\text{ m}}$$
$$v = 5.32\frac{\text{m}}{\text{s}^2}$$

3. (a) The electric field is given by both

$$E = \frac{F}{q} \text{ and } E = \frac{V}{d}$$

This lets us write

$$\frac{F}{q} = \frac{V}{d} \text{ or}$$
$$F = \frac{qV}{d}$$
$$F = \frac{(1.6\times10^{-19}\text{C})\times(0.12\text{V})}{0.5\text{m}}$$
$$F = 3.8\times10^{-20}\text{N}$$

Because the charges have the same magnitude and only the opposite direction, the voltage is the same except for the opposite direction, and the distance is the same, the charges will experience equal forces.

(b) Because we know the force and distance, we can use either $W = \Delta K$ to become

$$W = \Delta K$$
$$Fd = \frac{1}{2}mv^2$$
$$\frac{2Fd}{m} = v^2$$
$$v = \sqrt{\frac{2Fd}{m}}$$
$$v = \sqrt{\frac{(2)\times(3.8\times10^{-20}\text{N})\times(0.5\text{m})}{9.1\times10^{-31}\text{kg}}}$$
$$v = 2.0\times10^5\text{m / s}$$

for the electron and

$$v = \sqrt{\frac{(2)\times(3.8\times10^{-20}\text{N})\times(0.5\text{m})}{1.67\times10^{-27}\text{kg}}}$$

$$v = 4.8\times10^{3}\text{ m / s}$$

(c) for the proton. The speed of the proton is only about 2% the speed of the electron. The charges will experience a magnetic force once they enter the magnetic field given by $F = qvB$. The force is perpendicular to the direction of motion so that the charges also obey the circular motion equation $F = mv^2/r$. A sketch would look like:

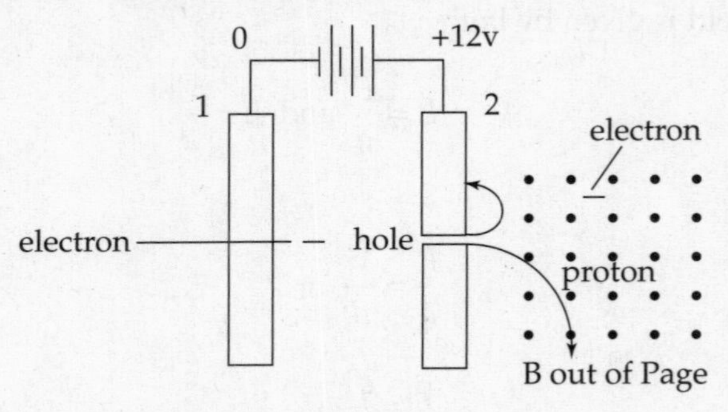

Setting these two forces equal to each other we have:

$$qvB = \frac{mv^2}{r}$$

$$qB = \frac{mv}{r}$$

$$r = \frac{mv}{qB}$$

A proton will have a radius given by:

$$r = \frac{(1.6\times10^{-27}\text{kg})\times(4.8\times10^{3}\frac{\text{m}}{\text{s}})}{(1.67\times10^{-19}\text{C})-(0.002\text{T})}$$

$$r = 2.3 \times 10^{-2}\text{ m}$$

An electron will have a radius given by:

$$r = \frac{(9.11\times10^{-31}\text{kg})\times(2.0\times10^{5}\frac{\text{m}}{\text{s}})}{(1.67\times10^{-19}\text{C})\times(0.002\text{T})}$$

$$r = 5.5 \times 10^{-4}\text{ m}$$

The charges will each hit the wall one diameter's distance away so we can double the above two radii and see them strike the plate.

The proton strikes 4.6×10^{-2} m below the hole. The electron strikes 1.1×10^{-3} m above the hole. The charges will strike 0.047 m apart.

(d) The time it takes to complete one circle would be given by $v = \frac{2\pi r}{T}$ if it traveled the full circle, but it only travels half way around so that we have

$$v = \frac{\pi r}{T}$$

$$T = \frac{\pi r}{v}$$

The proton will take

$$T = \frac{(\pi) \times (2.3 \times 10^{2}\ \text{m})}{4.8 \times 10^{3}\ \frac{\text{m}}{\text{s}}}$$

$T = 1.5 \times 10^{-5}$ s for a proton and

$$T = \frac{(\pi) \times (5.5 \times 10^{4}\ \text{m})}{2.0 \times 10^{5}\ \frac{\text{m}}{\text{s}}}$$

$T = 8.6 \times 10^{-9}$ s for an electron

4. (a) The equivalent resistance of the circuit can be found by first calculating the resistance of the two resistors in parallel and then adding that value to the resistor in series. For parallel circuits:

$$\frac{1}{R_p} = \frac{1}{R_1} + \frac{1}{R_2}$$

$$\frac{1}{R_p} = \frac{1}{40\,\Omega} + \frac{1}{60\,\Omega}$$

This means R_p is 24 Ω. The total resistance of the circuit is thus

$R_s = R_1 + R_2 = 24\ \Omega + 20\ \Omega$ or 44 ohms.

(b) Because $P = IV$ and $V = IR$ we can substitute in to obtain

$$P = IV$$

$$P = \frac{V}{R} V$$

$$P = \frac{V^2}{R}$$

$$P = \frac{(12\text{V})^2}{44\ \Omega}$$

or $P = 3.27$ W

(c) The energy transferred in 15 minutes is given by:

$$E = Pt$$

$$E = (3.3\text{W}) \times 15\text{ min}\left(\frac{60\text{s}}{\text{min}}\right)$$

$$= 2970\text{ J}$$

(d) The energy added goes into heat. This is given by $\Delta Q = mc\Delta T$. We also need to know the density of water is 1g/mL so that 1000 mL equals 1000 g. Therefore our heat equation becomes

$$\Delta T = \frac{\Delta Q}{mc}$$

$$\Delta T = \frac{2970\text{ J}}{(1000\text{ g}) \times \left(1\frac{\text{cal}}{\text{g°C}}\right)}$$

$$\Delta T = 2.970\frac{\text{J°C}}{\text{cal}}$$

Because 1 cal is 4.186 J we can simplify this to

$$\Delta T = 2.970\frac{\text{J°C}}{\text{cal}}\left(\frac{1\text{cal}}{4.186\text{ J}}\right) = 0.71\text{°C}.$$

This means the final temperature of the mixture is 10.7 °C

5. (a) Any image that can be formed on a screen is a real inverted image.

(b) The thin lens equation states $\frac{1}{s_i} + \frac{1}{s_o} = \frac{1}{f}$. Picking the first two data points 15 and 61.5, we obtain

$$\frac{1}{s_i} + \frac{1}{s_o} = \frac{1}{f}$$

$$\frac{1}{61.5\text{cm}} + \frac{1}{15\text{cm}} = \frac{1}{f}$$

$$0.083\text{cm}^{-1} = \frac{1}{f}$$

Thus, $f = 12$ cm

if an image is placed 6 cm away from the lens we have

$$\frac{1}{s_i}+\frac{1}{s_o}=\frac{1}{f}$$

$$\frac{1}{s_i}=\frac{1}{f}-\frac{1}{s_o}$$

$$\frac{1}{s_i}=\frac{1}{12\text{cm}}-\frac{1}{6\text{cm}}$$

$$\frac{1}{s_i}=\frac{-1}{12\text{cm}}$$

or $s_i = -12$ cm. This is a virtual upright image.

(c) A ray-tracing diagram would look something like this.

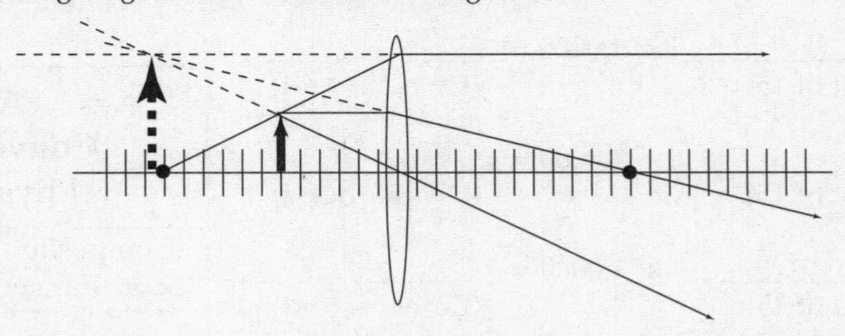

The diameter of the lens only affects the ability to gather light. It would decrease the brightness of the image but not change the location of the image.

6. (a) For a double slit experiment, $X_n = \frac{n\lambda L}{d}$. Solving for d we get:

$$d = \frac{n\lambda L}{X_n}$$

$$d = \frac{(3)\times(4.7\times10^{-7}\text{ m})\times(2\text{m})}{0.42\text{m}}$$

Or $d = 6.7 \times 10^{-6}$m

(b) For the third-order maximum the path difference is 3λ. This would simply be

$(3) \times (4.7 \times 10^{-7}\text{m})$ or 1.4×10^{-6}m

(c) The third-order maximum for the new laser is

$$X_n = \frac{n\lambda L}{d}$$

$$X_3 = \frac{(3)\times(6.3\times10^{-7}\text{ m})\times(2\text{m})}{6.7\times10^{-6}\text{ m}}$$

or 0.56 m. This means the two third-order maxima are 0.56 m – 0.42 m or 0.14 m apart.

HOW TO SCORE PRACTICE TEST 2

SECTION I: MULTIPLE-CHOICE

____________ × 1.3043 = ____________

Number of Correct (out of 70) — Weighted Section I Score (Do not round)

SECTION II: FREE RESPONSE

(See if you can find a teacher or classmate to score your essays using the guidelines in Chapter 3.)

Question 1 ____________ (out of 15) × 1.0000 = ____________ (Do not round)

Question 2 ____________ (out of 15) × 1.0000 = ____________ (Do not round)

Question 3 ____________ (out of 15) × 1.0000 = ____________ (Do not round)

Question 4 ____________ (out of 15) × 1.0000 = ____________ (Do not round)

Question 5 ____________ (out of 10) × 1.5000 = ____________ (Do not round)

Question 6 ____________ (out of 10) × 1.5000 = ____________ (Do not round)

Sum = ____________

Weighted Section II Score (Do not round)

AP Score Conversion Chart Physics B Exam

Composite Score Range	AP Score
112–180	5
85–111	4
57–84	3
40–56	2
0–39	1

COMPOSITE SCORE

____________ + ____________ = ____________

Weighted Section I Score + Weighted Section II Score = Composite Score (Round to nearest whole number)

ABOUT THE AUTHORS

Steve Leduc has been teaching at the university level since the age of 19. He earned his Sc.B. in theoretical mathematics from MIT at 20, and his M.A. in mathematics from UCSD at 22. After his graduate studies, Steve co-founded Hyperlearning, Inc., an educational services company that provided supplemental courses in undergraduate math and science for students from the University of California, where he lectured seventeen different courses in mathematics and physics. He has published four math books, *Differential Equations* in 1995, *Linear Algebra* in 1996, The Princeton Review's *Cracking the GRE Math Subject Test* in 2000 and *Cracking the Virginia SOL Algebra II* in 2001, as well as a physics book, The Princeton Review's *Cracking the SAT II Physics* in 2000. Through Hyperlearning, Steve has directed the creation and administration of the most successful preparation course for the medical school entrance exam (the MCAT) in California, where he has taught mathematics and physics to thousands of undergraduates. He currently owns two-to-the-eleventh power CDs and has seen *Monty Python and The Holy Grail*, *The Lord of the Rings* trilogy (extended versions), and *Blade Runner* about two-to-the-eleventh power times.

John J. Miller earned his B.S. in physics at Illinois Institute of Technology and M.Ed. at University of Illinois at Chicago. He has been teaching physics since 1990. He currently teaches at New Trier Township High School in Winnetka, Illinois. In addition to teaching physics, the two other major areas of his life that keep him in balance are his family, (wife Becky, and three children, Gregg, Katie, and Jackie), and his martial arts (he's practiced tae kwon do since 1982 and jujitsu since 1986).

INDEX

F

G

H

I

J

K

L

M

N

O

P

Q

R

S

T

U

V

W

Y

Z

Completely darken bubbles with a No. 2 pencil. If you make a mistake, be sure to erase mark completely. Erase all stray marks.

1. YOUR NAME: ______________________________
(Print) Last First M.I.

SIGNATURE: ______________________________ DATE: ___/___/___

HOME ADDRESS: ______________________________
(Print) Number and Street

____
City State Zip Code

PHONE NO. : ______________________________
(Print)

IMPORTANT: Please fill in these boxes exactly as shown on the back cover of your test book.

2. TEST FORM

3. TEST CODE

0	A	0	0	0
1	B	1	1	1
2	C	2	2	2
3	D	3	3	3
4	E	4	4	4
5	F	5	5	5
7	G	7	7	7
8		8	8	8
9		9	9	9

4. REGISTRATION NUMBER

0	0	0	0	0	0
1	1	1	1	1	1
2	2	2	2	2	2
3	3	3	3	3	3
4	4	4	4	4	4
5	5	5	5	5	5
7	7	7	7	7	7
8	8	8	8	8	8
9	9	9	9	9	9

5. YOUR NAME

First 4 letters of last name					FIRST INIT	MID INIT
A	A	A	A		A	A
B	B	B	B		B	B
C	C	C	C		C	C
D	D	D	D		D	D
E	E	E	E		E	E
F	F	F	F		F	F
G	G	G	G		G	G
H	H	H	H		H	H
I	I	I	I		I	I
J	J	J	J		J	J
K	K	K	K		K	K
L	L	L	L		L	L
M	M	M	M		M	M
N	N	N	N		N	N
O	O	O	O		O	O
P	P	P	P		P	P
Q	Q	Q	Q		Q	Q
R	R	R	R		R	R
S	S	S	S		S	S
T	T	T	T		T	T
U	U	U	U		U	U
V	V	V	V		V	V
W	W	W	W		W	W
X	X	X	X		X	X
Y	Y	Y	Y		Y	Y
Z	Z	Z	Z		Z	Z

6. DATE OF BIRTH

Month	Day		Year	
JAN				
FEB				
MAR	0	0	0	0
APR	1	1	1	1
MAY	2	2	2	2
JUN	3	3	3	3
JUL		4	4	4
AUG		5	5	5
SEP		7	7	7
OCT		8	8	8
NOV		9	9	9
DEC				

7. SEX

MALE

FEMALE

The Princeton Review.

FORM NO. 00001-PR

Section 1

Start with number 1 for each new section.
If a section has fewer questions than answer spaces, leave the extra answer spaces blank.

1. A B C D E
2. A B C D E
3. A B C D E
4. A B C D E
5. A B C D E
6. A B C D E
7. A B C D E
8. A B C D E
9. A B C D E
10. A B C D E
11. A B C D E
12. A B C D E
13. A B C D E
14. A B C D E
15. A B C D E
16. A B C D E
17. A B C D E
18. A B C D E
19. A B C D E
20. A B C D E
21. A B C D E
22. A B C D E
23. A B C D E
24. A B C D E
25. A B C D E
26. A B C D E
27. A B C D E
28. A B C D E
29. A B C D E
30. A B C D E
31. A B C D E
32. A B C D E
33. A B C D E
34. A B C D E
35. A B C D E
36. A B C D E
37. A B C D E
38. A B C D E
39. A B C D E
40. A B C D E
41. A B C D E
42. A B C D E
43. A B C D E
44. A B C D E
45. A B C D E
46. A B C D E
47. A B C D E
48. A B C D E
49. A B C D E
50. A B C D E
51. A B C D E
52. A B C D E
53. A B C D E
54. A B C D E
55. A B C D E
56. A B C D E
57. A B C D E
58. A B C D E
59. A B C D E
60. A B C D E
61. A B C D E
62. A B C D E
63. A B C D E
64. A B C D E
65. A B C D E
66. A B C D E
67. A B C D E
68. A B C D E
69. A B C D E
70. A B C D E
71. A B C D E
72. A B C D E
73. A B C D E
74. A B C D E
75. A B C D E
76. A B C D E
77. A B C D E
78. A B C D E
79. A B C D E
80. A B C D E
81. A B C D E
82. A B C D E
83. A B C D E
84. A B C D E
85. A B C D E
86. A B C D E
87. A B C D E
88. A B C D E
89. A B C D E
90. A B C D E

Completely darken bubbles with a No. 2 pencil. If you make a mistake, be sure to erase mark completely. Erase all stray marks.

1. YOUR NAME: ______________________________
(Print) Last First M.I.

SIGNATURE: ______________________ DATE: ___ / ___ / ___

HOME ADDRESS: ______________________________
(Print) Number and Street

____
City State Zip Code

PHONE NO. : ______________________________
(Print)

IMPORTANT: Please fill in these boxes exactly as shown on the back cover of your test book.

2. TEST FORM

3. TEST CODE

0	A	0	0	0
1	B	1	1	1
2	C	2	2	2
3	D	3	3	3
4	E	4	4	4
5	F	5	5	5
7	G	7	7	7
8		8	8	8
9		9	9	9

4. REGISTRATION NUMBER

0	0	0	0	0	0
1	1	1	1	1	1
2	2	2	2	2	2
3	3	3	3	3	3
4	4	4	4	4	4
5	5	5	5	5	5
7	7	7	7	7	7
8	8	8	8	8	8
9	9	9	9	9	9

5. YOUR NAME

First 4 letters of last name				FIRST INIT	MID INIT
A	A	A	A	A	A
B	B	B	B	B	B
C	C	C	C	C	C
D	D	D	D	D	D
E	E	E	E	E	E
F	F	F	F	F	F
G	G	G	G	G	G
H	H	H	H	H	H
I	I	I	I	I	I
J	J	J	J	J	J
K	K	K	K	K	K
L	L	L	L	L	L
M	M	M	M	M	M
N	N	N	N	N	N
O	O	O	O	O	O
P	P	P	P	P	P
Q	Q	Q	Q	Q	Q
R	R	R	R	R	R
S	S	S	S	S	S
T	T	T	T	T	T
U	U	U	U	U	U
V	V	V	V	V	V
W	W	W	W	W	W
X	X	X	X	X	X
Y	Y	Y	Y	Y	Y
Z	Z	Z	Z	Z	Z

6. DATE OF BIRTH

Month	Day		Year	
JAN				
FEB				
MAR	0	0	0	0
APR	1	1	1	1
MAY	2	2	2	2
JUN	3	3	3	3
JUL		4	4	4
AUG		5	5	5
SEP		7	7	7
OCT		8	8	8
NOV		9	9	9
DEC				

7. SEX

MALE

FEMALE

The Princeton Review

FORM NO. 00001-PR

Section 1

Start with number 1 for each new section.
If a section has fewer questions than answer spaces, leave the extra answer spaces blank.

1. A B C D E
2. A B C D E
3. A B C D E
4. A B C D E
5. A B C D E
6. A B C D E
7. A B C D E
8. A B C D E
9. A B C D E
10. A B C D E
11. A B C D E
12. A B C D E
13. A B C D E
14. A B C D E
15. A B C D E
16. A B C D E
17. A B C D E
18. A B C D E
19. A B C D E
20. A B C D E
21. A B C D E
22. A B C D E
23. A B C D E
24. A B C D E
25. A B C D E
26. A B C D E
27. A B C D E
28. A B C D E
29. A B C D E
30. A B C D E
31. A B C D E
32. A B C D E
33. A B C D E
34. A B C D E
35. A B C D E
36. A B C D E
37. A B C D E
38. A B C D E
39. A B C D E
40. A B C D E
41. A B C D E
42. A B C D E
43. A B C D E
44. A B C D E
45. A B C D E
46. A B C D E
47. A B C D E
48. A B C D E
49. A B C D E
50. A B C D E
51. A B C D E
52. A B C D E
53. A B C D E
54. A B C D E
55. A B C D E
56. A B C D E
57. A B C D E
58. A B C D E
59. A B C D E
60. A B C D E
61. A B C D E
62. A B C D E
63. A B C D E
64. A B C D E
65. A B C D E
66. A B C D E
67. A B C D E
68. A B C D E
69. A B C D E
70. A B C D E
71. A B C D E
72. A B C D E
73. A B C D E
74. A B C D E
75. A B C D E
76. A B C D E
77. A B C D E
78. A B C D E
79. A B C D E
80. A B C D E
81. A B C D E
82. A B C D E
83. A B C D E
84. A B C D E
85. A B C D E
86. A B C D E
87. A B C D E
88. A B C D E
89. A B C D E
90. A B C D E